# Process Development Division



**Core Programming Topic at the 2011 AIChE Annual Meeting**

## Minneapolis, Minnesota, USA
## 16 – 21 October 2011

ISBN: 978-1-61839-733-1

**Printed from e-media with permission by:**

Curran Associates, Inc.
57 Morehouse Lane
Red Hook, NY 12571



Printed by Curran Associates, Inc. (2012)

For permission requests, please contact AIChE
at the address below.

AIChE
3 Park Avenue
New York, NY 10016-5991

Phone: (203) 702-7660
Fax: (203) 775-5177

www.aiche.org

**Additional copies of this publication are available from:**

Curran Associates, Inc.
57 Morehouse Lane
Red Hook, NY 12571 USA
Phone: 845-758-0400
Fax: 845-758-2634
Email: curran@proceedings.com
Web: www.proceedings.com

# TABLE OF CONTENTS

# (13a) Mechanistic Modeling of Fast Pyrolysis of Cellulose to Predict Bio-Oil Composition

**Vinu Ravikrishnan** and Linda J. Broadbelt, Chemical and Biological Engineering, Northwestern University, Evanston, IL

Fast pyrolysis, a potential strategy for the production of transportation fuels from biomass, involves a complex network of competing reactions, which result in the formation of bio-oil, non-condensable gaseous species, and solid char. Bio-oil is a mixture of anhydro sugars, furan derivatives, and oxygenated aromatic and low molecular weight (LMW) compounds. Currently, the successful modeling of fast pyrolysis reactors for biomass conversion is hampered by lumped kinetic models, which fail to predict the bio-oil composition. Hence, a fundamental understanding of the chemistry and kinetics of biomass pyrolysis is important to evaluate the effects of process parameters like temperature, residence time and pressure on the composition of bio-oil.

We have recently developed a comprehensive mechanistic model to characterize the primary products of fast pyrolysis of pure cellulose. The model incorporates the following condensed phase reactions with ionic and electrocyclic fragmentation mechanisms for the formation of various LMW products in bio-oil: glucosidic bond scission, retro-aldol, retro-Diels-Alder, 1,2-dehydration, 1,3-dehydration, hydrolysis, cyclic Grob fragmentation and enol-keto transformation reactions. The kinetic rate coefficients for most of the above reaction steps were obtained from the literature, based on either experimentally determined Arrhenius parameters for cellulose and model compounds, or theoretically computed values of activation energies using quantum chemical calculations. A computational framework based on continuous distribution kinetics was constructed to solve the kinetic model. The model tracks 96 distinct polymer species and 25 LMW products, and incorporates over 350 reactions. The model predictions compare well with the experimental data (Patwardhan *et al.*, *J. Anal. Appl. Pyrolysis* **2009**, 86, 323-330) for the formation of important products like levoglucosan (60 wt.%), glycolaldehyde (6 wt.%), 5-hydroxymethyl furfural (3 wt.%), furfural (1 wt.%) and formic acid (6 wt.%). Levoglucosan was found to be formed by the initiation of ionic chain ends by mid-chain glucosidic bond scission, followed by end-chain ionic unzipping. Some of our efforts in extending the kinetic model of pyrolysis of pure cellulose to cellulose with mineral matter and hemicellulose will also be discussed in this presentation.

See more of this Session: Catalytic Conversion of Renewable Resources to Synthesis Gases and Pyrolysis Oils
See more of this Group/Topical: Sustainable Engineering Forum

# (13b) Mixed Rare Earth Oxides (REOs) for Desulfurization and Tar Removal From Gasifier Effluents

**Rui Li**, Joseph Bridges, Weishi Kong and Kerry.M. Dooley, Chemical Engineering, Louisiana State University, Baton Rouge, LA

The removal of $H_2S$, $NH_3$ and tars from syngas generated in biomass or coal/biomass gasification plays an important role in further processing of syngas. We have developed materials capable of both $H_2S$ adsorption and tar reforming at high temperatures (~600-650°C). These are mixed rare earth oxides (REOs) such as $CeO_2/La_2O_3$, prepared by templated sol-gel methods, and then impregnated with transition metals (Mn, Fe, etc.). Desulfurization tests have been carried out using model syngas mixtures. Successive desulfurization-regeneration cycles some of these materials (e.g., $Mn/Ce/LaO_x$) demonstrated reasonable sulfur capacities in the presence of water and $CO_2$, along with regenerability. Naphthalene and propane have been used as models for tar reforming/cracking, and the effects of $H_2S$, CO and $CO_2$ on the model reactions were examined. $H_2S$ and $CO_2$ are strong inhibitors, while CO is not. Fresh, used and regenerated sorbents and catalysts have been characterized by BET, TPR, XRD, XANES and EXAFS.

See more of this Session: Catalytic Conversion of Renewable Resources to Synthesis Gases and Pyrolysis Oils
See more of this Group/Topical: Sustainable Engineering Forum

# (13c) Multi-Scale Study On the Pyrolysis of Sustainable Biomass Feedstock

**Jessica D. Murillo**, Chemistry, Environmental Sciences, Tennessee Technological University, Cookeville, TN, Joseph J. Biernacki, Chemical Engineering, Tennessee Technological University, Cookeville, TN and C. Pat Bagley, College of Agricultural and Human Sciences, Tennessee Technological University, Cookeville, TN

**Multi-Scale Study on the Pyrolysis of Sustainable Biomass Feedstock**

Jessica D. Murillo

Department of Chemistry, Environmental Sciences, Tennessee Technological University, Cookeville, TN 38505

Joseph J. Biernacki

Department of Chemical Engineering, Tennessee Technological University, Cookeville, TN 38505

C. Pat Bagley

College of Agricultural and Human Sciences, Tennessee Technological University, Cookeville, TN 38505

Pyrolysis of lignocellulosic material for the production of biofuels and value-added products offers a short- to long-term renewable energy option to help curb fossil fuel use, gain energy independence, and mitigate global climate change. The main components of lignocellulsic material are cellulose, hemicellulose and lignin; however, there are many issues with using such a complex chemical feedstock for the production of renewable fuels. During pyrolysis the main components degrade via primary and secondary reactions, producing a spectrum of products, including volatiles, char, and bio-oil. To gain a more complete understanding of the pyrolytic behavior of lignocellulosic biomass, a deeper knowledge of the morphology and chemical structure is required.

Preliminary meso-scale experimental results from the pyrolysis of tall fescue hay, sorghum-sudangrass, Bermuda grass, switchgrass, and alfalfa hay, have paved the way for study at microscopic lengths. The physical and chemical characteristics of each grass hay was studied using X-ray diffraction, mid-infrared spectroscopy (MIR) and electron scanning microscopy to determine the microstructure and to quantify the content of lignin, cellulose, and hemicelluloses. Comparison studies on the reaction kinetics of grass samples and isolated pseudo-components were performed using thermogravimetric analysis (TGA).

See more of this Session: Catalytic Conversion of Renewable Resources to Synthesis Gases and Pyrolysis Oils
See more of this Group/Topical: Sustainable Engineering Forum

# (13d) Catalytic Cracking of Oak Pyrolytic Vapors Via Bench-Scale In-Situ Fixed-Bed Catalysis

**David J. Mihalcik**, Akwasi A. Boateng, Charles A. Mullen and Neil Goldberg, Eastern Regional Research Center, Agricultural Research Service, USDA, Wyndmoor, PA

Pyrolysis oils derived from lignocellulosic feedstocks by way of fast pyrolysis offer the possibility of a distinct alternative to the use of petroleum-based fuels. However, an inherent drawback associated with pyrolysis oil its high oxygen content which is due to the nature of the lignocellulosic biomass from which it derives. Extensive efforts are being put forth to treat pyrolytic vapors in situ using zeolite catalysts. For our experiment, the catalytic conversion of oak pyrolysis vapors was performed at 500°C using a 2.5 kg/hr fluidized-bed fast pyrolysis reactor that was fitted with a re-heated (425°C) split-stream fixed-bed catalytic cracker containing zeolite-type catalyst. The placement of the split-stream fixed-bed catalytic reactor was varied within the condenser train to determine the effect of the removal of increasing amounts of water and oxygenated components from within the pyrolytic vapor stream. The cracked pyrolytic vapors resulting from the fixed-bed catalyst were collected by way of immediate contact with a dry-ice acetone bath followed by secondary collection using a methanol spray condenser. Karl Fisher titration and elemental and quantitative GC/MS analyses were performed to determine relevant trends within the experiment. The recovered liquid products that passed through the fixed-bed catalytic reactor displayed a remarkable decrease in the amount of oxygen containing species and a significant increase in the presence of aromatic hydrocarbons.

See more of this Session: Catalytic Conversion of Renewable Resources to Synthesis Gases and Pyrolysis Oils
See more of this Group/Topical: Sustainable Engineering Forum

# (13e) A Dual-Stage Laminar Entrained Flow Reactor (LEFR) and Plug Flow Reactor (PFR) System for Studying Biomass Pyrolysis and Gasification

**Mark W. Jarvis**, Calvin Mukarakate, David Robichaud and Mark Nimlos, National Renewable Energy Laboratory, Golden, CO

An understanding of pyrolysis vapor cracking kinetics and mechanisms are important for developing approaches for reducing tars during gasification and ultimately reducing the cost of producing transportation fuels from gasification and fuel synthesis. In order to reduce tars using predictive design approaches such as Computation Fluid Dynamics (CFD), it is necessary to have robust chemical reaction models containing mechanism that lead to the formation of these species. Existing reaction models only contain limited, highly generalized mechanisms for the formation of tars and are not robust enough to be predictive. In particular, the initial steps of gasification are inadequately characterized. These steps are primarily cracking reactions that break down the pyrolysis vapor species into smaller molecules ultimately leading to syngas and tars. In this study, we have undertaken an experimental investigation of these types of cracking reactions. The solid phase vaporization reactions were separated from the gas phase cracking reactions by conducting experiments in a two-stage reactor. In the first stage, particles of biomass were pyrolyzed in a Laminar Entrained Flow Reactor (LEFR) at a low temperature (450°C) and in the second phase the vapors were cracked at variable residence times (0.2 to 1.5 s) and temperatures (430° to 680°C) in a Plug Flow Reactor (PFR). The tar species and other product gases were measured using a Molecular Beam Mass Spectrometer (MBMS). The data from the MBMS showed trends in the tar species concentrations as a function of residence time and temperature, which will allow the development of more accurate and robust chemical kinetics models of gasification. Eventually, these kinetic models will be used in CFD models of gasification in order to optimize it for reduced tar formation and increased syngas yield.

See more of this Session: Catalytic Conversion of Renewable Resources to Synthesis Gases and Pyrolysis Oils
See more of this Group/Topical: Sustainable Engineering Forum

# (13f) Reforming of Residual Tars and Oils From Biomass Gasification

**Lyman Frost**[1], Elango Elangovan[1] and J. Hartvigsen[2], (1)Ceramatec, Inc, Salt Lake City, UT, (2)Electrochemical, Ceramatec, Inc, Salt Lake City, UT

One of the barriers to wider use of many kinds of biomass is the fact that they generate refractory tars and oils when gasified. These compounds cause difficulty when they are passed to downstream components in the system by causing plugging and drop in performance. The compounds can be condensed and removed but this involves extensive heat exchange, increased cost, and removal to a hazardous waste landfill.

Working as a subcontractor to Emery Energy, a Salt Lake City company with expertise in biomass gasification, Ceramatec will be applying its non-thermal plasma reformer technology to eliminate any refractory tars and oils that exit the Emery gasifier. Preliminary laboratory work indicates that it is possible for the plasma reformer to operate on the gas exiting the gasifier and convert these tars and oils to additional hydrogen and carbon monoxide. This work is being done at Western Research Institute in Laramie, Wyoming under a US Department of Energy award to Emery Energy.

This paper will present an overall process structure of the gasification process and the non-thermal plasma reformer in particular. Operational characteristics and laboratory results related to the reforming of the refractory tars and oils will be presented. The larger pilot sized facility at Laramie will also be described. Data is expected to be available from the larger pilot sized facility in Laramie by the time for the conference. The resultant gas stream is of a quality that could be used by a high temperature fuel cell.

See more of this Session: Catalytic Conversion of Renewable Resources to Synthesis Gases and Pyrolysis Oils
See more of this Group/Topical: Sustainable Engineering Forum

# (13g) Low Temperature Gasification of Impregnated Biomass Feedstock

**Foster Agblevor**, Biological Engineering, Utah State University, Logan, UT, Ofei Mante, Biological Systsems Engineering, Virginia Polytechnic Institute and State University, Blacksburg, VA, Allen Aradi, Afton Chemical Inc, Richmond, VA and Tze-Chi Jao, Afton Chemical, Richmond, VA

Hybrid poplar biomass feedstocks were impregnated with catalysts and gasified at three different temperatures in a bubbling fluidized bed reactor. The impregnated biomass were gasified in a nitrogen atmosphere at 400 °C, 600° C, and 800 °C. The gas yield increased with increasing temperature as expected and the char and liquid yields decreased with increasing temperature. Te catalytic effect was observed at 400 C when the impregnated gasification was compared with the control that did not contain any catalyst. At 400 C, the impregnated sample had a gas yield of 51wt% compared to 41 wt% for the non-catalytic gasification at 450 C. The char yield was 11 wt% compared to 14 wt% for the non-catalytic run at 450 C. The char yield decreased from 11 wt% to 6.5 wt% when the gasification temperature was increased from 400 to 800 C. Although there was 88 wt% gas yield at 800 C, the thermal effect was very strong and the catalyst was not effective in cracking the lignin degradation products. Thus, this catalyst needs considerable improvement to effectively crack the lignin fraction of the biomass pyrolysis products.

See more of this Session: Catalytic Conversion of Renewable Resources to Synthesis Gases and Pyrolysis Oils
See more of this Group/Topical: Sustainable Engineering Forum

# (21a) Thoughts On Unit Ops Lab: What Was, Is, and Ought

**Matthew J. Cline** and B. Erik Ydstie, Chemical Engineering, Carnegie Mellon University, Pittsburgh, PA

The responsibilities of engineering educators continuallyevolve with, adapt to, and even influence industrial and societal conditions. Facultiescontinuously struggle over what, and how, to best prepare students. Prof.Ogden at Bucknell asserted in 1934 that a certain minimum amount ofsemi-plant equipment must be used in order to present more effectively theapplications of the theory of chemical engineering as actually practiced inindustrial operations. The Society for the Promotion of EngineeringEducation, now the American Society for Engineering Education, studied chemicalengineering laboratory instruction and equipment for the next year. The natureof that study (and of subsequent studies), along with the attendant findingsand recommendations, illustrate the status quo to a large extent. Thehistorical development of chemical engineering laboratories, in particular theteaching of methods for developing and scaling industrial production processes, will be traced. Juxtaposed with the underlying motivations for more recentcalls for change, one will see why undergraduate chemical engineeringlaboratory equipment and teaching practices continue to shift focus away frompilot-plant scale operation and process design. The ongoing changes to theUnit Operations lab at Carnegie Mellon University demonstrate increased focus,in particular, to engineering practice skill development and to student-led,research or development projects of a wide variety of topics. Furthermore,each project may have different desired educational outcomes. This *is* and,we assert, *ought* to be. The likely consequences designers of chemicalprocesses and producers of chemical products may see will be discussed.

See more of this Session: Design, Construction and Operation of Unit Operations Labs and Pilot Plants
See more of this Group/Topical: Process Development Division

# (21b) Scale-up of Membranes for Separation of Hydrogen From Syngas for Carbon Capture

**David H. Anderson**, Carl R. Evenson, John D. Faull, Darcie D. Bailey and Doug S. Jack, Eltron Research & Development, Inc., Boulder, CO

In partnership with the Department of Energy (DOE) and Eastman Chemical Company, Eltron Research & Development Inc. is developing a process technology that separates hydrogen from carbon dioxide in syngas, enabling clean power generation and carbon capture in coal-fired Integrated Gasification Combined Cycle (IGCC) power plants. The process is based on membranes that permeate pure hydrogen and retain the $CO_2$ at high pressure to aid its eventual sequestration. In addition, membrane separation can achieve nearly complete carbon capture depending on process configuration. Scale-up is in progress for the membrane technology including pilot plant testing on coal-derived syngas. The status of the scale-up steps and recent results of the testing will be presented.

See more of this Session: Design, Construction and Operation of Unit Operations Labs and Pilot Plants
See more of this Group/Topical: Process Development Division

# (21c) Integrating Membrane Separations Into the Chemical Engineering Laboratory

**Daniel Anastasio**, Chemical, Materials, and Biomolecular Engineering, University of Connecticut, Storrs, CT and Jeffrey R. McCutcheon, Department of Chemical, Materials and Biomolecular Engineering, University of Connecticut, Storrs, CT

Crossflow reverse osmosis (RO) and forward osmosis (FO) systems were constructed to introduce senior-level chemical engineering students to key facets of membrane separations. During the RO experiments, students examined several commercial RO membranes, measuring water flux and salt rejections at various pressures, flow rates, and feed water salinities. During FO experiments, students examined a single commercial FO membrane, measuring water flux and feed conductivity at various draw concentrations, flow rates, and temperatures. Ultimately, students used the data they collected to calculate each membrane's hydraulic permeability constant and a variable indicative of membrane selectivity (salt rejection for RO and salt permeability for FO). The data presented is taken directly from the laboratory reports of undergraduate seniors at the University of Connecticut. These data were compared to the manufacturer's specifications. Boundary layer theory was utilized in both experiments to predict theoretical permeability and/or flux values after accounting for concentration polarization. Together, these versatile systems have granted students a thorough understanding of membrane separations and mass transport phenomena in the context of a capstone undergraduate laboratory course.

See more of this Session: Design, Construction and Operation of Unit Operations Labs and Pilot Plants

See more of this Group/Topical: Process Development Division

# (21d) Project Based Learning In Senior Unit Operation Laboratory At the University of Michigan

**Henry Y. Wang**, Pablo LaValle and Matthew A. Robinson, Department of Chemical Engineering, University of Michigan, Ann Arbor, MI

The University of Michigan has one of the oldest unit operations laboratories in this nation having been part of our core curriculum in Chemical Engineering since the founding of the department. All students must work on several types of process equipment such as a reactor, distillation column and evaporator to enhance their understanding of the fundamental technical knowledge from various lecture courses. During the past few years, we have modified the existing equipment in the laboratory and integrated them to form a simulated biodiesel production facility. In addition to the advantages of providing the students a clear understanding of the interconnectedness of these unit operations in a real production process, this approach also allows the laboratory to become an invaluable teaching tool for various soft skills fundamental to our profession. Skills such as team work, time management, the importance of producing practical information in a timely manner for other groups or teams are emphasized as the students work in an interdependent group environment to develop the specification and design of a large industrial scale facility based on the information gathered from an actual pilot plant or laboratory environment. Additionally, the students are encouraged to produce by the end of the semester a single group report on the viability of the process integration and improvement for the entire plant. We shall share our experience in this transformation of the senior unit operation laboratory in this talk.

See more of this Session: Design, Construction and Operation of Unit Operations Labs and Pilot Plants
See more of this Group/Topical: Process Development Division

# (21e) Using Pilot Plants In a Capstone Unit Operations Laboratory Course to Create a Chemical Manufacturing Experience

**John F. Sandell**, David W. Caspary and Jason M. Keith, Chemical Engineering, Michigan Technological University, Houghton, MI

Unit Operations courses are typically considered "capstone" courses in many undergraduate chemical engineering programs. At Michigan Technological University, each semester of the senior year has a required 3 credit unit operations focused course. This paper details the content for the second of these two courses which is a unique, manufacturing focused Chemical Plant Operations course utilizing two fully automated pilot plants. The pilot plants include a two-story, 30 gallon jacketed, batch polymerization reactor (PDMS reactor) and a three-story, distillation column (Solvent Recovery Unit or SRU). Both pilot plants are part of the Process Simulation and Control Center (PSCC) at Michigan Technological University. The course incorporates participation by numerous industrial speakers as well as utilization of an Emerson Process Management Delta V™ control system, Foundation™ fieldbus and Hart-enabled process instrumentation, an OSIsoft PI™ data acquisition system, Six Sigma quality control tools, and ISA S-88 methods. Lecture material includes quality management, the application of statistical process control, and current trends in quality manufacturing. Students complete two "project cycles" requiring the completion of proposals prior to each run followed by the submission of comprehensive written and oral reports.

See more of this Session: Design, Construction and Operation of Unit Operations Labs and Pilot Plants
See more of this Group/Topical: Process Development Division

# (21f) Distributed Control System Implementation for Undergraduate Labs

**Peyton C. Richmond**, Chemical Engineering Department, Lamar University, Beaumont, TX

A Honeywell Experion DCS system was recently acquired and configured as part of a multi-year plan to expand and remodel the Lamar University Chemical Engineering Department's Unit Operation and Process Control Labs. The first phase is complete: a new "Control Center" with 10 Flex Stations was established and the Department's two Process Control Experiments were converted from independent single loop control to DCS based control. Those Control Experiments were used by our undergraduates for the first time during the spring 2011 semester. The next phase will involve extending the DCS system to the Unit Operations laboratory which is located in a separate building. This presentation will include some of configuration choices made and the lessons learned during the acquisition and configuration of this industrial DCS system for academic use.

See more of this Session: Design, Construction and Operation of Unit Operations Labs and Pilot Plants
See more of this Group/Topical: Process Development Division

# (21g) "I Thought PTFE Tape Was Optional": Teaching Practical Engineering Skills As a Component of Unit Operations

**Eric P. Codner**, Chemical and Biological Engineering, University of Wisconsin - Madison, Madison, WI

Undergraduates are increasingly skilled at computer simulation, seemingly at the expense of mechanical aptitude and knowledge of process fabrication and operation techniques. While this may not be a serious problem in large organizations, many beginning engineers enter small companies in which they are the sole technical resource, and a lack of practical skills may substantially hinder their professional progress.

The unit operations course at the University of Wisconsin – Madison has traditionally required students to design and construct several experiments during the term. Over the past four years our approach has undergone considerable development, resulting in the following initiatives:

- Students in the unit operations and fluid mechanics laboratories are now required to assemble portions of the equipment prior to performing experiments, in order to familiarize them with tools, materials, and equipment commonly found in process environments.
- Many experiments are structured to include a troubleshooting or process improvement component, in which students analyze the operation of the experiment and make determinations concerning maintenance, backup operation, or repair.
- There is increased emphasis on measurement uncertainty and statistics, and most measurements are obtained using data acquisition systems. A number of the experiments employ PLCs for measurement and control purposes.

This paper discusses the current state of the senior laboratory curriculum, student response to the changes, and ongoing challenges such as minimizing equipment damage and successfully training teaching assistants.

See more of this Session: Design, Construction and Operation of Unit Operations

# EFFECTIVE APPLICATION OF PROJECT RISK MANAGEMENT

*Stephen L. Cabano*

*Pathfinder, LLC*
*Cherry Hill, NJ*

## Biography

*Steve Cabano is President of Pathfinder, LLC, a project management consulting firm specializing in the planning, development and execution of domestic and international capital plant construction projects. With more than 25 years of project engineering and project management experience in the industrial and commercial facilities industry, he has provided cost, scheduling, procurement, and similar project related services for owner and government clients in the Petroleum, Petrochemical, Chemical, Environmental, Pharmaceutical, Industrial and Commercial industries in the United States and throughout the world.*

## Introduction

In today's complex capital project environment, a project manager has a myriad of risks to consider and address. The traditional risks include: managing to the project cost and schedule commitments, addressing regulatory and environmental concerns, availability of internal resources vs. contracting out for services, among others. In today's increasingly active capital project execution environment, a whole new set of potential risks have arisen. Several of these include availability of competent engineering and construction contractors, experience of engineering and construction supervisors, material and equipment availability, the multi cultural nature of today workforce, etc.

## Basis of Observations

This paper will address capital project risk management issues from an owner's perspective concerning projects in the process industry. For the purposes of this paper, we define the process industry as the oil, gas, chemical, petrochemical and related industry sectors. We will also discuss the application of risk management tools and techniques across an industry accepted, 5 Phase "Best Practice" Capital Project Delivery Process as outlined below in Figure 1:

# EFFECTIVE APPLICATION OF PROJECT RISK MANAGEMENT

**Figure 1**

***The Project Management Process***

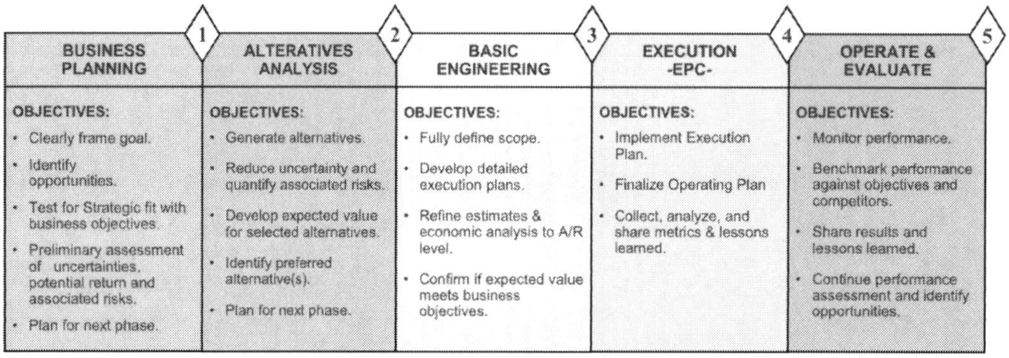


## The Risk Management Process

The Project Management Institute (PMI) identifies Risk Management as one of the basic Project Management Bodies of Knowledge. (See Figure 2)

**Figure 2**

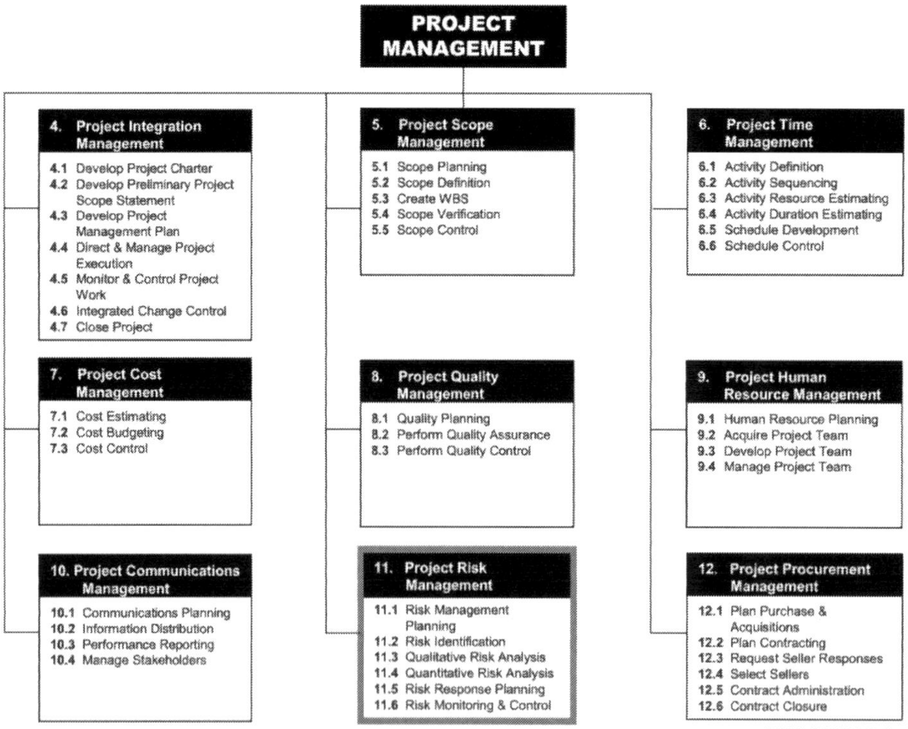


*2008 PMBOK Guide*

## EFFECTIVE APPLICATION OF PROJECT RISK MANAGEMENT

PMI defines Risk Management in the following 6 categories:

- Risk Management Planning
- Risk Identification
- Qualitative Risk Analysis
- Quantitative Risk Analysis
- Risk Response Planning
- Risk Monitoring and Control

Let's take each one of these categories and identify the issues related to each in today's project execution environment.

### *Risk Management Planning*

For a Risk Management approach to be effective there needs to be a company-wide established process which is applied, utilized and adhered to by all project representatives, as well as a project specific risk management plan integrated into the project's execution plan.

Company-wide processes should outline the corporate tools, resources, and deliverables by phase. A series of support "think lists"/risk considerations should be established for each project phase to assist the project team in identifying project risks. An example of this is outlined below in Figure 3:

**Figure 3**

***Risk Management Think List by Project Phase***

| OPPORTUNITY (1) | ANALYSIS (2) | DEFINITION (3) | EXECUTION (4) | OPERATE (5) |
|---|---|---|---|---|
| • What are the business risks?<br>• What market influences will impact this opportunity?<br>• How will these risks impact the company? | • What are the opportunities and threats for each alternative evaluated?<br>• How do these risks impact cost, schedule, operations, reliability, etc.?<br>• What is the benefit of implementing a new technology? What are the risks? | • What are the project risks associated with the selected alternative?<br>• What are the business risks associated with the selected opportunity?<br>• How will the identified risks impact the overall project?<br>• Are there mitigation options that should be incorporated into the base scope/execution plan? | • What new risks have been uncovered during detailed engineering, procurement, construction and Start-up?<br>• How will these new risks impact the commitment made to management?<br>• Are these late risks manageable and will they impact operations? | • What will impact the effective handover to operations?<br>• What risks can be mitigated through application of effective Lessons Learned?<br>• Are there risks that will affect operability/ maintainability and therefore operating costs and/or reliability/ utilization factors? |

# EFFECTIVE APPLICATION OF PROJECT RISK MANAGEMENT

The Project Execution Plan (PEP) should also have a specific Risk Management sub-section dedicated to each of the defined sections of the PEP. This sub-section should define the risks associated with that component of the PEP and describe how the team will handle the defined risks. The sections of a "Best Practices" PEP Include:

### *Section 1 - PROJECT CHARTER*

- Strategic and Business Objectives
- Project Objectives
- Scope of Work
- Support of Local Communities/ Workforce/ Enterprises

### *Section 2 - PROJECT PLANNING BASIS*

- **Risk Management Plan**
- Organization Plan
- Master Plan/ Schedule
- Contracting Plan
- Funding and Cost Estimate Plan

### *Section 3 - PLAN FOR IMPLEMENTATION*

- Design Management and Value Improving Practices(VIP) Plan
- Constructability Plan
- Procurement and Materials Management Plan
- Construction Management and Labor Plan
- Mechanical Completion/Commissioning Through Start Up Plan
- Regulatory and Environmental Management Plan
- Project Controls Plan
- Quality Management Plan
- Safety Management Plan
- Informational Management and Communication Plan
- Document Management Plan
- Operations Management Plan
- Plan to Capture Lessons Learned
- Special Factors Plan (Modularization, Project Suspension/ Restart Plan etc.)

# EFFECTIVE APPLICATION OF PROJECT RISK MANAGEMENT

In a Best Practices PEP, the risk sub-section includes a write-up on the Risk Management Plan, the identified risk, what impact the risk will have on the project (+/-), how the team intends to address the risk, who is responsible for implementing the mitigation approach, when the mitigation is intended to be completed and what benefits (quantifiable/tangible of cost/schedule improvements) should be expected as a result of implementing the recommended mitigation. This is often called the risk register and becomes part of the PEP, but is also updated. Risks are mitigated as other risks become apparent.

***Risk Identification***

As an initial step in the Risk process, the team needs to "know what it does not know". It needs to brainstorm project issues that can both negatively and positively impact the project and then determine how probable the risk occurrence will be, as well as the impact to the project if the risk occurs. This comes from experience. Having "been there, done that". It also requires research of Lessons Learned from previous projects (inside and outside of the company) which will, at the very least, stimulate thought provoking issues that the team can evaluate for possible applications to their project(s).

So, based on the above discussion of limited experienced resources across the industry, how do we know what we do not know? How do organizations today share the limited project experience across projects to gain the knowledge and input for all projects rather than targeting just a few projects?

A traditional step includes implementation of a Risk Identification workshop in coordination with the Project Execution Planning workshop. As the team defines the execution approach that best suits the project and business objectives, various risks would be unfurled and documented as part of the PEP. The risks would then be assigned to responsible parties. These parties, together with the project team, would define mitigation strategies for each of the identified risks. This approach works well when the project team is made up of highly experienced resources representing their respective disciplines. Today's teams do not necessarily have this cross section of experience and must rely on other risk identification techniques.

Today, we rely heavily on checklists of questions and Lessons Learned related to various aspects of the project scope and execution strategy. These checklists only go so far – as the individual asking the questions may not see the inherent risk in how these questions are answered.

There is one technique that has proven effective but takes a mature organization to assure appropriate implementation. What some organizations have realized is that, with the limited pool of experienced project management resources available, an effective approach to gaining the best use of their experience is not necessarily to put them on a single project, but to have them act as an independent review resource that all projects must go through at key junctures in the project. The experienced individuals are asked to "Put their black hats on" and see what issues they see in the scope definition, execution plan, etc. of the project as it moves through the companies. These individuals need to have the intestinal fortitude to drill down into details at key junctures in the project to better prepare the project team for the future project phases. This function is often part of a corporate-wide Project Management Organization (PMO), which is a shared service across the organization.

As projects become more internationally integrated and technologically complex, some of the consistent risks identified in today's project environment include:

- Limited experienced planning, oversight and execution resources
- Competing business demands
- Significant competition for limited pool of qualified Engineering and Construction contractor resources
- Inadequate scope and/or site condition definition
- Escalation variability
- Supplier/Fabricator shop loading (over-commitment impacting delivery schedules)
- Labor Productivity – dwindling supply of qualified craft workers
- Extreme sensitivity to economic conditions/ capital investment reticence
- Discipline to follow established strategies, plans, processes and procedures (schedule pressures)
- Ill-defined contractual requirements (scope of work and services)
- Insufficient recognition of host country requirements/project environment (work rules, labor availability/qualifications, etc.)
- Logistical challenges (world-wide procurement, customs/duties, transportation, infrastructure at location, etc.)
- Learning curve costs (just-in-time recruiting, on-boarding, orientation)
- Staff turnover
- "First time" consortium/"partnership"/"alliance" arrangements – integration, coordination, systems and procedure compatibility
- Others...

These are but a few examples of the risks and challenges facing business and project ventures, particularly in mega-projects and/or international/multi-national situations/endeavors.

## EFFECTIVE APPLICATION OF PROJECT RISK MANAGEMENT

Risks today that are typically overlooked are often related to the execution environment more so than with the technology and processes involved. As stated previously, lack of experienced project management resources is a huge risk issue today. Availability of competent international suppliers that can deliver the quality and timeliness required is another major concern, as is availability of contractor teams that have worked together in the past and are not thrown together solely for the project at hand. Multi-cultural execution teams also represent some unique challenges (in some cases).

### *Qualitative Risk Analysis*

Identifying project risk is only part of the battle; it is time and cost prohibitive to mitigate all possible project risks. With this in mind, the team must prioritize project risks and select the most effective Risk Response for each risk based on business and project goals and objectives. The prioritization process must include a measurement of the impact the risk will have, if it occurs, as well as the probability that the risk will occur. A typical risk matrix is included below as Figure 4:

**Figure 4**

# Risk Matrix

**Scope of use:** Utilized to assess, evaluate and categorize risks associated with capital project planning and execution, and Major Projects Contracts, Purchase Orders.

Project Specific

**Consequence** (Note: For purpose of selecting impact of event, utilize most severe outcome of Project Specific and Corporate consequences analyzed)

**Likelihood** How often would it be expected to have similar causes or circumstances aligned for the event with the defined consequences to re-occur? Please make reference to site, company and industry historical data to help predict the re-occurrence frequency of such an event.

Project Specific

| Cost/Financial Impact | Schedule | Scope | Severity | **Very Low** --- < 5% | **Low** --- =>5% and < 20% | **Medium** =>20% & < 50% | **High** --- =>50% & < 80% | **Very High** --- =>80% |
|---|---|---|---|---|---|---|---|---|
| | | | | Event will almost surely not happen | Event will quite unlikely happen | Event will quite likely happen | Event will very likely happen | Event will almost surely happen |
| Cost Δ > $10MM or Cost Δ > 5%*Capex; which ever is smaller | > 6 months | Scope or Quality does not meet performance or business expectations | Very High | Moderate | Significant | Significant | Critical | Critical |
| 1MM <Cost Δ ≤10MM; or 2%*Capex< Cost Δ ≤ 5%*Capex; which ever is smaller | > 3 months but ≤ 6 months | Scope or Quality unacceptable to Project Manager | High | Negligible | Moderate | Significant | Significant | Critical |
| 100K <Cost Δ ≤1MM; or 1%*Capex< Cost Δ ≤ 2%*Capex; which ever is smaller | > 1 month but ≤3 months | Significant impact on scope or quality of major project deliverable and/or component | Medium | Negligible | Moderate | Moderate | Significant | Significant |
| 10K <Cost Δ ≤100K; or 0.5%*Capex< Cost Δ ≤ 1%*Capex; which ever is smaller | > 1 Week but ≤ 1 month | Impact on scope or quality of minor project deliverable and/or component | Low | Negligible | Negligible | Moderate | Moderate | Significant |
| 0 <Cost Δ ≤10K; or 0%*Capex< Cost Δ ≤ 0.5%*Capex; which ever is smaller | Negligible | No noticeable scope or quality impact | Very Low | Negligible | Negligible | Negligible | Negligible | Moderate |

Consequences

This too is a function where the above defined shared PMO risk function can be of benefit. The experienced project representative can bring some realism to the quantitative assessment of the identified risks. Inexperience can lead to downplaying what might seem to be minor issues, which if thoroughly analyzed, could have major impact to project success potential.

### *Quantitative Risk Analysis*

The team must also measure the impact a risk might have on the project cost and schedule. The main issue here is that one risk may impact a number of project activities. As an example, site logistics or poor transportation to and from the site which might include limited road accessibility, multiple bridges and overpasses, poor road conditions, etc., will not only impact delivery of equipment and materials but also may cause construction productivity issues (people cannot get to the site in a timely manner), emergency response timing, local pubic relations issues, etc. These impacts all need to be taken into consideration when quantitatively measuring the impact of an identified risk.

There are a number of excellent risk analysis tools on the market that, when used properly, can assist the project team in measuring the impact risks have on the project cost and schedule. We will not go into their functionality in this paper, but many use a Monte Carlo simulation approach to develop risk profiles for both cost and schedule. This type of analysis assists the team in selecting which Risk Response effort provides the most project/business benefit for the investment. The project team needs to be well-versed in utilizing these tools, as they can lead to a misunderstanding of the impacts of risk if applied incorrectly. Owner management also needs to define what level of cost and schedule probability the project must "shoot for", as this will affect how much time and cost contingency are applied to the project.

If a selected risk mitigation strategy is applied, the cost, time and resources required to implement this strategy must also be considered, quantified and included in the base cost estimate and schedule. This is often overlooked and leads to false risk analysis readings.

### *Risk Response Planning*

Probably the most important aspect of Risk Management is what the team will do to address the risk. Identification and measurement are critical but the real test is in how the team addresses the risks. There are a number of risk response options which are outlined in Figure 5 below:

**Figure 5**

**Responding to Risk**

| RESPONSE TYPE | EXAMPLES |
|---|---|
| Avoidance | Choosing only high probability options, eliminating high risk approaches on parallel programs, screening subcontractors. |
| Prevention | Focusing on preventive measures, extensive potential problem analyses, extensive bench testing, mockup, and simulation work |
| Reduction of Impact /Mitigation | Focus on contingency planning, large contingency funding, frequent progress reviews |
| Compromising | Tending toward low risk options, choosing low return options, pursuing redundant paths |
| Transfer | Through subcontracting, warranties, performance bonds, insurance |
| Assumption | Any decision to proceed implies assumption of risk |

Often risk registers are developed but not implemented; risk responses are identified, but too late. Many times the symptom is addressed but not the root cause (example: poor productivity in the field - the owner authorizes more resources but the true causes were engineering quality and/or late material deliveries).

In the case of engineering quality, the risk may be avoided by the selection of high caliber technical personnel within contractual terms.

### *Risk Monitoring and Control*

Once the Risk Response approach has been implemented, management and project team representatives need to be assured that the selected response is doing what it intended to do. Project Control plays a key role here. The team needs to be assured that the money, resources and time allocated to a risk mitigation option are successfully achieving the benefits expected. If not, the team needs to reevaluate the option and develop an alternate approach.

One of the main issues we have today is the weakness in project controls across the industry. Of all the 'brain drain' suffered across the project disciplines, the degradation of the project control capability has been the greatest. This is a risk in itself. The team needs to be assured that it has a strong in-house, project control capability to assure the team can measure the effectiveness of the mitigations options.

# EFFECTIVE APPLICATION OF PROJECT RISK MANAGEMENT

## Why Is Risk Management so Important Today?

Today's projects are more complex, involve more planning and execution entities, are executed in a more challenging environment and involve multi-national resources. These issues, combined with a less experienced workforce are a recipe for disaster. Identification, measurement and response planning can aid in eliminating and/or minimizing the negative impact that risks may have on project success.

## Who is responsible?

Risks are shared by the owner, contractors, suppliers, and other project stakeholders. Ultimately, the owner bears the burden of accountability for all project risks. At the end of the day, it's the owner's money. Product delivery is an owner issue (especially if they have made sales agreements and commitments to their customers). Quality becomes an owner issue – as the owner has to operate the facility for the next 20 – 30 years.

## In Conclusion

A Project Manager in today's world cannot effectively manage a project without also managing the project risks. How aggressively one tackles the risk is dependent on the business/project drivers for the project, scope definition, execution team experience, etc., and is all part of the project execution strategy defined in the planning process. Risk management is one of the most critical skill sets a Project Manager needs to successfully manage projects in today's environment. Mastering this knowledge area goes a long way toward assuring achievement of project goals, objectives and success factor requirements.

# (34b) The Legal System As a Driver for Innovation In Energy Technology

**Richard D. Jordan**, IP, Dickinson Wright PLLC, Washington, DC and Mary Ternes, McAfee& Taft, Oklahoma City, OK

This presentation will provide an overview of how the legal system acts as a primary driver for innovation in energy related technology, considerations for funding from public and private sources, and considerations for determining how to best protect intellectual property developed in the innovation effort. The presentation will be made by invited policy and legal experts. These experts will discuss policy and procedural considerations in which governmental agencies, such as the Environmental Protection Agency, "intervene" for the good of society to require industry to provide "cleaner" technologies. These legal drivers in turn cause the industry to invest a great amount of funds to innovate in order to meet the new legal requirements. Panel experts will discuss: 1) a few of the policy issues that are considered by governmental institutions in implementing the new laws and regulations, 2) government grants or funds that are made available to stimulate innovation in such a capital intensive market, and 3) considerations on how to protect valuable intellectual property that will be created in getting the new innovations to market.

See more of this Session: Law, Risk, and Ethics of Technology and Project Management
See more of this Group/Topical: Management Division

# (34d) Patenting Basics

**Vivek Koppikar**, Patent Attorney, Dale City, VA

This partially invited session will provide a fundamental review of the patent procurement process and patent law in general, and is designed for anyone who has any type of interest in the patenting process and/or deals with patents in their work in industry and/or academia. Little or no background knowledge in patents is needed by those intending to attend this session. The session will begin with the very basics of the patenting process as well as providing them with basic terminology. Next, the session will provide a basic synopsis of the patent procurement process with the US Patent and Trademark Office. The session will then focus on portions of a patent most relevant in patent disputes and patent litigation. Case studies involving various patent scenarios will be presented. The session will wrap up with a question and answer session where attendants can ask basic questions dealing with the patent procurement process and patent law in general. This session will provide a foundation for two following patent sessions focusing first on drivers for innovation in energy technology and, next, on litigation for engineers. Abstracts are solicited for case studies from lawyers and engineers with various patent experience

See more of this Session: Law, Risk, and Ethics of Technology and Project Management
See more of this Group/Topical: Management Division

**Ethical Project Management**

**Dr. Eldon R. Larsen**
**Marshall University**

## ABSTRACT

Project managers are often faced with situations that present ethical dilemmas. These situations can be anything from very simple, like something in a conversation, to something very complex, like how to accept or not accept a gift. This paper presents some principles and suggestions regarding ethics in project management. It also highlights the main concepts in the Project Management Institute's ® Code of Ethics and Professional Conduct.

## INTRODUCTION

Many people have written about ethics. The objective of this paper is not to re-hash past studies. Rather, I attempt to present significant concepts and principles through both quotations and discussion that will help the reader gain a vision of what it means to be ethical in one's behavior. This applies very much to project managers, as they are frequently making decisions that have ethical questions surrounding them.

## CARDINAL RULES OF ETHICAL BEHAVIOR

Following are a few quotations in which are embedded basic principles and rules of ethical behavior. For several of the quotations are given "sub-quotations" that are closely related, but are expressed in entirely different ways.

- "Therefore all things whatsoever ye would that men should do to you, do ye even so to them" (Matthew 7:12)
    - "You can easily judge the character of a man by how he treats those who can do nothing for him." (James D. Miles)
- "And above all things have fervent charity among yourselves: for charity shall cover the multitude of sins." (I Peter 4:8)
    - "When I do good, I feel good. When I do bad, I feel bad. That's my religion." (Abraham Lincoln)
- "Thou shalt love thy neighbor as thyself." (Leviticus 19:18)
    - "Laws control the lesser man. Right controls the greater one." (Chinese proverb)
- "To thine own self be true, and it must follow, as the night the day, thou canst not then be false to any man." (William Shakespeare, *Hamlet)*

- "To know what is right and not do it is the worst cowardice." (Confucius)

## FOOTNOTES TO THE CARDINAL RULES

- Though you legally can do something to someone else, it might not be alright to do it to him.
- Ethical people normally follow the laws, but are not constrained from going beyond the bare minimum of the law to do what is right
- Good ethical behavior is at the heart of excellent leadership—you do what is right because it is right, not because it benefits you monetarily or otherwise
    - "Keep true, never be ashamed of doing right; decide on what you think is right and stick to it." (*George Eliot)*
- "If moral behavior were simply following rules, we could program a computer to be moral." (Samuel P. Ginder)

## ETHICAL PROJECT MANAGEMENT AND PRINCIPLES

Ethical project management involves right behavior personally, legally, socially, and environmentally. Inconsistencies in one area can affect behavior in another area. Consistency of ethical behavior in all areas strengthens a person's integrity.

We normally consider the goal of project management to be to achieve project objectives. Some questions we should ask ourselves, however, are the following:

- How well do we meet the objectives, and within what constraints?
- Does it matter how we get there?
- Are some objectives more important than others?
- How do we treat team members? Stakeholders?

How might ethical considerations affect the answers to the above questions?

Many organizations have a code of ethics by which they expect their participants to abide. However, it is important to consider how the code was developed. I suggest that you should always questions a code of ethics which is based solely upon the premise that right or wrong is determined either by opinion or by statute. Truth is true whether people agree with it or not. And even making a statute does not change that truth is still true, and false is false. True principles will always be true, no matter how hard we might try to ignore them, change them, or go against them.

The fact that true principles will always be true can present us with a *principles dilemma*. One of the challenges we face is when principles seem to conflict with each other. For example, is it right to tell a lie in order to save an innocent person's life? Many people would probably say "Yes." This leads us to realize that one principle might be more important than another, even though they are both true (Don't kill. Don't tell a lie.). It is often our conscience and our experience that tell us what to do in these circumstances. In fact, good ethical behavior is closely tied to a sensitive conscience.

## PRINCIPLES AND VALUES

Principles and values are not exactly the same thing, although they can be very closely related. Most values are built upon some principles. Values are like roadmaps that guide us in our decision-making processes. Therefore, it is important for us to identify and understand our values—companies and organizations need to understand their own core values. Once we better understand our values, we can more easily make an ethical decision through some kind of structured analysis.

Cultural norms can affect what is considered acceptable or not. While principles (truth) will always be true, the application of the principles might change significantly, depending upon the situation. Values (how we perceive life) sometimes vary with culture, religion, upbringing, etc., and therefore change the situation. They relate to perspective. So, it is important to understand cultures when considering what decision we face is ethical. We don't have to break principles when we adjust for different values—they are not the same thing.

Project management ethical considerations often involve the following questions.

- Are you obeying the law? Are you following guidelines provided by your organization and the customer's organization? Which one should take priority?
- Are you treating other people the way you would want to be treated, and would you do it either in private or in public and still have personal peace?
- What is driving your actions? What if everyone did what you are considering doing? Would that be good?
- Is your choice the right thing to do, according to the principles you know to be true?

There are also some legal considerations when dealing with ethical questions. For example:

- "The Foreign Corrupt Practices Act – FCPA, 1977, states with respect to bribery that it is a crime for a bribe to be made to a foreign official or political party for the purpose of obtaining or retaining business or for directing business to another person even if this flows through an intermediary or consultant.
- "However, not all payments are defined as "bribes." Payments for "routine" government action are allowed under the FCPA including the dispensing of permits, licenses and police protection as noted in the example."[1]

Following is an ethical, legal situation to illustrate some of the challenges in making ethical decisions:

- Country A allows companies to perform drug testing on employees, but allow employees to refuse to be tested
- In Country B, companies can mandate that employees and contractors submit to drug testing, and refusal can be grounds for dismissal
- A Country A contractor is asked to perform work for a company in Country B, that asks the contractor to submit to a drug test. He refuses. Right choice or not?

Even though the contractor has the right to refuse drug testing according to the policies of his own home country A, because he is doing work for company in country B he should follow the more stringent policies of country B.

**The Project Management Institute's Code of Ethics and Professional Conduct**

The Project Management Institute's Code of Ethics and Professional Conduct[2] includes many fine points of ethical behavior for project managers. For this reason, it is useful to summarize them here.

1. Take responsibility for the decisions we make or don't make, and for the resulting consequences
2. When making decisions, take into account the best interests of society, public safety, and the environment
3. Keep commitments
4. If we make a mistake, take ownership and responsibility, and correct it. If others make mistakes, communicate appropriately as quickly as possible
5. Protect proprietary or confidential information
6. Be informed regarding policies, laws, regulations related to our work, profession, and volunteer activities
7. When substantiated by facts, report unethical or illegal conduct to appropriate management and, if necessary, to those affected by the conduct
8. Be informed about the customs of others and avoid being disrespectful
9. Listen to other people's opinions and views and try to understand them
10. When a conflict occurs, deal directly with the people involved
11. Be professional, even if this conduct is not reciprocated
12. Negotiate in good faith
13. Do not use power to benefit personally at the expense of others
14. Do not act abusively
15. Respect the rights of others

16. Demonstrate transparency in decision making
17. Be impartial as much as possible
18. Allow equal access to all who are authorized to have information
19. Make opportunities equally available to qualified candidates
20. Actively and fully disclose potential conflicts of interest to appropriate stakeholders
21. Refrain from engaging in decisions or influencing others when we have a conflict of interest, unless or until: we have made full disclosure to affected parties; we have an approved mitigation plan; and we have obtained consent from affected stakeholders to proceed
22. Do not hire/fire, reward/punish, or award/deny contracts based on personal considerations like favoritism, nepotism, bribery, etc.
23. Do not discriminate against other people based on things they cannot control or based on their religion
24. Apply your organization's rules without favoritism or prejudice
25. Seek the truth
26. Communicate truthfully
27. Provide timely, accurate information
28. Make implied or explicit commitments in good faith
29. Strive to create a non-threatening environment for people to tell the truth

## SUMMARY

Ethical project management involves many facets of application, all of which should be based upon foundational principles of good management and human relations. The basic rules or principles of ethical behavior have existed for thousands of years and will continue to exist for thousands more. Truly successful project management applies these ethical principles fairly according to the situation and the needs of the people involved.

## REFERENCES


1. http://www.bia.ca/articles/pj-project-managers-ethical-dilemma.htm
2. http://www.pmi.org

# (56a) Solvent Selection Techniques for Pharmaceutical Processes

**Kevin G. Joback**, Molecular Knowledge Systems, Inc., Bedford, NH

Solvents are heavily used throughout many pharmaceutical processes. Solvents are used for cleaning, for forming azeotropes, in crystallizations and liquid extractions. The proper selection of solvents is thus critical to the development of a successful process.

In this paper we will present an overview of various techniques for selecting solvents. We will start by examining the thermodynamic, process design and environmental sources of the physical property constraints needed to guide the selection process. We will then present some of the computational techniques used for the selection process. Finally, an example will be shown to demonstrate how these techniques can be used to select crystallization solvents.

See more of this Session: Solvent Design for Pharmaceutical Processing
See more of this Group/Topical: Process Development Division

# (56b) Efficient, Reliable and Predictive Solvent Design for Pharmaceutical Processes

Elisa Conte, CAPEC, Department of Chemical and Biochemical Engineering, Denmark Technical University, Lyngby, Denmark, Rafiqul Gani, Department of Chemical and Biochemical Engineering, Denmark Technical University, Lyngby, Denmark, Peter A. Crafts, Process R&D, AstraZeneca Pharmaceuticals Ltd., Cheshire, United Kingdom and **Sascha Sansonetti**, CAPEC, Department of Chemical and Biochemical Engineering, Denmark Technical University (DTU), Kgs. Lyngby, Denmark

Solvents play a very important role in the processing of pharmaceutical products. For example, use of pure solvents and antisolvents for crystallization and in organic synthesis steps are critical in pharmaceutical process design. In fact, the identity of the solvent or antisolvent is a key issue for efficient solute recovery in the case of crystallization and solid reactant conversion in the case of reaction. From an environmental point of view, the solvents also need to be "green" and improve process sustainability. Also, in pharmaceutical formulations, solvents are needed. In many situations, solvent blends could enhance the solubility of the active pharmaceutical ingredient (API) and show a 'solubility peak' (solubility higher than the solubility in the two pure solvents). Therefore, it is important to have a method for solvent design method that is able to cover pure solvents as well as solvent blends.

In this presentation a systematic method to identify/design pure solvents, solvent blends that show solubility peaks and antisolvents that cause a large decrease in solubility when added to the API-solvent system is presented.

The method is based on the theory of the conceptual segments introduced by Chau-Chyun and Song (2004) with the NRTL-SAC model: hydrophobic, polar and hydrophilic segments. First, a Marrero and Gani (Marrero and Gani, 2001) type of group contribution model is developed and applied to estimate the conceptual segment values in the absence of experimental data. The advantage here is that the procedure is now independent of the need for experimental data. Next, the API solubility in *apriori* generated list of pure solvent candidates is evaluated and the pure solvents are ranked according to their performance (for example, solvent power). Once the most promising candidates have been identified, mixture (blend) design is performed to identify the solvents that mixed with the best candidates give the solubility peaks. This is done by identifying the combination of conceptual segments that maximizes the excess (mixing) properties. The same procedure is also applicable for design of antisolvents and blends of solvents-antisolvents. In this case, however, a solubility valley is desired. That is, the method uses the excess properties of mixing to identify an enhancement of the API solubility for solvent-solvent blends (the formation of a solubility peak) or a reduction of the API solubility for solvent-antisolvent blends (the formation of a solubility valley). For organic synthesis or API solubility, the same procedure is also applicable.

A case study dealing with the crystallization of a well-known pharmaceutical product, ibuprofen, is highlighted to illustrate the application of method for identifying solvents, antisolvents and to design solvent-antisolvent blends for use in cooling crystallization followed by drop-down crystallization. Other examples involving paracetamol and hydrocortisone are also briefly presented.

**References**

Chen C.-C. and Song Y. Solubility Modeling with a Nonrandom Two-Liquid Segment activity coefficient model. *Ind. Eng. Chem. Res.* 2004, 43, 8354-8362.

Marrero J. and Gani R. Group contribution based estimation of pure component properties. *Fluid Phase Equilibr.* 2001, 183-184, 183-208.

See more of this Session: Solvent Design for Pharmaceutical Processing
See more of this Group/Topical: Process Development Division

# (56c) Determining Solubility of APIs and Intermediates From Automated Parallel Experiments and Modeling

**Naresh Chennamsetty**, Olav Lyngberg and Jun Qiu, Bristol-Myers Squibb Company, New Brunswick, NJ

The design of efficient pharmaceutical processes such as reaction and crystallization is critically dependent on choosing an optimal solvent system. Selecting the solvent system requires the knowledge of the solubility of APIs or intermediates in several solvents and solvent mixtures. However, even with the use of automation and material sparing techniques, the direct experimental measurement of solubility in all possible solvents and solvent mixtures and as a function of temperature is often impractical and inefficient due to the limited quantities of material available during the early stages of process development. To address this issue, earlier we have presented an approach based on combined automated parallel experiments and modeling to estimate the solubility in a large number of solvents and solvent mixtures. Here we explore the value of different solubility models, their experimental input requirements, output capabilities, and the overall success of an integrated empirical-modeling solubility workflow. The solubility modeling is typically performed using the NRTL-SAC (Nonrandom Two-Liquid Segment activity coefficient) model [1]. We compare the performance of NRTL-SAC with two other alternative models: PC-SAFT (Perturbed-Chain Statistical Associating Fluid Theory) [2] as implemented by Scienomics and regressed-UNIFAC (UNIversal quasichemical Functional-group Activity Coefficients) as implemented by Dynochem. Unlike the standard UNIFAC [3], the regressed-UNIFAC [4] from Dynochem treats the solute as a single functional group. Furthermore, we evaluate the predictive capabilities of these three different models and discuss their unique strengths, limitations, and accuracy. We also discuss the FTE (Full Time Equivalent) savings obtained through our combined automated parallel experiment and modeling approach in the pharmaceutical process design.

**References**

[1] Chen CC, Song Y, Solubility Modeling with a Nonrandom Two-Liquid Segment activity coefficient model. Ind. Eng. Chem. Res. 43, 8354-8362 (2004)

[2] Gross J, Sadowski G, Perturbed-Chain SAFT: an equation of state based on a perturbation theory for chain molecules. Ind. Eng. Chem. Res. 40, 1244-1260 (2001)

[3] Fredenslund A., Jones RL., Prausnitz JM., Group-Contribution Estimation of Activity Coefficients in Nonideal Liquid Mixtures", AIChE J., 21(6), 1086-1099 (1975)

[4] Crafts PA., The role of crystallization and solubility modeling in the design of active pharmaceutical ingredients, (in *Chemical product design : toward a perspective through case studies*, Ng KM, Gani R, Dam-Johansen K., Eds.), (2007).

See more of this Session: Solvent Design for Pharmaceutical Processing
See more of this Group/Topical: Process Development Division

# (56d) Solvent Design for Optimal Kinetics: From Quantum Mechanical Predictions to Experimental Validation

Heiko Struebing[1], Zara Ganase[1], Panagiotis G. Karamertzanis[1], Alan Armstrong[2], Amparo Galindo[1] and **Claire S. Adjiman**[1], (1)Centre for Process Systems Engineering, Dept of Chemical Engineering, Imperial College London, London, United Kingdom, (2)Dept of Chemistry, Imperial College London, London, United Kingdom

The performance of pharmaceutical processes depends on the flowsheet structure, the design of the equipment and the operating conditions, to a large extent, on the choice of solvents. The performance of the reaction steps is strongly affected by solvent choice: the rate of a given reaction in two different solvents can vary over several orders of magnitude as the solvent affects the solubility of the reactants and the rate constant – itself a function of the reaction's free energy surface. In practice, solvents are often selected in the early stages of process development, and considered as fixed for the purpose of reactor design. The search for optimal solvents is often limited to a restricted set of options, potentially leading to sub-optimal designs. This is partly due to the complexity of predicting solvent effects on reactions.

In this work, we propose a systematic methodology for solvent design for reactions based on quantum mechanical calculations and optimization-based computer-aided molecular design (CAMD). The first step of the design methodology involves the computation of the reaction rate constants for the reaction(s) of interest in 7 different solvents. This is done using conventional transition-state theory. The free energy barrier in each solvent is obtained through quantum mechanical calculations on the reactants and transition states, in which the solvent is captured via a continuum solvation model. Specifically, we use the M06-2X functional with the SMD solvation model, as implemented in Gaussian 09. In the second step of the design methodology, we use the quantum mechanical data to build a predictive model of solvent effects on the rate of reaction. For this purpose, we use the solvatochromic equation and group contribution approaches. This yields a "surrogate" simplified model which can be used to predict the reaction rate constant in thousands of solvents at very low computational cost. In a third step, we embed this model in a mixed-integer CAMD formulation, in which we specify the size of the solvent design space and any additional design constraints (e.g., limits on the solvent melting point). By solving this optimization problem, we obtain a prediction of the solvent which maximizes the reaction rate constant. We verify the rate constant obtained for the best solvent with the surrogate model by comparing to a quantum mechanical calculation. If there is a mismatch, we return to the second step and improve the surrogate model by including the newest rate constant obtained by quantum mechanics. We iterate until convergence is achieved and verify the performance of the final predicted solvent experimentally.

The proposed approach is applied to a Menschutkin reaction, a classical $S_N2$ reaction often studied in the context of solvent effects, where a solvent that maximizes the reaction rate constant is sought. The results of the quantum mechanical calculations are compared to experimental measurements of the kinetics by $H^1$ NMR, for a range of solvents. Good quantitative agreement is obtained. The application of the design methodology leads to the identification of an improved solvent for the reaction, a finding which is verified experimentally. Such an approach is readily applicable to the selection of solvents in systems with competing reactions, and is particularly promising where selectivity is an issue.

See more of this Session: Solvent Design for Pharmaceutical Processing
See more of this Group/Topical: Process Development Division

## Practical Applications of QbD for a Parenteral Drug Product

*Nancy J. Harper, Venkat Koganti and Gautam R. Ranade, Pharmaceutical Research and Development, Pfizer, Groton, CT*

This presentation addresses the application of Quality by Design principles to two common unit operations encountered in the manufacture of parenteral products. Examples will be described for 1) the application of mixing models to facilitate the scale-up of the API dissolution step, and 2) an approach for establishing a sterilization design space based on chemical kinetics and sterilization theory.

### API Dissolution During Solution Compounding

In this example, mixing conditions to achieve complete dissolution of an API are dependent on several process variables, including API particle size, temperature, agitation rate, agitation time, and tank/agitator configuration. The fluid dynamics of the mixing vessel are scale-dependent, but can be modeled to enable predicted mixing conditions on scale-up. The approach taken in this example was to develop a lab-scale mixing vessel that was geometrically similar to the production vessel. The key aspect ratios of the production equipment (impeller diameter/tank diameter, liquid height/tank diameter, distance between impeller/tank diameter) were reproduced in a lab-scale mixer. Based on fluid dynamic principles, maintaining a similar power per unit volume (P/V) during mixing at either scale will result in similar solid-liquid mass transfer coefficient and therefore similar mixing times to achieve dissolution. Determinations of P/V were made using computational fluid dynamic (CFD) modeling of the mixing systems. Dissolution studies were performed in the lab-scale mixer to evaluate the combined influences of agitation rate and batch temperature on dissolution time, using worst-case API (a batch with particle size parameters near the maximum control limits). Using CFD modeling to determine agitation rates that will achieve the same P/V in the commercial tank, along with a constant scaling factor statistically determined during development of the model for this tank geometry, predicted agitation rates were determined that would achieve complete dissolution within the same time interval as observed in lab-scale. The suitability of the predicted agitation rates was determined for one full-scale batch, intentionally manufactured at the lower limit of the temperature range, representing worst-case conditions for dissolution. Successful dissolution within the target time interval served to confirm the model predictions and verify the minimum agitation rate and time for commercial production.

### Terminal Sterilization

The molecule for this example is heat sensitive, and when exposed to pharmacopoeial (overkill) sterilization conditions (15 minutes at 121 °C), the extent of degradation exceeds an acceptable impurity level. Therefore a temperature below 121°C was evaluated for the sterilization cycle. Both microbiological lethality and degradant formation are directly dependent on cumulative thermal exposure, and therefore sterilization conditions are well suited for the development of a design space. To establish the design space, sterilization temperature and exposure time were explored in terms of conditions that simultaneously provide 1) a minimum $\log_{10}$ reduction of 8 (assuring sterility), and 2) maximum degradant formation according to the specified limit (assuring appropriate purity).

This approach to the establishment of a design space is unique in that it allows the use of first principles (chemical kinetics and sterilization theory) to define acceptable boundaries for process parameters. While some experimentation is required (i.e., determination of degradation rate constants and sterilization D-value), these studies are scale-independent and can be conducted at the lab bench. The experimentally determined constants are incorporated into theoretical equations that are used to define the boundaries of the design space.

The degradation rate constants were determined in development studies by traditional chemical kinetic experimentation utilizing drug product exposed to elevated temperatures in the range of sterilization conditions (above 100°C). Rate constants were determined at worst-case conditions of pH and headspace oxygen content based on targeted in-process control limits, thus representing worst-case stability conditions within the formulation design space. The Arrhenius equation was then used to define a time/temperature profile that results in a pre-defined maximum limit of degradation. For sterilization requirements, the D-value of a thermally resistant spore-former (*Geobacillus stearothermophilus*) was determined in the drug product solution. Based on sterilization theory, a time/temperature profile was derived that would result in an 8 log reduction of microorganisms. Superposition of these two time/temperature profiles then defines the design space, shown in Figure 1.

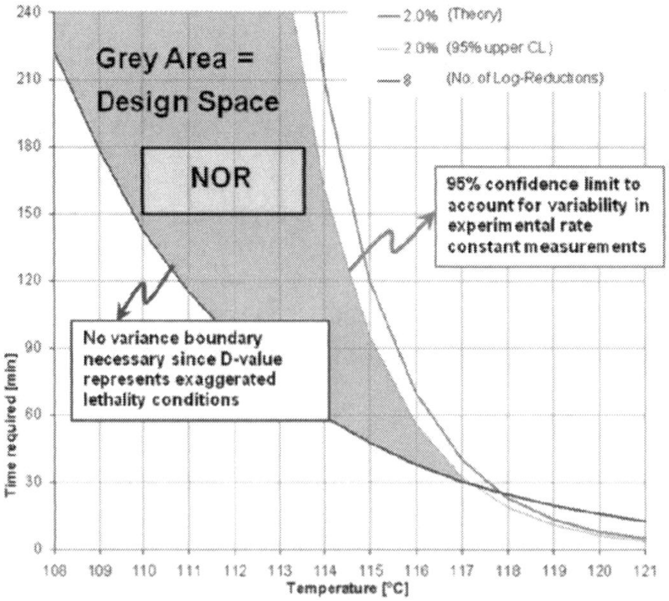


Figure 1. Moist Heat Sterilization Design Space and Normal Operating Ranges

The design space definition includes the 95% confidence limits for degradation. Statistical variance considerations for the lethality input (log-reduction) were not applied since the lethality experiments (D value generation) were performed with highly heat-resistance spores (*Geobacillus stearothermophilus*) to reflect exaggerated conditions.

Acknowledgements

Mixing Model: Fred Carroll, Rich Ferraina, Mark Berry, Yang Liu, Ken Norris, Robert Leasure, Jeff Gaudio (Pfizer); Rick Falk, Yogesh Waghmare (Bend Research)

Sterilization: EFPIA Mock P2 Injectables group: Heinz Wiederkehr (Roche), Thomas Backensfeld (Bayer Schering), Ritva Haikala (Orion Pharma), Christian DeMuynck (Nycomed), Brian Withers (Abbott)

# (68b) Rational Design and Scale-up of Batch Lubrication Processes for Immediate Release Formulations

**Joseph Kushner IV**, Daniel O. Blackwood, Bruce C. MacDonald, Mark Polizzi, Francis M. Moore and Bruno C. Hancock, Pharmaceutical Development, Pfizer Worldwide Research and Development, Groton, CT

**Purpose:** To provide stepwise guidance for lubrication of pharmaceutical powders through the development of a unifying model which captures lubrication effects during blending and compression.

**Methods:** Literature models of powder properties following lubrication blending over a range of scales and conditions were reviewed. Additionally, a new model was developed by lubricating pharmaceutical powders in a tablet press feed frame. Lubricated powders were compressed on a development-scale rotary tablet press (Kilian T100) with varying throughput and feed-frame paddle rotation speeds. The resulting tablets were measured for tensile strength and a feed frame lubrication sensitivity model was developed. The two models were then combined to create an integrated lubrication understanding framework.

**Results:** Lubrication process design and scale-up guidelines were developed that addressed the selection of the appropriate amount of lubricant, the impact of equipment design (scale & principle of operation), and the need to select processing conditions that will result in the consistent manufacture of product possessing the desired extent of lubrication. To provide quantitative guidance to this end, a new unified model was developed which captured the effect of lubrication on pharmaceutical powders during blending and compression.

**Conclusions:** The proposed stepwise approach outlines the current state-of-the-art for conventional tablet blend lubrication processes, and embodies the principles of Quality-by-Design and risk-based drug product development. As such it is recommended as the starting point for scientists undertaking the development of immediate release tablet formulations.

See more of this Session: Application of Quality by Design In Drug Product Process Development I

# (68c) Automation and Control of Drug-On-Demand Technology

**Laura Hirshfield**, Arun Giridhar, Gintaras V. Reklaitis, Venkat Venkatasubramanian and Michael T. Harris, Chemical Engineering, Purdue University, West Lafayette, IN

After the implementation of initiatives such as PAT and QbD, there has been a drive towards efficient manufacturing in the pharmaceutical industry. The concept of "mini-manufacturing" of drugs, as opposed to mass production of drugs, has shown promise in this area. We describe one such "mini-manufacturing" process, "drug-on-demand," which uses drop-on-demand printhead technology to deposit active pharmaceutical ingredients (API) onto edible substrates. This method also allows for the layering of different APIs and the creation of personalized dosage forms, in which the amount of API can be varied depending on the patient. In this work, we focus on the automation and control strategy for the drug-on-demand pilot facility done as part of the Engineering Research Center for Structured Organic Particulate Systems. A proper control strategy is necessary so that the system can run at optimal setpoints and adjust in real-time, allowing for an efficient manufacturing process and a precise on-spec drug product. The control strategy for our "drug-on-demand" setup allows us to execute an automated, optimized, and controlled print cycle while closely monitoring drop size, drug morphology, and drop deposition pattern.We summarize our achievements and current research in this area.

See more of this Session: Application of Quality by Design In Drug Product Process Development I
See more of this Group/Topical: Topical I: Comprehensive Quality by Design in Pharmaceutical Development and Manufacture

# Dynamic simulation of the dissolution of pharmaceutical tablet integrated with continuous manufacturing process

Presenter: Lukasz Mioduszewski

Research Participants: Fani Boukouvala, Marianthi Ierapetritou, Rohit Ramachandran

ENGINEERING RESEARCH CENTER FOR
STRUCTURED ORGANIC PARTICULATE SYSTEMS
RUTGERS UNIVERSITY
PURDUE UNIVERSITY
NEW JERSEY INSTITUTE OF TECHNOLOGY
UNIVERSITY OF PUERTO RICO AT MAYAGÜEZ

## Scientific Goals

- Develop a dynamic model of dissolution of pharmaceutical tablets as part of a dynamic flowsheet simulation of a continuous tablet manufacturing process
- Correlate operating and design parameters of upstream processes to properties of tablet dissolution through the integration of processing units
- Develop a model for tablet dissolution using Fick's Second Law with a non-constant diffusion coefficient

## Introduction

### *Flowsheet simulation:*

❑ Flowsheet simulation is used for **process design**, **dynamic optimization** and **flexibility analysis** of complex large-scale integrated systems.

❑ Application of simulators for solid processes is still very primitive compared to state of the art for fluid processes.

❑ Through dynamic flowsheet simulation of an integrated tablet manufacturing line, operating conditions and design of the system can be linked to properties of final product

❑ The effects of possible perturbations during operation can be and the necessary control scheme to alleviate them can be identified

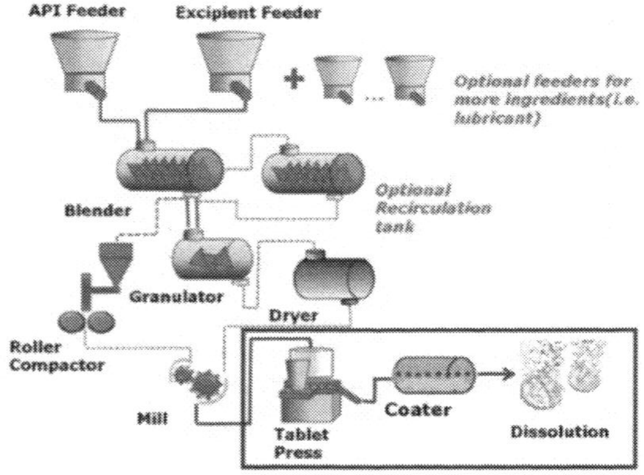


### *Tablet Dissolution:*

A tablet is ingested and passed through the esophagus to the stomach, which is an aqueous environment. This is the first place where the tablet may dissolve. The rate of dissolution is a key target for controlling the duration of a drug's effect.

## *Methods:*

### *Representation of tablet:*

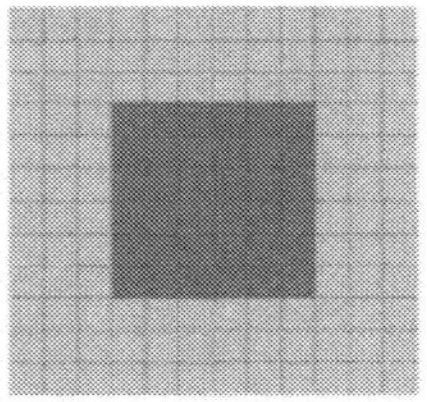

We initialized the void fractions with the equation for porosity. We assigned the center of the tablet a value of void fraction 20% larger than that for the outer tablet volume

### *Governing equations dissolution:*

Fick's Second Law

$$\frac{\partial c_i}{\partial t} = -\nabla \cdot (-D_i \nabla c_i) + s_i \quad i = \mathrm{A,B}$$

where

Source term

$$s_i(\mathbf{x}) = \begin{cases} k_i(c_i^{sat} - c_i(\mathbf{x})) & \forall \mathbf{x} : \phi_i(\mathbf{x}) > 0 \\ 0 & \text{otherwise} \end{cases}$$

Diffusion coefficient

$$D_i = D_{i0}\left(1 - \sum_{j=1}^{n} \phi_j\right)^{\alpha_i}$$

Evolution equation

$$\frac{\partial \phi_i}{\partial t} = -\frac{s_i}{\rho_i}$$

**Where:**
$C_i$ is the concentration of component i
$D_i$ is the diffusion coefficient
$s_i$ is the source term
A,B are components (API and Excipient)
$k_i$ is the dissolution rate constant
φ is the phase volume fraction (porosity)
α is diffusivity exponent
ρ is density

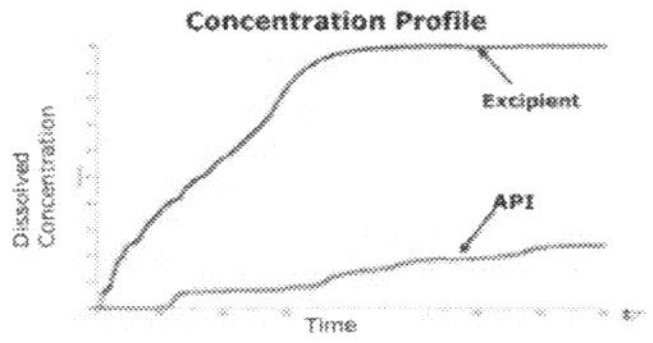


### *Integration of tablet press and dissolution*

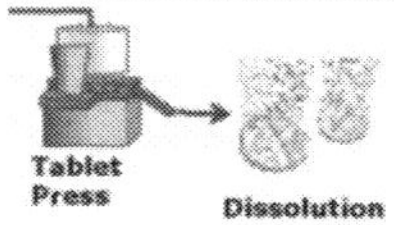


Correlation between compression force (F) of tablet press to porosity of tablet

$$\varphi = 0.5921 * \mathrm{EXP}(-0.0002 * F)$$

**Developed from empirical analysis (Nokhodchi, A. Ford, J. Rowe, P. Rubinstein, M.)**

Correlation between mean particle size and Diffusion coefficient

$$D_{i0} = kT / 6\pi\eta R_H$$

Boltzmann constant; Temperature; Viscosity; Particle radius

Brownian Motion

## *Results:*

### *Graphical Profiles of Results:*

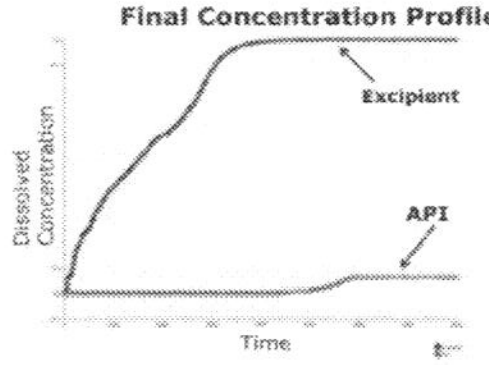


The concentration profiles for both API and Excipient increase until entire quantity is dissolved

The porosity tends to zero for both API and Excipient. Profile for API tends to zero slowly because API volume fraction is initially higher than that of Excipient. Also, the API is located in the middle of the tablet, therefore its dissolution takes longer

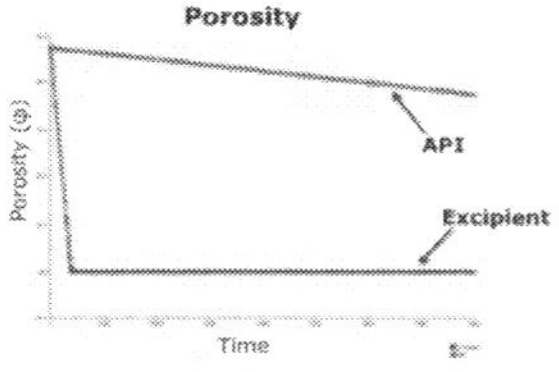


The spatial concentration profile of the tablet is shown. The concentration is maximum in the middle of the tablet and tends to zero when approaching the edges of the tablet

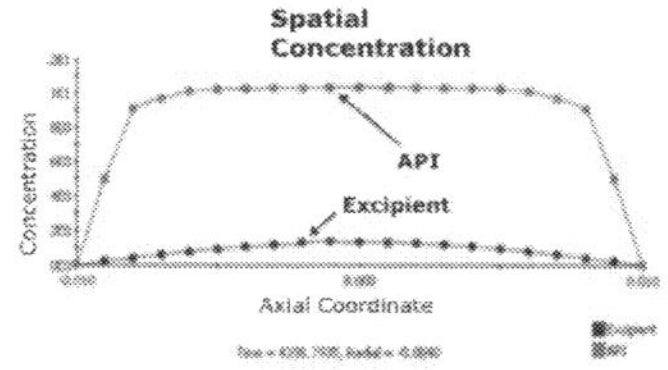


## Conclusions

1. Dissolution model is the last step in the flowsheet simulation and determines many important properties of the final product
2. Tablet press and dissolution models were connected using Brownian motion and empirical data available for hydroxypropylmethylcellulose in regard to the compaction properties and void fractions
3. Flowsheet simulation of integrated system can help identify optimum design, operating conditions and sequence of tablet manufacturing process in order to produce high quality product with consistency

## Future Work

1. Conducting experiments with the actual tablet press and obtaining real data for a specific pharmaceutical component
2. Modifying model parameters for a specific drug
3. Modifying the empirical equations in the model to customize the flowsheet for particular compounds

# (68e) Innovative Techniques for Planning Experiments Across Mulitple Unit Operations

**Kelly Canter**, Pharmaceutical Sciences, Pfizer Inc., St. Paul, MN and Leonard Perry, University of San Diego, San Diego, CA

Many well-established analytical tools and methodologies exist for developing and characterizing a product for manufacture. However, the application of these tools all too often can fall short of their objective or full potential to gain process understanding – even in the hands of a well trained chemometrician or statistician. Another essential and uncommon taught skill is in the planning of experiments over the lifecycle of a products development and subsequent manufacture. The selection of the proper experimental approach and planning activities that precede the actual experimentation are critical to successfully establish process understanding. The application of proper risk management tools along with established experimental planning principles aide significantly in effectively establishing a product's experimental development plan. There are well documented sources that aid in the planning and selection of designed experiments to achieve a desired outcome, especially when looking at a single unit operation at a time. Yet minimal attention is given to the planning and sequential approach required for experimentation across multiple units operations. This paper highlights a set of tools and guidelines for assessing technical issues and establishing an appropriate experimental plan when looking across a products manufacturing process. A pharmaceutical case study involving the experimental strategy taken to develop the design space for a Quality-by-Design filing is provided, and the trade-offs between resource limitations and statistical consequences is included.

See more of this Session: Application of Quality by Design In Drug Product Process Development I
See more of this Group/Topical: Topical I: Comprehensive Quality by Design in Pharmaceutical Development and Manufacture

# (68f) Real-Time Process Management and Knowledge Management for Continuous Pharmaceutical Manufacturing

Girish Joglekar, **Arun Giridhar**, Gintaras V. Reklaitis and Venkat Venkatasubramanian, Chemical Engineering, Purdue University, West Lafayette, IN

The manufacture of pharmaceutical drug products has benefited significantly from a QbD approach, including the enabling of continuous production over batch processing. Developing and improving processes for continuous manufacture requires the combined use of process data with process and formulation knowledge. Process data can be generated across multiple unitops simultaneously, and correcting deviations in one critical process parameter or critical quality attribute may require manipulations of a totally different process unit. Such integrated control must be done in a manner that is both easy to scale to different sizes and easy to transfer to different materials, requiring an integrated approach in gathering and managing process knowledge and using it flexibly for new processes.

In this work, we present the results and case studies from the Engineering Research Center for Structured Organic Particulate Systems (ERC-SOPS), in developing a real-time process management system for continuous pharmaceutical manufacture, that makes use of an ontological informatics framework to manage process knowledge of multiple types, including material properties (such as stress-density relationships), process parameters (such as process setpoints), equipment parameters (such as die and punch sizes in a tablet press), and process control knowledge (mathematical models for PID and model predictive control, process signatures for fault diagnosis, mathematical models to plan future setpoint changes, etc). The models taken together operate on a wide range of time scales, and the knowledge management framework is designed to be flexible for all uses. Further, we show how the system architecture allows for the use of various control systems and modeling programs in a flexible manner. Finally, we present case studies in tablet manufacturing and drop-on-demand systems, and summarize our current results.

# (71a) An Integrated Approach to Separations Process Research

**Kyle P. Kostroski**, Refining & Logistics Technology, BP Products North America Inc., Naperville, IL

The importance of separations processes is demonstrated by their ubiquity in the chemical and petrochemical industries. Many separations processes are also technologically mature, however, and it is often tempting for companies to forego rigorous separations research by limiting capital spending on benchtop- and pilot-scale programs. For obvious economic reasons, experimental work has been replaced by the use of process simulators. While this is understandable, it comes with a caveat: simulators are only as good as the data they use to make predictions. To avoid the so-called "garbage in, garbage out" problem associated with simulation, an integrated approach to separation process research might be necessary.

This talk will focus on how bench-scale and pilot-scale experimental facilities can be leveraged cost-effectively to do the key experiments required for enhancing process simulations, informing process economic evaluations, and enabling process scale-up. Development of a solvent extraction process for hydrocarbon separation will serve as a case study. Three efforts were undertaken in parallel: bench-scale solvent screening, pilot-plant validation and determination of scale-up parameters, and integration of experimental data into process simulations. The results of this study indicated that neither simulation results nor experimental data alone were adequate to assess the overall performance of the extraction process. However, by integrating key bench and pilot plant data into the simulation, a more accurate prediction of separation performance was achieved.

See more of this Session: Best Practices In Pilot Plant Separation
See more of this Group/Topical: Process Development Division

# (71b) Drying Aqueous Biphenol Salt for Polyetherimide Monomer Synthesis

**James M. Silva**[1], David Woodruff[1], Thomas Guggenheim[2] and Albert Stella[1], (1) General Electric Global Research Center, Niskayuna, NY, (2)SABIC Innovative Plastics, Mt. Vernon, IN

Polyetherimide synthesis via displacement polymerization is important for process simplification and product flexibility. A key step in the synthesis of biphenol-based polyetherimides is the production of biphenol (BP) disodium salt with minimal moisture content. To achieve sufficient control over the salt composition, the BP disodium salt is synthesized as an aqueous solution. Salt drying is accomplished by introducing an aqueous salt solution into a vessel containing a hot organic solvent. This removes most of the water from the aqueous BP salt solution as an overhead product. The process of adding the aqueous BP salt solution into hot solvent is called "sprayover".

During the laboratory and semi-pilot development phases of this program, we found that the sprayover vessel must be maintained at or above a certain minimum target temperature in order to avoid agglomeration of the BP salt and caking on the vessel walls. We utilized the sprayover vessel target temperature at the 50-liter and 50-gallon scales. However, as aqueous BP salt is sprayed into the hot solvent a large amount of solvent is lost to the overhead stream. Such a high solvent evaporation rate causes serious solvent inventory control problems for a sprayover vessel with a short-path (non-reflux) condenser configuration.

By utilizing a partial reflux condenser in the overheads system, we avoided agglomeration and caking in the sprayover vessel and achieved stable operation at the target temperature. We were initially concerned that the refluxing solvent, which is saturated with water, would yield a higher water concentration in the sprayover vessel (compared with the short-path condenser configuration). However, application of the phase rule and ASPEN Plus® simulations of the solvent-$H_2O$ system showed that the equilibrium water concentration in the sprayover vessel depends only on the sprayover vessel temperature and pressure, and is independent of the condenser configuration. Measured solvent moisture levels were comparable for both condenser configurations.

The calculated heat duty to operate a sprayover vessel at the target temperature is

6% higher with a partial reflux condenser than with a short-path condenser. However, the calculated amount of solvent in the distillate at the target temperature is 33X lower with a partial reflux condenser than with a short-path condenser.

Sprayover tests in 50-liter and 50-gallon sprayover vessels demonstrated that with a partial reflux condenser, the solvent can be contained in the system at high target temperatures. In these tests, solvent losses were minimal, the vessel walls were clean after sprayover at both scales, and the salt product polymerized normally.

See more of this Session: Best Practices In Pilot Plant Separation
See more of this Group/Topical: Process Development Division

# (71c) Design of Reactive Distillation Column for Non-Parameterized Components - Which Experiments Are Essential

**Oliver Litzmann**[1], Thomas Grtzner[2], Gnter Wozny[1] and Jens-Uwe Repke[3], (1) Chair of Process Dynamics and Operation, Berlin Institute of Technology, Berlin, Germany, (2)Lcdrn, Lonza AG, Visp, Switzerland, (3)Institut fr Thermische Verfahrenstechnik, Umwelt- und Naturstoffverfahrenstechnik, TU Bergakademie Freiberg, Freiberg, Germany

**Design of reactive distillation column fornon-parameterized components - which experiments are essential**

Litzmann, O., TU Berlin, Berlin; Lorenz, H.-M., TU Berlin, Berlin; Wozny, G., TU Berlin, Berlin; Grtzner,T., Lonza AG, Visp/CH; Repke, J.-U., TU Freiberg, Freiberg_

Designing reactive distillation processesrequires a great number of fitted parameters depending on the model used. Theseparameters are known for test-systems like MTBE or TAME but not for systemswhich have never before used in reactive distillation. As fitting of parametersrequire a lot of time and money for trials it is crucial to know the impact ofan improved model to the prediction accuracy. If this is known it is possibleto carry out fewer experiments with and fit a set of sufficient parameters inorder to describe the RD process as good as needed to strivethe goal.

Application to a new industrial substancesystem

The system analysed consists of an organic acid, analcohol, the corresponding ester and water whereupon some components cannot bedescribed by standard UNIFAC methods. The esterficationtakes place in a reactive distillation column designed for this system. The reactionis catalysed by the heterogeneous catalyst Amberlyst 36.

In order to describe the results of thecolumn, models are used which take different phenomena into account and usedifferent sets of parameters which have been measured independently.

Fig. SEQ Fig. * ARABIC 1: Comparison of the composition forecast by pseudo homogeneous and adsorption based model

For instance the reaction can be described either byequilibrium or by kinetics which can be corrected by adsorption effects in addition.This kinetic have been measured in a batch reactor and is described with significanthigher accuracy by an adsorption based kinetic model (see Fig.1) Theportability of these model improvement and similar effects to the reactivecolumn simulation will be presented and discussed by means of comparison withthe experimental results from the lab scale plant.

See more of this Session: Best Practices In Pilot Plant Separation
See more of this Group/Topical: Process Development Division

# (71d) Design and Fabrication of Pilot Scale Gas/Liquid Separators

**Oliver Orrell**, Zeton Inc, Burlington, ON, Canada

Separation processes are an important component of chemical and petrochemical plants. In large facilities, gas/liquid, liquid/liquid, and gas/liquid/liquid separators can benefit from engineered internals such as coalescing plate packs, weirs, and baffles. In smaller pilot facilities, these internals can be difficult to employ due to size constraints. This session will discuss various approaches of designing and fabricating pilot scale separators that are effective without the benefit of internals available to their full-scale counterparts. The main focus will be on a recent redesign of a small scale three phase separator that has been proven to have less liquid carryover than earlier designs.

See more of this Session: Best Practices In Pilot Plant Separation
See more of this Group/Topical: Process Development Division

# (71e) Pilot Plant Development of Molecular Separations by Organic Solvent Nanofiltration

**Andrew G. Livingston**, Andrew T. Boam, Andrea Meniconi, Fui-Wen Lim and Marcin Makowski, Evonik Membrane Extraction Technology Limited, London, United Kingdom

Organic Solvent Nanofiltration (OSN) separates molecules present in solutions of organic solvents. It is the same in principle as the well known RO/nanofiltration used in water treatments, except that the separations are based in organic solvents. This opens up a whole range of separations applications in organic processes – concentration, purification, catalyst retention and solvent recovery are just a few of these.

In OSN, a pressure gradient is applied across a solvent stable nanoporous film, inducing transport of liquid across the film. Depending on relative permeation rates, molecular species in a feed liquid stream can be concentrated and separated. Typically OSN will use a small fraction (less than 10%) of the energy of thermal methods such as evaporation and distillation, and can work at temperatures around ambient. Challenges for application of OSN technology in the refining, chemicals and pharmaceutical industry have been: (i) the lack of membranes with wide solvent resistance, and which retain their nanostructure with exposure to aggressive liquids such as dipolar aprotic solvents (DMF, DMSO, THF); and (ii) acceptance of the introduction of a new process technology to large scale chemical processes.

With respect to the first challenge, this presentation will describe the recent development of two new classes of commercially available OSN membranes and modules, and their applications to processes typically used in chemical manufacturing routes. The DuraMem$^{TM}$ series OSN membranes are made through post-formation processing including crosslinking of polymer membranes, which makes them stable in polar solvents such as acetone, DMF and THF. The PuraMem$^{TM}$ S series are thin film composite membranes which show exceptional performance in apolar solvents. Both these membranes are available for lab testing as flat sheets, and as spiral wound cartridges for scale up.

To meet the second challenge, the development of organic processes where OSN is used as a molecular separation step proceeds from lab bench scale to kilo/pilot scale, and then to process scale. The laboratory bench testing programme must yield key performance data to show that OSN is a viable technical solution, and that the technology is economically viable. Suitable process models are employed to predict the performance of the process, and to decide the size of the appropriate pilot scale tests that will provide convincing data for full scale plant design. Next, it is essential to move to a pilot plant testing programme utilising spiral wound membrane modules. The spiral modules must remain stable in the organic solution being processed, and so it is very important that the construction of these modules is the same for the pilot programme as will be used for the scaled up process. Membrane lifetime is a key factor that must be estimated from the data obtained in the pilot testing programme. Typically, at Evonik we have used spiral wound modules of 1.8" x 12" (0.2m2) through to 2.5" x 40" (2 m2) for the testing programmes; sometimes we also employ process size 4" x 40" modules for advanced tests. Where the application of OSN is in the pharmaceutical industry, compliance with GMP requirements and the need for demonstrating that leachable materials from the spiral element will not affect the overall process, is also essential.

The paper will describe examples of pilot testing programmes that have been carried out and which have led to process scale installations of OSN technology. OSN membrane modules are being successfully applied in cGMP manufacturing processes in the pharmaceutical industry (GSK, Johnson and Johnson), and for larger scale chemicals processing including homogenous catalyst retention at Evonik, and these applications will be described.

See more of this Session: Best Practices In Pilot Plant Separation
See more of this Group/Topical: Process Development Division

# (71f) Pilot Scale Liquid-Liquid Extraction Process Development for Emulsifying Viscous Organic-Aqueous Mixtures

**William Hollar**[1], Babatunde Oguntade[1], Pankaj S. Gautam[2] and Zeljko Kuzelevic[1], (1)Sabic Innovative Plastics, Mount Vernon, IN, (2)Process Technology, SABIC Innovative Plastics, Mt. Vernon, IN

The tendency of specific two-phase liquid-liquid systems to emulsify makes proper pilot scale process development extremely important for successful process scale-up. This paper will focus on a case study of a polymer solution – water mixture that exhibited strong emulsification tendencies and the challenges associated with piloting and scale-up. Its behaviors and unique features will be compared to a system with low emulsification tendency. Efforts to measure liquid-liquid mixture properties and correlate them to input process conditions and equipment design will be detailed. A brief description of ongoing process improvement efforts will be included.

See more of this Session: Best Practices In Pilot Plant Separation
See more of this Group/Topical: Process Development Division

# (84a) Generic Model and Data Based Framework for Analysis and Development of Crystallization Processes

**Kresten Troelstrup Meisler**[1], Noor Asma Fazli Abdul Samad[1], Krist V. Gernaey[2], Nicolas von Solms[3] and Rafiqul Gani[1], (1)CAPEC, Department of Chemical and Biochemical Engineering, Technical University of Denmark (DTU), Kgs. Lyngby, Denmark, (2)PROCESS, Department of Chemical and Biochemical Engineering, Technical University of Denmark (DTU), Kgs. Lyngby, Denmark, (3)CERE, Deparment of Chemical and Biochemical Engineering, Technical University of Denmark (DTU), Kgs. Lyngby, Denmark

Crystallization is an efficient process for separation of compounds, which are solid in their pure form at the given separation conditions. Crystallization is often applied in the production of salts and active pharmaceutical ingredients (API), where the crystallization step is an essential part of the manufacturing process for many pharmaceutical products. In recent years the monitoring and analysis of crystallization operations has received increased attention due to the growing need to control more sophisticated production lines as well as the increased possibilities of monitoring such processes with the advance of measurement techniques. This presentation addresses the need for a generic model and data based framework for the generation of population balance based models for crystallization operations in one to three dimensions with the analysis of data-driven crystallization kinetic models. The crystallization models are valid for batch as well as continuous operations.

A generic multi-dimensional model-based system for use in crystallization modeling (for cooling as well as solvent and/or anti-solvent based operation) has been set up and its use to study a wide range of chemical (crystallization) systems as well as different crystallizer operation phases (scenarios) has been demonstrated previously (1). This system has now been combined and extended with systematic methods for collection of data and model-based analysis of known crystallization kinetic models and their use in simulating crystallization operations. This combination provides a framework for monitoring, analysis and optimization of crystallization operations. Measurements can be taken and used for offline analysis and parameter regression. With an established kinetic model it is possible to translate the predicted model results into a form, which is directly comparable (and visualized) to the data for evaluation of the model. Furthermore, if there is model – data mismatch, the measured data can be used for online parameter estimation.

The models used for the crystallization usually include several phenomena such as growth, nucleation, agglomeration and breakage in order to give a satisfactory description of the crystallization operation. The generic framework handles population balances for these phenomena and has been expanded into covering multiple dimensions accounting for temperature and concentration gradients in the crystallization operation.

The expanded model framework combined with the systematic approach to establish the kinetic models for use in general crystallization operations in combination with monitoring tools will be presented using case studies involving API crystallization.

**Works Cited**

1. *A generic multi-dimensional model-based system for batch cooling.* **Noor Asma Fazli Abdul Samad, Ravendra Singh, Gürkan Sin, Krist V. Gernaey, Rafiqul Gani.** 2011, Computers and Chemical Engineering, Vol. 35, pp. 828–843.

See more of this Session: Crystallization Process Development I - Techniques and Tools
See more of this Group/Topical: Process Development Division

# (84b) Quality-by-Design of the Crystal Size Distribution of An Anti-Arrhythmic Drug Using a Crystallisation Process Informatics System (CryPRINS)

**Zoltan K. Nagy**, Chemical Engineering Department, Loughborough University, Loughborough, United Kingdom, Ali N. Saleemi, Chemical Engineering, Loughborough University, Loughborough, United Kingdom and Gerald Steele, Pharmaceutical Development, AstraZeneca R&D Charnwood, Loughborough, United Kingdom

A crystallization control strategy is proposed, which combines the supersaturation control and direct nucleation control to obtain *in situ* consistent seed production and minimize undesired secondary nucleation, to achieve enhanced robustness in the product quality. Focused beam reflectance measurement (FBRM) is used for direct nucleation control where the number of nuclei is maintained at a desired level, during the beginning of the operation by correcting the number of counts by heating the system to produce controlled fine dissolution. After the number of counts is stabilized, and consistent seed is generated, the approach switches to a closed loop supersaturation control using concentration measurement based on ATR-UV/Vis probe. The supersaturation control prevents multiple nucleation events and at the same time maintains enough supersaturation for crystal growth. Supersaturation is maintained at the desired level by variation of temperature or of solvent/anti-solvent ratio. The benefits of this approach include suppression of secondary nucleation, better final crystal size distribution and production of desired polymorphs as well as significant decrease in the variability of crystalline product properties. The approach is evaluated using several pharmaceutical compounds (small organic molecules such as paracetamol, sufathiazole and a large molecule anti-arrhythmic drug). The quality of the crystals obtained both from lab scale and pilot plant are compared in order to check the robustness of the approach. The approach is implemented in an object oriented software environment, the Crystallisation Process Informatics System (CryPRINS), which enables the application of various process analytical tools (PAT) in different feedback control strategies, providing industry standard OPC communication interface to any distributed control system. The CryPRINS allows straightforward application of the proposed feedback control approaches at both laboratory as well as industrial scale from the same software environment providing an automated scale-up system through adaptation of the operating profiles via feedback control.

See more of this Session: Crystallization Process Development I - Techniques and Tools
See more of this Group/Topical: Process Development Division

# (84c) Model-Based Decision Support for Design and Operation of Pharmaceutical Crystallisation Processes: Efficient Workflows for Validation Against Experiments and Scale-up

**Sean K. Bermingham**, Process Systems Enterprise, London W6 7HA, United Kingdom and Ugo Cocchini, GlaxoSmithKline, Stevenage SG1 2NY, United Kingdom

This paper describes GSK's assessment of available tools and techniques for model-based design and optimisation of crystallisation processes and considers the potential for using the same models to quantify the design space of these processes. In order for model-based decision support to be of practical value to the pharmaceutical industry, it is essential to have workflows that allow model development, experimentation, parameter estimation, process optimisation and scale-up to be done in a period of 4-8 weeks.

The batch cooling crystallisation of an API from a solvent was selected as a **case study**. The process was seeded and investigation of crystal images revealed that the dominant mechanisms are growth, attrition and to a lesser extent agglomeration. The key challenges from the process development perspective are to be able to predict attrition and agglomeration as a function of operating conditions, crystalliser/agitator type and equipment scale.

A number of **experiments** were conducted to investigate the final PSD and solute concentration over time as a function of seeding conditions (amount and PSD of the seeds) and cooling profile. Slurry samples were taken at critical time points. Solid samples were isolated by vacuum filtration and analysed for PSD by Malvern Mastersizer, whilst assay was performed on the liquid filtrate to measure residual solute concentration.

A **model** of the batch cooling crystallisation set-up used for the experimental work was developed using a commercially available tool. The developed model is based on a population balance framework that supports both steady-state and dynamic applications. For the **model validation**, the unknown kinetic parameters were estimated against the solute concentration and PSD measurements of the performed experiments. This was done for all experiments simultaneously.

The validated model was subsequently combined with flow information from a CFD model to construct a so-called Multizonal model (see Figure below) that allowed prediction of local wall temperatures, primary and secondary nucleation rates, growth (and dissolution) rates as well as agglomeration rates. This enables **scale-up** on a sound physical basis.

The value of a model-based approach to process development is to reduce the number of experiments required to understand and predict the crystallisation behaviour. The approach used here allows one to predict the behaviour at the same scale (e.g. to troubleshoot an existing process or to reduce batch time whilst satisfying PSD and purity constraints) as well as at different scales or different configurations at the same/similar scale (e.g. to aid scale-up).

See more of this Session: Crystallization Process Development I - Techniques and Tools
See more of this Group/Topical: Process Development Division

# (84d) Development of a Task Based Crystallizer Design

Anamaria Soare, Andrzej Stankiewicz and **Herman J. M. Kramer**, Process & Energy, Delft University of Technology, Delft, Netherlands

Crystallisation is one of the oldest and economically most important separation technologies in chemical industry. Many interacting physical phenomena occur in crystallization processes, involving nucleation, growth, attrition, agglomeration and classification of crystals. The operating conditions and the spatial staging of the phenomena jointly determine the properties of the crystal particles in a non-uniform manner. The complex interplay between the various physical phenomena poses many challenges to the design and operation of crystallization processes.

A task based design procedure is developed aiming at improved product quality and more flexible process operation and design of industrial crystallization by creating predictive models based on first principles, improving the design of crystallizers by applying new concepts and implementation of advanced model based control. In this novel functionality driven design approach the crystallization process is decomposed into fundamental building blocks. A task is defined as an intentional transformation of matter from a given state to a desired state driven by physical phenomena. The aim of this novel task based design approach is to generate procedures to structure these fundamental tasks in a way that allows for the optimization of complex transformation process from feed stream to final product. In that perspective several experimental setups dedicated to single crystallization tasks were designed and tested. It has been shown, at laboratory scale, that it is possible to isolate a number of crystallization phenomena such as nucleation, growth and supersaturation generation in dedicated equipment. This not only enabled the optimization of the process conditions to get the best possible results for these crystallization tasks, it also enabled in theory the design of a new generation of crystallizers with superior flexibility in product quality.

Based on the task based design approach new concepts were developed applying alternative methods for the nucleation of crystals, crystal growth and supersaturation generation.

The results show a number of new promising technologies that can be applied in a task based crystallizer approach to achieve better and flexible control of the product quality in industrial crystallization processes. Especially the laser induced nucleation and ultrasound induced nucleation might be effective technologies to control the number of seed crystals, fed into a growth crystallizer such as the airlift reactor.

See more of this Session: Crystallization Process Development I - Techniques and Tools
See more of this Group/Topical: Process Development Division

# (84e) Systematic Approach for Chiral Separation Via Diastereomeric Salt Formation

**Venkata S. Sistla**, Jan von Langermann, Heike Lorenz and Andreas Seidel-Morgenstern, Physical and Chemical Foundations of Process Engineering, Max Planck Institute for dynamics of complex technical systems, Magdeburg, Germany

In order to prevent the possible complications, L- and D-enantiomers of pharmaceutical substances must be separated from each other. Direct separation is very difficult as enantiomers have identical physical and chemical properties. Unlike enantiomers diastereomeric salts possess different physical and chemical properties. The difference in properties endorsed diastereomeric salt resolution (classical resolution) to be one of the strong separation techniques for pure enantiomers. During the process enantiomers are treated with an optically active resolving agent to form diastereomeric salts that can be separated via crystallization. Often in industry, classical resolution is performed with limited data on phase behavior and metastable zone width data for diastereomeric salts. The separation process can be planned effectively and yield can be improved, if the above data is available.

In the present work the diastereomeric salt separation of DL-phenyl glycine is presented. Pure salts of both D-and L-phenyl glycine with camphor-D-sulphonic acid were synthesized in the laboratory and characterized with the help of different analytical techniques like $^{1}H$ NMR, XRPD, DSC and Raman spectroscopy. Further, basic thermodynamic and kinetic data like solubility phase diagram and metastable zone width were measured in different solvents. On this basis different possibilities for the resolution of both salts were derived and executed in the laboratory. The comparison of results and selection of best process out of all above possible separation processes will be discussed.

See more of this Session: Crystallization Process Development I - Techniques and Tools
See more of this Group/Topical: Process Development Division

# (84f) Thermodynamics-Based Approach to Crystallization Process Development

**Christianto Wibowo**, ClearWaterBay Technology, Pomona, CA

Just as vapor-liquid equilibrium-based methods are routinely used to design distillation processes, it is only natural that in designing a crystallization process one should start with solid-liquid equilibrium (SLE), which provides the foundation for further development efforts. Basically, the SLE behavior imposes thermodynamic limitations as to which components can or cannot crystallize out under a given operating condition. With the knowledge of SLE behavior, one can easily identify the feasible operating region to get the desirable product as well as the theoretical yield under any given condition. While it is also crucial to take into account non-equilibrium behavior such as kinetic and mass transfer effects, beginning with SLE ensures that one gets off on the right foot.

This presentation outlines an integrated approach that emphasizes the understanding of SLE phase behavior in the development of crystallization processes. Visualization using phase diagrams greatly enhance the understanding of the SLE phase behavior, leading to an easier analysis. The use of an SLE model, especially for multicomponent systems, allows convenient representation and interpretation of experimental data. To illustrate the approach, case studies that are based on or inspired by actual industrial projects are presented.

See more of this Session: Crystallization Process Development I - Techniques and Tools
See more of this Group/Topical: Process Development Division

## A debottlenecking study for a dehydration column

*Ahmed A. Youssef and Mohan Khadilkar*

*SABIC Innovative Plastics*, Mt. Vernon, USA*

### 1. Abstract

This study presents a case study of the debottlenecking of a distillation column used in the dehydration of a process stream. Due to recent developments on the process, limitations on the water stripping performance arose. Therefore, a debottlenecking study was conducted to alleviate the load on the distillation column, allow improved capacity and prevent the operation under the flooding regime. A thorough investigation of the system's hysteresis and operating conditions (flow rates, temperatures, pressures, compositions, etc.) was carried out. This showed, for instance, an increase in the pressure drop across the column height. An ASPEN Plus®* model was developed, validated and utilized to identify the associated bottlenecks, their interactions and their impact on the performance of the water removal column. A set of recommendations with respect to the design and operating parameters including adjustments to the feed temperature are presented to showcase the potential solutions of the problem.

* ASPEN Plus is a trademark of AspenTech.

### 2. Introduction

The utilization of a distillation step in order to accomplish a dehydration process is a common practice in industrial operations. Depending on the nature of the process and the operating conditions of the distillation columns, a number of issues can impact their performance. Such malfunction can be manifested in, but not limited to, a decrease in the separation efficiency and irregular heat effects around the condenser and/or the reboiler for example. Some of the typical root causes can be sought by investigating the degree of flooding in the column, the pressure drop across its sections, the feed stream composition and the column's internals (packing, trays, distributors, etc.). In the current study, a detailed description of the system in use, the "symptoms" of the observed malfunction and the route for tackling the problem, along with the conclusions and recommendations will be outlined.

### 3. Case description

#### 3.1 Background

The dehydration section in discussion consists of two distillation columns. The feed to the first column (C-41) is a "wet" adduct solution containing a given amount of water governed by the operating conditions. Water is removed in C-41 (also will be referred to as the water stripper) as overheads while the "dry" adduct solution separates out in the bottoms stream. The water stripper's distillate proceeds to the second column (C-48 and also called the azeotrope column) for further processing. The azeotrope column's feed is, basically, a vapor phase and both of its overheads and bottoms streams exiting the dehydration section in discussion. The second column is required to yield further concentration of C-41's overheads.

A Process Flow Diagram (PFD) highlighting the dehydration section is given in Figure 1. It is noteworthy that C-41 operates with no condenser while C-48 has a condenser and a reboiler. The water stripper is 6' 6" in diameter and contains three packing sections. The packing material is Mellapak plus of varying characteristics. Internally, C-41 also has a liquid and a gas distribution plates. As for the azeotrope column, it is 4' in diameter and houses two Mellapak packing sections.

There are several controllers around the dehydration section that maintain its desired operation. These include the temperature of the section's feed, the temperature at the bottom of C-41 that correlates to the steam consumption directly, and the azeotrope column's reflux ratio.

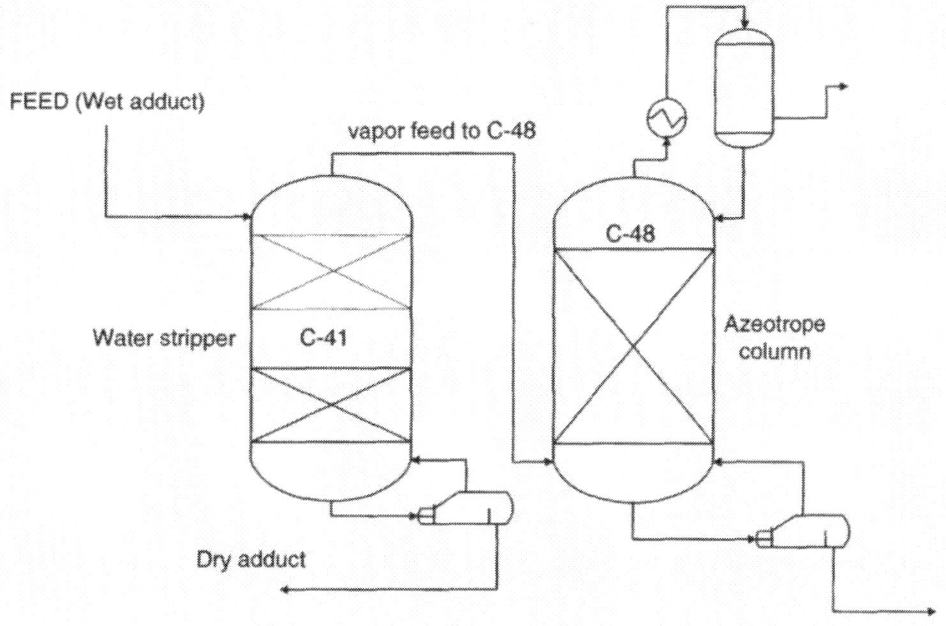


Fig. 1 – Dehydration section PFD

### 3.2 Problem description

A new technological addition developed within Sabic Innovative Plastics was introduced to the process. This resulted in a significant increase in the percent water fed to the dehydration section. At low operational rates, the dehydration section was able to maintain similar performance compared to prior to the implementation of the new modification. That state of operation continued for ~9 months. However, as the plant operation was ramped up towards its maximum, a sharp increase in the pressure drop across the water stripper column was observed (see Figure 2) corroborated with an increase in the normalized reboiler steam usage (see Figure 3).

Amid fears of a flooding condition inside the C-41 column, adjustments were made to cut the amount of water in the feed. Despite that effort, the continuous increase of the operation capacity showed no improvement in the above trends.

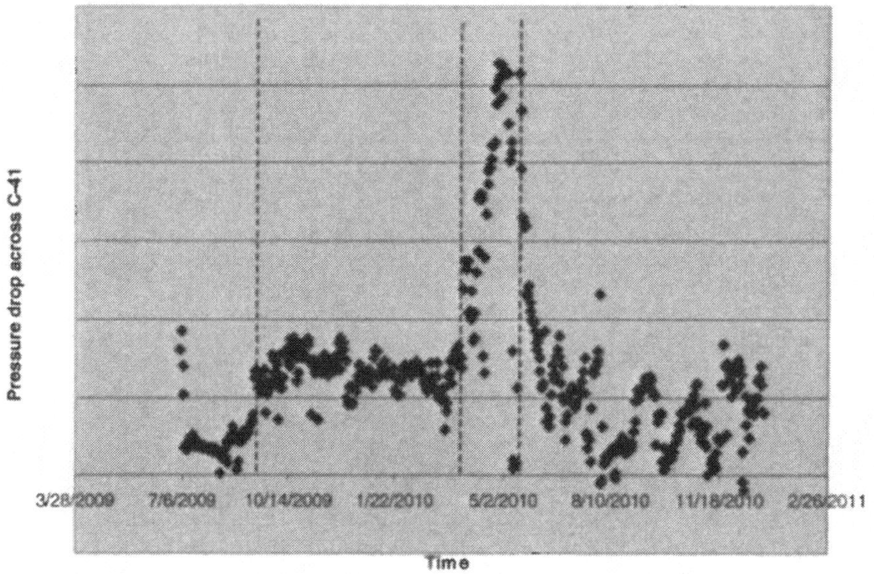


Fig. 2 – Pressure drop across C-41

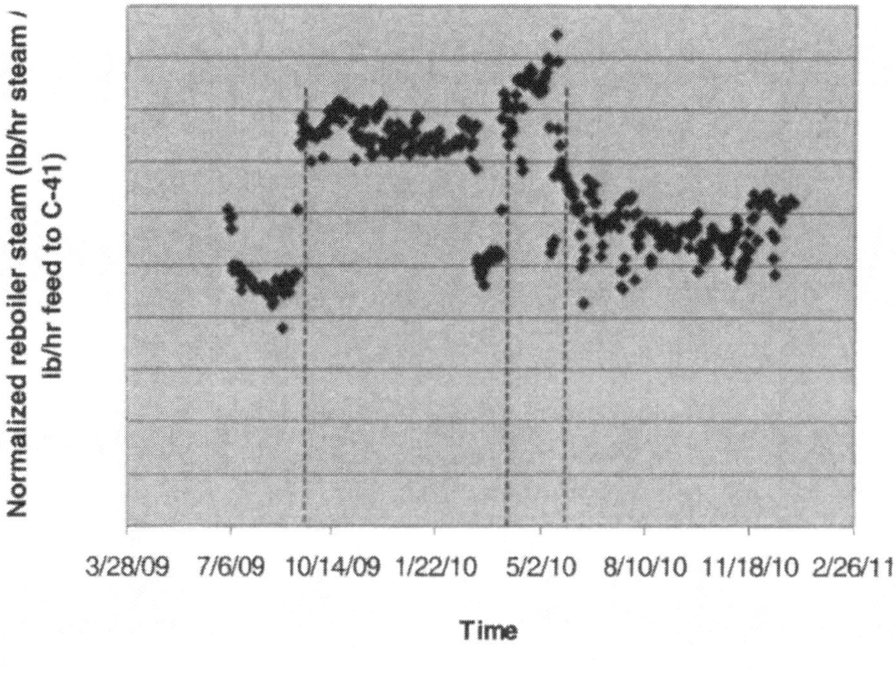


Fig. 3 – Normalized C-41 reboiler steam to feed flow rate

Hence, the above indicators drove a halt on the utilization of the new technology and necessitated the restoring of the original conditions until a better understanding of the situation is reached and a debottlenecking study is completed.

It is noteworthy that the modification to the process limits the plant to 70% of its capacity. The debottlenecking would result in a saving in the raw material usage that translates to ~ $2 MM/y. In addition, the modified process yields a product of higher quality.

In order for this to be accomplished, a mass balance was performed around the dehydration section (see Figure 4). The latter was limited by the availability of the composition data available from plant sampling.

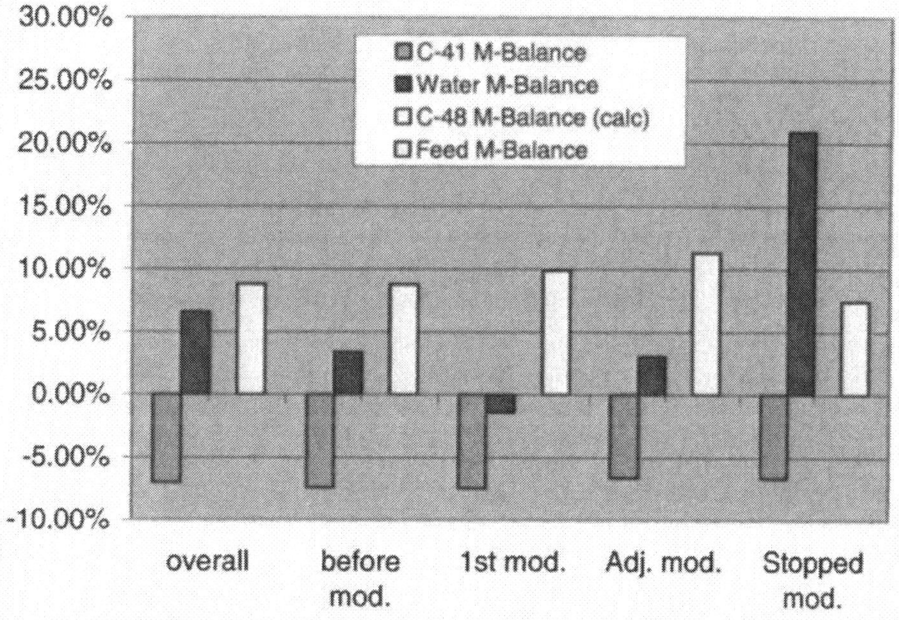


Fig. 4 - Mass balance closure around the dehydration section

## 4. Root Cause Investigation

The water content in the dehydration column bottoms is a critical indicator of the efficiency of separation. Historical data for the water content during the period of 2009 through 2010 when the modification was employed is shown in Figure 5.

Throughout the time frame of concern, and in spite of the change in the production of the plant, minor effect on the water level in the bottoms of C-41 is observed. Potential modifications in operating conditions can cause such trend as will be highlighted later.

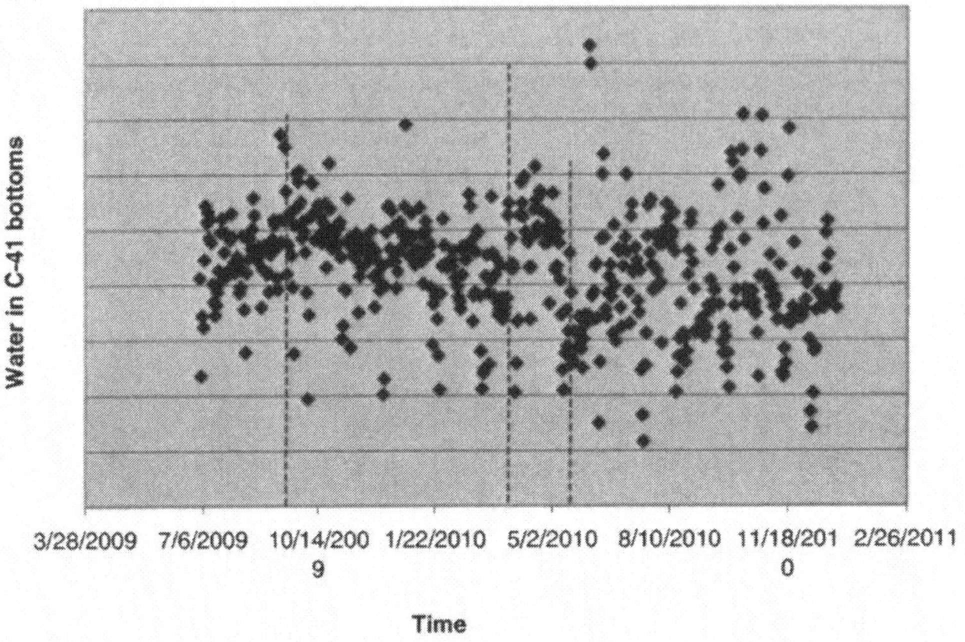


Fig. 5 - Water content in C-41 bottoms

One of the key relationships around the dehydration section is the reboiler steam requirement for C-41 when plotted against the pressure drop in the column. This is particularly important since it correlates one of the main variables (bottoms temperature) that govern the water content in the bottom stream of the water stripper (hence, its efficiency) to the pressure drop, which is an indicator of any malfunction in the operation of the column and its capacity. This relationship is shown in Figure 6. It is clear from Figure 6 that as the reboiler steam usage in the water dehydration column increases (potentially as a result of higher feed flow rate and/or higher bottom's temperature), the pressure drop increases accordingly.

The second key component of the factors that would impact the performance in the dehydration section is the inlet temperature. Higher feed temperature increases the amount of flash yielding to lower work done by the column yielding more separation efficiency and taking the system away from flooding regime [1].

However, plotting the feed temperature vs. water content in bottom showed no obvious trend in the range of feed temperature evaluated for the current study on the water content in the bottom of the water stripper (Figure 5).

The latter can be attributed to the small range of variability in the feed temperature employed within the dehydration section, which does not allow for a wide range of data to estimate the impact of changing the feed temperature.

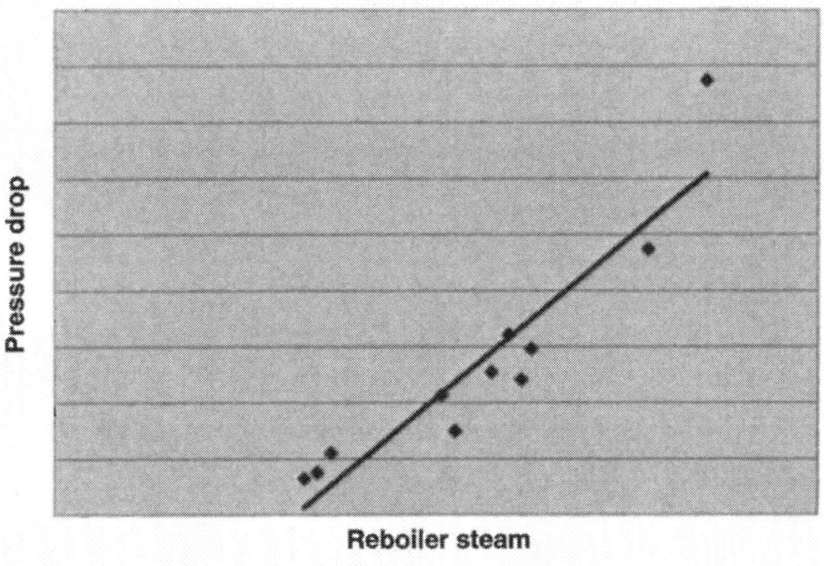


Fig. 6 - Reboiler steam vs. pressure drop in C-41

Moreover, a comparison of such operating conditions against a sister plant site of SABIC Innovative Plastics shows that a much broader range of feed temperature was employed.
The question of whether additional heating capacity can be utilized for the feed stream needs to be answered. The dehydration section's current capability allows flexibility in supplying hotter feed to the C-41 column. However, the benefits from such practice need to be evaluated which will be accomplished in the next section via the simulations using the developed ASPEN model.

## 5. ASPEN model development

The above representation implies that the list of factors that influence/indicate the dehydration section performance include:

- Feed stream properties
- Feed temperature
- Pressure drop in the columns
- Packing details
- Steam usage in the reboiler

### 5.1. ASPEN Model setup

The flow diagram of the dehydration model is shown in Figure 7. The model input and specifications for the two columns can be listed as follows:
Water stripper (C-41):

- Radfrac with no condenser
- Input: Feed flow rate and composition, pressure profile.
- Design spec: Bottoms' water content (vary distillate to feed ratio).
- Packing – 3 sections

- No specific efficiency (Murphree) specified.

Azeotrope column (C-48):

- Operating with a fixed Reflux Ratio as input.
- Input: Pressures and distillate to feed ratio.
- Design spec: Bottoms' temperature.

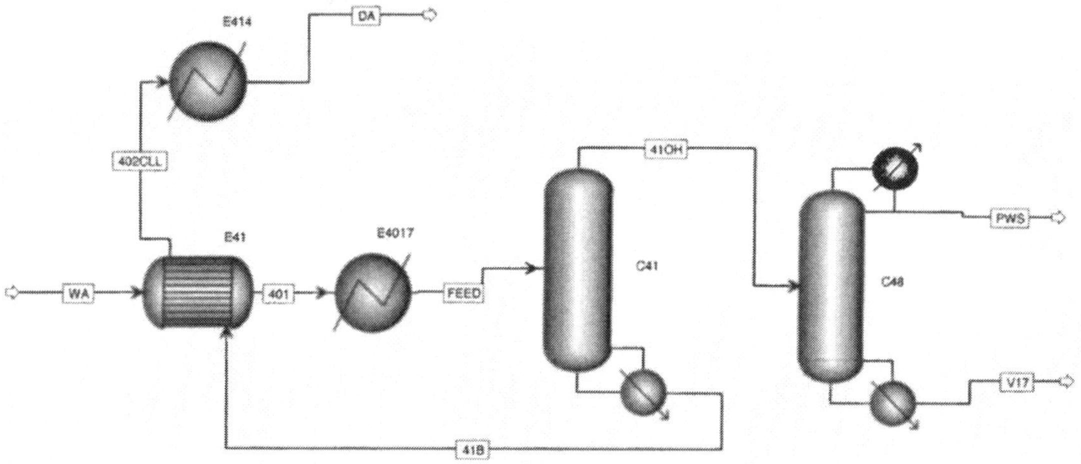


Fig. 7 – ASPEN model diagram

## 5.2. Model validation and sensitivity

As discussed above, the data period under investigation where the system modifications were utilized in the plant includes a variety of plant productivities as well as a wide range of operating conditions. This is crucial for the selection of the validation periods.

A total of 10 validation periods were utilized.

The current dehydration section model does:

- Predict the C-41 top temperature well as well as the C-48 top temperature
- Match reasonably the C-41 bottom temperature
- Meet the flow rate at the C-48 overheads
- Predict the composition in C-48 overheads fine
- Agree with the C-41 reboiler duty (more deviation at higher loads/water content) well

On the other hand, the ASPEN model showed a mismatch of the split across C-41, hence, underestimate the C-48 feed stream and its bottoms. The latter may be attributed to the fact that the mass balance demonstrated in Figure 5 was not fully closed but rather at ~6.5%. This small deviation can attribute to the mismatch noted above. Another explanation of the mismatch can be due to an entrainment effect in C-41 that can also lead to a certain difference in the reboiler duty at higher capacities.

Higher feed temperature is known to yield flashing. This relieves the load of the C-41 column allowing better separation. This is, predictably by the model, desirable. A sensitivity analysis on that variable is shown in Figure 8.

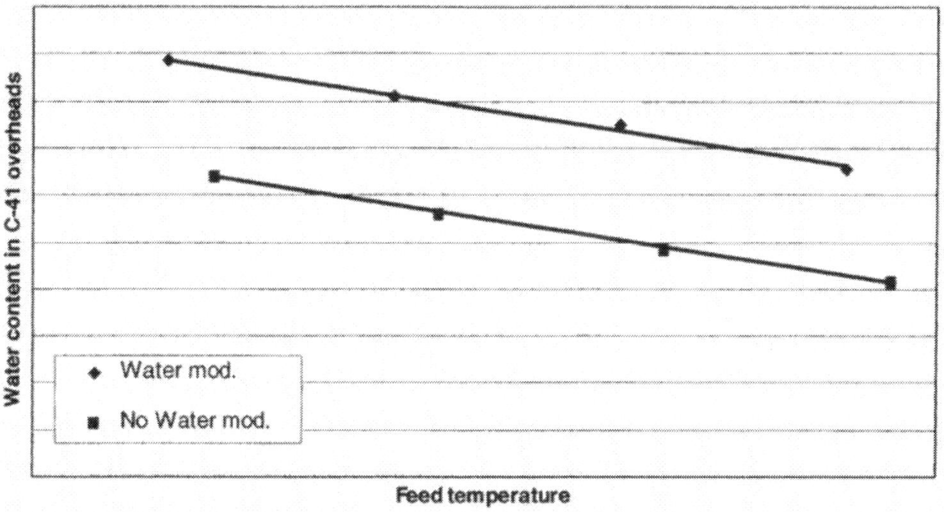


Fig. 8 –ASPEN predicted sensitivity analysis of the feed temperature

## 6. A discussion on flooding and pressure drop

The investigation of the water stripper's internals revealed that in addition to the packing, liquid and gas distributors also are present. A big relief and a decrease in the pressure drop can be gained from the removal of the gas distributor and hence it is recommended to check the condition of the gas distributor and evaluate the removal to ensure the benefits. This practice has proved harmless in similar columns operating under similar conditions in our sister plants.

A column operated under flooding conditions before gas distributor removal showed a drastic decrease in the top temperature afterwards (~70 °F step).

Corresponding top temperatures in our current case does not exhibit more than ~10 °F difference between extreme cases with the most extreme modifications in the process.

## 7. Conclusions and recommendations

In conclusion, the study for debottlenecking the dehydration section in our plant was accomplished by analyzing the plant data and evaluating optimized operating conditions with an ASPEN model.

The main recommendations from this work can be listed as follows:

- Design an EVOP for a gradual increase of the water content in the feed of the system to better assess the impact on the dehydration section performance.
- Consider the removal of the gas distributor in the water stripper; a practice that proved harmless on the operation of the column yet efficient in tackling pressure drop issues.
- Including additional compositional sampling in extra positions around the system.

## 8. Acknowledgments

The authors are grateful for the support provided by the Manufacturing team. In addition, special thanks go to the managers of the Center of Excellence for the initiation of the project.

*Sabic Innovative Plastics is a trademark of SABIC Holding Europe B.V..

## Reference

[1] Erik Delsman, Internal Sabic Innovative Plastics technical memo report, 2009-SABICIP-0102.

# (140b) Fast NMPC Applied to Industrial High Purity Propylene Distillation

**Gilvan A. G. Fischer**, Henrique Lage Refinery, Petrleo Brasileiro S.A. Petrobras, So Jos dos Campos, Brazil and Lorenz T. Biegler, Chemical Engineering, Carnegie Mellon University, Pittsburgh, PA

Distillation is one of the most used separation processes. Distillation columns are largely studied on process control because they are directly related with product quality, energy consuming and control difficulties due to nonlinearities. With increasing competitiveness, there is a demand to push the process to most profitable operating point, generally closer to the boundary of the admissible operating region, without violating the operating limits and satisfying product specifications. To guarantee the achievement of these requisites the controller must explicitly consider process nonlinearities and constraints. These problems motivate the use of nonlinear model predictive control (NMPC) with rigorous dynamic models. The NMPC is able to improve operation stability and increase profit.

In this work a detailed propylene distillation model of an industrial unit is derived. The model is represented by tray-by-tray equations consisting of mass balances, energy balances, phase equilibrium, hydraulic and summation equations. This DAE system contains the rigorous description of the plant, bound constraints for the differential, algebraic states and for the manipulated variables. Here, we use a simultaneous collocation-based approach. The infinite dimensional problem was transcribed to a finite element discretization at Radau collocation points.

This industrial-size distillation first principle model incorporates the discretized dynamic as algebraic constraints and has thousands (40,000) of variables which is challenging to solve. One advantage of this dynamic model is that it has a sparse structure. In order to reduce computational complexity, the NLP solver must have the ability to exploit this sparse structure. Another advantage is that the dynamic model is completely algebraic and first and second order derivative information can be obtained directly from modeling platforms. To solve the resulting NLP problem we use the interior-point solver IPOPT (Wächter & Biegler, 2006). To handle the bound constraints IPOPT add logarithmic barrier terms to the objective function. IPOPT applies Newton's method to the KKT conditions of system to solve each barrier problem. The factorization of the KKT matrix is the most expensive step at each iteration in the algorithm. For large systems, this factorization step can take a significant amount of computation time.

A barrier for these optimization-based strategies of NMPC is long time feedback delay due to on-line solution. In order to solve this problem have been developed advanced-step NMPC strategies (Zavala & Biegler, 2007) that use the sensitivity equation to get the fast updated solution. The control action is extracted from the approximate solution vector and applied on plant. The asNMPC has the same nominal stability properties of the ideal NMPC (Zavala et al, 2008). Success on the application of this strategy for an air separation unit (Huang et al, 2009) leads to the study of other industrial problems.

Industrial data from a propylene unit localized at REVAP (Henrique Lage Refinery, Petrobras-Brasil) were used to determination of system parameters for the propylene distillation model. The main uncertainty sources are the tray efficiencies and heat transfer coefficients. In this work, we are verifying the performance of the MHE/asNMPC maintaining the composition on desired level and maximizing the valuable product on a modeled industrial case, dealing with measured disturbances as plant feed, unmeasured disturbances as well as the influence of the model-plant mismatch.

References

A. Wächter and L. T. Biegler, On the Implementation of a Primal-Dual Interior Point Filter Line Search Algorithm for Large-Scale Nonlinear Programming, Mathematical Programming 106(1), pp. 25-57, 2006

Zavala, V. M.; and Biegler, L.T. The Advanced Step NMPC Controller. Proceedings of the IEEE Workshop on Advanced Process Control Applications for Industry - Vancouver, Canada. 2007

V. Zavala, C. Laird, L. Biegler, A fast moving horizon estimation algorithm based on nonlinear programming sensitivity, Journal of Process Control 18 (2008) 876–884.

R. Huang, V.M. Zavala and L.T. Biegler. Advanced step nonlinear model predictive control for air separation units. J. of Process Control,19(4) 678-685, 2009.

See more of this Session: Applications of Optimization In Industrial Practice
See more of this Group/Topical: Process Development Division

# MULTIOBJECTIVE OPTIMIZATION OF A SPOUTED BED REACTOR

**Ghanim M. Alwan**[1], **Stoyan Nedeltchev**[2], **Shreekanta Aradhya**[3] and Muthanna Al-Dahhan[3], (1)Chemical Engineering, University of Technology, Baghdad, Iraq, (2)Department of Chemical and Biological Engineering, Missouri University of Science and Technology, Rolla, MO, (3)Chemical and Biological Engineering, Missouri University of Science and Technology, Rolla, MO

Abstract:

Gas-solid spouted beds are either cylindrical bed with cone base or the whole bed is in a cone shape where the gas enters as a jet. The gas forms a spout region that carries the solids upward in a diluted phase which forms a fountain at the top of the bed where the solids fall down and move downward in the annular region. These beds have found wide and efficient applications as reactors, dryers, coaters, etc. Various design and operating variables affect the performance of these beds. It seems that the performance of the gas-solid spouted bed benefit from solids uniformity structure along with lower pressure drop (PD). Therefore, the focus of this work is to perform multiobjective optimization on gas-solid spouted bed, to maximize the uniformity index (UI) of the solid particles across the spouted bed and to minimize the PD along the bed. Hence, UI and PD are considered as the objective functions. Four decision variables were selected which affect the objective functions of the process. These decision variables are: gas velocity, pressure, particle density and particle diameter. The dynamics of the spouted bed reactor are highly interactive nonlinear processes. Thus, the optimization technique is a powerful tool to obtain the desired design parameters and set of operating conditions. This would guide the experimental work and reduce the risk and cost of the design and operation. It also yields the best range of the operating conditions which could improve the efficiency of the spouted bed reactor. Various optimization algorithms are available. However, the reliability of the optimized solutions depends on the formulation of the objective functions and the selected optimization technique.

In this work, genetic algorithm has been used which is considered the best stochastic search technique for highly nonlinear objective functions. The accuracy and the reliability of the optimization search could be enhanced by using adaptation operators. It has been found that maximum UI could be obtained with high density particles, while minimum PD could be obtained with lower particle density. The results of the optimization have been compared with the experimental data using sophisticated optical probe and high accuracy pressure transducers.

Keywords: Genetic algorithm, Optimization, Particle Technology, Spouted bed

## Nomenclatures:

C: Concentration of solid particles,[-]

dp: Diameter of solid particle, [mm]

P: Pressure into spouted bed, [Kpa]

UI: Uniformity index of solid particles,[-]

Vg: Superficial velocity of gas, [m/s]

## Greek letters:

*f*s: Density of solid particles,[Kg/m^3]

ÂP: Pressure drop a cross the spouted bed,[Kpa]

€: Porosity of bed, [-]

Ø: Sphersity of the solid particle, [-]

## Introduction:

A spouted bed is a special case of fluidization.It is an effective means of contacting gas with coarse solid particles. There is increasing a pplication of spouteds as reactors such as; coating, desulfurization, CO2 capture, combustion and gasification of coal and biomass (Limtrakuletal.2004).The spouted bed is a kind of high performance reactor for fluid-solid particles reaction, also it is a hybrid fluid-solid contacting system (Wang etal.2001).

The conical spouted bed reactor may be an alternative to the fluidized bed reactor for obtaining olefins by polyolefin pyrolysis, and moreover, it is also suitable for obtaining high yields of wax, which is a raw material that is handled easily in commercial cracking units, either by steam cracking or by catalytic cracking in FCC units (Aguado etal.2002).

Conversion and coating of particle nuclear fuel is performed in spouted bed reactor, in order to achieve uniform particle coating and prevent agglomeration of the particle bed. Spouted bed reactors are specialized fluidized bed reactors with unique fluidization characteristics and particle handling advantages (Wendt etal.2007).

Figure (1) illustrates the spouted bed schematically.The bed is made up of three distinct regions: the dilute core called the spout, the surrounding annular dense region called the annulus, and the the dilute fountain region above the bed surface. Solid particles are carried up rapidly with the fluid (usually gas) in the spout to the fountain and fall down onto the surface of the annulus by gravity.In the annulus, the particles move slowly downward and, to some extent, inward as aloose packed bed(Epstein and Grace, 1984.

Plasma when coupled with spouted bed reactor of gas-solid reaction brings in several advantages such as high rate of heat & mass transfer (Sathiyamoorthy, 2010).However, it is better to develop the design of the spouted bed reactor to overcome the large pressure drop and instability of the operation and also improve the uniformity of the products resulting from the chemical or physical treatment due to the elimination of the back mixing (Prachayawarakorn etal.2005).

Spouted bed coating reactor helps in uniform exposure of surface to the precursor by systematic circulation of particles within the reactor. The spouted bed reactor is characterized of its stable solid circulation which is responsible to enhance heat and mass transfer (Sathiyamoorthy etal. 2010).

Most studies, concering solid behavior and hydrodynamic correlations of bed reactors, have been done at ambient temperature. The lack of studies at higher temperatures is due difficulties associated with measuring techniques under these conditions. In extrapolating hydrodynamic parameters derived at ambient temperature to higher temperatures, only the physical property changes of gas and solid phases, such as density and viscosity are taken into consideration (Radmanesh, etal.2005).

The mimicking with the spouted bed nuclear particles coating reactor was investigated in the present work using the steel and glass beads as the solid particles and the air as the spouted gas at ambient temperature. Steady state study was done to select the effective process variables. The uniformity of the solids particle was considered as the new key parameter in the scale-up study of the present spouted reactor. It seems that the performance of the spouted bed reactor benefit from solids uniformity structure along with lower pressure drop.

Therefore the objective was to maximize the uniformity index of the solid particles and minimize the pressure drop across the bed. Optimization technique is a powerful tool to obtain the desired design parameters and set of operating conditions. This would guide the experimental work and reduce the risk and cost of the design and operation. It also yields the best range of the operating conditions which could improve the efficiency of the spouted bed reactor.

Multiobjective optimization technique was needed to solve the conflicted optimization problem of the system. Two advanced optimization techniques were implemented with the present spouted bed reactor which were Genetic and Pattern Search Algorithms.

## Experimental set-up:

The experimental set-up was designed and constructed in the best way to collect the data as explained in Figure(2).The cylindrical spouted bed made of Plexiglas was employed in this study which was the major part of the experimental set-up. The bed is (3 inches in diameter and 36 inches in height).At the bottom of the bed, there is a 60° cone-shaped distributor (3 inches in height), on which 20 holes (0.25 inch in diameter) are perforated horizontally. The distributor was made of stainless steel (316L) with thickness of(0.25 inch).The spouting nozzle (0.25 inch in diameter) locates in the center of the distributor.

The spouted gas was air from the air compressor and the gas flow rate was controlled by the pressure regulator and measured by the flow meter (Figure 2). The solid particles used were steel and glass beads with different average diameters and properties (Table 1).

Holes are drilled at vertical intervals of (1.86 inch) along the column wall with diameter of (0.5 inch), in which the optical probe was placed at different radial positions: 1.5, 1.25, 1.0,0.75,0.5,and 0.25 inch also at axis positions:7.5, 5.5 and 3.5 inches above the distributor.

The newly optical probe (manufactured by the Institute of Chemical Metallury, Chinese, and Academy of Sciences) can be used to measure both solids concentration and solids velocity and their fluctuation. The Particle Velocity analyzer (PV6) which manufactured by the 'Institute of Chemical Metallurgy, Chinese, Academy of Science')consists of; photoelectric converter and amplifying circuits, signal pre-processing circuits, high-speed A/D interface card and its software PV6, is adapted to the optical probes.

The pressure drop across the bed is measured by the pressure transceducer (Type: PX309-002G5V) with the range of (0-2.0 psig) manufactured by omega company . The pressure into the spouted bed adjusting within the desired values by using the inverted circular stabilizer, 60 mm in diameter was installed at the top of the bed column. This was preventing the spout fountain from swaying.

The structure of the fountains depended upon the value of the air velocity which entered to the spouted bed. Figure (3-a) illustrates the fountain in the spouted bed for different spouted gas superficial velocities. The Computed Tomography (CT) as a radio active technique based on gamma ray was used in the present work to monitor the solid holdup distribution into the radial axis of the spouted bed reactor (Figure 3-b).Also Computational Fluid Dynamic (CFD) simulation on the spouted bed was used to appear the solid hold up into the axial direction(Figure 4). These techniques were used to confirm the measurements of the optical probes. The red color indicates to high concentration of sol ids (solid bulk) while the blue color indicates to very low concentration of particles (gas bulk).

The effective process variables are affecting the solid concentration and pressure drop into the reactor are; gas velocity, pressure, particle density and particle diameter ((Wendt etal.2007). The operating conditions of the process are explained in Table (2).

The calibration of the optical probe needed to about (30-40 miutes).The interval time of each measurement was within 3.0 minutes.The warming time of the experimental rig was 30.0 minutes.However each experiment needed to about two hours and for validity the experiment was repeated twice.

## Optimization Technique:

Deterministic algorithms for function optimization are generally limited to convex regular functions. However, many functions are either not differentiable or needed a lot of difficult mathematical treatment: discretiation, decomposition, sensitivity computing…etc.Therefore, stochastic sampling methods could used instead to optimize such functions.

In the present work two optimization techniques were used:

-Genetic Search Algorithm

-Pattern Search Algorithm

The genetic algorithm (GA) has been applied to investigate the optimum conditions of the process.GA is search algorithm based on mechanics of natural selection and natural genetics.GA is based on Darwins' theory of 'survival of the fittest'.Thereare several genetic operators.Such as; population, selection, crossover, mutation…etc.Each chromosome represents a possible solution to the problem being optimized, and each bit (or group of bits) represents a value of variable of the problem (gene).A population of chromosomes represents a set of possible solution.these solutions are classified by an evaluation function, giving better values, or fitness, to better solution(Gupta and Srivastova, 2006).

Pattern Search algorithm was used to confirm the optimal results obtained by the Genetic algorithm also when implemented with the genetic algorithm process to switch to Hook-Jeeves algorithm.The Pattern Search algorithm was used to refine the decision variables (Palonen etal.2009).In the present work, the Genetic & Pattern search were applied with the aid by the Tool Box of MATLAB language (version 7.10.0.499-R2010a).

## Result and Discussion:

The stability conditions and the solid particles concentration distribution were studied in the spouted bed reactor. Also the uniformity index of the solid particles and the pressure drop across the bed were focused in the present work.

Flow Regimes:

The flow regimes in the present spouted bed reactor were studied to limit the stable gas velocity. (Xu etal. 2009) were previously mentioned to that as follows:

-Vg<0.785 (m/s)          packed bed flow regime

- 0.785 <Vg<0.887 (m/s)          unstable spouting flow regime

- 0.887<Vg (m/s)          stable spouting flow regime

All the parameters of the reactor process are difficulty to be measured; also the efficiency of the spouted reactor is dropped at the unstable conditions. A stable spouted bed is observed when the particles in the bed fluidize homogenously and they are not discharged from the apparatus.

## Uniformity Index:

In the present work, the uniformity index relation was developed and implemented regarding to the Nedeltchev and Zhang relations (Nedeltchev etal.2000 and Zhange tal.2011). For steady state condition, the equation can be written as follows:

UI= (Cav-Cmin)/ (Cmax-Cmin) (2)

Where;
Cav: Average solid concentration, [-]

Cmin: Minimum solid concentration [-

Cmax: Maximum solid concentration [-]

UI: Uniformity index of the solid particles, [-]

The results of CT and CFD techniques confirmed the validity of equation (2).Figure (5-a) illustrates the effect of the velocities of air on the uniformity index (UI)of the steel and glass beads.The UI decreased with increasing the air velocity (Vg) as a result of increasing the disperation of the solid particles due the kinetic energy of the solid particles was increased.Also two regions appeared in these curves which represented the transition region from the packed bed flow regime at Vg of 0.74 m/s to the stable spouting flow regime at Vg of 0.95 and 1.0 m/s.The axial mixing is more vigorous than the radial mixing with the increase of spouting air velocity (Zhang etal.2007).The static bed height has important effect on the uniformity of the final mixtures. with increasing the static bed height, the degree of mixing axially decreased then increased, but in the lateral direction decreased (Hao etal.2008).Also with the increase in the spouting gas velocity the coating thickness linearly decreased (Sathiyamoorthy etal.2010).TheUI of the steel beads is higher than that of glass beads (Figure 5-a) due to high scattering occurred with the particles of low density. So the low concentration of solid particles was obtained for all positions with the glass beads system this affected directly on the values of UI. Figure (5-b) explains the effect of the spouted bed pressuer on the UI.As the pressure increased the stochastic motion and the scattering of particles will be increased, then the uniformity of solid particles decreased. The density of the solid particles has the positive effect on the UI of the spouted bed reactor (Figuer6-a).Since the increasing of the particles density tend to get the bed more strength and resistance to the motion and dispersion against the vortex of the fountain. So it gives the solid particles more stability and uniformity.The denser material is continued to spout in the centeral region while the less dense particle formed atorriodal vortex around the central spout (Halow etal.2010).The solid beads diameter has negative effect on the UI of the solid particles as shown in Figure (6-b).As the diameter of the particles decreased, the free area (mean free path) into the bed increased. Then the particles of small size are helpful to raise the mixing speed (Hao etal.2008).This gets the smaller particles more uniformity in the radial distribution. However, the uniformity index is good indication for improve the mass, heat and the conversion of reactants into the nuclear spouted bed reactor.

Pressure Drop:

Figures (7&8) explain the effect of the process variables (air velocity, bed pressure, density and diameter of solid particles) on the pressure drop across the reactor bed.The pressure drop increased with increasing the velocity of air for both steel&glass beads due to increasing the kinetic energy and the interaction of the solid particles (Correa etal.1999).

In the case of steel beads the pressure drop is higher than that with glass beads because of high strength and friction with steel beads (Figure 7-a).The effect of the spouted bed pressure on the pressure drop across the bed was very low. Then the pressure drop increased slightly with increasing the inlet pressure of the air to the spouted bed (Figure 7-b).The dense particles(steel) create more resistance and friction against the air flow and then tend to raise the pressure drop across the bed (Figure 8-a).The porosity of the bed increased with the large particles diameter, so that the strength of the solids to the air flow then reduced.This tend to reduced the pressure drop across the spouted bed (Figure 8-b).These conclusions also confirmed by (Zhong etal.2006).

## Optimization Problem:

The spouted bed is a hybrid fluid solid contacting system (Wang etal.2001).So that the dynamic modeling for optimization requires model that describe the essential dynamic characteristics of the process under study. For the present complex nonlinear spouted bed process, a simple and effective model may be difficult to derive. Process measurements are the most important source of information of a specific process.While first principles provide general information about the behavior of the process, process measurements are required to identify the process model (Roffel and Betlem, 2006).

In the present work, it is recommended to formulate the objectives i.e. the uniformity index of the solid particles (UI) and the pressure drop ($\Delta P$) across the spouted bed reactor with the effective (decision) process variables. The decision variables are; air velocity, bed pressure, density and diameter of particles. The objective of optimization search is to maximize UI and minimize $\Delta P$.

Two advanced nonlinear regression optimization algorithms were used as: Newton-Quasi and Hook-Jeevs pattern moves to formulate the developed nonlinear objective functions with the aid of the computer program (Statistica version6).

The empirical correlations are:

Maximize $UI=0.184*Vg\text{^}-0.214*P\text{^}-0.301*\int s\text{^}0.316*dp\text{^}-0.267$ (3)

With correlation coefficient of(R=0.99).

Minimize $\Delta P=0.037*Vg\text{^}0.381*P\text{^}0.05*\int s\text{^}0.407*dp\text{^}-0.221$ (4)

With correlation coefficient of (R=0.94)

Subject to Inequality constraints of;

$$0.74\leq Vg\leq 1.0$$

$$101.0\leq P\leq 312.0 \qquad (5)$$

$$2400.0\leq \int s\leq 7400.0$$

$$1.09\leq dp\leq 2.18$$

From equations (3 and 4), one can conclude that the spouted bed is highly nonlinear interacted process. Equation (3) confirms that the air velocity, reactor pressure and particle diameter have the negative effect on the uniformity index (UI) while the density of solids has the positive effect as described in the previous section. Also the density of the solid particles seem the more effective variable which have higher power of (0.316).

From equation (4), one can conclude that the air velocity, reactor pressure and density of the solid particles have positive effect on the preesure drop across the bed, while the diameter of particles has negative effect. The density of solid particles also is the more effective variable.

In extrapolating hydrodynamic parameters derived at ambient temperature to higher temperatures, only the physical property changes of gas and solid phases, such as density and viscosity are taken into consideration (Radmanesh, etal.2005).

However, the spouted bed reactor is a hybrid fluid-solid contacting system.So that the optimization problem of the present work represented by the two conflicted equations(equations 3&4) with inequality constraints (equation 5).

## Optimization Search:

Optimization technique is a powerful tool to develop the design of the spouted bed reactor to overcome the large pressure drop, the instability of the operation and also improve the uniformity index of the solid particles.

In the present work, the stochastic techniques (Genetic and Pattern search) were used to obtain the optimal solutions of the process depended upon the process models of the system (equations 3, 4&5).The objective of the present work is to maximize the uniformity index (UI) of the solid particles and minimize the pressure drop ($\Delta p$) across the bed.

Two optimization searches were used:

-Single objective search:

In this search, the optimization technique was applied separately for each objective functions (equations 3&4) with inequality constraints (equation 5).Several experiments were applied to investigate the optimal solution of the system.Also the operators of Genetic algorithm and pattern search were tunned to get the best solution. Tables (3&4) explain the best operators of the two algorithms for the two objectives. Figures (9&10) illustrate the outputs of the algorithms excutions.GA was implemented with the pattern search by using the hybrid function to refine the decision variables (Table3).The pattern search technique (Figure 10) confirms the Genetic algorithm and get approximately the same optimum values as shown in Table (6).The results of the optimization searches are agree since the values of the objectives and the continuous &discrete variables within the the limits of the operating conditions. Also the experimental results confirm the optimal results that the maximum UI obtained with high density material (steel beads) while the minimum ($\Delta P$) obtained with low density (glass beads).This confirms the mathematical model of the process (equations 3&4) that the density of solids was the effective variable on UI and $\Delta P$.Also, the accuracy of the optimization techniques are high due to it can capture the discrete variables (density and and diametr of solid particles) with the integer values and the velocity of air (Vg) at the stable region (Table 6).The objective function of UI needed to (55) numbers of generation which is more than those for the objective function of $\Delta P$ which operated with (51) generations.This is due to that the function of UI (equation 3) is more complex than $\Delta P$ function (equation 4).However, the maximum UI was obtained with the steel beads system while the minimum $\Delta P$ obtained with the glass beads system for the same operating conditions. The accuracy of the optimization searches could be increased by the adaptation of the Genetic and Pattern search algorithms operators because of highly interacted and nonlinears objective functions.

-MultiObjective Search:

In this technique the two objective functions were used together.Also several experiments were done to select the best operators of GA as shown in Table (5).Since the objective functions are conflicted, so that 26 chromosomes were created (Figure 11-a).Each chromosome represents the possible solution to the optimization problem, and each bit represents the value of the variables of the problem.This means that the optimization search created 26 sets for the process design and operating conditions. This means that we can operate the spouted bed reactor with others new materials and operating conditions. optimization search can reduce the number of experimental runs, also the consumed cost for design and operation. The designer or operator can select the best solution regarding to the limits of the operating conditions. Figure (11-b) illustrates the output of the algorithm excution.The optimum values of UI are between (0.447 to 0.755) while for $\Delta P$ within (0.895-1.672).The best optimum results are approximately closed that obtained by the single search (Table 6).This indicates to the good selection of the algorithm operators (Table 5).Due to the two objectives functions are conflicted then the number of generations is equal to 109.This high when compared with the single objective techniques which have numbers of generation within (51 to 55) as shown in Table (3).The optimum results by the multiobjective search were stay in the limits of the operating conditions of the process (equation 5).Also this technique confirms the resuls obtained by the single objective techniques as explained in Table (6).This indicates to high accuracy of the optimization techniques used in the present work.However, the multiobjective search is more accurate than the single objective technique since it treated with the overall process variables of the system and could give the globle optimum results. The improvement was exactly obtained by using the optimization techniques.The maximum UI was obtained by the experimental work conditions were 0.485 and 0.55for steel&glass beads respectively as shown in Figure (5).The value of UI could be increased to 0.75 by the optimization techniques within the limits of the operatingconditions.Also the improvement for $\Delta p$ function was occurred when compared the experimental results (Figure 7) with the optimum result of 0.89Kpa obtained within the same limits of the operating conditions. Table (7) explains the improvement of the optimization technique on the objectives of the spouted bed reactor when compared the best experimental results with that obtained by the optimum search within the limits of operating conditions. The increasing of UI will be obtained by the steel beads and the decreasing of $\Delta P$ obtained by the glass beads. This means that the density of solid particles have the major effect on the UI and $\Delta P$.

The success of the optimization search depended on the best construction of the objective correlations with the decision variables and on the good selection of the optimization algorithm (Palonen etal.2009).In the present work, the stochastic GA technique was successful with the interacted non linear system.The accuracy of the optimization search can be increased by adaptation the algorithm operators.

## Conclusions:

1-Spouted-gas bed is a hybrid fluid-solid contacting reaction system. The spouted reactor was highly interacted nonlinear process system.

2- The uniformity index of the solid particles get good indication for enhances the conversion of reactants and coating of particles. The reducing of the pressure drop across the bed within the stable flow dropped the risk of the dissipated pumping energy.

3-The effective variables are; gas velocity, pressure, particle density and particle diameter.

4-The density of solid particles was the effective variable on the uniformity index of solid and the pressure drop across the spouted bed.

5-It was better to correlate the objectives directly to the decision variables of the process. The empirical model based optimization could be implemented to find the optimum parameters.

6- It is better to develop the design of the spouted bed reactor to overcome the large pressure drop, the instability of the operation and also improve the uniformity of the products. The optimization technique was the powerful tool for generate developed designs and sets of operating conditions for the spouted bed.

7-The optimal results can reduced the number of experimental runs and the cost consumed for design and operation. Also it limits the best range of operating conditions which improve the efficiency of the spouted reactor.

8-Maximum uniformity index obtained with high density beads (steel) while minimum pressure drop obtained with lower particles density (glass).

9-Good validity obtained when compared the optimal solutions with the experimental results.

10-The success of optimization search related to the formulation of the objective function also the selection of the suitable searching technique.

11-Genetic algorithm was the suitable stochastic search for the nonlinear functions. The accuracy of the search could be increased with the adaptation of the algorithm operators.

## References:


1-Limtrakul, S., Boonsrirat, A.and Vatanathm, T.'DEM Modeling andSimulation of a Catalytic Gas-Solid Fluidized Bed Reactor: a Spouted Bed as aCase Study', Dept.ofChem.Engng.kasetsar University, Bangkok, Thailand, Elsevier Ltd, (2004).

2-Wang, Z., Chen, P., Li, H., Wu, C.andChen, Y.'Study on the Hydrodynamics of a Spouting-Moving Bed', Ind.Engng.Chem.Res, 40, pp.4983-4989, (2001).

3-Aguado,R.,Olazar,M.,Prieto,R. and Bilbao,J.,'Kinetic Study of Polyolefin Pyrolysis in a Conical Spouted Bed Reactor',Ind.Eng.Chem.Res.41(18),pp.4559-4566,(2002).

4-Wendt, D.S, Bewley, R.L. and Wides, W.E.,'A Spouted Bed Reactor Monitoring System For Particulate Nuclear Fuel', proceeding of space nuclear conference, Boston, Massachusetts, June 24-28, paper 2023, (2007).

5-Epstein, N., and Grace, J.R.,'Spouting of Particulate Solids', Hand Book of Powder Science and Technology, Chapter 11,2nd, Van Nostrand Reinhold Co., N.Y.pp.509-536,(1984).

6-Sathiyamoorthy, D.,'Plasma Spouted-Fluidized Bed for Materials processing, Journal of Physics, Conference Series 208, (2010).

7-Prachayawrakorn, S., Reuengnarong, S.and Soponronnarit, S.'Characteristics of Heat Transfer in Two-Dimensional Spouted Bed', School of Energy and Materials, University of Technology, Bangkok, Thailand,(2005).

8-Radmanesh,R.,Mabrouk,R.,Chaouki,J.,and Guy,C.,'Effect of Temperature on Solids Mixing in a Bubbling Fluidized Bed Reactor', International Journal of Chemical Reactor Engineering,Vol.3,A16,(2005).

9-Gupta, A.K. and Srivastava, R.K.,'Integral Water Treatment Plant Design Optimization: a Genetic Algorithm Based Approach',IE Journal,vol.87,September,(2006).

10-Palonen, M., Hasan, A. and Siren, K.,'A Genetic Algorithm for Optimization of Build-ing Envelope and HVAC system Parameters', Eleventh International IBPSA Conference, July 27-30, Glasgow, Scotland, (2009).

11-Xu,J.,Bao,X.Wei,W.,Bi,H.T.,Grace,J.R.,and Lim,C.J.'Chaotic Characteristics of Pressure Fluctuation in a Gas Spouted Bed', the Canadian Journal of Chemical Engineering, vol.87, April, (2009).

12-Nedeltchev,S.,Ookawara,S.,and Ogawa,K.,'Time-dependent Mixing Behaviors in Lower andUpper Zones of Bubble Column', J.chem.Engng.of Japan, Vol.33, No.5, pp.761-767, 2000).

13-Zhang, Y., Zhong, W.and Jin, B.,'New Method for the Investigation of Particle Mixing Dynamic in a Spout-Fluid Bed', Powder Technology, 24, January, (2011).

14-Zhong, W., Zhang, M., Jin, B.and Xiao, R.,'Experimental and Model Investigations on Gas Mixing Behaviors of Spout-Fluid Beds,' International Journal of Chemical Reactor Engineering, vol.5, Article A32, (2007).

15-Hao, H., Guoxin, H.and Fengchao, W.,'Experimental Study on Particles Mixing in an Annular Spouted Bed, Energy and Conversion Management, 49,257-266, Elsevier, (2008).

16-Halow, J.S, Holsopple, K.andFinng, E.A,'Segregation in a Spouted Fluidized Bed, AIChE, Annual Meeting, Salt Lake City, U.S.A, (2010).

17-Correa, N.A., Freire, J.T and Correa, R.J,'Improving Operability of Spouted Beds Using a Simple Optimization Control Structure', Brazilian.J.Chem.Engng.vol.16, no.4, Dec.Saopaulo, (1999).

18-Zhong, W., Chen, X.and Zhang, and M.'Hydrodynamic Characteristics of Spout-Fluid Bed: Pressure Drop and Minimum Spouting/Spout-Fluidizing Velocity', Chemical Engineering Journal, 118, 37-46, (2006).

19-Roffel, B. and Betlem, B.,'Process Dynamics and Control: Modeling for Control and Prediction', Jhon-Wiley, Sons Ltd, 3rd edition, (2006).

Table (1)-Properties of the particulate materials.

| Material | dp(mm) | ∫s(Kg/m^3) | € | Ø |
|---|---|---|---|---|
| Steel beads | 1.09 | 7400.0 | 0.42 | 1.0 |
| Glass beads | 1.09 | 2450.0 | 0.42 | 1.0 |
| Glass beads | 2.18 | 2400.0 | 0.41 | 1.0 |

Table (2)-Operating conditions of the system.

| Air velocity(m/s) | Pressure(Kpa) | Temperature(K) |
|---|---|---|
| 0.74 | 101 | 298 |
| 0.95 | 101 | 298 |
| 1.0 | 312 | 298 |

Table (3)-The best operators of the single objective GA.

| Operator | Objective(UI) | Objective(ΔP) |
|---|---|---|
| Population type | Double vector | Double vector |
| Population size | 80 | 100 |
| Creation function | Feasible population | Feasible population |
| Scaling function | Rank | Rank |
| Selection function | Roulette | Stochastic uniform |
| Crossover function | Scattered | Scattered |
| Crossover fraction | 0.8 | 0.7 |
| Mutation function | Adaptive feasible | Adaptive feasible |
| Migration direction | Forward | Both |
| Migration fraction | 0.1 | 0.2 |
| Hybrid function | Pattern search | Pattern search |
| Number of generation | 55 | 51 |
| Function tolerance | 1.0E-6 | 1.0E-6 |

Table (4)-The best operators of the single objective Pattern Search.

| Operator | Objective(UI) | Objective($\Delta$P) |
|---|---|---|
| Poll method | Gps positive basis 2N | Gps positive basis 2N |
| Polling order | Consecutive | Consecutive |
| Search | Complete search | Complete search |
| Initial mesh size | 1.0 | 1.0 |
| Scale | On | On |
| Expansion factor | 2.0 | 2.0 |
| Contraction factor | 0.5 | 0.4 |
| Penalty factor | 100 | 100 |
| Cache tolerance | 1.0E-4 | 1.0E-4 |
| Mesh tolerance | 1.0E-6 | 1.0E-6 |

Table (5)-The best operators of the multiobjective GA.

| Operator | Type and values |
|---|---|
| Population type | Double vector |
| Population size | 80 |
| Selection function | Tournament |
| Crossover function | Scattered |
| Crossover fraction | 0.8 |
| Mutation function | Adaptive feasible |
| Migration direction | Forward |
| Migration fraction | 0.2 |
| Hybrid function | Fgoalattain |
| Number of chromosome | 26 |
| Number of generation | 109 |
| Function tolerance | 1.0E-6 |

Table (6)-Optimal solutions by the optimization Techniques.

| | Objective(UI) | | Objective(ΔP) | | MOGA | |
|---|---|---|---|---|---|---|
| Decision variables | SOGA | PS | SOGA | PS | Max. | Min. |
| Gas velocity(m/s) | 0.89 | 0.889 | 0.89 | 0.88 | 0.89 | 0.8 |
| Bed pressure(Kpa) | 101 | 101 | 101 | 101 | 101 | 101 |
| Density of solid(Kg/m^3) | 7400 | 7400 | 2400 | 2400 | 7400 | 2400 |
| Diameter of particle(mm) | 1.09 | 1.09 | 2.18 | 2.18 | 1.09 | 2.18 |
| Value of objective | 0.76 | 0.77 | 0.88 | 0.89 | UI=0.45-0.75<br>ΔP=0.89-1.76 | |

PS: Pattern Search

SOGA: Single Objective Genetic Algorithm

MOGA: Multi Objective Genetic Algortim

Table (7): Comparison between the best experimental and the optimal results.

| Materials | Max.UI(exp) | Optimal UI | Min.ΔP(exp) | Optimal ΔP |
|---|---|---|---|---|
| Steel beads | 0.53 | 0.75 | 1.82 | 1.67 |
| Glass beads | 0.42 | 0.45 | 0.94 | 0.89 |

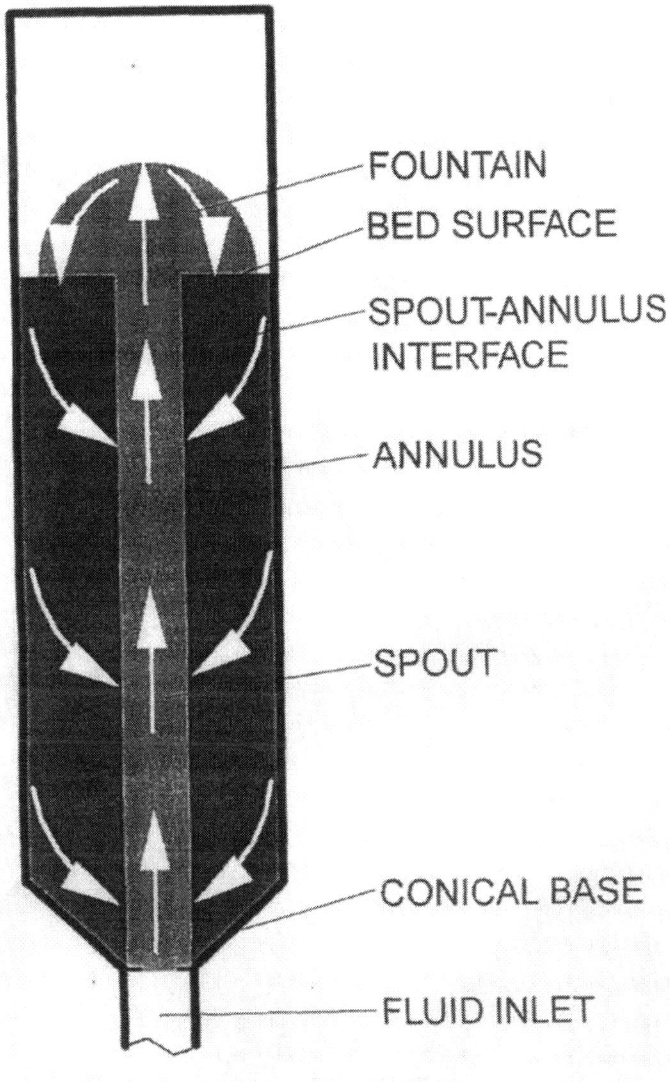


Figure(1):Spouted bed.

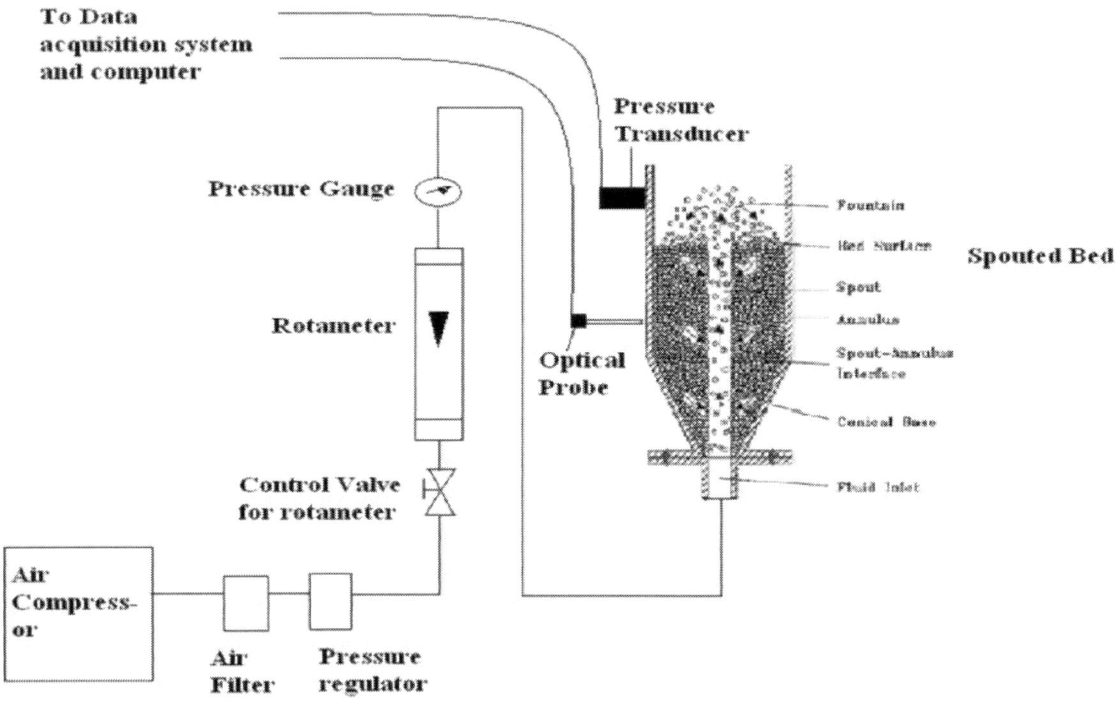


Figure (2): Experimental set-up

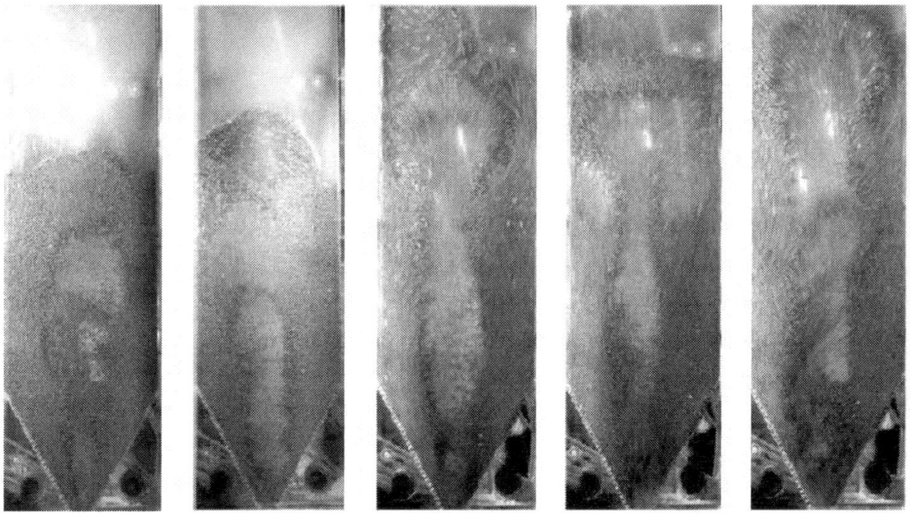

(a)

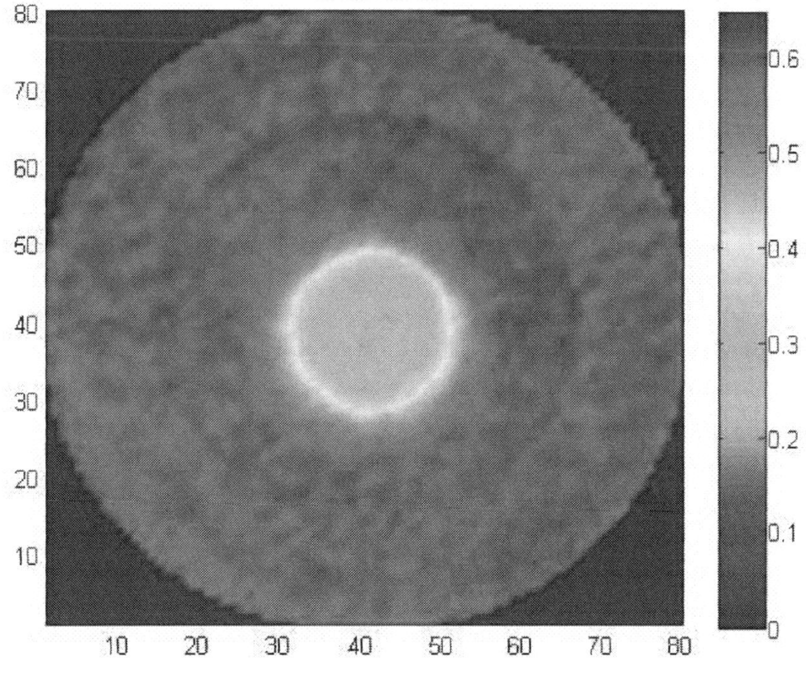


(b)

Figure (3): (a)-Flow regimes into the spouted bed for different spouted gas velocities.
(b)-Computed Tomography of solid holdup distribution into the spouted reactor.

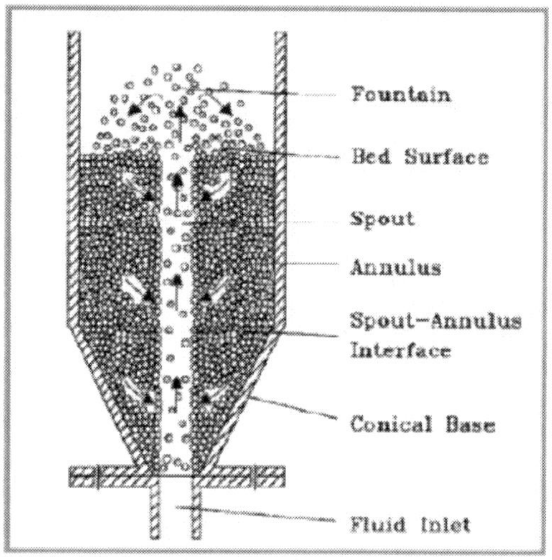


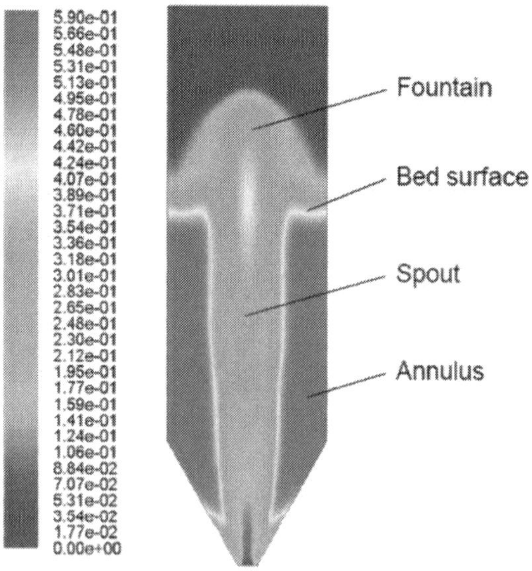


Figure(4):CFD simulation on the spouted bed appears the solid particles hold up.

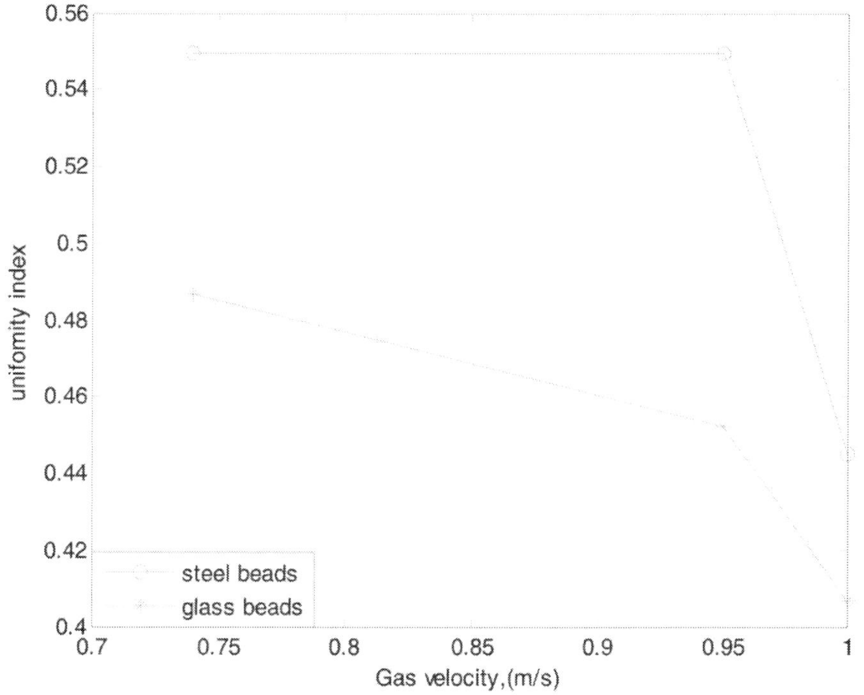


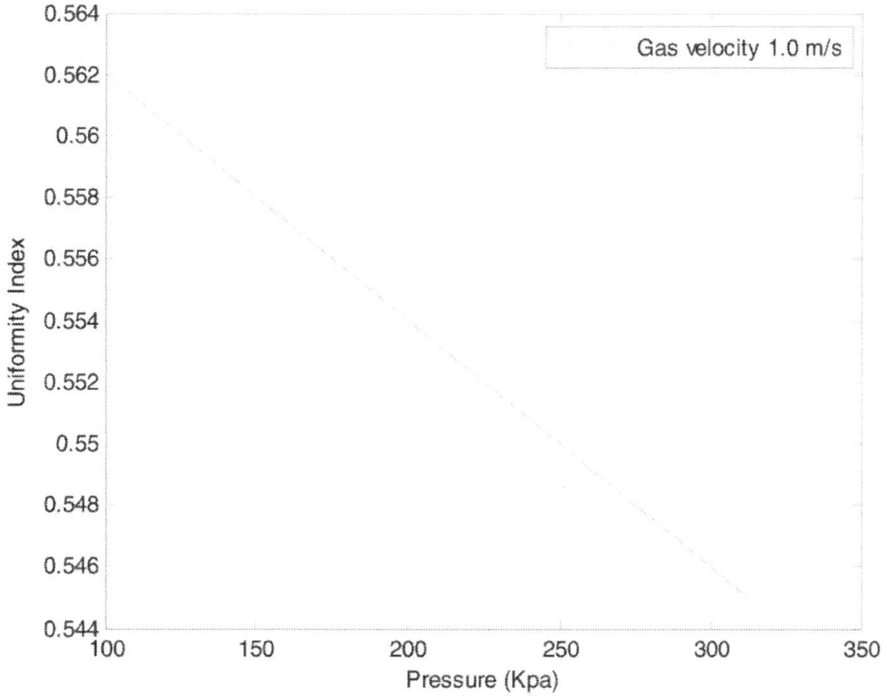


Figure (5): Uniformity index against :(a)gas velocity,(b)bed pressure.

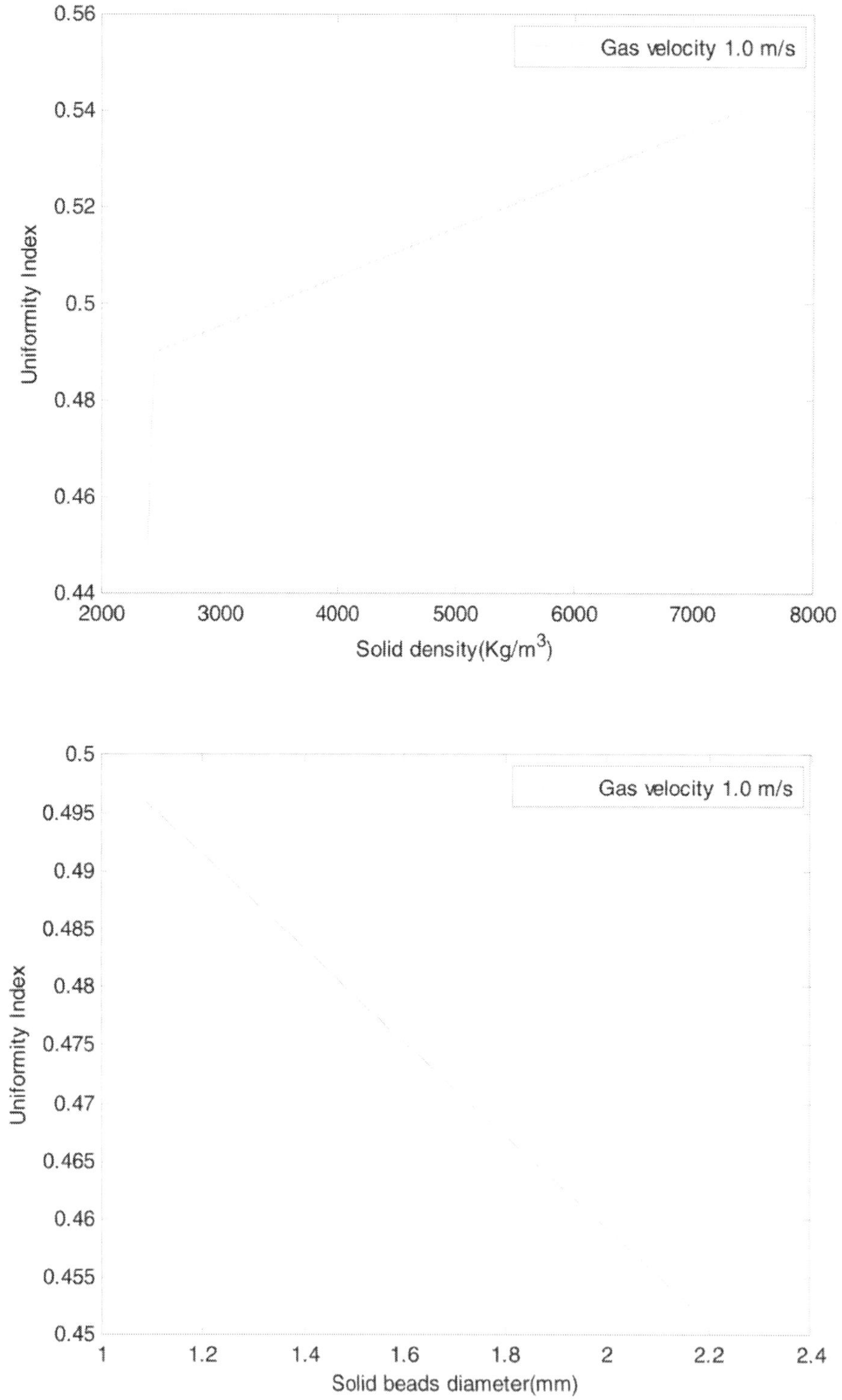


Figure (6): Uniformity index against ;(a)solid density,(b)solid diameter.

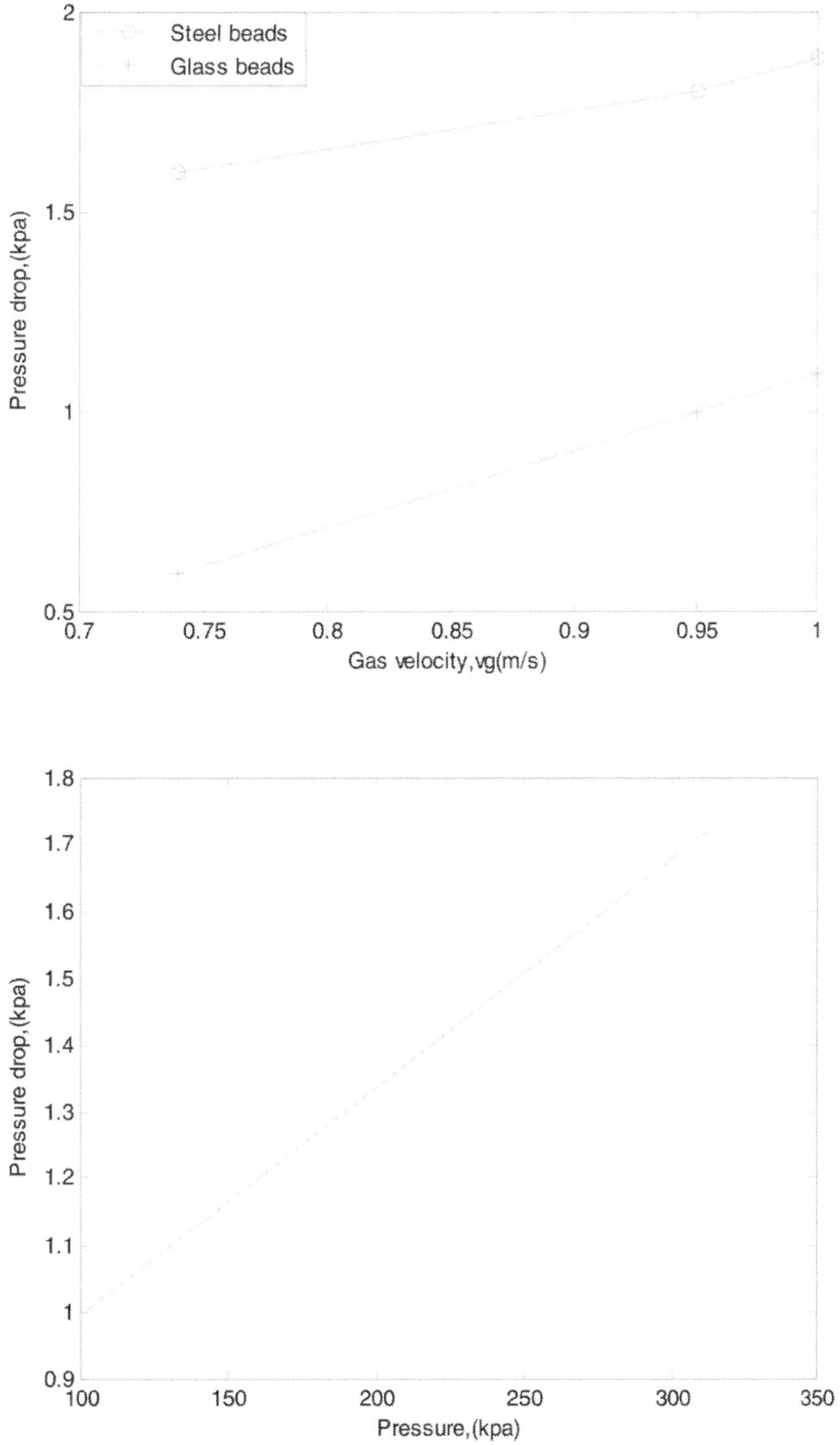


Figure (7): Pressure drop against ;(a)gas velocity,(b) bed pressure.

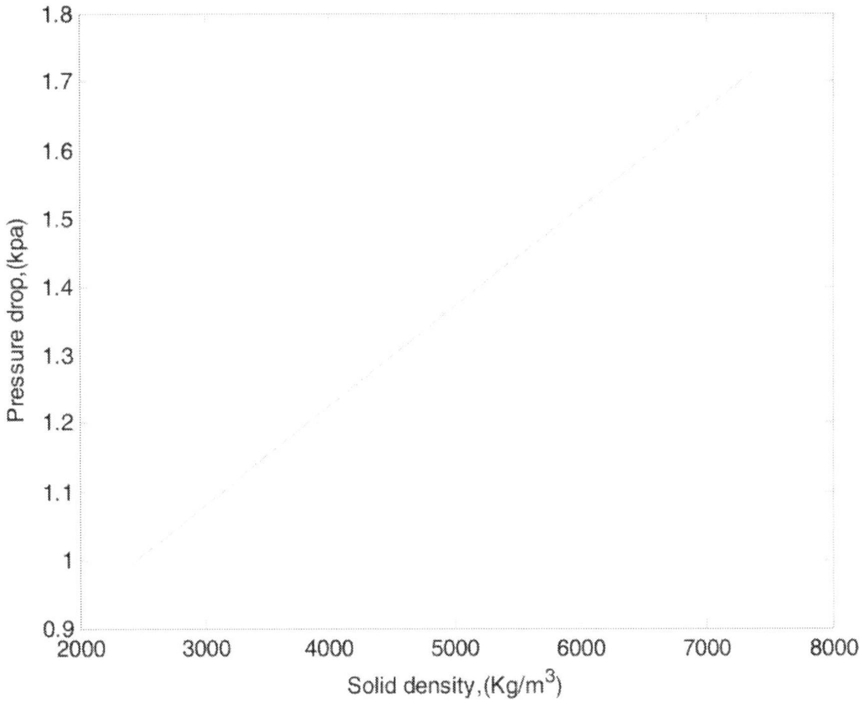


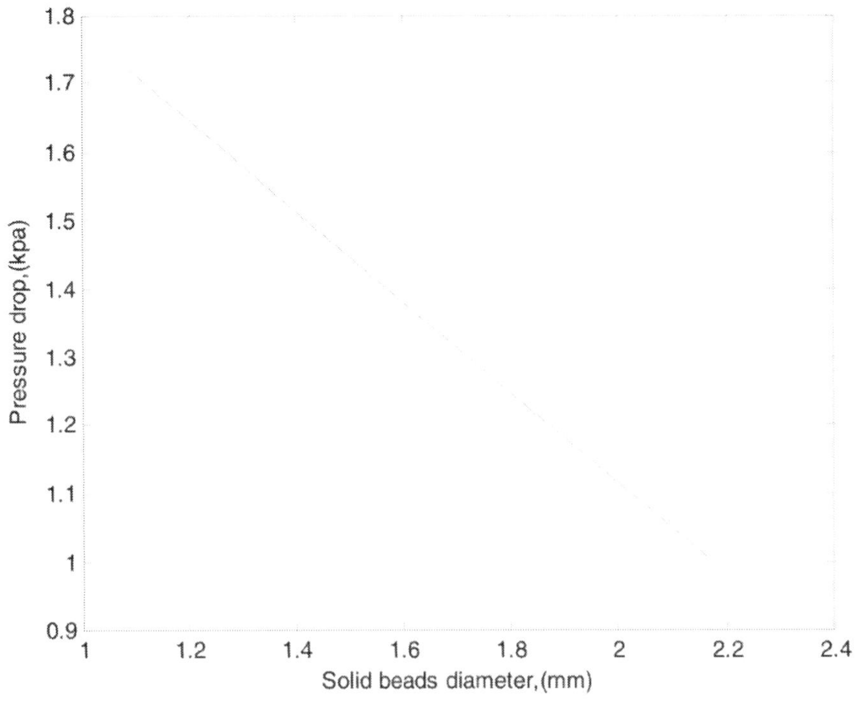


Figure (8): Pressure drop against ;(a)solid density,(b)solid diameter at vg=1.0.

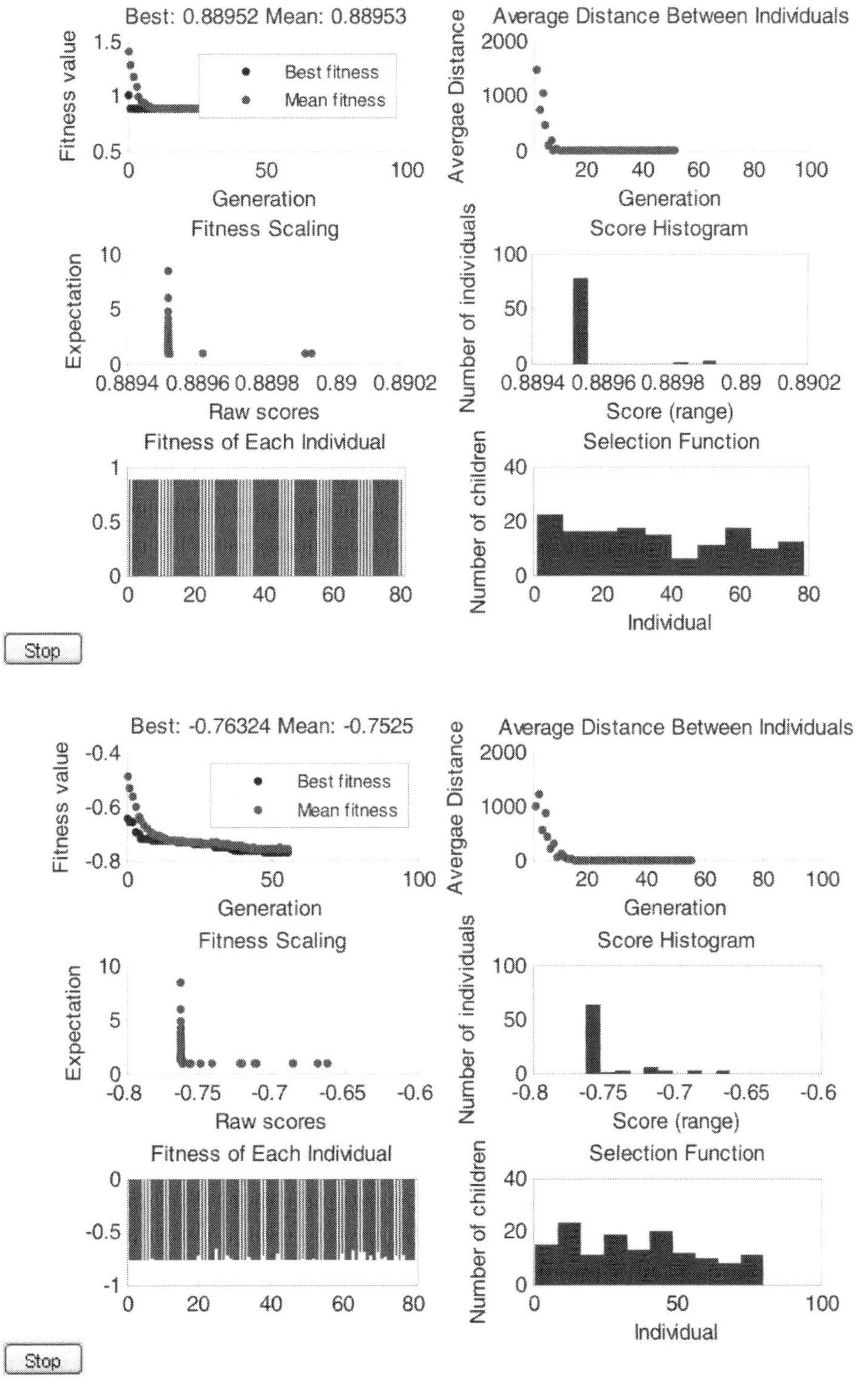


Figure (9): Output of single objective genetic algorithm for ;(a) pressure drop, (b) uniformity index.

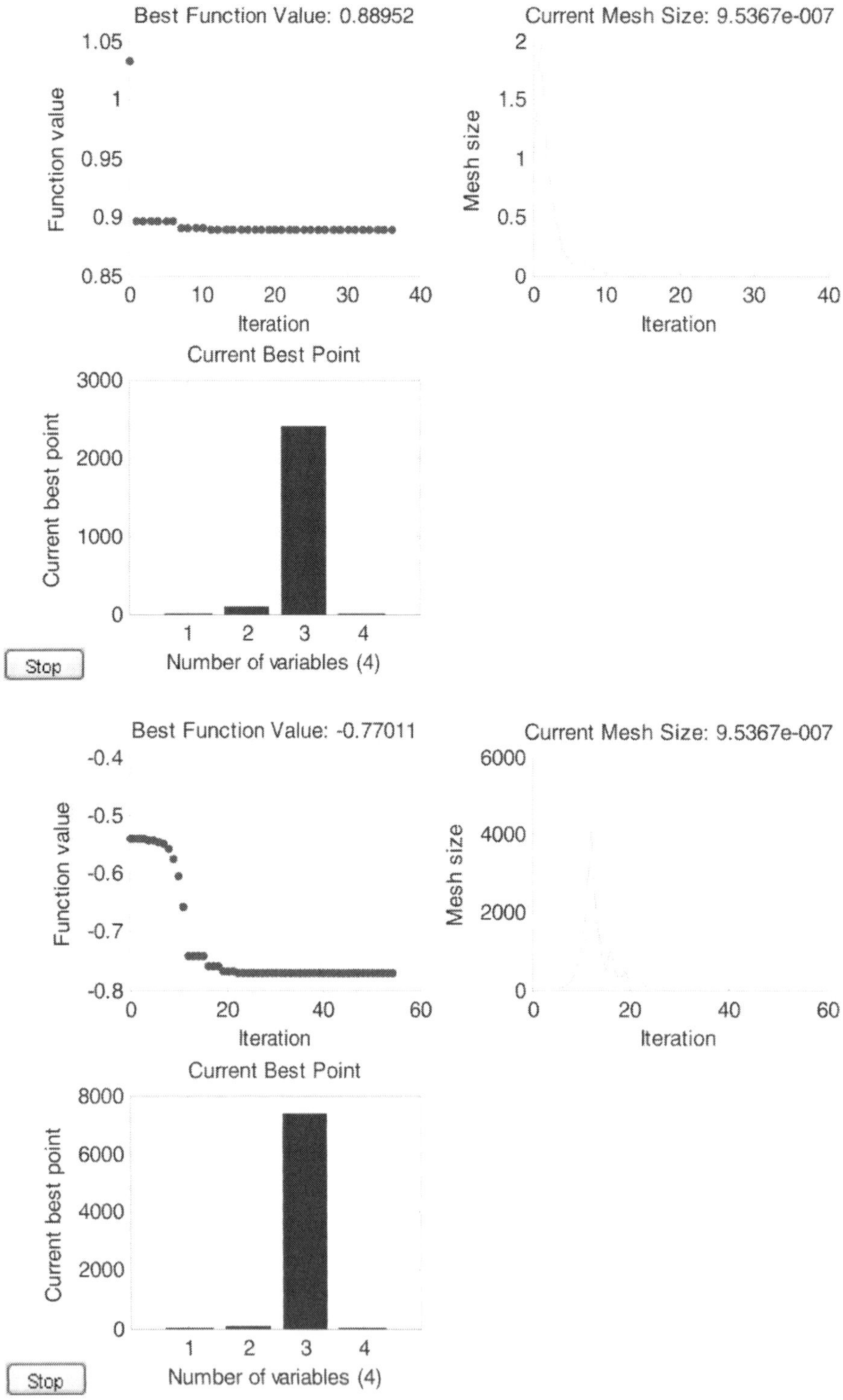


Figure (10): Output of single objective search pattern algorithm for ;(a) pressure drop,(b)uniformity index.

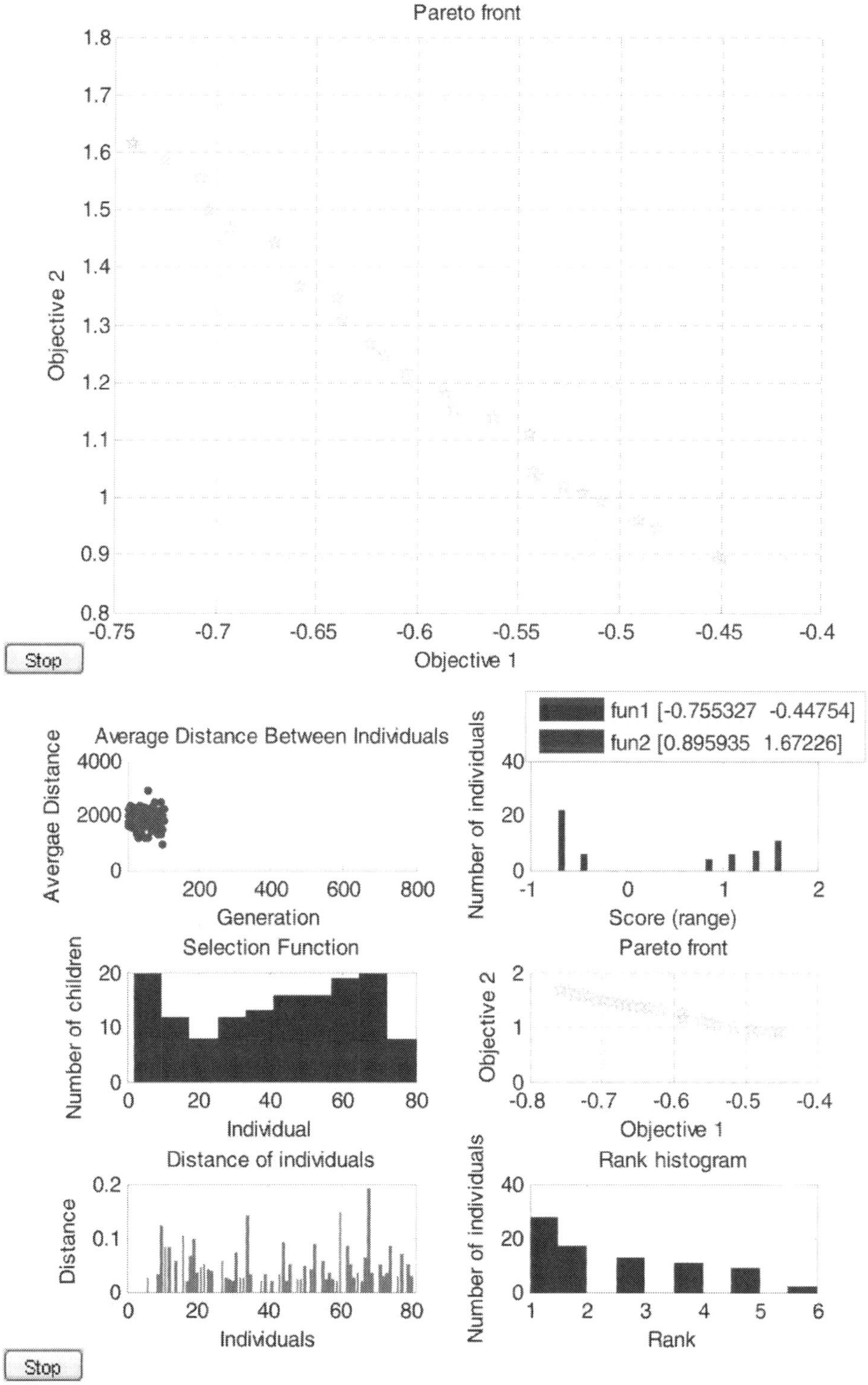


Figure (11): Output of multiobjective genetic algorithm ;( a) Pareto front, (b) genetic parameters.

# (140d) A Multifaceted Experimental Design Investigation of Nodule Formation In Electroformed Parts In Compact Disc Plating Operations

**Z. Otero Gephardt**, Chemical Engineering, Rowan University, Glassboro, NJ and John Heinzel, NSWCCD Code 985, Naval Sea Systems Command - Philadelphia, Philadelphia, PA

A multifaceted experimental design combining chemical mixture and operational parameters was developed to identify the factors affecting nodule formation in electroformed parts in compact disc plating operations. The mixture variables investigated were the bath wetting agent (surface tension reducer), bromide salt (anode corrosion) and grain refiner (stress reducer) concentrations. The anode nickel level and plating head rotational speed were the operating variables studied. A disc quality score to account for and quantify nodule size and location was developed, and optimal operating parameters and bath mixture concentrations were identified. The analysis indicated that high level interactions among mixture variables and operating parameters have a significant effect on the disc quality score. All plating bath components and, to a lesser extent, rotational speed had a significant effect on disc quality for the variable ranges investigated. The concentrations of wetting agent and stress reducer in the plating bath had the most significant impact on nodule formation. The location of the nodules on the disc was not influenced by plating bath or operating variables.

See more of this Session: Applications of Optimization In Industrial Practice
See more of this Group/Topical: Process Development Division

# MOLECULAR MODELS USING VBA TO OPTIMIZE THE PROCESS OF PETROLEUM REFINING

*Tatiane G. Vilas, University of São Paulo, São Paulo, Brazil*

*Jadson P.A. Silva, Petrobras Distribuidora, Rio de Janeiro, Brazil*

*Cláudio O. Nascimento, University of São Paulo, São Paulo, Brazil*

## Abstract

This work makes the combination between information from mass spectroscopy analysis and algorithm search in VBA (Visual Basic for Applications) for identify a specific molecule in groups of known molecules. The method proposed in this work is based on classifying compounds by their molecular weights, i.e., the compounds present in a mixture will be separated into classes by a mass difference as small as you want. You can thus see which group of molecules previously generated an oil compound belongs by its molecular weight. As these groups were generated by compounds which are of a known composition, it is known that composition of a given compound will have between the possible compositions. To generate molecules based on knowledge of components of oil, we used the formulas of general organic compounds, which are formed by homologous series. Homologous series are series of organic compounds with similar chemical properties whose components differ by a constant reactive molecular weight (CH2). These series are generated and their molecular weights are calculated and arranged in increasing order of mass to be generated by the groups developed a computational procedure in VBA.

## Introduction

Petroleum is a complex mixture of hydrocarbons with thousands of different molecular components. To determine the molecular composition of the oil components and products of the refinery is not a feasible task with the current analytical techniques because of the large number of components. Modeling the composition and behavior of these complex mixtures in refinery processes has over the years, evolved into a concept called lumping (Quann, 1992). The lumping models are quick and easy to formulate and solve, making them more dominant in the process control, design and optimization. In lumping models, thousands of individual constituents in a complex raw material are grouped into a broad but measurable category of classes of compounds or boiling point range (Quann, 1996).

The lumping methods have been applied to model almost all the reaction processes of refining, such as catalytic cracking, hydrotreating, etc. These methods are helpful when the main requirement is more comprehensive information, such as yield or octane number of gasoline. However lumping models cannot provide detailed molecular information, such as paraffins, olefins, benzene, sulfur, which are important for refining operations (Hu, 2001).

The molecular basis for refining model is to characterize the composition of the oil. Modern analytical techniques have been used, such as gas chromatography (GC), high-performance liquid chromatography (HPLC) and mass spectrometry (MS), nuclear magnetic resonance spectroscopy (NMR) to characterize the oil fractions. Techniques allowing obtaining the molecular composition of petroleum streams (Aye, 2005).

In this work we developed a method to describe the molecular composition of samples of asphalt. Using data obtained from analytical techniques to obtain molecular properties, along with a database generated by homologous series is possible to systematize the identification of components of oil.

**Metodology**

The method proposed in this work is based on classifying compounds by their molecular weights, i.e., the compounds present in a mixture will be separated into classes by a mass difference as small as you like. It is possible thus know which group of molecules before generating a given compound of oil by its molecular weight. As these groups were generated by compounds, which are of a known composition, that composition of a given compound will be known among the possible compositions.

To generate molecules based on knowledge of components of oil, we used the general formulas of organic compounds, which are formed by homologous series. The series is homologous series of organic chemical compounds with similar chemical properties whose components differ by a constant reactive molecular mass ($CH_2$). These series were generated and their molecular weight were calculated and arranged in order of increasing mass groups to be generated by a computational routine developed in VBA language. Below, in Table 1, are some examples of functions from organic compounds present in oil, such as simple hydrocarbons, nitrogen compounds, sulfur compounds, oxygenates, aromatics and saturated compounds closed chains among others, determined to make up the database of molecules used in this work.

**Table 1.** Organic compounds homologous series.

| Organic Groups | Organic Compounds General Formula | Carbon number |
|---|---|---|
| Alkanes or Paraffins | $C_nH_{2n+2}$ | |
| Alkadienes | $C_nH_{2n-2}$ | n>=2 |
| Benzene rings | $C_nH_{2n-6}$ | n>=6 |
| Olefins | $C_nH_{2n}$ | n>=2 |
| Alkines | $C_nH_{2n-2}$ | n>=2 |
| Alcohol single bonds | $C_nH_{2n+2}O$ | |

| | | |
|---|---|---|
| Alcohol one double bond | $C_nH_{2n}O$ | |
| Carboxylic acids with single bonds, acyclic | $C_nH_{2n}O_2$ | |
| Carboxylic acids with one Double Bond, acyclic | $C_nH_{2n-2}O_2$ | n>=3 |
| Ketones with single bonds, acyclic | $C_nH_{2n}O$ | n>=3 |
| Ketones with one double bonds, acyclic | $C_nH_{2n-2}O$ | n>=4 |
| Ether with single bonds, acyclic | $C_nH_{2n+2}O$ | n>=2 |
| Ether with one double bonds, acyclic | $C_nH_{2n}O$ | n>=3 |
| Ester with single bonds | $C_nH_{2n}O_2$ | n>=2 |
| Ester with one double bond | $C_nH_{2n-2}O_2$ | n>=4 |
| Amines single bonds | $C_nH_{2n+3}N$ | |
| Amines with one double bond | $C_nH_{2n+1}N$ | n>=2 |
| Amides single bonds | $C_nH_{2n+1}ON$ | n>=3 |
| Mercaptans | $C_nH_{2n+2}S$ | |
| Carbon disulfides | $C_nH_{2n+2}S_2$ | n>=2 |
| Benzothiofenes | $C_nH_{2n-10}S$ | |
| Di-Benzothiofenes | $C_nH_{2n-16}S$ | |
| Polysulfides | $C_nH_{2n+2}S_3$ | |

With the help of mass spectrometry samples of asphalt were analyzed, which were determined the molecular weight of compounds. These compounds had their masses determined by comparison with the similar series through a computer system developed in VBA. And thus classified into groups according to a criterion of difference in mass, as illustrated in the diagram below.

**Figure 1**. Illustrative scheme to search for groups of molecules

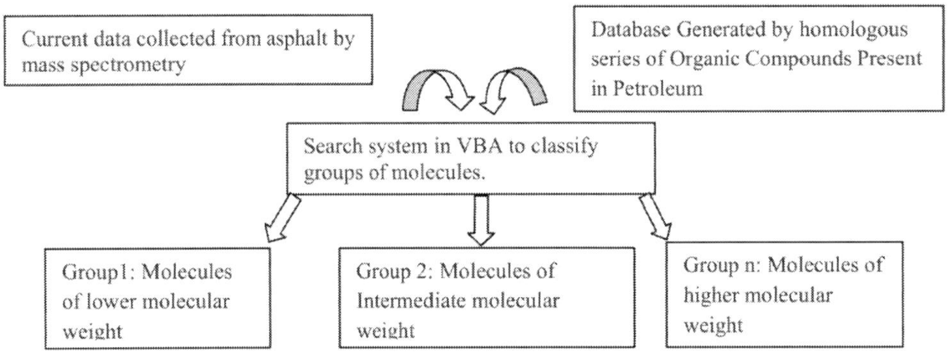


The search system was generated with algorithms in VBA language, the operation is as follows: since the homologous series was generated and terminated their molecular weight compared with the molecular weight obtained from mass spectrometry. Each theoretical molecular weight (generated by homologous series) will be compared with those obtained from experimental analysis, and groups are formed when the difference between the masses is less than a predetermined threshold.

**Figure 2**: Schematic illustration of the operation of the search system

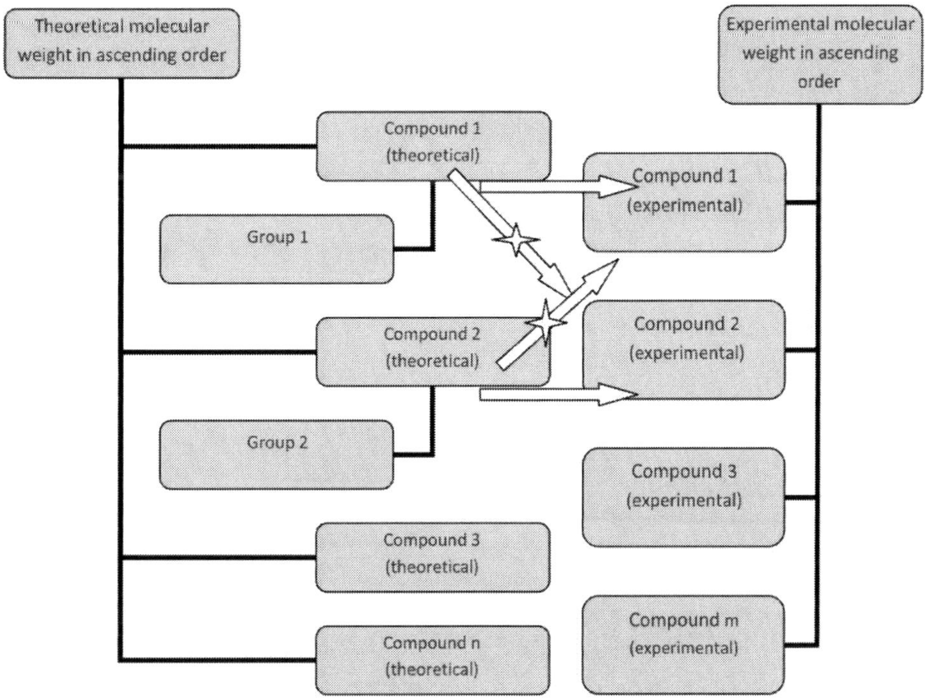


## Results and Discussion

For the classification of the components of asphalt sample was generated a database of molecular weight of the homologous series, whose examples are found in the methodology above. Sample analysis were made of asphalt in the mass spectrometer and obtained their molecular weight with high accuracy. To systematize the classification and separation of the groups, we used a search algorithm developed using VBA (Visual Basic for Applications).

**Table 2**: Asphalt compounds

| Groups generated by homologous series | Molecular weight of the sample (g/mol) |
|---|---|
| 276, 1582 (C20 H20 O) | 276, 1145 |
| | 276, 1383 |
| | 276, 1508 |
| | 276, 1747 |
| | 276, 1872 |
| 276, 2089 (C18 H28 O2) | 276, 2321 |
| | 276, 2447 |

| | |
|---|---|
| | 276, 2685 |
| | 276, 2811 |

**Table 3**: Asphalt compounds with high molecular weight

| Groups generated by homologous series | Molecular weight of the sample (g/mol) |
|---|---|
| 332, 2191 (C24 H28 O) | 332, 2008 |
| | 332, 2134 |
| | 332, 2372 |
| | 332, 2498 |
| | 332, 2517 |
| | 332, 2948 |
| 332, 3079 (C23 H40 O) | 332, 3074 |
| | 332, 3436 |

Tables 2 and 3 above show the groups generated by the proposed method, such groups are of structural isomers. Grouping structural isomers reduces the complexity of the system for thousands perhaps millions of molecules. Structural isomers with the same number of carbon are compounds that have similar physical and chemical properties which allow correlating them (Hu, 2002). These correlations, together with kinetic properties and thermodynamics are possible to obtain molecular models, to be attached to the models of existing processes in the refinery.

## Conclusions

Current oil has a large number of different molecular components, but to represent the molecules or to build only a small number of molecular groups is required. Having a set of atomic elements such as C, H, O, S, N, which are the molecules of oil, this set can be thought of as a structure group, with a molecular formula to promote the representation of the molecule. The molecular formula is limited and can only provide molecular weight of molecules and cannot distinguish between a large set of stoichiometric isomers.

From the standpoint of modeling the refining process requires a set of molecular information such as presence of: double bond, branching, sulfur, oxygen, nitrogen, or molecular arrangement simplified to provide a convenient representation in the fundamental chemical processes.

**References**

Aye,M.; Zhang,N.(2005). A Novel Methodology in Transforming Properties of Refining Streams into Molecular Information. Chemical engineering Science, 60:6702-6717.

Hu,S.; Towler,G.; Zhu,F.(2002). Combine Molecular Modeling with Optimization to Stretch Refinery Operation. Ind. Eng. Chem.,Res.41:825-841.

Hu,S.; Zhu,X.X.(2001). A General Framework for Incorporating Molecular Modelling into overall Refinery Optimisation. Applied Thermal Engineering 21:1331-1348.

Quann,R.; Jaffe,S. (1992). Structure-Oriented Lumping: Describing the Chemistry of Complex Hydrocarbon Mixtures, Ind. Eng. Chem.,Res.31:2483-2497.

Quann,R.J.; Jaffe,S.B. (1996). Building Useful Models of Complex Reaction Systems in Petroleum Refining. Chemical Engineering Science,51:1615-1635.

# (140f) Optimization and Thermodynamic Analysis of Mixed Refrigerant System In Ethylene Plants

**Karthik Krishnadevarajan**[1], Jian Zhang[1] and Qiang Xu[2], (1)Dan F Smith Department of Chemical Engineering, Lamar University, Beaumont, TX, (2)Dan F. Smith Department of Chemical Engineering, Lamar University, Beaumont, TX

## Optimization and Thermodynamic Analysis of Mixed Refrigerant System in Ethylene Plants

Karthik Krishnadevarajan, Jian Zhang and Qiang Xu

Dan F Smith Department of Chemical Engineering

Lamar University, Beaumont, TX 77710

### Abstract

Refrigeration is one of the most important and crucial system in chemical plants. A refrigeration system generally works by removing heat from low temperature streams through vapor compression cycles at the expense of mechanical work. Since a refrigeration system can cool down a process stream below the ambient temperature, it is indispensable to cryogenic cooling and used for separation process in many chemical industries, such as large scale productions of ethylene, oxygen, nitrogen and liquefied natural gas.

Usually refrigeration systems employ single component as refrigerant as long as it is thermodynamically desirable and operationally feasible. But a multi component mixture can also be used as refrigerant, because a mixed refrigerant system provides a continuum of temperatures with smaller temperature differences even at the lower temperatures. This leads to a low energy loss in the system. Mixed refrigerant system has many inherent advantages than the traditional single component refrigerant system.

In an ethylene plant, usually refrigeration is provided by cascading two or three single-component refrigeration systems. This process is not only expensive but also inefficient because it needs more facilities in comparison to single component refrigeration systems. The process streams in ethylene plant vary from 40 to -140 $^{o}$C. For a single component refrigerant it is impossible to provide constant refrigeration at a particular temperature. Mixed Refrigerant is used in order to improve refrigeration process by using four components as mixed refrigerant.

The present work deals with developing a mixed refrigerant system for an ethylene plant. The process is developed using a rigorous simulation model. Also the process is optimized for the optimal design and operating parameters of the system. The design and operating parameters include compressor work and total flow rate of the mixed refrigerants. This model has few challenging factors. First, the minimum temperature difference in the heat exchangers is taken as 2 $^{o}$C. Second, the heat ------duty of all the heat exchangers in the process is fixed based on the plant data. Finally, the process is non linear so optimization is more complex. To demonstrate this work, process streams thermodynamics are also analyzed using thermodynamic analysis. The base case results are compared with optimized results based on the rigorous simulation and optimization methods.

See more of this Session: Applications of Optimization In Industrial Practice
See more of this Group/Topical: Process Development Division

# (140g) Real Time Optimization of Industrial Gas Plants and Networks

**Tong Li**, Air Liquide, Newark, DE

Industrial gases are widely used in many areas including the chemical, refining, metallurgy, environmental, and semiconductor industries. The industrial gases used in large quantities typically include oxygen, nitrogen, argon, hydrogen, and carbon monoxide (CO).

The air gases – oxygen, nitrogen and argon – are usually produced through cryogenic distillation, when a large amount of them are needed in high purity. In a cryogenic air separation unit (ASU), air is first liquefied at a very low temperature and then separated into different components through distillation. This is a very energy intensive process. For example, the Air Liquide Group runs about 400 ASUs worldwide, and its electricity consumption in 2010 was more than one thousandth of the total world electricity consumption. Steam methane reforming (SMR) is one important process to produce hydrogen and CO. In this process, natural gas is reacted with steam in a heated furnace at the presence of catalysts to produce syngas, which is a mixture of hydrogen, CO, methane, and other gases. This mixture is then further processed to produce pure hydrogen, CO, or certain mixtures. These products are often transported to customers through a pipeline, which connects multiple SMR plants and customers. SMR processes also consume a lot of energy, mainly in term of natural gas. Around half of Air Liquide's annual energy bill is for the purchase of natural gas.

It can be seen that improving the efficiency of these industrial gas plants and pipelines is important in improving an industrial gas company's profit and competitiveness while reducing its energy consumption and carbon footprint. Real time optimization (RTO) is an important technique to achieve such a goal. A real time optimizer takes market dynamics such as changes in energy price and client demands into account in real time, and calculates the most profitable mode of operating the plants by using the plant models with energy consumption equations and process constraint equations. Then the calculated optimal set points are downloaded to the plant control system to ramp the plant or sent to operators as references.

In this work, two RTO applications in industrial gas production are studied: one ASU plant and one SMR pipeline. The important aspects of the applications are addressed including the general IT structure and functions of its different software components, important steps in completing such projects, and major challenges and corresponding solutions in mathematical optimization.

See more of this Session: Applications of Optimization In Industrial Practice
See more of this Group/Topical: Process Development Division

# (149a) Additives Used for Process Optimization - a Case Study On Ibuprofen

Thomas Vetter, Institute of Process Engineering, ETH Zurich, Zurich, Switzerland, **Marco Mazzotti**, Institute for Process Engineering, Swiss Federal Institute of Technology Zurich, ETHZ, Zurich, Switzerland and Jrg Brozio, Pharmaceutical & Analytical Ddevelopmentt, Novartis Pharma AG, Basel, Switzerland

Quality (purity, bio-availability) and processability of crystals generated in a crystallization or precipitation process are of key importance in the pharmaceutical industry. While a high product quality is demanded by both the consumer and federal regulations, a good processability is in the interest of the pharmaceutical company itself, as it makes downstream processing easier and therefore saves costs. Both the quality and the processability largely depend on the crystal size distribution. Therefore, influencing the crystal size distribution to obtain a desirable product is a highly promising approach. To do so, polymeric additives are used in this work to alter nucleation and growth kinetics of the model substance Ibuprofen.

Polymeric additives have two distinct advantages in comparison to tailor-made additives that have been proposed earlier to alter crystallization kinetics[1,2]. First, unharmful and already approved polymers can be used as additives in the crystallization process, such that an expensive and time-consuming approval procedure is avoided. Second, due to the immense size difference between the solute and additive molecules, the additive molecule is unlikely to be incorporated into the crystal lattice of the solute molecule, therefore ensuring crystal purity[3,4].

In this work, metastable zone width measurements and desupersaturation experiments are presented for the solute/additive combination Ibuprofen/Pluronic F127. The experiments are carried out in a 50:50 (w/w) mixture of ethanol and water.The data is obtained with state of the art measurement tools, such as attenuated total reflectance Fourier transform infrared spectroscopy (ATR-FTIR) and focused beam reflectance method (FBRM) measurements. Spectra obtained from the ATR-FTIR measurements are used together with a calibration model and chemometric methods to measure the evolution of the Ibuprofen concentration in solution. FBRM is used to measure nucleation.

By fitting a population balance equation model to the experimental data, the nucleation and growth kinetics are estimated. Knowing the influence of different temperatures, supersaturations and concentrations of Pluronic F127, on the crystallization kinetics of Ibuprofen, the crystallization process can be designed to yield a desirable crystal size distribution.


[1] Weissbuch, I., Addadi, L., Lahav, M., Leiserowitz, L., 1991. Molecular Recognition at Crystal
Interfaces. Science, 253 (5020), 637–645.

[2] Weissbuch, I., Addadi, L., Lahav, M., Leiserowitz, L., 1995. Understanding and Control of
Nucleation, Growth, Habit, Dissolution and Structure OF 2-dimensional and 3-dimensional
Crystals using tailor-made Auxiliaries. Acta Crystallographica Section B-Structural Science,
51, 115–148.

[3] Simonelli, S. P.; Mehta, S. C.; Higuchi, W. I., 1970. Inhibition of Sulfathiazole Crystal
Growth by Polyvinylpyrrolidone. Journal of Pharmaceutical Sciences, 59 (5), 633-637.

[4] Vetter, T.; Mazzotti, M.; Brozio, J., 2011. Slowing the growth rate of Ibuprofen crystals using
the polymeric additive Pluronic F127. Crystal Growth & Design, submitted.



See more of this Session: Crystallization Process Development II - Case Studies
See more of this Group/Topical: Process Development Division

# (149b) Crystallization-Based Integrated Process for the Resolution of a Pharmaceutically Relevant Intermediate

**Jan von Langermann**[1], Heike Lorenz[1], Malte Kaspereit[1], Martin Hedberg[2], Matthew J. Jones[2], Bjrn Herschend[3], Kerstin Larson[3], Mozzafar Shakeri[4], Jan-Erling Bckvall[4], Achim Kienle[1] and Andreas Seidel-Morgenstern[1], (1)Max Planck Institute for dynamics of complex technical systems, Magdeburg, Germany, (2) Pharmaceutical Development, Process Engineering, AstraZeneca AB, Sdertlje, Sweden, (3)AstraZeneca AB, Sdertlje, Sweden, (4)Department of Organic Chemistry, Arrhenius Laboratory, Stockholm University, Stockholm, Sweden

2,6-Pipecoloxilidide(PPX) is an important intermediate for the synthesis of mepivacaine,ropivacaine and bupivacaine,which are frequently used local anesthetics of the amide type for severalpurposes in medicine[1]. Based on its chiral nature, anenantioseparation of PPX is needed to exclude additional negative side effectsfrom the counter-enantiomer. Bycombining three different techniques, chromatography, enantioselective crystallizationand racemization, an improved process with a higher yield was achieved incomparison to the use of a single full-separation process. All 3 unitoperations were executed within a jointly used solvent system with fullrecycling (counter-enantiomer of PPX and solvents) including a minimumrequirement of solvent exchanges / removals. The target molecule was obtainedwith an enantiopurity of >99.5% in multi-gram scale.

Within this contribution the design of theintegrated process is presented with a focus on the final enantioselectivecrystallization process. The relevant physical chemical properties of PPX arepresented, e.g. binary and ternary phase diagrams, polymorphism study, andfinally the successful design of an effective crystallization processes onindustrial scale.

[1] Kleemann, A.; Engel, J.; Kutscher,B.; Reichert, D. *PharmaceuticalSubstances - Synthesis, Patents and Applications of the most relevant APIs*, 5th ed.; Georg Thieme Verlag KG: Stuttgart, New York, **2009**.

See more of this Session: Crystallization Process Development II - Case Studies

# (149c) Modeling, Screening and PAT: The Development of a Pharmaceutical Crystallization

**Brendon G. Ricart**, Chemical Process Development and Commercialization, Merck & Co., Rahway, NJ and Dimitrios Zarkadas, Chemical Process Development & Commercialization, Merck and Co. Inc., Rahway, NJ

**Modeling, Screening and PAT: The Development of a PharmaceuticalCrystallization.**

Crystallization isa critical unit operation in pharmaceutical processing because it determinesthe crystal form, particle size distribution and purity of the drug substance.Maintaining a consistent process upon scale-up requires an understanding ofgrowth and nucleation kinetics, relevant mixing time-scales, and local shearforces in varying geometries. Thedevelopment of a robust crystallization process for an HCl salt of Compound 1had been particularly difficult due to polymorphism issues, purity and particlesize/shape concerns and poor bulk density in existing processes. Form controlwas particularly difficult due to several undesirable hydrates andsolvates. A previous process produced amixture of crystal forms, solids with bulk density was less than 0.1 g/cm$^3$,and required two processing steps. Therequirements for optimal drug product formulation were crystals smaller than 20microns, aspect ratios between 1:5 and 1:3 and maximum obtainable bulk density.

A new single-step process was developed, which made use ofsynergistic solubility in a binary system, and desupersaturatedin a ternary solvent system which consistently controlled crystal form. Single solvent screens failed to yield anacceptable solvent, but an optimal mixture of toluene/methanol was indentifiedin binary screens. Extensive processmodeling using commercially available software and *de novo* algorithms enabled processing parameters for mixing, acidaddition rate, and filtration rate to be accurately estimated, leading to awell-controlled growth-dominated crystallization. Process performance was further optimized byon-line particle size tracking and HPLC monitoring. Using this comprehensive approach, a robustcrystallization was developed for a challenging salt form of Compound 1, producingover 90% yield, consistently meeting particle size specifications, doublingbulk density to 0.17 g/cm$^3$, and improving purity from ~96 to >99%.

# (149d) Process Enhancements for a pH Sensitive Crystallization

**John G. Gaertner**, James Marek, Samrat Mukherjee, Shailendra Bordawekar and Mina Dimitri, Process Research & Development, Abbott Laboratories, North Chicago, IL

A monomodal agglomerate particle size distribution and the desired morphology were achieved for an API crystallization that was pH sensitive, leading to a robust, efficient crystallization process. Irregular crystals were formed during growth in a variable pH environment. Switching from a simultaneous dual fed batch addition to a single feed addition, together with characterization/optimization of crystallizer mixing and pH enabled reduction of pH gradients. Statistically designed experiments were employed to demonstrate achievement of the desired particle size and morphology over a wide operating range. Benefits included higher crystallization yield, a 3x increase in agglomerate size, a reduction of the seed charge from 11% to 3%, widening of the design space for the range of seed particle sizes (eliminating the need for special seed milling), and reducing feed charge time by half. The process was successfully scaled up 100-fold to the commercial reactor.

See more of this Session: Crystallization Process Development II - Case Studies
See more of this Group/Topical: Process Development Division

# (149e) API-Crystal Preparation In A Continuously Seeded Tubular Crystallizer

Rafael J. P. Eder[1], Julia Grill[1], Elisabeth Schmitt[1], **Simone Schrank**[2], Eva Roblegg[2], Maximilian Besenhard[3], Stefan Radl[1], Heidrun Gruber-Woelfler[1] and Johannes G. Khinast[3], (1)Institute for Process and Particle Engineering, Graz University of Technology, Graz, Austria, (2)IPS/Pharmaceutical Technology, University of Graz, Graz, Austria, (3)Research Center Pharmaceutical Engineering (RCPE), Graz, Austria

We present a continuously seeded tubular crystallizer system, which allows the crystallization of pharmaceutical substances under controlled conditions (1-5). Ethanolic suspensions of acetylsalicylic acid, ibuprofen as well as acetaminophen seeds are fed into a tubular crystallizer, where they are mixed with hot, highly concentrated ethanolic solutions of the corresponding model-substances. Cooling generates supersaturation upon which the seeds grow to form product crystals while being conveyed through the tubing. Due to the tubular appearance and the small inner dimensions of the crystallizer in the few millimeter range it is possible to adjust the temperatures along the tubing according to the needs of the crystallization and to achieve narrow residence time distributions of the crystal slurry in the pipe. Hence, the seeds entering the feed end simultaneously grow in a similar environment (concentration, temperature etc.) and have approximately the same amount of time for growth. Thus, narrow product crystal size distributions (CSDs) can be obtained.

The effects of various critical process parameters on product characteristics have been investigated. Residence times of the suspensions in the tubing have been changed by altering the length of the crystallizer or the flow rates respectively. Further parameters that have been varied independently were the seed loading, seed properties as well as the cooling trajectories along the tubular crystallizer. Moreover, the spatial orientation of the tubular crystallizer coils were subject to variations (horizontal or vertical) in order to assess if particle segregation phenomena lead to different results regarding the crystal size distribution. Furthermore, a segmented flow of slurry and air-bubbles was implemented to completely rule out particle size segregation phenomena and to reduce the residence time distribution of the crystals to a minimum. All experiments clearly resulted in significant crystals growth despite the short residence times in the few minute range. Moreover, crystal masses increased by a few g/min and steady-state conditions were obtained instantaneously. Suppression of nucleation events by avoiding rapid cooling, thus high levels of supersaturation together with narrow residence time distributions of the slurry in the tubing allowed successful achievement of narrow CSDs.

Furthermore, simulations regarding the temperature gradients and the crystal growth have been performed in order to get a better understanding of the process. Modelled and experimental results were compared to each other and calculations where foundation for process improvements.

(1) R. J. P. Eder, S. Radl, E. Schmitt, S. Innerhofer, M. Maier, H. Gruber-Woelfler, J. Khinast. Continuously Seeded, Continuously Operated Tubular Crystallizer for the Production of Active Pharmaceutical Ingredients. *Crystal Growth and Design*. **2010**, 10 (5), 2247-2257

(2) R. J. P. Eder, E. K. Schmitt, J. Grill, S. Radl, H. Gruber-Woelfler, J. Khinast. Seed Loading Effect on the Mean Crystal Size of Acetylsalicylic Acid in a Continuous-Flow Crystallization Device. *Crystal Research and Technology*. **2011**, 46 (3), 227-237

(3) R. J. P. Eder, H. Gruber-Woelfler, J. Khinast. Kontinuierliche Kristallisation in einem Rohrkristallisator. *Chemie Ingenieur Technik*. **2009**, 81 (8), 1171

(4) R. J. P. Eder, E. K. Schmitt, J. Grill, R. Radl, H. Gruber-Woelfler, J. G. Khinast. Kontinuierliches Wachstum von API Impf- zu Produktkristallen in einem Rohrkristallisator. *Chemie Ingenieur Technik*. **2010**, 82 (9), 1469-1470

(5) R. J. P. Eder, E. K. Schmitt, J. Grill, M. Maier, S. Innerhofer, S. Radl, H. Gruber-Woelfler, J. G. Khinast. Crystallization of APIs in a continuously seeded tubular crystallizer. *Scientia Pharmaceutica*, **2010**, 78, 664

See more of this Session: Crystallization Process Development II - Case Studies
See more of this Group/Topical: Process Development Division

# (153a) Towards Sustainable Manufacturing: An Approach Evaluating Energy Reduction Techniques for System Improvement

Tamer O. Girgis, Department of Chemical Engineering and Materials Science, Wayne State University, Detroit, MI and **Yinlun Huang**, Dept. of Chemical Engineering & Material Science, Wayne State University, Detroit, MI

Manufacturing industries are facing major challenges due to economic globalization, increased energy costs, etc. Capital investments are highly dependent on the economy and driven by customer demands. Hitherto, a variety of technologies have been developed for energy efficiency improvement, such as those developed by the Department of Energy, which have been used for industrial system assessment. By these technologies, system performance improvement opportunities may be identified, such as those through equipment/process modification, control adjustment, workforce training, system management improvement, etc. However, how to quantify the sustainability performance of the technologies and how to use them in a systematic way is still a challenge.

In this paper, we introduce an energy technology integration approach for sustainable manufacturing. By this approach, the known energy reduction technologies will be assessed quantitatively by a set of suitable sustainability metrics. The process system to be improved will be assessed by the same set of sustainability metrics as well. The assessment results will be analyzed together and the most suitable set of energy reduction technologies will be selected to improve process system's sustainability to the expected level, for which the investment will be still under budget limit. The introduced approach can provide a scientific guidance to industry sustainability enhancement. The efficacy of the approach will be demonstrated through case studies on electroplating systems.

See more of this Session: Energy Sustainability, Challenges and Solutions
See more of this Group/Topical: Sustainable Engineering Forum

# (153b) Optimal Synthesis of Fuel Gas Networks for Energy Sustainability

M. M. Faruque Hasan and **Iftekar A. Karimi**, Dept of Chemical & Biomolecular Engineering, National University of Singapore, Singapore, Singapore

The world is facing two critical and inseparable issues in energy and climate change. The fossil fuels (oil, coal, and natural gas) that supply over 85% of the current primary worldwide demand will continue to be the major source for the near future. However, coal, oil, and natural gas emit 208, 164, and 117 pounds of $CO_2$ per one million Btu of energy respectively. Since the cleanest energy is the one that we never use, immediate measures to reduce their usage by conserving energy are extremely critical in arresting the dreadful impact on the environment due to the GHG (Green House Gas) emissions.

In the USA, the chemical industry consumes nearly 20% of the total industrial energy consumption (Worrell et al., 2000).[1] Energy contributes as much as 40% to the total operating cost of a chemical plant (Zargarzadeh et al., 2007).[2] Most plants buy electricity or fuel externally to satisfy their energy demands. Some even consume a portion of their raw materials, products, or byproducts as fuel. For instance, refineries use products such as vaporized LPG and fuel oil as fuel supplements. This is called FFP or Fuel From Product. On the other hand, base-load liquefied natural gas (LNG) plants use the feed (natural gas) as a fuel to run their gas turbine drivers and generators. This is called FFF or Fuel From Feed. Obviously, FFF and FFP directly reduce plant capacity and revenue, and measures to reduce their consumptions are certainly desirable.

Many chemical plants produce a variety of hydrocarbon gases with fuel value. To this end, an attractive option is to use the usually many impure, waste, surplus, byproducts, purge, or side streams with some heating value as fuel, instead of sending them to flare. In general, most plants have units that can consume such fuels. A fuel gas network (FGN) integrates and uses these fuel gases appropriately to make their best use. The FGNs are critical components of most chemical plants such as LNG and refinery. With the increasing demand of chemical and petrochemical products, decreasing natural resources, and stringent restrictions on GHG emissions, integrating Energy, Environment and Economics (IEEE) is the key to sustainability. The use of a well-designed and well-operated FGN is an excellent example of IEEE. It can yield "greener and leaner" operation through reduced emissions arising from reduced flaring and substantial savings arising from greater fuel efficiency. In addition to reduced fuel usage, waste, and pollution, the benefits of an FGN may also include higher profit margin, increased plant energy security, augmented product quality, and even stable operation.

However, a systematic approach to design and operate realistic FGNs has been missing in the literature. The mixing and re-distribution of fuels is essentially a generalization of the well known pooling problem. While the work on the pooling problem and its extensions and variations is extensive, limited work exists on the integration of "fuel streams" (versus refinery fuels) per se to reduce the energy cost as an issue. De Carli et al. (2002)[3] addressed the intelligent management and control of a fuel network. Wicaksono et al. (2006)[4] proposed a mixed-integer nonlinear programming (MINLP) model for integrating various fuel sources in an LNG plant. Wicaksono et al. (2007)[5] extended this to integrate JBOG (Jetty BOG) as an additional source. Hasan et al. (2008)[6] addressed the optimal synthesis of FGN and presented two models based on two superstructures, one with 1-stage and the other with 2-stage mixing. However, all these works focused primarily on mass and property integration to meet customer needs and demands. They assumed isothermal and isobaric systems and isothermal mixing. These are limiting assumptions, as pressure and temperature variations are significant and often necessary in most plants. Lastly, even those works that addressed FGNs neither accommodated various fuel quality requirements of the consumers rigorously nor provided a comprehensive model to optimize FGN synthesis.

In this work, we address the optimal synthesis of an FGN with many practical features such as auxiliary equipment (valves, pipelines, compressors, heaters/coolers, etc.), non-isobaric and non-isothermal operation, non-isothermal mixing, nonlinear fuel quality specifications, fuel/utility costs, disposal and treatment costs, and emission standards. We develop a nonlinear program (NLP) based on a superstructure that includes essential equipment such as compressors, valves, and utility heaters/coolers in addition to sources and sinks. Thus, it embeds plausible alternatives for heating/cooling, moving, mixing, and splitting. We successfully apply our model to three real-life case studies from LNG and refinery industries to demonstrate that an FGN can save 40-50% of total energy cost of a plant and reduce fuel-from-feed or fuel-from-product consumptions by similar amounts. We also address the issue of retrofitting new sources of fuel (e.g. integrating JBOG in an LNG plant) into an existing FGN. Finally, we highlight the utility of optimal FGN synthesis in reducing the energy cost of fuel-using and energy-intensive chemical plants. This work represents an important contribution towards conserving energy, preserving environment, and improving plant economics using advanced techniques of process systems optimization.

**References**


1. Worrell, E.; Phylipsen, D.; Einstein, D.; Martin, N. Energy Use and Energy Intensity of the U.S. Chemical Industry. *Environmental Energy Technologies division*. 2000. http://ies.lbl.gov/iespubs/44314.pdf.
2. Zargarzadeh, M.; Karimi, I. A.; Alfadala, H. E. Olexan: a tool for online exergy analysis. In Proceedings: *ESCAPE17*, Bucharest, Romania, May 27–30, 2007.
3. De Carli, A.; Falzini, S.; Liberatore, R.; Tomei, D. Intelligent Management and Control of Fuel Gas Network. In: Proceedings of the 28th Annual Conference of the *IEEE Industrial Electronics Society*, 2002, p. 2921–2926.
4. Wicaksono, D. S.; Karimi, I. A.; Alfadala, H. E.; Al-Hatou, O. I. Optimization of Fuel Gas Network in an LNG Plant. Presented in *AIChE Annual Meeting*, San Francisco, CA, USA, Nov. 12–17, 2006.
5. Wicaksono, D. S.; Karimi, I. A.; Alfadala, H. E.; Al-Hatou, O. I. Integrating Recovered Jetty Boil-Off Gas as a Fuel in an LNG Plant. In Proceedings: *ESCAPE17*, Bucharest, Romania, May 27–30, 2007.
6. Hasan, M. M. F.; Karimi, I. A. Optimal Synthesis of Fuel Gas Networks. *INFORMS Annual Meeting*, Washington D. C., USA, Oct. 12-15, 2008.


See more of this Session: Energy Sustainability, Challenges and Solutions
See more of this Group/Topical: Sustainable Engineering Forum

# (153c) Nationwide Energy Supply Chain Analysis for Transportation Fuels

**Josephine A. Elia**[1], Richard C. Baliban[2] and Christodoulos A. Floudas[2], (1) Chemical Engineering, Princeton University, Princeton, NJ, (2)Chemical and Biological Engineering, Princeton University, Princeton, NJ

Hybrid energy processes are an attractive avenue towards a more sustainable energy future due to their flexibility in converting multiple types of inputs into a consistent range of products. For the transportation sector, Fischer-Tropsch synfuels can be generated from a diverse of feedstock inputs to replace petroleum-based products [1]. Using the novel hybrid coal, biomass, and natural gas to liquids (CBGTL) process, the United States can utilize domestically available resources to produce gasoline, diesel, and kerosene to fulfill the fuel demands for the country [2-6]. Further, the incorporation of different types of biomass as feedstocks helps mitigate greenhouse gas emissions and further the efforts for renewable energy production [7]. Such hybrid processes, however, require the cooperation between multiple economic domains to achieve efficient resource allocation. Thus, an integrated and systematic approach that evaluates the interconnection between key stakeholders is crucial to attain a cost-competitive energy supply chain.

We developed an optimization framework to solve a large-scale, nationwide energy supply chain problem that takes into account the varying degrees of resource availabilities and the demand profile for the United States [4]. The supply chain begins at the feedstock source locations, ends at the demand locations, consists of optimized CBGTL facilities using the process synthesis with simultaneous heat, power, and water integration method [6]. These facilities are the 108 optimal plant topologies analyzed in [6] with distinct combinations of six coal feedstock types, three biomass feedstock types, one natural gas composition, three plant capacities, and two carbon management options (i.e., with or without carbon sequestration). A superset of candidate facility locations across the country is derived from a filtered list of United States counties in the lower 48 states based on the availability and transportation cost of feedstocks to each county.

The CBGTL facilities and the candidate locations, along with information on the nationwide configuration for the feedstock, fuel demands, and modes of transportation obtained from published government-based databases [4], serve as parameter inputs to the formulated large-scale mixed-integer linear optimization (MILP) model that represents discrete and continuous decisions in the energy supply chain. The model is solved using CPLEX to obtain the optimal supply chain topology with the minimum overall cost of fuel production for the entire network, the facility selections and locations, and the interconnections between feedstock sources, facility locations, and demand locations.

Ten case studies are presented to illustrate the performance of the optimal supply chain network under three major scenarios, namely the supply chain network with i) unrestricted topology, where the selections of the facility locations are not restricted to any geographical constraints, ii) imposed degrees of plant distribution, where a minimum number of selected facilities in the network is introduced, and iii) restricted topology, where the selection of the plant locations are restricted to be in proximity to current energy infrastructures. Facets of the model solution provide a quantitative basis to evaluate and examine trade-offs in investment decisions under these different scenarios.

[1] Takeshita, T.; Yamaji, K. Important roles of Fischer-Tropsch synfuels in the global energy future. *Energy Policy* 2008, *36*, 2773-2784.

[2] Baliban, R. C.; Elia, J. A.; Floudas, C. A. Toward novel hybrid biomass, coal, and natural gas processes for satisfying current transportation fuel demands, 1: Process alternatives, gasification modeling, process simulation, and economic analysis. *Industrial & Engineering Chemistry Research* 2010, *49*, 7343-7370.

[3] Elia, J. A.; Baliban, R, C.; Floudas, C. A. Toward novel hybrid biomass, coal, and natural gas processes for satisfying current transportation fuel demands, 2: Simultaneous heat and power integration. *Industrial & Engineering Chemistry Research* 2010, *49*, 7371-7388.

[4] Elia, J. A.; Baliban, R. C.; Xiao, X.; Floudas, C. A. Optimal energy supply network determination and life cycle analysis for hybrid coal, biomass and natural gas to liquid (CBGTL) plants using carbon-based hydrogen production. *Computers and Chemical Engineering* 2011, In press.

[5] Baliban, R. C.; Elia, J. A.; Floudas, C. A. Optimization Framework for the Simultaneous Process Synthesis, Heat and Power Integration of a Thermochemical Hybrid Biomass, Coal, and Natural Gas Facility. *Computers and Chemical Engineering* 2011, In press.

[6] Baliban, R. C.; Elia, J. A.; Floudas, C. A. Simultaneous Process Synthesis, Heat, Power, and Water Integration of Thermochemical Hybrid Biomass, Coal, and Natural Gas Facilities, 2011, In preparation.

[7] Somerville, C.; Youngs, H.; Taylor, C.; Davis, S. C.; Long, S. P. Feedstocks for Lignocellulosic Biofuels. *Science* 2010, *329*, 790-792

See more of this Session: Energy Sustainability, Challenges and Solutions
See more of this Group/Topical: Sustainable Engineering Forum

# (153d) Development of An Integrated CO2 Capture, Transportation and Storage Infrastructure for the UK and North Sea Using An Optimisation Framework

Paolo Prada[1], **N.V.S.N. Murthy Konda**[2] and Nilay Shah[2], (1)E&P Division, Eni S.p.A., Milan, Italy, (2)Centre for Process Systems Engineering, Department of Chemical Engineering, Imperial College London, London, United Kingdom

Carbon Capture and Storage (CCS) is widely regarded as an effective medium-term $CO_2$ abatement technology. However, high cost of the capture technologies, at their current state of the art, and the huge upfront costs of the $CO_2$ transport pipelines and storage facilities are hindering the large-scale development and full-chain development at commercial scale. Thus, it is of utmost importance to minimize the associated costs, as much as possible, to facilitate optimal design of CCS networks – this emphasizes the need for rigorous optimization based approach such as the one we employed in this study.

Despite the aforementioned challenges, especially within the context of the UK, given the presence of large number of $CO_2$ emitters and the vicinity of large storage sites in the Southern North Sea, CCS potential for the UK is significant. Thus, it is worthwhile to investigate the economic feasibility of the development of an integrated CCS infrastructure (which may benefit from shared transport and storage components, and economies of scale) together with the possible physical layout and evolution of such a large-scale infrastructure.

In this study, we investigated the optimal deployment of such an infrastructure based on overall cost (including capital and operating costs) minimisation. This is formulated as a Mixed Integer Linear Programming (MILP) problem. The scope of this work includes 33 large UK $CO_2$ emitters and 6 Southern North Sea sinks. Relevant cost estimates and integrated infrastructure layouts are generated through 'snapshot' model for several scenarios with varying $CO_2$ reduction targets from 5 to 105 $MtCO_2$/year.

Our results show that, for all the scenarios, practical integrated infrastructures can be developed. An initial 'seed' infrastructure is required in the Southern Yorkshire and Humber region and captured $CO_2$ is routed towards Southern North Sea depleted gas fields via the Theddlethorpe terminal. With increasing reduction targets, the infrastructure evolves and incorporates Thames estuary area emitters whose $CO_2$ is separately routed offshore via the Bacton terminal. In the first stage, the CCS system is dominated by coal power stations, primarily Drax, while CCGTs' contribution becomes significant above 75 $MtCO_2$/year reduction targets. An optimal overall cost of 81.5 $\$/tCO_{2,Total}$ is achieved at 30 $MtCO_2$/year after an initial decline primarily due to increasingly shared transport and storage infrastructures. As the reduction target increased beyond 30 Mt/year, more expensive $CO_2$ sources are included in the network – subsequently, the overall cost continually increases up to a maximum of 99.8 $\$/tCO_{2,Total}$ when the capture target increased to 105 $MtCO_2$/year.

Sensitivity analysis performed on key parameters – i.e., scaling factors and gas price – revealed that not only the overall costs and but also the timing of retrofit of gas-based power generation are affected. On the other hand, oil refineries, cement works and smaller capacity CCGTs are discarded as cost-optimal emitters to capture in all cases.

The MILP model is thus shown to be an effective framework for the optimal planning and, thereby facilitating, large-scale deployment of CCS infrastructure. (Disclaimer: This work is done while the 1st and 2nd authors were at Imperial College London and the content/opinions/results discussed do not necessarily represent those of the associated institutions)

See more of this Session: Energy Sustainability, Challenges and Solutions
See more of this Group/Topical: Sustainable Engineering Forum

# (153e) Energy Balances for the Production of Cellulosic Ethanol

**Kyriacos Zygourakis** and Venetia Rigou, Chemical and Biomolecular Engineering, Rice University, Houston, TX

Cellulosic ethanol has the potential to meet a substantial fraction of the US needs for liquid transportation fuels. To remove the perceived major barriers to the large-scale commercialization of this technology, a lot of attention has been focused lately on overcoming the recalcitrance of plant cell walls to deconstruction, on the development of more efficient fermentation methods and on genetically engineering innovative energy crops that maximize fuel production. However, the commercial success of cellulosic ethanol will depend to a large extent on our ability to design energy efficient plants.

We report here some recent results from a systematic analysis of the energy requirements of cellulosic ethanol production. A comprehensive process model for the production of ethanol from cellulosic biomass (corn stover, switchgrass etc.) was developed and implemented using the SuperPro simulator package. We paid particular attention to the operation of the utilities section of the plant, a combined heat and power (CHP) unit that generated the steam and electricity needed for the purification of ethanol and the other plant units.

A detailed parametric analysis was carried out to study how the energy balances of each plant section and the overall energy efficiency of the plant were influenced by changes in (a) the composition of the biomass feedstock, and (b) the conversion levels of the pretreatment, hydrolysis and fermentation stages. We carefully analyzed the requirements of the utilities section of the ethanol plant where byproduct streams of the production process were utilized for energy generation. One of the key questions asked was whether these streams were a sufficient fuel source for meeting the energy requirements of the plant or whether we needed to provide extra fuel in the form of additional biomass or natural gas.

Our calculations indicate a significant trade-off between ethanol production and external energy inputs. Increases in the cellulose and hemicelulose contents of the feedstock or increases in the conversion of the fermentation stage would result (as expected) to higher ethanol production rates. This increase, however, would be offset by an increase in the requirement for extra energy inputs in the form on natural gas or biomass fed directly to the utilities section. Therefore, any changes designed to increase ethanol production may be accompanied by lower overall efficiencies and net energy ratios for the biorefinery. These results cast some doubt on the ultimate effectiveness of efforts aimed at engineering innovative energy crops with high cellulose/hemicellulose content in order to maximize fuel production.

See more of this Session: Energy Sustainability, Challenges and Solutions
See more of this Group/Topical: Sustainable Engineering Forum

# (153f) High Photon Flux Photoelectrochemical Reactor: Modelling and Operation

**Klaus Hellgardt**[1], C.-K. Ong[2], G. Kelsall[2] and S. Dennison[2], (1)Chemical Engineering, Imperial College London, London, United Kingdom, (2)Imperial College London, London, United Kingdom

Production of hydrogen via sustainable technologies using renewable resources is an important goal for future global energy generation and storage. Photoelectrolysis is an attractive approach in this context, both, for its simplicity and the opportunity of employing relatively cheap materials. Considerable effort has been expended on the development of materials for photoelectrolysis - less so on the development of photoelectrochemical reactors. Here we model and demonstrate the operation of a high photon flux photoelectrochemical reactor and identify some of the key criteria for the rational design of an efficient, practical device. Mathematical models have already been developed to account for fluid flow and current distribution. An additional model rationalises the minority carrier distribution within the photoelectrochemical film itself. Process intensification through solar concentration will be demonstrated with a flat plate device, leading to a 5-fold increase in hydrogen production rate per unit active semiconductor area, or, conversely a significant reduction in device cost.

See more of this Session: Energy Sustainability, Challenges and Solutions
See more of this Group/Topical: Sustainable Engineering Forum

# (166a) Laboratory Activities for Introducing Students to Biomaterials

**Jennifer Vernengo** and Kevin Dahm, Chemical Engineering, Rowan University, Glassboro, NJ

Because of the increasing prevalence of musculoskeletal, cardiovascular, and neurodegenerative diseases, there is a necessity to engineer biomaterials that can be used to treat these painful and debilitating disorders. This paper presents two experiments that introduce freshman and senior-level undergraduate Chemical Engineering students to the design and evaluation of biomaterials. By designing ceramic-polymer composites for bone substitution and hydrogels for cartilage replacement, students develop new knowledge in biomaterials science that also contributes to ABET learning outcomes for undergraduate Chemical Engineering courses. Assessment results will be presented, including those of pre and post-tests, which indicate the effectiveness of the learning activities

See more of this Session: Labs and Experiments In the ChE Curriculum
See more of this Group/Topical: Education

# (166b) Integration of the Unit Operations Laboratory with a Focus On Biofuel Production

**Jim Pfaendtner** and Danilo Pozzo, Chemical Engineering, University of Washington, Seattle, WA

This presentation will highlight our reconfiguration of the senior chemical engineering unit operations (UO) lab (CHEM E 437) at the University of Washington. Using the themes of biotechnology and renewable energy, we have developed a new pedagogical model, which emphasizes 1) collaboration between teams and 2) developing instincts for thinking about the impact of single unit operations on the operation of an entire production facility. The production of renewable energy is one of the most important technological problems that we face today. This challenge also offers us an opportunity to motivate and shape the early careers of Chemical Engineering undergraduate students. The most innovative component of the new laboratory is the complete integration of new and existing experimental stations. The senior UO lab was integrated to model a bio-fuel production plant where student groups work on individual operations that make up a complete process. This full-plant view of the laboratory allows students, for the first time, to evaluate the effects of their decisions on upstream and downstream plant operations. Furthermore, it also provides a common framework to promote active discussion and engagement among student groups. The transformation of the course included the development of completely new modules for fermentation of biomass and the modification of existing equipment and modules for the treatment, separation and extraction of product and waste streams. The new fermentation modules utilize internet-based remote monitoring technologies to track the development of fermentations while students are outside of the laboratory. Fully interconnected units now define a common goal of reducing costs and improving productivity and replace the original independent and unrelated experiments. The new structure also allows us to easily incorporate design concepts, such as cost analysis and environmental compliance, into the laboratory. The objective of the re-designed course is to provide a realistic structure that is congruent with what students will experience after graduation. The new laboratory structure is also designed to foster leadership, creative thinking, composure under uncertainty and the critical review of information. Furthermore, with the new structure, we also continue to meet the original learning objectives of instructing students on the basics of experimental planning and reporting. We will additionally present evaluation and assessment of the student experience from the first round of the new lab and our updated plans for continuing to improve the student experience in our course.

See more of this Session: Labs and Experiments In the ChE Curriculum
See more of this Group/Topical: Education

# (166c) A Drug Delivery Experiment Using Alginate Microspheres

**Stephanie Farrell** and Jennifer Vernengo, Chemical Engineering, Rowan University, Glassboro, NJ

Controlled release drug delivery has recently become a major research and development focus area of today's pharmaceutical industry. New drug delivery systems are continually being designed to dispense a drug at a predetermined rate, to a target area in order to maintain a therapeutic concentration of drug at a desired location in the body. Rowan University is currently working with the Engineering Research Center for Structured Organic Particulate Systems (ERC-SOPS) of Rutgers University to develop educational materials involving drug delivery technology. We describe a drug delivery module in which students investigate factors affecting drug release from hydrogel drug delivery systems. Calcium alginate microspheres and other geometries are produced containing a model drug, limonene (yellow food dye) or bovine serum albumin. Using a guided inquiry approach, students investigate the effects of surface area, degree of cross linking, drug molecular weight, and stirring speed. Students learn introductory concepts related to experimental design and data analysis, mass transfer, and biomaterials. Calculus and chemistry concepts are also reinforced.

See more of this Session: Labs and Experiments In the ChE Curriculum
See more of this Group/Topical: Education

# (166d) Developing Tools for Teaching Chemical Engineering Unit Operation Design

**Rachael H. Elder**, Department of Chemical and Biological Engineering, University of Sheffield, Sheffield, United Kingdom

The teaching of chemical engineering unit operation design in a project week environment is essential for a synoptic course and enhanced student learning. This mini project, funded by the Engineering Subject Group of the Higher Education Academy, has two distinct aims: improving student awareness of resources available for Chemical Engineering Design and development of a pilot Staff Toolkit to assist lecturers new to Chemical Engineering Design. The second year design week is used as an example of how different resources can be introduced and used in a design exercise.

Experience from supervising third year undergraduate design projects over the last couple of years has highlighted the significant amount of time spent on directing students to resources where they can find information. This not only detracts from supervision time but also reduces the useful learning content of the project. Students should be aware of a wide range of resources and not just those relevant for the design they complete during their design projects. The rationale behind this work is that by providing a more thorough grounding in the second year design week, students will be better prepared for their IChemE accredited design project and for life as a Chemical Engineer.

Second year design week involves a task based design, focussing on prescribed unit operations in a defined process, so is ideal for directing the students to the wealth of new online design resources relating to both underlying theory and practical design of unit operations. The synoptic nature of the project week makes it a vital part of the curriculum to ensure student understanding. This paper discusses some of the different resources available and the way they have been implemented into the design week.

Student awareness of online resources before and after design week was evaluated via questionnaires and focus groups. The impact of the online resources introduced in the design week on the cohort's third year design project was also evaluated. The results show that the students found the design week and online resources extremely useful.

Providing a teaching resource in the form of a pilot staff toolkit is another key aim of the project as it will be an invaluable teaching aid, especially given the increasing number of academics lacking real industrial design experience. The pilot of the staff toolkit will be demonstrated and discussed.

See more of this Session: Labs and Experiments In the ChE Curriculum
See more of this Group/Topical: Education

# (166e) Alternative Lab Reports - Engineering Effective Communication

**Daniel Lepek**, Department of Chemical Engineering, The Cooper Union, New York, NY and Richard Stock, Chemical Engineering, The Cooper Union, New York, NY

For many chemical engineering undergraduate programs, the laboratory sequence allows students to experience the application of chemical engineering outside of the classroom. Typically, the results of these experiments are analyzed and written up in a classic laboratory report format. Although writing laboratory reports is a valuable skill, today's engineers are expected to report their results and express their findings in different forms of oral, written, and visual communication.

We have introduced a new initiative at The Cooper Union in the senior chemical engineering laboratory sequence that provides the students with a range of skill-sets that prepares them to communicate successfully on a professional level in different contexts and environments. As a baseline, students were expected to submit 2 laboratory reports a semester, written on the level of a scholarly journal article. To develop and stretch their communication skills, students used their laboratory experiences to submit executive memos to company CEOs, formulate grant proposals to government agencies. and created posters that convey their results in a visual setting. Students also made oral presentations tailored to a variety of key audiences of different scientific and corporate backgrounds.

The Cooper Union's CONNECT (Cooper's Own No Nonsense Engineering Communication Training) Program, also initiated as part of this laboratory sequence in 1996, partnered this initiative, providing students with feedback and specialized training in both written and oral communication as well as graphics.

In this presentation we will illustrate student work using these report alternatives, indicate how we assessed their effectiveness, present examples of the feedback and specialized training stimulated by the students' work and outline the lessons learned as we move on to our next year.

See more of this Session: Labs and Experiments In the ChE Curriculum
See more of this Group/Topical: Education

# (166f) Wiki Implementation In a Senior Unit Operations Laboratory

**James P. Abulencia**, Department of Chemical Engineering, Manhattan College, Riverdale, NY

Laboratory reports have traditionally been the primary deliverable for students enrolled in the unit operations laboratory course. These reports are typically comprised of several sections that emulate the format of a traditional scientific publication. Because these documents can be very lengthy, it is time consuming for educators teaching this course to read and assess the work. To this end, a wiki system was implemented wherein students can collaboratively contribute to the body of knowledge pertaining to a particular experiment. This wiki is comprised of four parts: 1) “Background”, which focuses on industrials applications, 2) “Theory and Equations”, which highlights important definitions, principles, and equations pertaining to the experiment, 3) “Technical Notes”, which provides a venue for sharing experimental details to subsequent groups, and 4) “Other” for facts/details that do not directly belong under the three aforementioned headings. Students are assessed on the impact of their contribution, as well as a final exam based on the material in the wiki. The benefits are as follows: 1) students are placed in an environment that encourages collaborative learning, 2) a final exam ensures that all students understand the experiments (not only the group leaders), and 3) the jigsaw approach provides student satisfaction on the personal level for contributing to the class’ knowledge.

Note to reviewer: A survey will be distributed at the conclusion of the course, recording student opinion on their experience with the wiki. Thus, the results are not currently available. However, students have articulated a positive experience overall.

See more of this Session: Labs and Experiments In the ChE Curriculum
See more of this Group/Topical: Education

# (166g) Web-Lab: Enhancing the Undergraduate Engineering Experience

**Cynthia Collins**, Department of Chemical and Biological Engineering, Rensselaer Polytechnic Institute, Troy, NY and **Kristala Jones Prather**, Chemical Engineering, Massachusetts Institute of Technology, Cambridge, MA

Web-Lab is an innovative SBE program that web enables laboratory experiments. A collection of bioreactors are located miles away in a central facility, and students, regardless of their physical location, can access and remotely interact with the bioreactor in real time over the internet. Multiple schools can access these web-based experiments and incorporate them into their curriculum, either as a lab or to supplement a lecture course. This is the objective of the SBE Web-Lab Initiative—to advance engineering education by breaking beyond the confines of the classroom and enhancing the typical classroom experience.

The upside of the web-based laboratories is to reduce the cost incurred per student to deliver state-of-the-art laboratory exercises and thereby enable more students to participate in a modern laboratory experience. Web-based laboratories will improve undergraduates' education by better preparing engineers with the latest skills and tools to compete in the global workforce.

See more of this Session: Labs and Experiments In the ChE Curriculum
See more of this Group/Topical: Education

# (166h) An Integrated Curricula Approach to a Great Capstone Lab Experience and Its Potential Implementation In South America

**Rocio Tijaro**, Ingenieria, Universidad Arturo Prat, Iquique, Chile, Mario Oyanader, Chemical Engineering, Tenneessee Tech University, Cookeville, TN and Joseph Biernacki, Chemical Engineering, Tennessee Technological University, Cookeville, TN

Traditional Unit Operation Labs (UOL) consist of a series of well define experiments that are run every semester with very little modification. As a matter of fact, a steady operation of the many units (Absorption Column, CSRT, Heat Exchanger, etc...) is very much preferred so the students have the chance to validate performance of unit the set up. In addition, a traditional UOL course usually have 3 to 4 credits hours, whereas in a Lecture-Lab integrated curricula the number of credit drops significantly to 1 to 2 credit hours. In this case, making a UOL core course an actual learning tool can be an opportunity for creativity and enhancement. In this paper, it is described how Chemical Engineering core courses converge into one single project with an open-ended problem. Particular attention is given to the strategic planning of the course, logistic associated, ABET requirement and external assessments. Other aspects are also includes as for example the details of the expected deliverables of the course, final presentations, and statistical analysis performed on the data collected. After five years of implementation at Tennessee TECH University, the pros and cons of the approach are evaluated for testing in a different environmental setting as it may be the North to the South American culture. The author's particular intentions are to document potential barriers as well as opportunities in a more global engineering education. The base case for the analysis is the new UOL concept as of a progressive program.

See more of this Session: Labs and Experiments In the ChE Curriculum
See more of this Group/Topical: Education

# (229a) Evaluation of Alternative Fuels to Replace Coal In Cement Manufacturing

**Steve R. Duke**, Chemical Engineering, Auburn University, Auburn, AL, Anton Schindler, Civil Engineering, Auburn University, Auburn University, AL and Don Stafford, US East Business Unit, Lafarge North America, Alpharetta, GA

A multiyear DOE-supported study was completed to identify, determine, and understand fundamental burn characteristics and properties of alternative fuel sources to replace coal for energy generation, with emphasis on impacts in cement processing. Portland cement manufacturing involves the combustion of solid fuels along with raw materials to produce clinker. Temperatures exceeding 1,400°C are needed to carry out reactions for cement chemistry. Energy needs for a typical manufacturing plant are 12,000 MMBTU/day. Potential alternatives to fossil fuels must provide adequate and economical energy but they must also result in products of combustion that have negligible impact on cement quality and emissions.

An operating Portland cement manufacturing plant (Lafarge North America's Roberta Plant in Calera, Alabama) served as facility for a series of trial burns to replace coal at various substitution rates with bio-based and waste fuels. More than 20 three-day trial burns were conducted from 2007 to 2010 replacing a portion of the coal with scrap tires, waste plastics, woodchips, switch grass, poultry litter, forests trimmings, scrap railroad ties, soy waste products, or waste liquids. Each trial burn contained one or more of the alternative fuels. An elaborate sampling and analysis plan was carried out for each trial burn with close cooperation between the Lafarge plant, the Lafarge testing laboratory, Auburn University, and outside laboratories.

Results will be presented from a comprehensive evaluation of the raw materials, fuels, clinker, cement, emissions, and plant operations from the test burns. The technically viability for each fuel as supplements for coal for direct burn in a manufacturing plant will be discussed. Alternative Solid Fuel (ASF) handling and process operations were established and evaluated for the use of each of the fuels. Replacement ratios (on a BTU basis) were up to 18% for the waste and bio-based fuels and the synergistic burns where two or more of the alternative materials were used together. Impacts on product quality and emissions were negligible and in some cases advantageous.

See more of this Session: Innovations for Energy Intensive Industrial Processes and Building Technologies
See more of this Group/Topical: International Congress on Energy 2011

# (229b) Kinetics Study of Calcium Carbonate Decomposition At High CO2 Environment

Yong-Hua Duan[1], **Jian Zhang**[1], Delong Xu[2], Hui Li[3], Le-le Yang[1], Xiao Fan[1], Wen-bin Yang[1], Yongxiang Ren[3] and **Yong Min**[3], (1)Material Institute, Xi'an University of Architecture and Technology,, Xi'an, China, (2)Institute for Powder Eng, Univ. of Archi. & Tech., Xi'an, China, (3)Xi'an University of Architecture &Technology, Institute of Powder Engineering, Xi'an, China

With the concerns of "global warming", CO2 sequestration from cement industry are becoming significant important for the "green processing" for the environments. This is because the cement processing released about 36% of CO2 comparing to the 12% from regular coal-power plant. CaO/CaCO3, a major component and raw material from cement, was investigated in order to design suitable processing to sequestrate the CO2 during the cement production. It was found that the raw material particle size, surface area as well as chemical components were important factors, which will effect on the CO2 absorption efficiency, cycle times, particle activities, and so on. Various analytical equipments such as SEM, XRD, FTIR, TGA, were used to monitor the processing and analysis materials during the processing.

It was found that there are two competitive mechanisms involved during the processing i.e., chemical reaction and physical diffusion. In order to investigate the dominated effect, we selected the different particles as model compounds for the analysis. As a result, we found the small particle (particle size less than 5 micron) is dominated by chemical reaction mechanism, i.e., CO2 can completely diffusion into the core of the particle and result in a complete chemical reaction to form CaCO3; and the larger particle is dominated by the physical diffusion mechanism, i.e., only partial CO2 can diffuse into the CaO particles and formed a "core-shell" particles, where the centre core is CaO and out-shell is CaCO3. In addition, the small particle always accomplished with a high cycle times and more complete chemical reaction, which was recommended to use in the industry.

Based on the cement processing, various chemical looping processes were used to demonstrate the CaCO3 Calcinations / CaO Carbonation Cycle such as, vacuum processing, solids flow processing, fluidization processing, and cocking gas recycling processing. The energy efficiency, raw material activities, and corresponding thermodynamic date will be presented during the presentation.

See more of this Session: Innovations for Energy Intensive Industrial Processes and Building Technologies
See more of this Group/Topical: International Congress on Energy 2011

# (229c) Application of Ionic Liquids In the Aluminum Electrorefining

**Zheng Yong**, Lu Xingmei and Zhang Suojiang, State Key Laboratory of Multiphase Complex Systems, Institute of Process Engineering, Chinese Academy of Sciences, Beijing, China

**Application of Ionic Liquids in the Aluminum Electrorefining**

Yong Zheng, Xingmei Lu *, Suojiang Zhang *

*State Key Laboratory of MultiphaseComplex Systems, Institute of ProcessEngineering, Chinese Academy of Sciences, 100190, Beijing, P. R. China*

**Corresponding author: xmlu@home.ipe.ac.cn,sjzhang@home.ipe.ac.cn*

The Hall-Heroult process for aluminum production has beenwidely employed in the past 100 years. According to this technique, the purityof primary aluminum is usually lower than 99.8% and needs to be improved further.At present, there are three main industrial methods applied in the aluminumrefining: zone melting, three-layer electrolysis and segregation process. However,these methods suffer from some inherent problems, such as high temperature and energyconsumption. Labeled as novel green solvents, ionic liquids have exhibited aseries of attracting properties and excellent performance in low-temperature aluminumelectrodeposition [1,2].The application of ionic liquids in aluminum electrorefiningis worth of investigation.

In our work, ionic liquid 1-butyl-3-methylimidazoliumheptachloroaluminate ([Bmim][$Al_2Cl_7$]) was prepared and used aselectrolyte in the aluminum electrorefining. The electrorefining processes were examined under different temperatures,potentials and current densities. Experimental results shownthat the main impurity elements, including Si, Fe, Tiand Na could be efficiently separated from primary aluminum in [Bmim][$Al_2Cl_7$]. Compact and smoothaluminum deposits with high purity (>99.9%) were obtained at 100 $^0$C, -0.5~-1 V and 10~15 mA/cm$^2$(see Figure 1 and 2). Furthermore, the current efficiency of aluminum electrorefining increased to 98%. And the energyconsumption was lower than 5 kWh/kg-Al. As for [Bmim][$Al_2Cl_7$],it could be recycled and reused after purification. In general, ionic liquids area class of promising low-temperature electrolytes which open a new way to thealuminum refining.

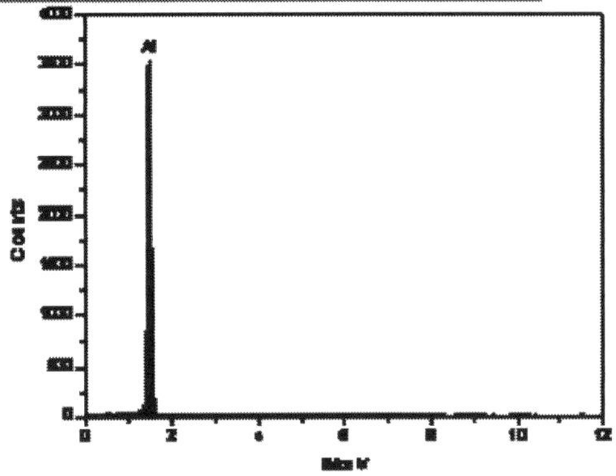

Figure 1. SEM image of the Al deposits

Figure 2. EDAX profile for the Aldeposits

## Reference

[1] A. Campion and P. Kambhampati, *Chem. Soc. Rev.*, 1998, **27**, 241.

[2] M. M. Zhang, V. Kamavaram andR. G. Reddy, *JOM*, 2003, **11**, 54.

See more of this Session: Innovations for Energy Intensive Industrial Processes and Building Technologies
See more of this Group/Topical: International Congress on Energy 2011

# (229d) Novel Zeolite Membranes for Energy-Efficient Air Dehumidification and Conditioning

**Rong Xing**[1], Yuxiang Rao[1] and Wei Liu[2], (1)Energy and Environment Directorate, Pacific Northwest National Laboratory, Richland, WA, (2)Pacific Northwest National Laboratory, Richland, WA

Traditional vapor compression air conditioners have very high power consumption for cooling of hot and humid air due to condensation of a large amount of water inside the cooler. We have developed a novel zeolite membrane supported on a thin (~50 micrometer thick) porous metal sheet to remove water vapor from an incoming humid air stream without changing its temperature, i.e., through vapor-phase membrane separation. The membrane dehumidifier reduces air humidity levels well below its due point under the cooling conditions so that only sensible cooling occurs. In this way, the energy consumption for air conditioning in hot and humid climate can be reduced by 50%. The high permeance and high H2O/air selectivity are critical performance parameters. Compared to previous membranes reported in the literature, our membrane exhibits a high H2O/air separation factor (> 100), a permeation flux about 2 or 3 orders of magnitude higher than the literature result, and excellent stability. This presentation will provide systematic discussion on impacts of membrane separation conditions, such as temperature, feed air humidity, and permeate pressure, on the membrane separation performances. The new membrane technology has great potential to be developed as a commercially-viable membrane dehumidifier by obtaining exceptional membrane separation performances and substantially reducing the zeolite membrane fabrication cost at the same time.

See more of this Session: Innovations for Energy Intensive Industrial Processes and Building Technologies
See more of this Group/Topical: International Congress on Energy 2011

# An Osmotic Membrane Dehumidifier

**Arthur S. Kesten** and Jack N. Blechner, Nanocap Technologies LLC, Longboat Key, FL

## OSMOTIC MEMBRANE DEHUMIDIFIER

It has long been known that relying on a vapor compression system to cool air down to the dew point in order to dehumidify is expensive to build and uses far too much energy to operate. The ability to dehumidify independently of cooling provides substantial advantage, provided that it can be done using inexpensive materials, uses little energy to operate, and is easily packaged with a separate cooling system.

An osmotic membrane dehumidifier has been developed that uses a flexible, semi-permeable membrane to facilitate capillary condensation of water vapor and the transport of condensed water through the membrane into a salt solution by osmosis. Here a humid gas stream is brought into contact with a semi-permeable membrane, which separates the gas stream from an osmotic (e.g., salt) solution. Some of the pores of the membrane are small enough to permit capillary condensation. Liquid formed within these pores connects with liquid formed in adjacent pores, collectively forming continuous paths of liquid. These 'liquid bridges' extend across the thickness of the semi-permeable membrane and provide paths by which water can travel across the membrane. Because the membrane is so thin, water concentration gradients across the membrane can be large. This can provide a large driving force for water transport between the humid air and the osmotic fluid. The flexibility of the polymeric membrane allows for considerable design flexibility that enhances the potential for retrofit with any cooling system.

## BACKGROUND

Conditioning of a gas stream, such as air, in warm weather generally involves removal of moisture and decrease of temperature to make the gas stream suitable for its intended environment. Current dehumidification technology is commonly based on the conventional, refrigerant vapor compression technology (DX technology) or on desiccant substrate capture (DS technology). DX technology requires cooling humid supply air, such as the air within a room and/or outside air, to the water vapor condensation point, with external heat rejection on the compression side. This usually requires the supply air to be cooled below comfortable temperatures, and, thereafter, either reheated or mixed with warmer dry air to raise its temperature to an acceptable level before directing it into the space being dehumidified. Twenty to thirty-five percent of the energy expended in cooling the high humidity air is utilized to remove the latent heat of condensation from the air. Cooling and dehumidification of the air are thus coupled. That makes it impossible to control comfort parameters independently, making the DX cycle less efficient, from an overall system perspective, than a technology that would allow independent control of sensible and latent heat.

In applications where the outside air has both high humidity and temperature and the functional use of the interior space generates high water vapor levels (e.g., populated convention halls, exercise rooms, school buildings, etc.), it may not be possible for the DX technology to maintain the air introduced into the interior space at the correct humidity and temperature for maintaining comfort. The air delivered is cool but 'muggy', since further cooling to remove additional water would result in the air being uncomfortably cool.

In stand-alone dehumidification using a conventional compression cycle, heat reject is in direct contact with room air. As a consequence, the room air becomes more comfortable from a humidity side, but may be too warm. Again, the comfort parameters are coupled.

DS systems are generally applied in central air, ducted systems. Water vapor capture on a solid phase substrate is efficient and rapid. However, removal of the water vapor from the pores requires energy input. It also requires removing the substrate from the high humidity air stream and placing it in an exhaust, water-reject stream, before adding the re-evaporation energy. Alternatively, the substrate remains fixed and the treated air and exhaust streams flow directions interchanged as is done in a parallel bed, desiccant dryer system. DS technology requires, in steady state operation, the addition of this energy at a rate equal to or greater than the latent heat of condensation of water in the desiccant substrate.

The advantage of DS technology is that humidity levels in the outside air and/or re-circulated air can be adjusted independently of the cooling step. The disadvantage is the requirement to move the substrate and treated air stream relative to each other for capture and rejection of the water vapor.

In a liquid desiccant system, direct contact equipment, such as packed-bed towers, allows for simultaneous heat and mass transfer between the air and the liquid. The essential issue here is the requirement for intimate contact of desiccant and air. The conventional technique of using a packed-bed adsorption tower is not energy efficient due to its low contact surface area per unit volume. Moreover, evaporation of some liquid desiccant in a packed-bed tower may cause contamination of the conditioned air stream.

Osmotic dehumidification provides a more compact and efficient option. The process is described in Refs. 1 and 2. In the earlier work, a capillary condenser with pore sizes around 2 to 3 nanometers was used to condense water vapor in air at moderately high relative humidity. The condenser (originally designed and fabricated by Mobil Oil) had been fabricated from a ceramic substrate using a molecular template which was subsequently calcined at high temperature to leave behind a fairly uniform porous structure. A semi-permeable membrane was used to separate the capillary condenser from an osmotic fluid. An osmotic driving force, resulting from a water concentration gradient, transports the condensed water through the condenser and into the osmotic fluid. This type of device is effective in promoting water transport at fluxes in excess of 1 liter/square meter-hour. Under some circumstances, however, the pore structure of the capillary condenser can become unstable and degrade over a relatively short period of time in a humid environment. In addition, the capillary condenser is typically made from a rigid material and is therefore limited to those applications where a rigid body is acceptable.

## DEMONSTRATION – PROOF OF CONCEPT

The feasibility of using commercially available membranes for capillary condensation and subsequent transport of condensate into a salt solution was demonstrated first in a laboratory scale experiment. A glass container, filled to the top with a salt solution, was covered with a membrane. A narrow tube attached to the side of the container extends up above the container. Humid air was blown across the top of the membrane and the rate of removal of water from the air was measured by the change in liquid level in the tube. Membranes tested were those used commercially for water purification. Salt solutions that were very effective were lithium chloride and magnesium chloride. The latter salt is inexpensive, benign and readily available. Results suggest that 1 square meter of the membranes that worked most effectively removes about 0.7 liters/hr of water vapor at relative humidity levels between 80 and 90%. The rate of condensation per unit of surface area is considerably smaller than for a uniform pore size capillary condenser. However, the polymer membranes are stable, inexpensive, flexible, and offer simple methods for providing large surface area interfaces between humid air and films of liquid salt solution.

## VALIDATION AND SCALABILITY – BENCH TESTS

Subsequent experiments were performed using a larger apparatus in which shallow trays were filled with a salt solution and each covered with a membrane. Trays were arranged in a cafeteria style with narrow spaces between the trays for air flow over the membranes. Air was blown over the membranes and then through an exhaust manifold. An anemometer was used to measure air flow and humidistats were placed upstream and downstream of the trays to measure temperature and humidity. A circulating pump was used to move the salt solution through the trays from inlet to outlet manifolds.

Tests were performed on arrays of trays and on single trays with various spacing and air flow. A single tray, for example, covered with a membrane of 0.13 square meters reduced relative humidity of an air stream flowing at 15 cfm (0.4 cubic meters per minute) by 7%. At 20 cfm, the relative humidity was reduced by 5%. An array of 12 of these trays (occupying a volume of less than 0.02 cubic meters) would remove about 0.5 liters/hr of water from a humid air stream. Reducing volumetric flow rates results in increased relative humidity reduction. However, capillary condensation with current membranes is limited by pore size and relative humidity is generally not reduced below 65-70%.

A mathematical model of the transport of air and removal of water vapor at membrane surfaces has been developed to help to optimize the design of the system. The tray system is but one of many potential configurations.

Ref. 1: Kesten, A.S. and Blechner, J.N., Versatile Dehumidification Process and Apparatus, U.S. Patent 7,758,671 B2, July 20, 2010

Ref. 2: Kesten, et al, Dehumidification Process and Apparatus, U.S. Patent 6,539,731 B2, April 1, 2003

# (229f) Ethanol/Water Biomass Feedstock Separation Through Inorganic A-Type Zeolite Membrane on Thin Porous Ni Sheet Support

**Yuxiang Rao**[1], Rong Xing[1] and Wei Liu[2], (1)Energy and Environment Directorate, Pacific Northwest National Laboratory, Richland, WA, (2)Pacific Northwest National Laboratory, Richland, WA

Bio-Ethanol/water separation by fractional distillation becomes a popular industrial process to remove water from ethanol-water mixture. However, the mixture of 95.6wt% ethanol and 4.4 wt% water is an azeotrope and cannot be further purified by distillation. Alternatively, this can be achieved by membrane separation technology since it is not works on vapor-liquid equilibrium but on pervaporation mode.

In this work, real ethanol feeds (rectifier bottom, 40 wt%, PH=3.99 and rectifier overhead, 95wt%, PH=4.54) from Pacific Ethanol Inc. were chosen to test the performance and stability of A type zeolite membrane (2~3um), which supported on novel thin porous Ni sheet substrate (~50um). The preliminary results shows that NaA zeolite membrane was unstable at operation condition (~100C), mainly due to the strong acidity of real ethanol-water biomass feedstock (PH≈4).

However, after neutralized with 0.1M NaOH solution to PH=8.5 or by pass a guard bed filled with weak base ion exchange resin (Amberlite@IRA-93), this novel membrane shows high performance (Water flux >2 kg/m$^2$ h, water/ethanol selectivity > 5000) and stability (>100hrs). ICP analysis result also shows that sodium and sulfur ions will remains in the feed side during the separation process, mainly due to their large hydrated diameter (~7.2 Å for $Na^+$ and ~8.6 Å for $SO_4^{2-}$) comparing with the pore size of NaA zeolite membrane. (~4Å)

See more of this Session: Innovations for Energy Intensive Industrial Processes and Building Technologies
See more of this Group/Topical: International Congress on Energy 2011

Process Intensification: A Historical Perspective.

Thomas A. Kenat, Ph.D., P.E.
KenaTech Process Engineering
Medina, Ohio

Extended Abstract:

While the term "Process Intensification" is fairly new, the concept has been around for many decades. Whenever processes are combined in unique or creative ways to increase yields, reduce equipment or operating costs, save energy, reduce undesirable byproducts, or gain other benefits, it could be called process intensification. In this presentation, the author draws on both the historical literature and some of his own unpublished research to show a few examples (some successful, some not so successful) that could have been called Process Intensification if the term had been around then.

A very early example of process intensification is the well-known method of producing pure ethanol. If ethanol is distilled from a water-ethanol mixture, the maximum purity is limited by the minimum-boiling azeotrope of 95.6% ethanol, which boils at 78.2 °C. The addition of a small amount of benzene makes it possible to boil off a ternary azeotrope boiling at 64.9 °C. This leaves behind anhydrous ethanol with only a few parts per million of benzene, useful in many chemical reactions but not fit for human consumption.

A process that has shown many examples of process intensification over the years is the manufacture of vinyl chloride, used primarily as the feedstock for vinyl plastics. The first process, invented in 1912 and commercialized in the 1920's, involved the reaction between acetlyene hydrogen chloride. The acetylene was produced from calcium carbide, made from coke and calcium oxide, and the HCl by burning hydrogen and chlorine gases together. This process, although highly polluting and not very energy efficient, was used through the 1930's and 1940's, and is still used extensively in China.

The first "balanced" process, which was more economical and energy efficient than the above process, involved the direct chlorination of ethylene to produce ethylene dichloride (EDC). The EDC was cracked in a furnace to produce vinyl chloride and HCl. The HCl was then combined with acetylene to produce more vinyl chloride, so that half was made from ethylene, and half from acetylene.

The second "balanced" process eliminated acetlyene as a feedstock. In the 1960's, B.F. Goodrich developed the widely licensed oxychlorination process, in which ethylene, HCl, and air were fed into a fluidized-bed catalytic reactor, where EDC was produced, and the heat of reaction was recovered as steam. In this balanced process, half of the EDC was produced by oxychlorination, and half was produced by direct chlorination of ethylene. The HCl from cracking all of the EDC was then recycled to the oxychlorination reactor.

When the author was working at BFGoodrich in process R&D, he was responsible for Oxychlorination catalyst testing and developing improvements to the vinyl chloride process. Two of those improvements were commercialized and became part of the company's vinyl process, which was widely licensed. The first was the Catoxid process, in which the chlorinated byproducts of vinyl chloride production (mostly higher chlorinated $C_2$ and $C_4$ compounds) were fed into a fluidized-bed catalytic reactor with air and burned. The effluent from this reactor, rich in HCl, was then fed directly into the oxychlorination reactor to recover the chlorine value, and the cooling jacket of the reactor produced steam. Not only did this process increase raw material and energy efficiency, but it also eliminated a major waste disposal problem.

The second new process was the Boiling Reactor, an improvement on the direct chlorination process. Normally, direct chlorination was run in a low-temperature liquid-phase reactor, with the heat of reaction removed with cooling water. The Boiling Reactor functions as the reboiler of the EDC distillation column, so that the heat of reaction is used to purify the EDC, rather than being dissipated in cooling water. This process was first run in glassware in the high-bay section of the Process R&D building, and then in a pilot unit attached to the EDC distillation column at the Calvert City vinyl chloride plant. The first commercial unit was installed in a Goodrich vinyl chloride plant near Houston.

Another process developed in our research was Chlorocracking, where we attempted to use the heat of chlorination to directly crack the EDC in a single fluid-bed reactor. This process proved that vinyl chloride could be made in a single-step process, but the high temperature resulted in the production of too many byproducts. The energy savings were not enough to make up for the higher raw material consumption. We also tried another direct process for vinyl, which unfortunately required the condensation of a large wet HCl stream. The problem with that process was that the only material suitable for that large condenser was tantalum, and the cost was prohibitive.

Since those days when BFGoodrich was in the business, the vinyl chloride process has undergone continued process intensification. In the Oxychlorination process, most companies are now using pure oxygen instead of air, which greatly reduces the size of the reactor or increases its capacity, thus reducing capital cost and also reducing effluents. Many companies have also come up with their variants on the Boiling Reactor design, to further increase yields and energy efficiency.

There have been ideas for many years that an ethane-based process could be developed, and many unsuccessful attempts have been made. It now appears that EVC (now INEOS) has a pilot plant in Germany that shows some promise in this area. Their timing could be just right. With the emerging technology of extracting natural gas from shale, low-cost ethane from natural gas could make this process even more economical.

# (255b) Enabling a Small Scale Gas-to-Liquid Process

**Jan J. Lerou**, Jan Lerou Consulting, LLC, Hilliard, OH

Minichannel process technology offers process intensification, in the form of enhanced heat and mass transfer, to a wide range of chemical reactions. This paper discusses the application of minichannel technology to the production of ultra-clean synthetic fuels. These fuels, which are one-to-one replacements for petroleum derived fuels, are produced by passing synthesis gas (syngas), a mixture of carbon monoxide and hydrogen, over a cobalt or iron catalyst; a process known as Fischer-Tropsch (FT) synthesis for its German inventors. The most efficient path to syngas is the conversion of natural gas via steam methane reforming (SMR). Both the SMR and FT benefit from the process intensification offered by minichannel technology, resulting in smaller, less costly processing hardware; thus, enabling cost effective production of synthetic fuels from smaller facilities, appropriate for stranded and associated gas resources, both on and offshore. The products from FT processes can be upgraded into diesel or synthetic paraffinic kerosene, or simply blended with crude oil for transport to the world market. Leading companies in this field included Chart Industries (USA), CompactGTL (UK), and Velocys, Inc. (USA).

See more of this Session: Process Intensification Plenary Session
See more of this Group/Topical: Process Development Division

# (255c) Micro-Structured Catalysts for Process Intensification

**Enrico Tronconi**, Gianpiero Groppi and Carlo Giorgio Visconti, Dipartimento di Energia, Politecnico di Milano, Milano, Italy

Laboratory of Catalysis and Catalytic Processes

**Abstract**

The first part of this talk deals with the investigation of open-celled foams (or sponges) as potentially enhanced catalyst supports. Ceramic and metallic open-celled foams are in fact micro-structured materials with interesting structural properties, which make them attractive as low pressure drop catalyst carriers: they come with high porosities (from 70 to 95%), and may have very high surface areas per unit volume. The adoption of foams as catalyst supports has been suggested for chemical processes (e.g. steam reforming of methane), as well as for the fast diffusion-limited processes of environmental catalysis (like e.g. in catalytic combustion, in DeNOx processes for aftertreatment of vehicle exhausts, or in catalytic partial oxidation of $CH_4$ for $H_2$ production). However, methods for making the foams catalytically active, and most of all engineering correlations of their transport properties, have been scarcely reported in the scientific literature so far.

After briefly touching the development of a procedure for coating open celled foams with thin layers of Pd-$Al_2O_3$, this talk will present measurements of gas-solid mass and heat transfer rates in metallic and ceramic foams, using CO oxidation and transient cooling experiments, respectively, and leading to generalized engineering correlations. In this respect, three types of catalyst support (foams, honeycomb monoliths and spherical pellets) will be compared, first on a generalized basis, then by simulating their application to the autothermal partial oxidation of methane to syngas: both cold start-up and steady-state behaviours are analyzed in the latter case. Finally, very recent experimental results concerning the overall (non-adiabatic) heat transfer performances of foams with different geometrical (porosity, ppi) and structural (FeCrAlloy, aluminium) properties will be illustrated.

The second part of the presentation will focus on a parallel investigation of conductive honeycomb monolith catalysts. We have shown in the past that there is potential for significant enhancement of radial heat transfer rates in industrial multitubular packed-bed reactors with external cooling if the random packings of catalyst pellets are replaced by structured catalysts with highly conductive honeycomb supports. This concept has been proposed on the basis of simulation studies, confirmed by heat transfer and reactive experiments at the lab scale, and claimed in patents. We present herein a proof-of-concept at an industrial scale based on a campaign (1600 hours) of o-xylene oxidation runs (c/o Polynt, Italy) in a single-tube pilot reactor (i.d. = 24.6 mm) loaded with washcoated ($V_2O_5/TiO_2$) monolithic catalysts with Al honeycomb supports (Corning Inc., USA) and operated at typical industrial conditions for PA (phthalic anhydride) production. The conductive monolithic supports afforded substantially reduced T-gradients in comparison with reference runs in the same pilot reactor loaded with conventional ring catalysts, the maximum T-difference with the salt bath being halved at the same hot spot temperature (440 °C) and the mean bed temperature being about 20 ° C higher. Temperature gradients were still moderate at the highest investigated o-xylene feed load (400 g/h at $Q_{air}$ = 4 $Nm^3/h$), which on the contrary represents an upper limit for current industrial PA packed-bed reactors. The twofold enhancement of radial heat transfer rates can be exploited e.g. either to increase the o-xylene feed load above 100 $g/Nm^3$ (and the PA productivity accordingly) within a retrofitting strategy, or to design new reactors with larger tube diameters. Ongoing related activities addressing the strongly exothermic Fischer-Tropsch Synthesis in fixed-bed reactors will be also reported.

See more of this Session: Process Intensification Plenary Session
See more of this Group/Topical: Process Development Division

# (255d) Intensification of Catalytic Reactions with Neoteric Solvent Media

**Bala Subramaniam**, Department of Chemical and Petroleum Engineering, Center for Environmentally Beneficial Catalysis, University of Kansas, Lawrence, KS

This talk will address the various unique ways in which solvent media such as supercritical fluids and gas-expanded liquids (GXLs) can be exploited to develop greener catalytic process concepts for $O_2$ and $H_2O_2$-based oxidations, hydroformylations and ozonolysis. The demonstrated advantages include process intensification at mild conditions by increasing dissolution of the limiting reagent (gaseous reactants such as $O_2$ or syngas or light olefins) in the GXL reaction phase;enhancing inherent safety by suppression of flammable vapors; and waste mitigation by reduced usage of organic solvents and by suppression of side reactions that generate undesired byproducts such as $CO_2$. Specific examples of novel process concepts to be discussed include the highly selective hydroformylation of higher olefins at mild conditions; a novel homogeneous ethylene oxide process virtually eliminating $CO_2$ formation as a byproduct; and a one-step spray reactor concept to produce polymer-grade terephthalic acid from *p*-xylene at high yields and purity.

See more of this Session: Process Intensification Plenary Session
See more of this Group/Topical: Process Development Division

## R&D Management from Beginning to End


*Mukund Karanjikar*
*Senior Associate*
*Technology Holding LLC*
*Salt Lake City, Utah*

*Jack Hipple*
*Principal*
*Innovation-TRIZ*
*Tampa, FL*

*Eldon Larsen*
*Professor of Engineering*
*Marshall University*
*South Charleston, WV*


### Extended Abstract


This workshop focuses upon managing R&D projects from beginning to end. Beginning refers to mental picture of an idea whereas end refers to solving the customers need and in the process generating profit for the organization. Using a case study, the speakers will delineate challenges involved at every stage of R&D project from a mental image of an idea to a profit making product. The talk also compares old paradigm of innovation management with the new proposed paradigm. The probability of success can be significantly improved using the new paradigm.


### Successful Innovation Management

We have seen many idea management flow sheets, gate management processes, and prioritization recommendations starting with an idea that already exists, but no discussion of where the idea came from or what was necessary to generate and frame it. Moreover attention is not paid to shaping and propagation of the idea. This has been especially true over the past decade when processes such as Six Sigma came into vogue whose main thrust was to eliminate deviation. New concepts and innovation are, by their very nature, deviant! In order to generate fresh and innovative ideas we must create a new thought pattern that may challenge the existing status quo in any number of ways. These can even include, at an extreme, replacing one's current products and businesses with totally new approaches and models.

The key elements of successful commercial innovation include:

1. Continuous flow of unmet needs, both from current customers and related to core competencies of the organization ("outside in" as well as "inside out"
2. Using clear mental models of problems and potential solutions
3. Perseverance for idea propagation
4. Clear definition of the business opportunity
5. An appropriate product development and management team
6. Efficient process scale up and continued input from the market
7. Successful project management to insure implementation on schedule and within budget

We will look at each one and discuss the important issues.

**1. Continuous Flow of Unmet Needs and Exploration of Uses for Core Competencies**

An organization needs to have a process for how it decides what to work on. This may sound simplistic, when this is left to intuition or accident; it is unlikely that the "best" projects will be chosen.

Steps to make sure that an innovation effort is capturing the best opportunities:

- Talk to groups that are not your current customers but that might be able to use the functional capability that you have as a core competence. Customer competitors are another valuable source. The customers of your customers are another excellent source. This insures that you have awareness several levels above your product and customers and alerts you to potential long term needs and trends that may not be obvious to your direct customer or may not be readily disclosed by them.

Figure 1: Level Diagram

In these discussions we want to make sure that we have a clear definition of what the "ideal result" that our customers are desiring. We can also look at parallel universes of technology and markets to explore opportunities.

**2. Using Clear Mental Models of Problems and Solutions**

It is one thing to "invent" a new product, but another to make sure that we understand how a potential customer is going to use the product as well as how the customer's customer is going to use their product. The days are gone when we can just make something, put it in a drum or tank car, and wait for the invoice to be paid. We must be intimately familiar with the product's use, its limitations (including regulatory issues if appropriate). We must understand the details of the physical use of the product and how it affects a customer's products and business.

It is also important to make sure that we do not have blinders on when thinking about use of our product. It is rare that there is only one way to accomplish a function. We must be constantly vigilant about competitive activities, not just from direct competitors, but from parallel approaches to meeting a customer's needs.

### 3. Perseverance for Idea Propagation

New business and product development is hard work and it is easy to get discouraged. An idea that starts our with a 10% probability of success can raise its odds to over 75% by the proper, early stage identification of what challenges must be overcome. If these are not clearly defined, then a new product or business development will flounder and struggle. The perseverance battle rests more on the shoulders of individuals than the organization or top management. Idea has to propagate and the best person to do so is the idea owner. Ideas may be facilitated top-down (by management) however propagation is always bottom up (by employees). A proprietary tool (YouSNAP®) is being proposed to better manage idea propagation. The five elements of the tool are:
a. You – Know yourself
b. Surroundings – Know your surroundings
c. Network – Develop your network
d. Approach – Decide an approach
e. Persistence – Be persistent and iterate

Employees need to understand a simple truth. Ideas are dime a dozen and never bought on face value. Frustration is part of innovation. One should know that failures will exist. Hence know yourself. Are you up to it? Then you should know your surroundings. Often times, a tendency exists in corporations to blame the corporation itself. What is corporation? It is made up of people like yourself trying to serve the customers. If there are non-cooperative people around you, the journey would be commensurately difficult. However the important thing is that you should know this before. This leads to having a robust network of supporters. Once you have done the necessary preparation, decide upon an approach. Finally be persistent. The presentation will discuss the tool in more interactive format.

### 4. The Business Opportunity

A new product is not a business unless we have a way of sustaining the product's sales, expanding its utility, and developing and enforcing an intellectual property portfolio. As we commercialize a new product, we must be constantly aware of, and be willing to address with resources and commitment, issues such as change in product form, packaging and shipping issues, product purity, and shelf life/stability. These issues are most adequately addressed by constant and continuous communication with the customer.

## 5. Product Development Teams

We frequently choose product team members based on "who's available". While this may be expedient, it may not serve the best interests of the new business development. The obvious component of team selection is function—marketing R&D, manufacturing, legal, sales, etc. but we frequently ignore the human side of team functioning. If team members spend too much time dealing with personal issues relating to other team members, that energy is not applied to successful new product launch. Tools that can assist in this include measurements such as the Myers Briggs family of instruments, the Hermann Brain Dominance instrument (HBDI), and the Kirton KAI assessment. Knowledge of these personality profiles will not change natural behavior, but can make team members aware of others' preferred styles and open discussion can eliminate hidden barriers to effective team functioning.

## 6. Efficient process scale up and continued input from the market

Often times, companies strive for launching the best product. It is important to realize that continued input from the market is critical in order to have that 'best product' in the market. It is necessary to efficiently scale up and launch a product sooner than later. It should be perfected using market feedback. Getting market feedback is the only way to close the R&D loop and serve your customers to the fullest.

## 7. Successful Project Implementation and Management

New product development does not end when the first semi-commercial sample has been OK'd by the customer. A new type of team is now required to oversee the details process design, plant construction (including all the sub issues of permitting, EPA regulations, etc.). Quick and efficient plant start up with appropriate contingency planning is also a part of this effort. Contingency planning, not only related to unforeseen plant construction and start up difficulties, but also to taking care of the customer's needs in these situations must be part of the plan.

Three keys to successful project management lead to successful project completion. These keys involve

- The importance of people
- The importance of defining and setting good objectives
- The importance of good planning

We will apply these three keys to successful project management. This is all about application, not just theory. Today, you can apply the lessons you will learn.

A case study will be discussed in the presentation.

**Process Optimization towards Sustainability Using Modified Particle Swarm Algorithm for a Multitude of Objectives**

*Mazdak Shokrian, Chemical Engineering, Oklahoma State University, Stillwater, OK and Karen High, School of Chemical Engineering, Oklahoma State University, Stillwater, OK*

**Extended Abstract**

Extreme purpose of finding the most cost effective design solution with the least waste generation possible, considering environmental and social impacts in chemical engineering design problems justifies developing a systematic methodology for incorporating sustainability concerns into design and optimization of chemical process. This paper presents an applicable methodology for incorporating sustainability concerns into design and optimization of chemical process systematically. The methodology incorporates economic, social and environmental metrics into the chemical process optimization at the initial design stage, which is totally different with the traditional end-of-pipe treatment methods.

First, a Particle Swarm Optimization algorithm was modified to enable handling a multitude of objectives imposed while considering sustainability concerns. In the multi-objective optimization problems, selecting the leaders (guides) in the swarm from a set of non-dominated solutions has a great impact on the convergence and diversity of the solution especially in handling high dimensional optimization problem. A quite novel approach was developed to adopt PSO into multi-objective problems for efficient convergence to a diverse final solution set especially while dealing with high number of objective. This algorithm keeps the record of best Pareto solutions to select the local guides (leaders). The method for selection of the members in archive and leaders ensure the diversity of Pareto front. This approach searches toward spars regions and also searches for multiple solutions while avoiding clustering around a specific solution. To avoid expelling successful leaders or good Pareto solutions in sparse regions from the archive an aggregated fitness is used. This accounts simultaneously for success rate and diversity of the members. Local best of each particle will be selected based on the concept of non-dominancy. In the case that the current and previous best position of a particle are both non-dominated, a strength fitness value based on the number of particles dominated by each could be used as well as other methods for selection between them. A unique equation for updating the position or velocity of the particles guides each particle to the nearest Pareto solution in decision space while considering the information of other Pareto solutions. The archive members are updated at each iteration using new non-dominated set obtained from new generation. It should be mentioned that the number of particles in the swarm is kept constant.

Secondly, a sustainability evaluator was linked to the optimization algorithm to quantitatively evaluate the objectives during optimization procedure. This tool uses a set of sustainability metrics that have been compiled from a variety of researchers by Ola Shadya [1] to quantify social and environmental concerns and economic of the process. The inputs into this tool are the mass and energy flows, raw material and product costs, and the operational conditions and the outputs of this tool are the economic, social and environmental metrics. In this work, to show the applicability of the methodology and the proposed optimization algorithm we tested the algorithm on the William Otto process which is a very common test problem and we focused on economic and process safety risk. Process safety risk was implemented by the index developed by Heikkila [2] and using data from the International Agency for Research on Cancer (IARC) to identify the carcinogenic risk. The economic metrics is the profit of the process to be maximized. Concerns of sustainability other than economic and resource usage, which are defined using cost, material and energy amounts, will be evaluated by their impacts. For example global warming impact for a process, is calculated by the potency factor of each material multiplied by its amount released into environment. The same concept of potency factors is applied to evaluate social impacts. Thus, all the concerns quantified in the sustainability evaluator are considered as an objective for optimization. Economic objective function in this work was profit and as well as equality and inequality constraints were obtained from the work of Dibella and Stevense [3].

Thirdly, an efficient constrained handling strategy was incorporated to cope with the optimization problem highly constrained with equality and inequality constraints. Among the common constraints handling method penalizing methods are not preferred here because there is no prior knowledge about the behavior of the problem and a proper amount to which each violating solution should be penalized is unknown. In this work, a simple and efficient constraint handling mechanism presented by Raquel and Naval (2005) [4] is used. In this method in order to find a non-dominated solution the concept of dominancy is redefined based on the violation from constraints. This method allows solutions with less amount of violation from constraints to be considered better.

Next, In order to ensure convergence to a mature final solution for termination of the algorithm a non-dominated based criterion introduces by Goel, T. and N. Stander [5] was used. In their approach the solution set of current generation is compared to the solution set of $y^{th}$ previous generation (y is the generation gap) and Consolidation ratio (CR), the ratio of the solutions in the solution set of $y^{th}$ previous generation which also present in the current solution set, is calculated. If CR is more than a threshold and also statistical criteria are met the algorithm is terminated. The generation gap acts as a filter for noise if this value is very small it may result in a pre mature solution. Once the Archive meets the stopping criterion the algorithm is terminated. This helps to avoid the problem of oscillation in the Pareto set and not incorporating a priori knowledge about the final solution or a fixed number of generation. In some situations, a fixed generation number would result in excessive generations or stopping the algorithm prematurely.

Finally, the multi-objective a highly constrained and non-linear mathematical model established by considering social and economic factors was solved by the modified Multi-Objective Particle Swarm Optimization, called MLMOPSO, algorithm. The proposed methodology was tested using the William Otto process. The result obtained from the implementation of the proposed methodology confirmed the strength of MLMOPSO optimization algorithm in highly constrained and non-linear optimization problem with high dimensionality in objective space. The new algorithm in addition to locating the optimal solution found by previous researchers dealing with only one objective (profit of the process) found more uncompromisable solutions due to presence of sustainability objectives.

**Keywords:** Multi-Objective, Optimization, Sustainability, Particle Swarm,

## References

1. Shadya, O.O., *Social, Economic and Environmental Metrices for the Sustainable Optimization of Chemical and Petroleum Processes*, in *School of Chemical Engineering*. December 2010, Oklahoma State University: Stillwater.
2. Heikkila A., *Inherent Safety in Process Plant Design. An Index-Based Approach*, in *Department of Chemical Technology* 1999, Helsinki University of Technology: Espoo. p. 132.
3. Di Bella, C.W. and W.F. Stevens, *Process Optimization by Nonlinear Programming*. Industrial & Engineering Chemistry Process Design and Development, 1965. **4**(1): p. 16-20.

4. Raquel, C.R. and J. Prospero C. Naval, *An effective use of crowding distance in multiobjective particle swarm optimization*, in Proceedings of the 2005 conference on Genetic and evolutionary computation. 2005, ACM: Washington DC, USA. p. 257-264.
5. Goel, T. and N. Stander, *A Non-Dominance-Based Online Stopping Criterion for Multi-Objective Evolutionary Algorithms*. International Journal for Numerical Methods in Engineering, 2010. 84(Compendex): p. 661-684.

# (329b) Distillation Design In Recovering Butanol Produced From Extractive Fermentation Processes

**Yihui Tom Xu**, DuPont Engineering Research & Technology, E. I.du Pont de Numours & Company, Wilmington, DE

Butanol is an important industrial chemical with a variety of applications, such as use as a fuel additive, as a blend component to diesel fuel, as a feedstock chemical in the plastics industry, and as a foodgrade extractant in the food and flavor industry. Each year 10 to 12 billion pounds of butanol are produced by petrochemical means. As the projected demand for butanol increases, interest in producing butanol from renewable resources such as corn, sugar cane, or cellulosic feeds by fermentation is expanding.

In a fermentative process to produce butanol, in situ product removal advantageously reduces butanol inhibition of the microorganism and improves fermentation rates and butanol concentrations. Technologies for in situ product removal include stripping, adsorption, pervaporation, membrane solvent extraction, and liquid-liquid extraction. In liquid-liquid extraction, an extractant is contacted with the fermentation broth to partition the butanol between the fermentation broth and the extractant phase. The butanol and the extractant are recovered by a separation process, for example by distillation. In the recovery process, the butanol can also be separated from any water, noncondensable gas, and/or fermentation by-products which may have been removed from the fermentation broth through use of the extractant.

This presentation will focus on design tools and methodologies for distillation processes recovering butanol from a butanol-containing organic phase obtained from an extractive fermentation process. The extractive fermentation product mixture contains butanol, water, extractant, and other byproducts and impurities created from the fermentation process. Design considerations for sustainability is addressed.

See more of this Session: Process and Product Development for Sustainability I
See more of this Group/Topical: Process Development Division

SK|KBR TECHNOLOGIES

## Value Maximization of C4 Stream in Petrochemicals Production:

# Introductions to C4AC and Ole-SIV

October, 2011

**Nicholas J. Kim**

SK KBR TECHNOLOGIES

# Contents

- ✓ SK-KBR
- ✓ Markets and Highlights
- ✓ C4AC
  - ➢ Process
  - ➢ Case Study
- ✓ Ole-SIV
  - ➢ Process
  - ➢ Case Study
  - ➢ Other Applications
- ✓ Summary

SK KBR TECHNOLOGIES
SK-KBR
SK energy
50:50
Joint Venture
KBR
License
Projects
Operating
Capital
Technology Development
– R&D Program
– Catalyst
– Process Design
– Commercial Units
SK-KBR
Technologies
Pte. Ltd.
Licensing Expertise
– Licensing Business Model
– Global Sales
– Market Presence
– BED Execution
Phase-I Technologies
• ATA (Advanced Transalkylation)
• C4AC (C4 Acetylene Converter)
• Ole-SIV (Olefin/Paraffin Separation)
Customers
(Licensees)
Phase-II Technologies
Additional technology platforms
will be added to the portfolio as
and when available
Privileged and Confidential

# Markets and Highlights

Butadiene and Butene-1

# Typical C4 stream

Area of interests

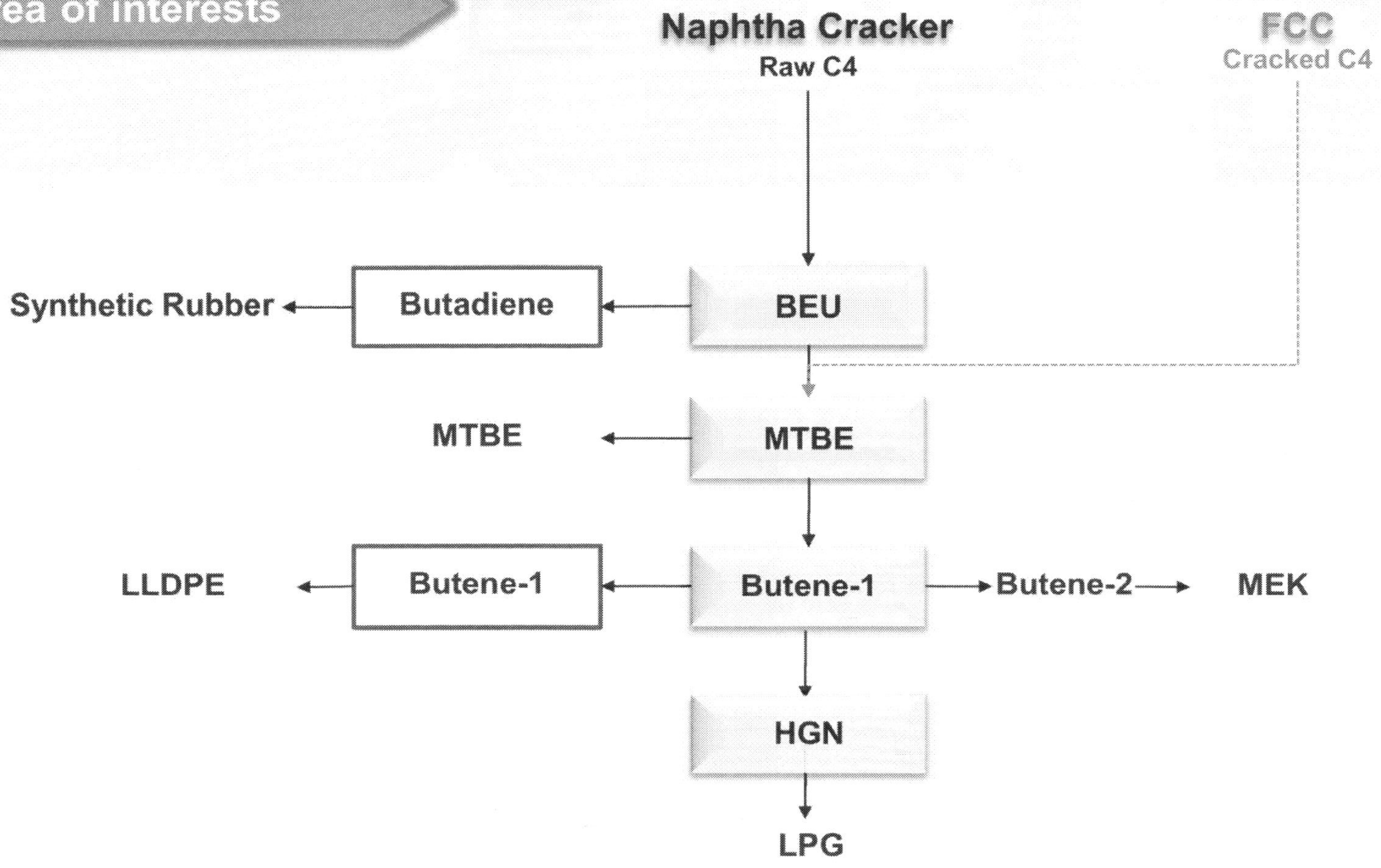

# Butadiene: talk of the town

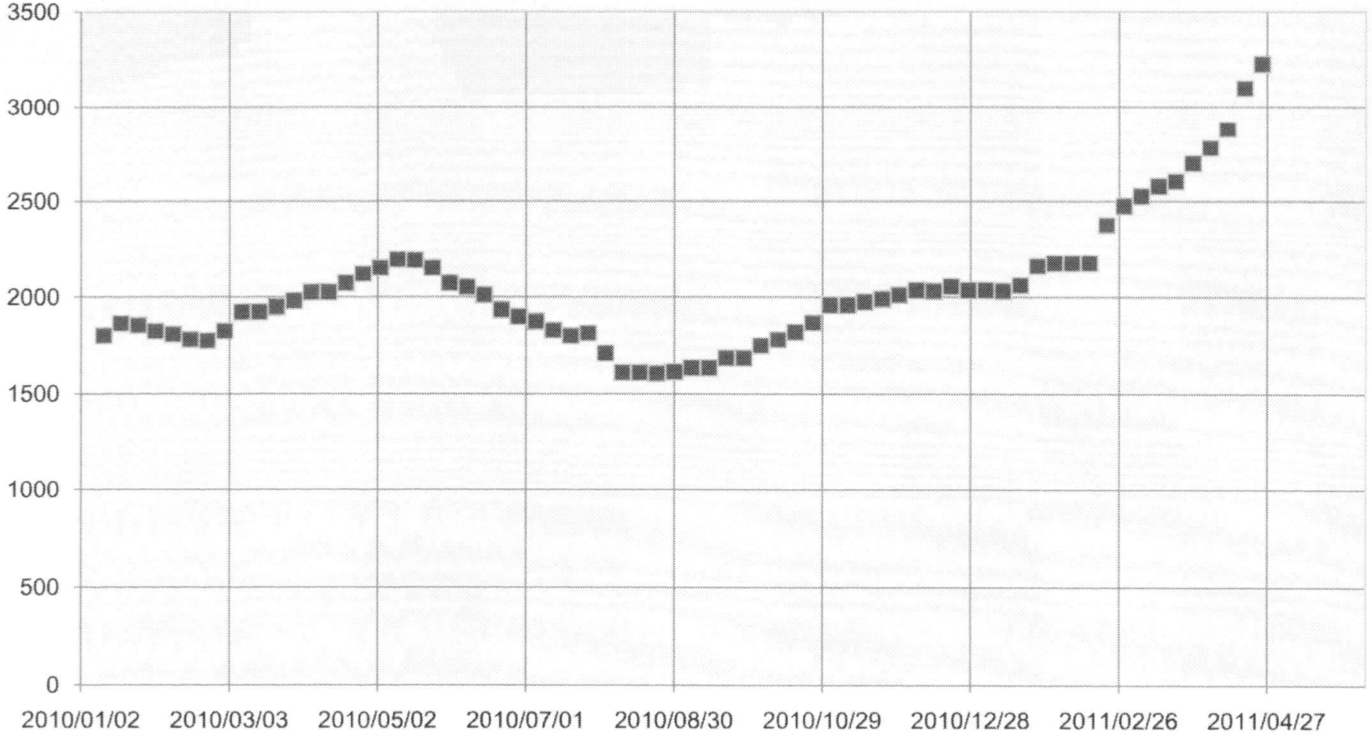


**Butadiene Spot Price *(USD/MT)***

# Butadiene outlook

SK KBR TECHNOLOGIES

## Demands

- ❑ Annual growth of 3.7% over the next 5 years (4.3% in Asia)
- ❑ Long-term demand at around 2.5% increase (3.2% in Asia)
- ❑ Growth driven mainly by China but India on deck
- ❑ Price likely stays high as $C_4$ supply permanently decreased

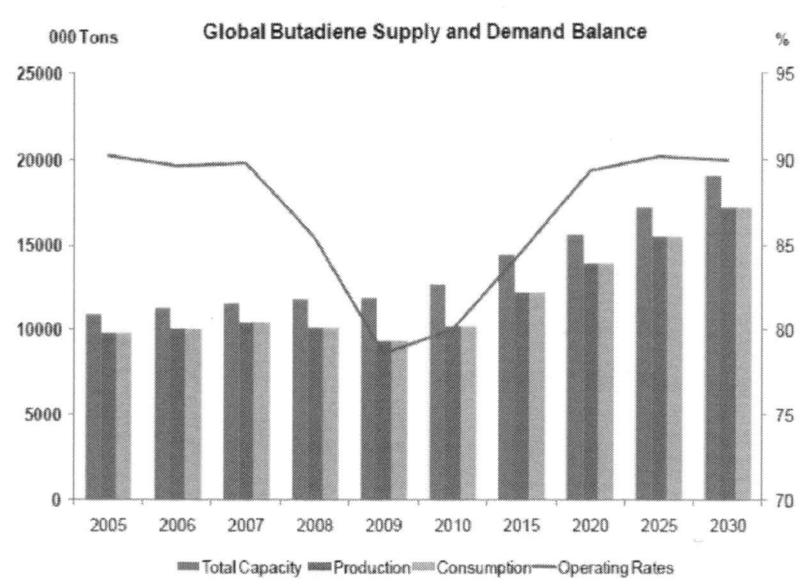


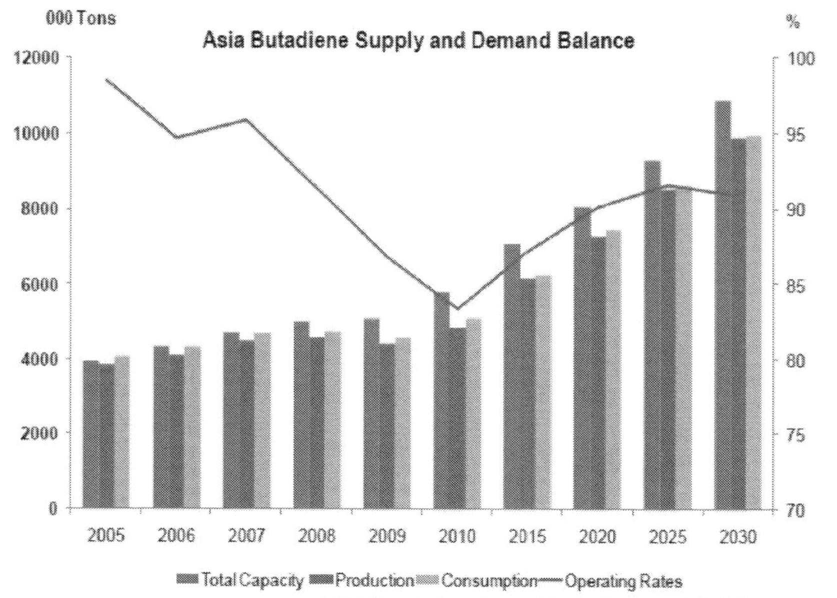


KBC at Indian Petchem 2010

Privileged and Confidential

SK KBR TECHNOLOGIES

# Butene-1 outlook

## Demands

- Polymer consumption per capita in emerging countries remains 1/5~1/3 of US/Japan
- Most polymer are expected to grow in double digits
- Despite capacity additions, LLDPE shortage is expected to prolong

*Polymer Growth vs. GDP for Last 5 years*

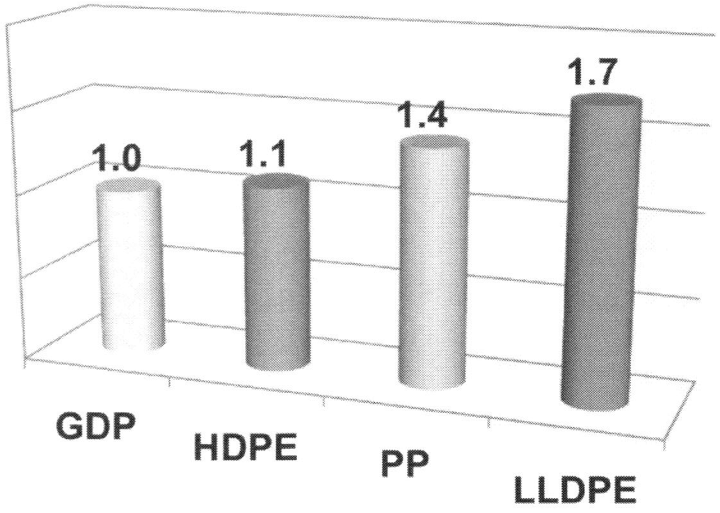


*Major Polymer Global Outlook*

| Product | 2010 | 2016 |
| --- | --- | --- |
| HDPE | Short | Short |
| LLDPE | Short | Short |
| LDPE | Short | **Surplus** |
| PP | **Surplus** | Short |
| PVC | Short | Short |

ONGC at Indian Petchem 2010

Privileged and Confidential

# Market highlight

**Opportunities & Treats**

- **$C_4$ shortage due to ethane/shale gas cracking**
- **Feedstock available in Asia**
- **In search for alternatives(esp. for BD) owing to the unprecedented price surge**

**Responses**

- **Seize the opportunity**
- **Value addition (optimization, integration, revamp…)**
- ***Competitiveness from technology***

# C4AC

## C4 Acetylenes Conversion to Butadiene

# Typical C4 stream of ethylene plant

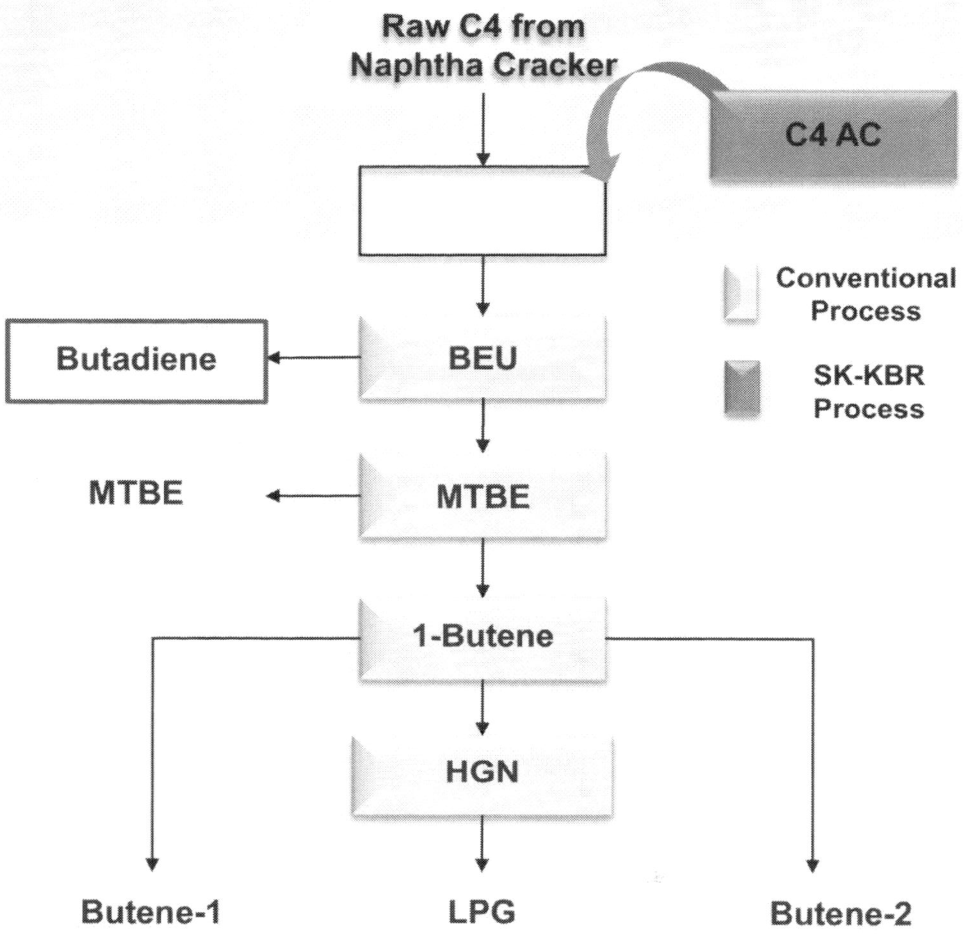


| Component | wt % |
|---|---|
| i-Butane | 0.47 |
| n-Butane | 1.70 |
| Cyclobutane | 0.06 |
| Methyl Cyclopropane | 0.04 |
| **1.3-Butadiene** | **47.24** |
| 1.2-Butadiene | 0.32 |
| **1-Butene** | **14.60** |
| **Isobutylene** | **25.42** |
| Trans-2-Butene | 4.66 |
| Cis-2-Butene | 3.78 |
| IsoPentane | 0.07 |
| C5+ | 0.03 |
| **MethylAcetylene** | **0.01** |
| **EthylAcetylene** | **0.24** |
| **VinylAcetylene** | **1.36** |

# Why acetylene converter?

- Why?
  - ✓ BD Loss in VA removal (Dilution with BD)
  - ✓ Acetylenes causing fouling → maintenance needs
- Solution: "selective hydrogenation" of C4 acetylenes (VA, EA)
  - ✓ BD yield Increase (VA to BD & BBR-1 and recovery of BD loss)
  - ✓ Reduce energy consumption

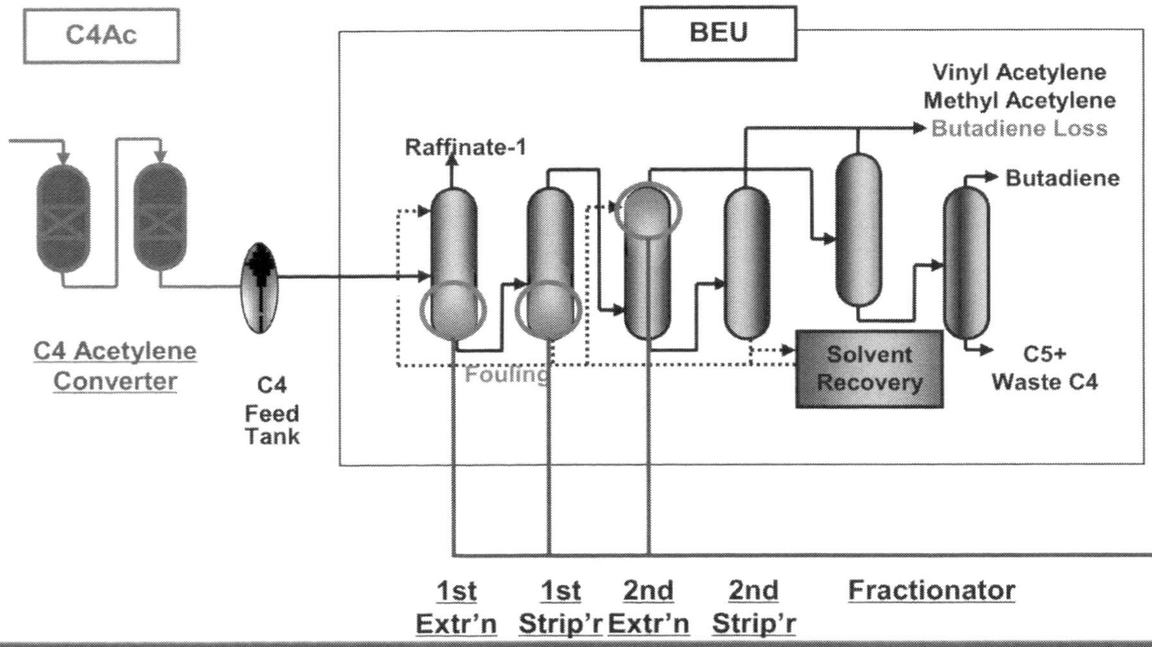


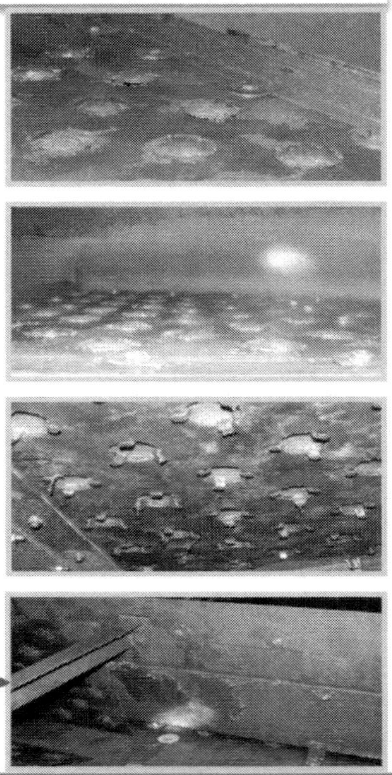

Privileged and Confidential

# Selective hydrogenation

- In general, acetylene conversion and butadiene yield are in trade off
- Selective hydrogenation is the key

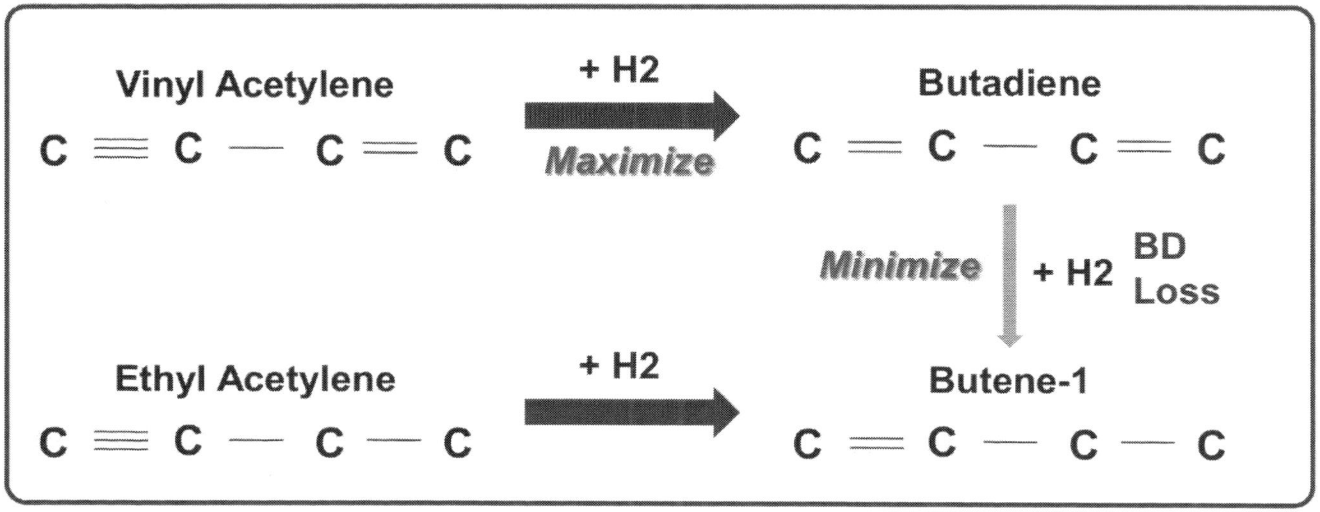


**Selective hydrogenation with proprietary catalyst(Olemax-456), co-developed with "Sud-Chemie"**

SK|KBR TECHNOLOGIES

# Single reactor scheme

**❑ Process performance of single reactor system**

- ✓ Conversion of VA : 50 ~ 80 % (depending upon process conditions)
- ✓ BD yield increase in BEU
- ✓ Less fouling

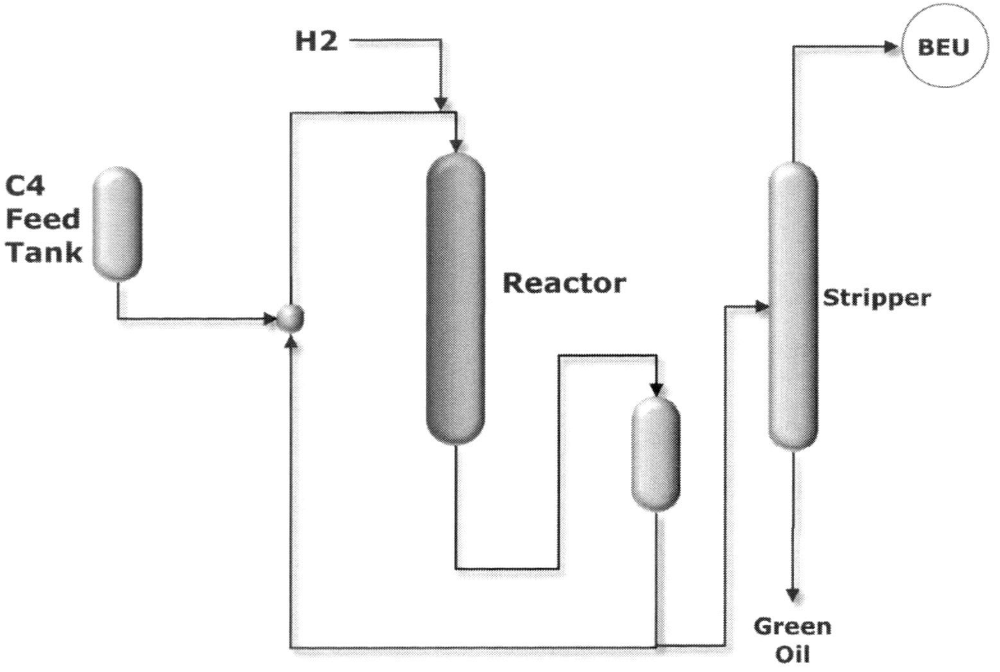


**❑ Operation conditions**

- ✓ Pressure:
  30 ~ 40 $kg/cm^2$,g
- ✓ Inlet temperature:
  30 ~ 60 °C
- ✓ H2/Acs. mol ratio:
  1.1 ~ 1.5
- ✓ Recycle/feed Ratio:
  1 ~ 2

SK KBR TECHNOLOGIES

## Series reactor scheme

- **Process performance of series reactor system**
  - ✓ Conversion of VA : nearly 100%
  - ✓ BD yield Increase in BEU
  - ✓ Less fouling
  - ✓ ***2nd EDC not necessary in BEU (savings in OPEX and/or CAPEX)***

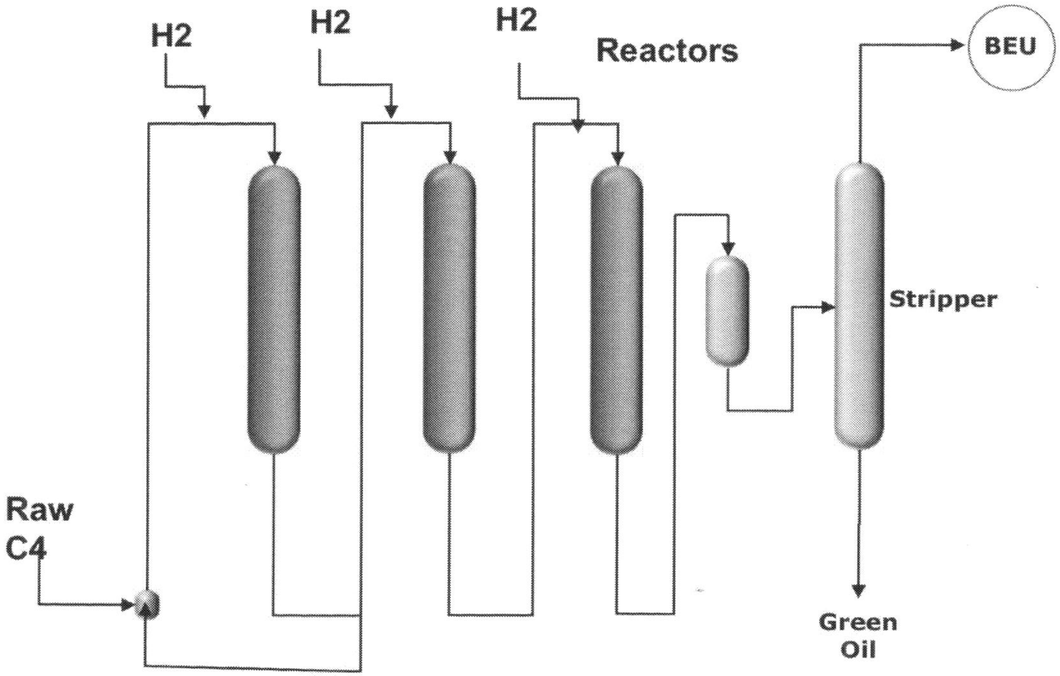


- **Operation conditions**
  - ✓ Pressure: 30 ~ 40 $kg/cm^2$,g
  - ✓ Inlet temperature: 30 ~ 60 °C
  - ✓ H2/Acs. mol ratio: 2.1 ~ 2.5
  - ✓ Recycle/feed Ratio: 1 ~ 6

Privileged and Confidential

SK KBR TECHNOLOGIES

# Equipment removal

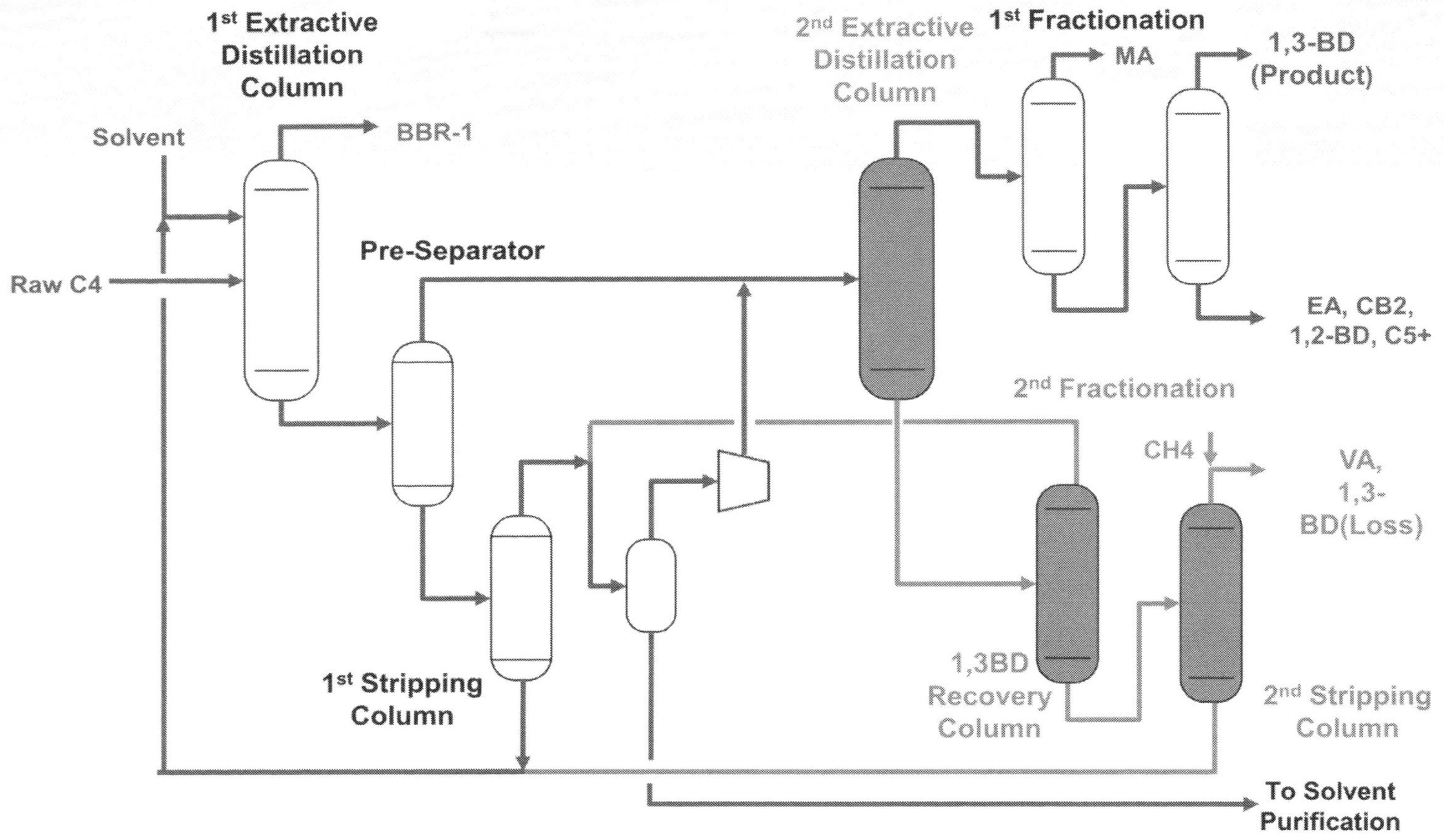


Privileged and Confidential

# Simplified BEU

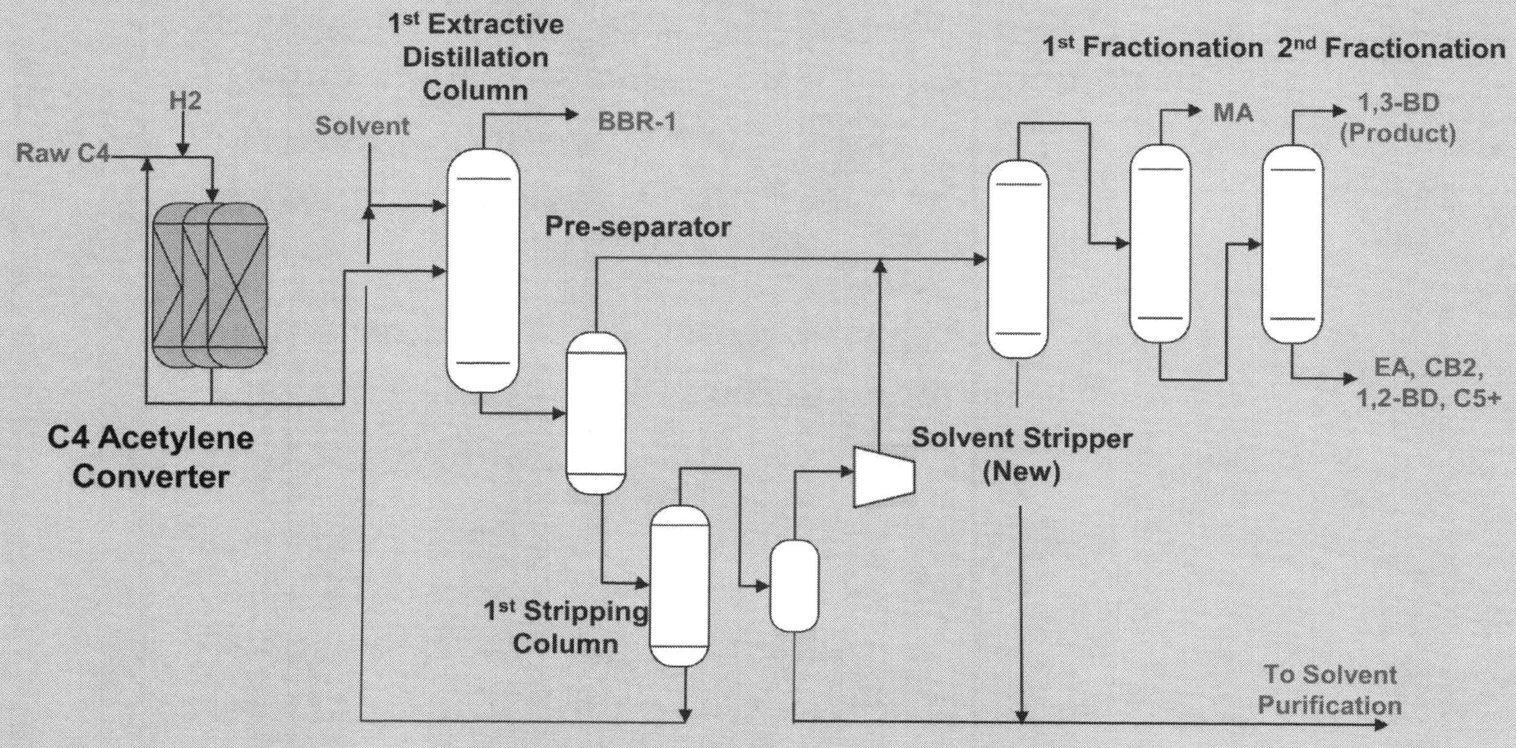


- ❑ Simplified BEU : 2nd EDC, 1,3BD recovery column, 2nd stripping column not necessary.
  - ✓ Lower Investment Cost(>30% ↓) & Energy Consumption(>20% ↓)
  - ✓ 1,3 BD Total Yield Increase(1~3% ↑)

# Summary

- Selective hydrogenation by proprietary catalyst
  - ✓ High activity → high throughput
  - ✓ High selectivity → low butadiene loss
- Catalyst regeneration
  - ✓ Short period of catalyst regeneration time
  - ✓ Virtually no performance degradation
- Economic benefits
  - ✓ Increase in butadiene yield
  - ✓ Saving in investment cost & utility consumption
  - ✓ Minimizing fouling and relevant maintenance

SK KBR TECHNOLOGIES

# Case study: C4 acetylene removal

| Component(wt%) | Feed (from customer) | Converter Out(Estimated) | | | |
|---|---|---|---|---|---|
| | | Single | | Series | |
| C3s | **0.140** | 0.17 | | 0.18 | |
| isobutane | **0.150** | 0.15 | | 0.15 | |
| n-butane | **0.520** | 0.52 | | 0.52 | |
| other C4 | **0.100** | 0.10 | | 0.10 | |
| trans-2-butene | **5.060** | 5.25 | | 5.42 | |
| 1-butene | **14.220** | 14.65 | | 15.01 | |
| cis-2-butene | **4.070** | 4.15 | | 4.20 | |
| iso-butene | **22.190** | 22.21 | | 22.23 | |
| 1,2-butadiene | **0.300** | 0.27 | | 0.24 | |
| **1,3-butadiene** | **51.440** | 51.44 | 0.0% | 51.12 | 0.6% |
| Methylacetylene | **0.080** | 0.0313 | 60.9% | 0.0032 | 96.0% |
| **Vinylacetylene** | **1.240** | 0.3718 | 70.0% | 0.0012 | 99.9% |
| **Ethylacetylene** | **0.190** | 0.1200 | 36.9% | 0.0173 | 90.9% |
| C5s | **0.300** | 0.42 | | 0.43 | |
| Green Oils(C6+) | **0.000** | 0.15 | | 0.39 | |
| SUM | **100** | **100** | | **100** | |

1. **@Converter Outlet, BD has Gain (VA→BD) or Loss(BD→BBR-1)** by **Hydrogenation** in order to convert VA. BD loss is recovered by BD Gain in BEU
2. **VA Removal** → Reduce BD Loss in VA Removal Column of BEU → **BD Gain in BEU**

SK KBR TECHNOLOGIES

# Case study: mass balance

| Components (KTA) | Without C4AC | | With C4AC | | | Performance | | |
|---|---|---|---|---|---|---|---|---|
| | BEU Feed | BEU Products (A) | Rxt. Inlet | BEU Feed (Rxt. Outlet) | BEU Products (B) | Change ,B-A (KTA) | Unit Price ($/Ton) | Savings (MM$/yr) |
| Raffinate-1 | 151.1 | 146.5 | 151.1 | 155.6 | 150.8 | 4.3 | 956 | 4.16 |
| 1,3-Butadiene | 167.8 | 161.1 | 167.8 | 167.0 | 165.4 | 4.3 | 2432 | 10.39 |
| Others(C3-, 1,2-BD, C5) | 2.4 | 0.0 | 2.4 | 2.7 | 0.0 | 0.0 | | 0.00 |
| C4 Acs | 4.9 | 0.0 | 4.9 | 0.1 | 0.0 | 0.0 | | 0.00 |
| Fuel Gas | | 14.1 | 0.0 | 0.0 | 5.3 | -8.8 | 1022 | -8.99 |
| Fuel Oil | | 4.1 | 0.0 | 0.0 | 5.1 | 1.1 | 675 | 0.72 |
| H2 | | | 0.5 | 0.0 | 0.0 | | 3623 | -1.74 |
| Utility Reduction | | | | | | | | 2.14 |
| Total | 326.3 | 325.7 | 326.7 | 325.5 | 326.6 | | | **6.69** |

- BD Total Recovery : 95.7% → 98.7%
- BD Purity : 99.7%

The above calculation is based on SK BEU process scheme

SK|KBR TECHNOLOGIES

# Case study: economics

## Investment Cost

**Total project cost: US$ 10.6 M (± 30%)**

- Aspen Icarus estimation
- Capacity : feed 37.5 ton/hr
- Series reactors x 4 (including standby reactor)
- ISBL
- Excluding:
  - ✓ license fee
  - ✓ catalyst (50 $m^3$, 34 ton)
  - ✓ dedicated regeneration facility

## Profit

**Profit: US$ 6.7 M/yr**

- BBR-1 increase (3.0 wt%)
- BD yield increase (2.7 wt%)
- Utility reduction (min. 20%)
- BEU fouling reduction → reduce maintenance cost

SK KBR TECHNOLOGIES

# Ole-SIV

## Paraffin/Olefin Separation by Adsorption

Privileged and Confidential

# Typical C4 flow of ethylene Plant

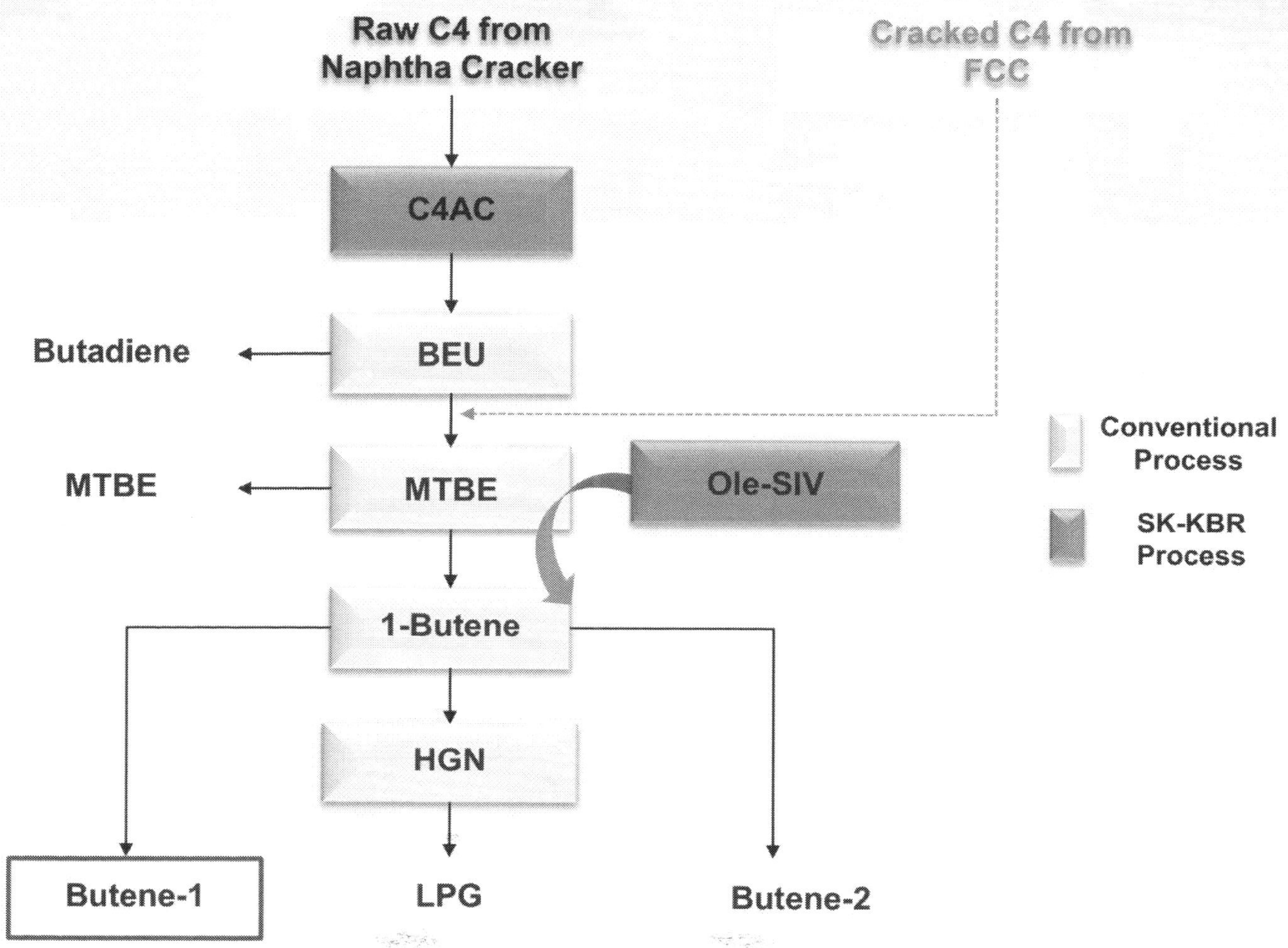


Privileged and Confidential

# Ole-SIV: separation by adsorption

- Adsorption technology to separate olefin/paraffin from mixed light gases
  - ✓ No vacuum operation
  - ✓ Vapor phase operation
  - ✓ Constant pressure operation
  - ✓ Continuous process

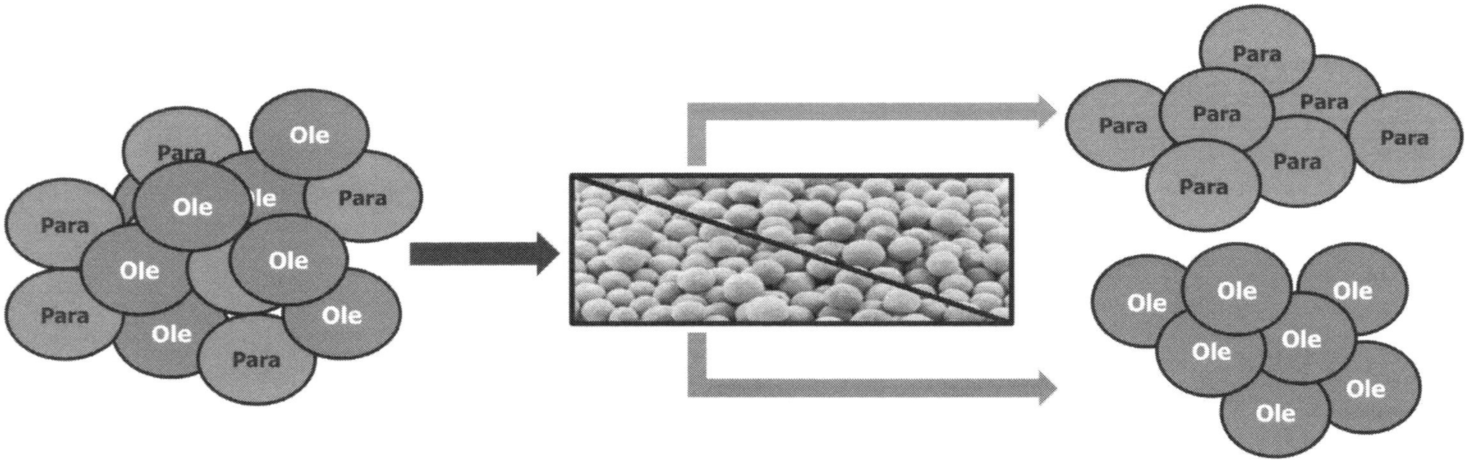

# Ole-SIV in butene-1 separation

## Conventional scheme

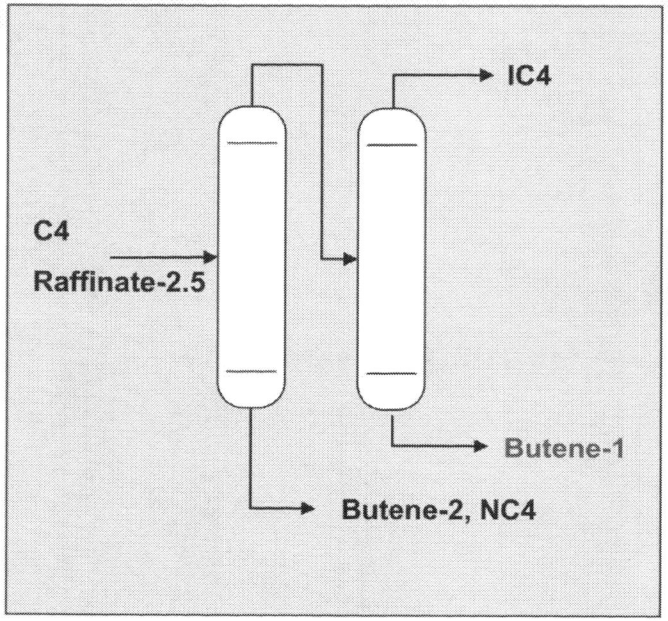


## New process scheme

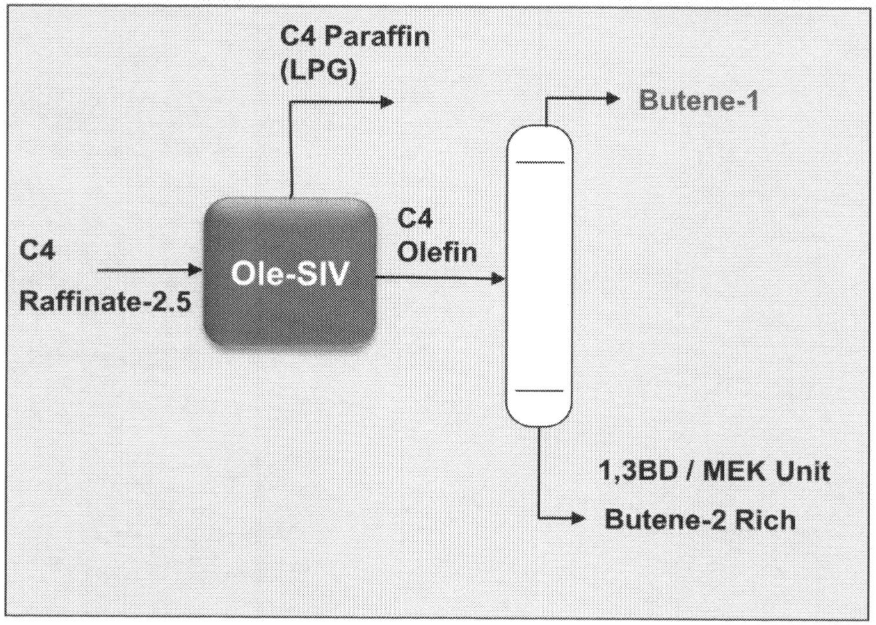

# Ole-SIV adsorption process

- ❑ **Adsorbent performance**
  - ✓ **High olefin selectivity**
  - ✓ **High adsorption capacity by patented material & structure**
- ❑ **Process steps**
  - ✓ **Displacement desorption technology**
  - ✓ **High performance desorbent by patented composition**

## Process Steps

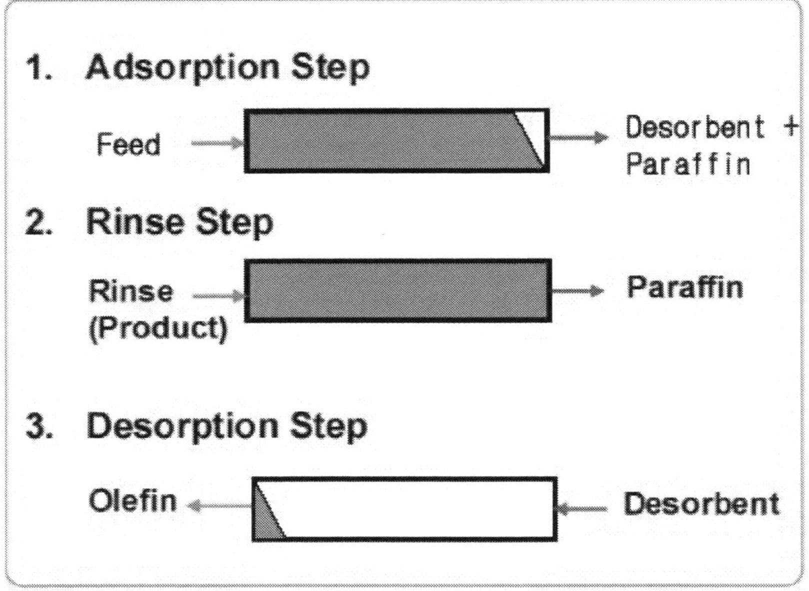

SK|KBR TECHNOLOGIES

# Ole-SIV process schematics

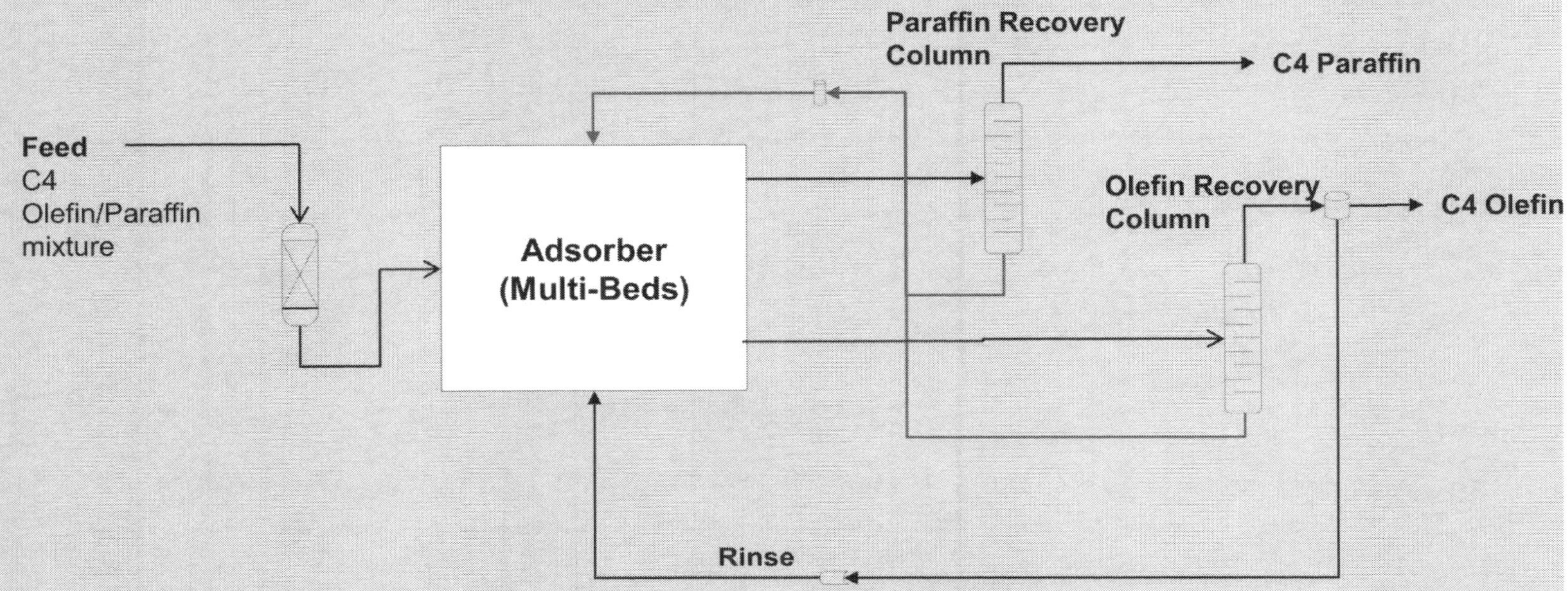


Privileged and Confidential

# Process configuration and sequence

SK KBR TECHNOLOGIES

## ❑ Five beds operation with one stand-by bed

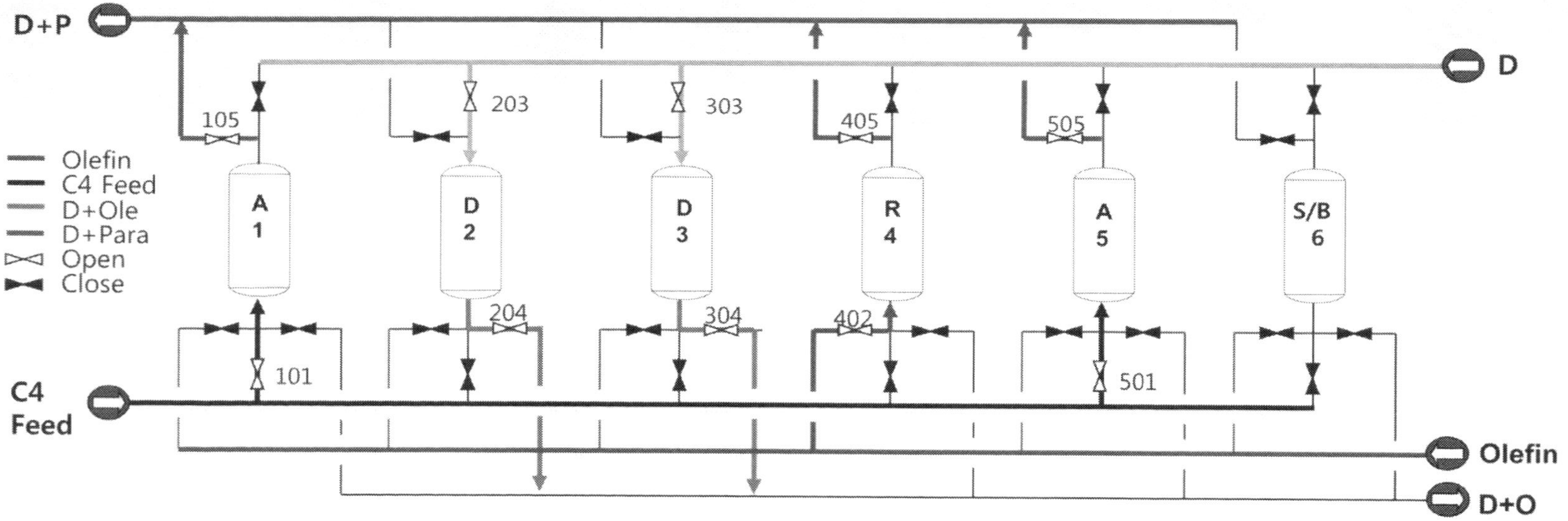


A: Adsorption Step | R: Rinse Step | D: Desorption Step

| | | | | | | | | | | |
|---|---|---|---|---|---|---|---|---|---|---|
| Bed#1 | A | | R | D | | A | | R | D | |
| Bed#2 | D | A | | R | D | | A | | R | D |
| Bed#3 | D | | A | | R | D | | A | | R |
| Bed#4 | R | D | | A | | R | D | | A | |
| Bed#5 | A | R | D | | A | | R | D | | A |
| Bed#6 | Stand-By | | | | | Stand-By | | | | |

Privileged and Confidential

SK KBR TECHNOLOGIES

# Performance and characteristics

| | |
|---|---|
| **High Performance** | ▪ **High Olefin Purity and Recovery**<br>- Optimal Results from Rinse Step and Proprietary Desorbent<br>- Either 99.0 wt% Recovery or 99.6 wt% Purity |
| **Flexibility** | ▪ **Processing Capacity of the Various Streams & Flexibility in Flow-rate Change**<br>- Feed Streams : C4 BBR-2.5 and 3, FCC C4, C3s, etc.<br>- Control of Purity and Recovery by Each Step Time |
| **Low Adsorbent Cost** | ▪ **No Noble Metal**<br>▪ **High Selectivity & Productivity** |
| **Low Investment & Utility Cost** | ▪ **Low Investment & Utility Cost in Comparison with Extractive-distillation commercialized on the same performances** |
| **Wide Applications** | ▪ **Applications**<br>- Revamp of Butene-1 Process or New Butene-1 Process<br>- Feedstock Pre-Treatment for 1,3-BD by ODH and MEK<br>- Propane/Propylene Separation |

SK KBR TECHNOLOGIES

## Summary

- Separating the olefin/paraffin in C4 stream
- Displacement desorption – no pressure swing, no vacuum, continuous process in vapor phase
- Performance
  - ✓ Recovery aim: up to 99.0 wt% with 94.5 wt% of purity
  - ✓ Purity aim : up to 99.6 wt% with 95.0 wt% of recovery
- Most economic option in butylenes extraction
- Saving in total investment cost up to 21%(new) and 30%(revamp)
- Energy saving up to 40%

# Case study: a revamp case

## ❑ Capacity increase : 25 kTA → 50 kTA

### By Copy Plant

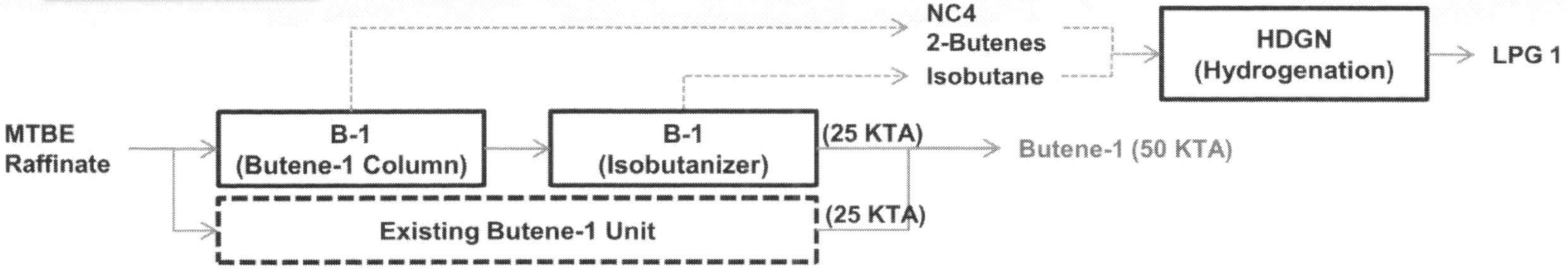


### By Ole-SIV

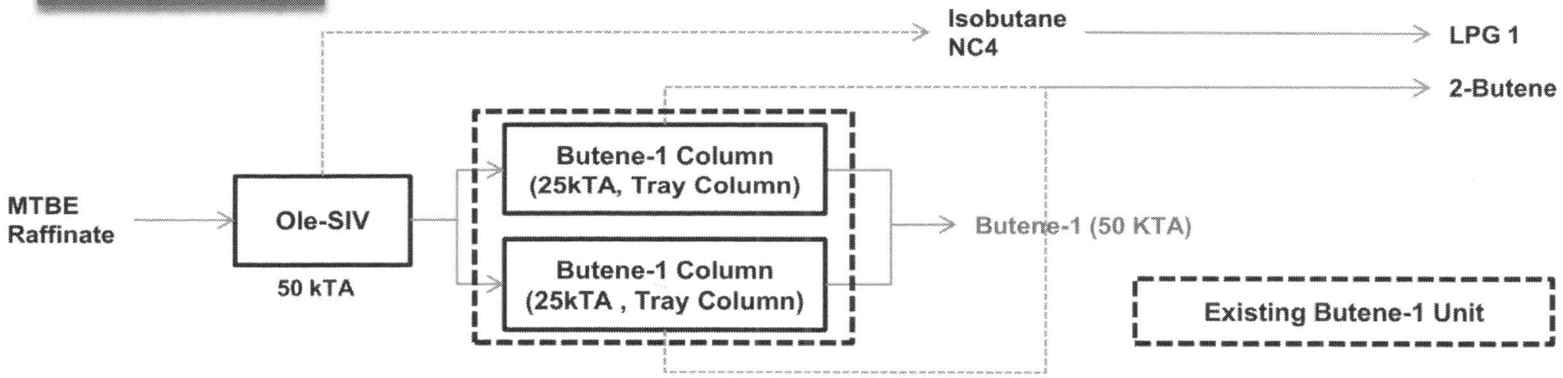

SK KBR TECHNOLOGIES

# Case study: economic gains

- Total investment cost reduction

| Feed Case | Conventional ($MM) | Ole-SIV ($MM) | Savings (%) |
|---|---|---|---|
| 50 kTA | 19.20 | 13.59 | **29.2%** |

- Energy savings

| Case | (MMKcal/hr) | Conventional | Ole-SIV | Savings (%) |
|---|---|---|---|---|
| 50kTA | Cooling Duty | -25.37 | -15.82 | 37.6% |
| | Heat Duty | 25.86 | 15.88 | 38.6% |

SK|KBR TECHNOLOGIES

# Other applications: FCC off-gas recovery

## ❑ Other applications

- ✓ $C_3$ splitter replacement
- ✓ **"Ethylene recovery from FCC off-gas"**

## ❑ Ethylene recovery from FCC off-gas

- ✓ 6 adsorbers, 3 distillation columns, 1 compressor
- ✓ Propylene refrigeration system ($T_{Lowest}$ = -20℃) as opposed to the conventional ethylene refrigeration system ($T_{Lowest}$ = -120℃)

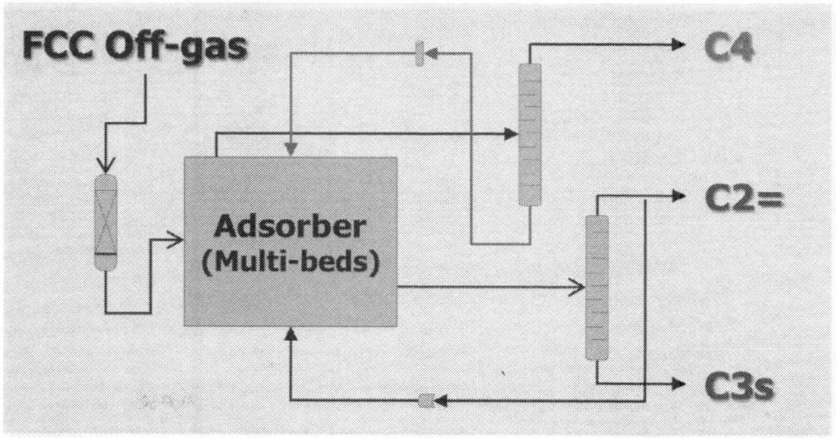

SK KBR TECHNOLOGIES

# Other applications: FCC off-gas recovery

- **SK confirms conditions and recovery**
  - ✓ 95% purity ethylene recovery for direct use in ethyl-benzene process
  - ✓ Additional propylene recovery from mixed C3s at FCC PRU
- **Plans for implementation in 2012~13 at SK Ulsan site**
  - ✓ Discussion on concurrent 3rd party's trail / JV option is welcomed

**FCC Off-gas**

| Component | wt% |
|---|---|
| $H_2$ | 2.4 |
| $N_2$ | 14.2 |
| Methane | 29.9 |
| Ethane | 20.4 |
| Ethylene | 29.4 |
| Propane | 0.7 |
| Propylene | 3.0 |

***335 KTA***

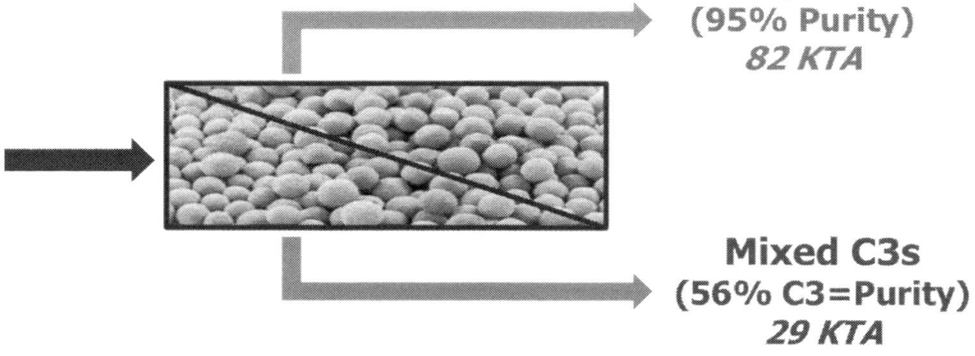

SK|KBR TECHNOLOGIES

# Summary

## C4AC and Ole-SIV

SK|KBR TECHNOLOGIES

# Summary

- **Butadiene in shortage and butylenes in demands**

- **C4AC**
  - ✓ Convert C4 acetylenes to butadiene: selective hydrogenation
  - ✓ Gain in butadiene yield
  - ✓ Saving in total investment cost & utility consumption
  - ✓ Minimizing fouling and relevant maintenance duties

- **Ole-SIV**
  - ✓ Most economic butylenes extraction by adsorption
  - ✓ Saving in total investment cost & utility consumption
  - ✓ FCC off-gas recovery under development for commercial trial

SK | KBR

TECHNOLOGIES

Privileged and Confidential

# DEVELOPMENT OF INHERENT SAFETY PERFORMANCE INDEX FOR SUSTAINABILITY STUDY ON THE PRELIMINARY PROCESS DESIGN

*Namjin Jang[1], Ikhyun Kim[1], En Sup Yoon[1]*

*[1]Seoul National University, Seoul, Korea*

## Abstracts

Sustainability study is mainly performed to design sustainable process in the chemical industries. To evaluate sustainability, Economic, environmental and social/safety indicators are mainly considered. Here, inherent safety index are considered as the safety indicator in the preliminary process design. In this paper, inherent safety performance index (ISPI) was proposed, which is consists of chemical hazard index (flammability, explosiveness, toxicity and reactivity) and process hazard index by unit (temperature, pressure, inventory). Also, the complexity, process safeness and operability of process are newly composed as the supplemental hazard index. In addition, the proposed index is performed with process simulators to get process operation date, thermodynamic properties, and chemical properties of mixture in MATLAB interface. The evaluation was done by using the six process routes for manufacturing methyl methacrylate (MMA) as a case study. To validate our proposed method, the several existed index with experts opinion was compared. The proposed methods can be guideline to decision makers of chemical process design when the safety and economic have conflicts because of government aspects and for practical engineering reasons. It can make easy to affect the process configuration and inherent safety in the preliminary process design than in the later phase of process design.

## Introduction

Development of chemical process is one of the core activities to launch the production of chemical materials in the chemical process industries. Especially, process design and synthesis is the essential engineering activity based on the mathematics, basic science, engineering science, economics, safety, reliability, and social consideration, etc. in the whole process design procedure. Also, it is more important than before due to increase of process complexity, handling of critical operation conditions, and handling of toxic materials, etc. [1].

Especially, sustainability have recently interested and studied in not only the society at large but also micro scale of process. Sustainability has become a driver for innovation since twenty years away from first definition of sustainable developments in the chemical industries. Also, Sustainability concept is used in from conceptual design of chemical process to life cycle of process in the overall chemical industries [2]. In this sense, the sustainable process is drawing attention as the upcoming frontier and alternative works. Generally, there are three major facets considered in sustainability, which is economy, environment/safety and society indicators. Several researches and studies have attempted to evaluate sustainability in order to identify sustainable design of the alternative process and systems [3, 4].

- The index of sustainability, IChemE
- AIChE sustainability index (SI), 2007
- Automating the Sustainability Metrics Approach in AIChE, 2004

Moreover, several major accidents were served to increase process safety in the process industries. Since then, the principles need to be emphasized repeatedly to draw the attention of those concerned in 1980s. After that, active research work on inherent safety design started in the 1990s in several countries. For example, the UK health and Safety Executive (EHS) encourages industries to use inherent safety design and is planning to issue a document "Successful Design for Health and Safety" suggesting possible usage of inherent safety design in several industries. CCPS (Center of chemical process safety in AICHE) in USA has published "Guidelines for Engineering Design for Process Safety" in 1993, and 1st revision of "An Introduction to Inherently safety design" in 2009 [5]. Also, several governments, academics, companies and associations have established guidelines, practices, and rules, etc. to encourage the development of inherently safer chemical processes and products [6].

Several studies have proposed tools for evaluation and estimation of inherent safety in the chemical process and plants. Gupta and Edwards (2002) [7], Palaniappan (2002) [8, 9], Khan and Amyotte (2004) [10] recently developed indices for inherent safer design. Most of indices adopted relative scoring method to measuring inherent safety design. The major two indices are Prototype Inherent Safety Index (PISI) by Lawrence in 1996 [11] and Inherent Safety Index (ISI) by Heikkila in 1999 [12]. Actually, ISI is focused on process units and structure in basic design step. However, it is hard to incorporate into process synthesis using computer aided process design tool. Also, several attempts are reported incorporation into process life cycle; chemical product design (research), basic process design, detail process design and construction, operation and management and decommissioning [13]. However, there have some considerations to improve applicability of inherent safety design index. Also, the existed indices did not reflect expert's recommendation and opinions.

In this paper, the inherent safety performance index (ISPI) is proposed to assess alternatives in the preliminary conceptual design phase. Also, we introduce inherent safety performance index with fuzzy set. Here, sub-indices of ISPI are modeled with fuzzy membership functions. Also, the fuzzy rules are written by using IF-THEN rules.

The evaluation was done by using the six process routes for manufacturing methyl methacrylate (MMA) as a case study. To validate our proposed method, the several existed indices with expert's opinions are compared.

## Inherent safety performance index

Inherent Safety Performance Indices (ISPI) is proposed based on the fuzzy logic methods (If – then rule) integrated with process simulators. First of all, the essential chemical and process properties are selected and categorized. Then, IF-THEN rule of fuzzy logic approach is written based on the expert experience and opinions. Finally, ISPI is applied to evaluate inherent safety using calculated property parameters in process simulators.

The proposed methods are composed of chemical hazard index, process hazard index and supplemental hazard index by modifying Inherent Safety Index (ISI, 1999).

The ISPI is calculated by Eq. (1), where the total inherent safety performance index (ISP) is multiply the sum of the chemical characteristic index (ISCI) and process characteristic index (ISPI) for each unit with supplemental characteristic index (ISSI).

$$IS_P = \sum_{i=1}^{n}(IS_{CI} + IS_{PI}) * IS_{SI} \quad (1)$$

where, i is equipment or stream of process, n is number of whole units or streams.

The chemical characteristic index ($IS_{CI}$) contains chemical characteristic sub-index and calculated by Eq. (2). These parameters are composed of flammability ($IS_{CF}$), explosiveness ($IS_{CE}$), Toxicity ($IS_{CT}$) and Heat of reaction ($IS_{CR}$). Also, the process characteristic index ($IS_{PI}$) contains process characteristic sub-index such as pressure ($IS_{PP}$), temperature ($IS_{PT}$) and inventory ($IS_{PI}$) and calculated by Eq. (3). Additionally, the kind of reaction ($IS_{CK}$: exothermic, endothermic, etc.) and kind of equipment parameter ($IS_{PE}$: tank, separator, distillation tower, heat exchanger, etc.) are multiplied with chemical characteristic index and process characteristic index, respectively.

$$IS_{CI} = IS_{CF} + IS_{CE} + IS_{CR} + IS_{CT} * IS_{CK} \quad (2)$$

$$IS_{PI} = (IS_{PT} + IS_{PP} + IS_{PI}) * IS_{PE} \quad (3)$$

The supplemental characteristic index ($IS_{SI}$) is composed of process safeness ($IS_{SS}$), complexity ($IS_{SC}$) and operability ($IS_{SO}$) and calculated by Eq. (4).

$$IS_{SI} = (IS_{SS} + IS_{SC} + IS_{SO}) \quad (4)$$

The calculation of ISPI is made on the basis of the worst case scenarios. Also, it is based on the results of calculation of process data in conceptual design phase. Therefore, it is very flexible according to both tendency of decision maker and engineering expert's opinions.

It is very important to understand that process may be inherently safe with respect to one criteria, but it can be possible unsafe in other view point [12]. So that, the inherent safety index needs to reflect expert's opinion or tendency of decision maker. And, fuzzy logic is one of the solutions for reflection of several uncertainties and rules to select possible process route among the alternatives.

In this step, we need to define more clearly inherent safety performance index including chemical, process, and supplemental characteristic sub-indices in following sections. The sub-index of ISPI is summarized in Table 1 and Table 2.

Table 1. Summary of Sub-index in the ISPI

| | Chemical Hazard Index | Process Hazard Index | Supplemental Hazard Index |
|---|---|---|---|
| Sub-index | Flammability<br>Explosiveness<br>Toxicity<br>Heat of reaction | Temperature<br>Pressure<br>Inventory | Safeness<br>Complexity<br>Operability |
| Descriptions of Chemical hazard sub- index | Flammability: NFPA rating (2007), HMIS and DOW F&EI<br>Explosiveness: (UEL – LEL) % depend on mixture, temperature and pressure<br>Toxicity (environmental): Overall toxicity index<br>Heat of the main/side reaction | | |
| Descriptions of process hazard sub- index | Temperature (K)<br>Pressure (atm)<br>Inventory: calculation (need to consider chemical characteristic)<br>– inventory = f (yield) | | |
| Descriptions of supplemental hazard sub- index | Safeness: Safety instrumental level: three level (SIL, ALARP)<br>Process complexity : INSIDE project<br>Process operability : range of operation, safety limits, equipment containments | | |

Table 2. The summarized score and fuzzy rule data of sub-index in ISPI

| Sub-index | Symbol | Score |
|---|---|---|
| Flammability | $IS_{CF}$ | 1 ~ 5 |
| Explosiveness | $IS_{CE}$ | 1 ~ 10 |
| Toxicity | $IS_{CT}$ | 1 ~ 9 |
| Heat of reaction | $IS_{CR}$ | 1 ~ 5 |
| Temperature | $IS_{PT}$ | 1 ~ 5 |
| Pressure | $IS_{PP}$ | 1 ~ 5 |
| Inventory | $IS_{PI}$ | 1 ~ 6 |
| Safeness | $IS_{SS}$ | 0 ~ 1 |
| Complexity | $IS_{SC}$ | Counts/100 |
| Operability | $IS_{SO}$ | 0 ~ 1 |

To assess inherent safety of design in the preliminary step, IF-THEN rules are adopted as the preference process in this paper. IF-THEN rules can be possible to reflect both expert's opinions or recommendation and tendency of decision maker's.

To generate IF-THEN rule as the preference process in the fuzzy logic, first of all, in such cases as expert opinion is applied with the application of an interpolation scheme. Of Course such an opinion may be quite subjective and imprecise. Generally, three different criteria for reasoning hazard are applied in this thesis, which is standard, easy, and hard. The "Standard" means typical hazard or risk encountered in many process industries. It can be represented by using IF-THEN rules in this thesis. Also, "Hard" means the high cost matrix however safer or acceptable high risk, whereas, the "easy" IF-THEN rules provides lower cost matrix or rules [13]. According to risk levels, the surfaces of fuzzy set are drawn.

Basically, inherent safety performance index, which are chemical hazard index, process hazard index and supplemental hazard index, are calculated same procedure. Also, whole scheme for inherent safety performance index with fuzzy logic are designed by performed hierarchical procedure. We built 4 FIS using 11 input parameters in fuzzy set. Also, weighting factors for the reaction types and equipment types is used in FIS. The hierarchical fuzzy system of ISPI with fuzzy sets is illustrated in Figure 1.

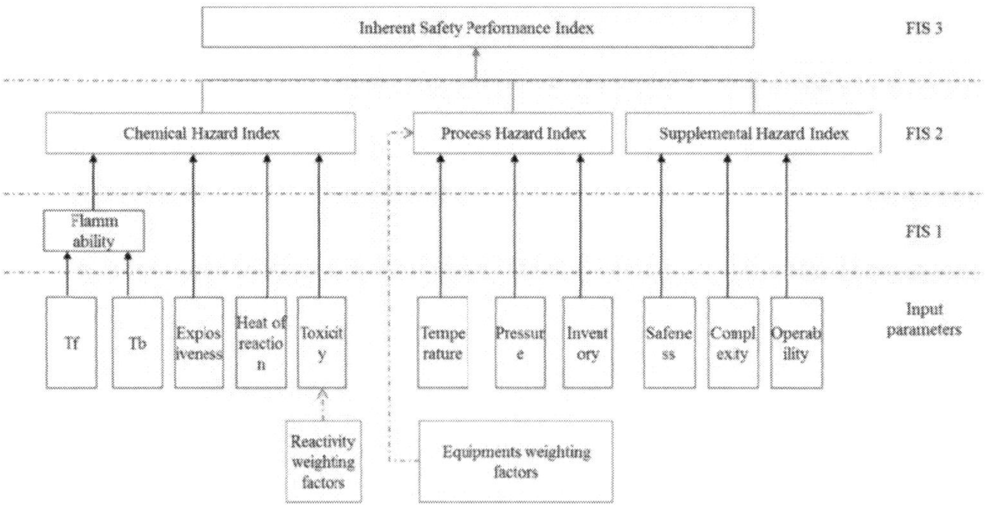


Figure 1. FIS tree of ISPI.

The scheme for calculation procedure of ISPI is briefly illustrated in Figure 2 as mention above. In this thesis, the ISPI analysis was performed on Matlab interface with process simulator data and database of chemical property.

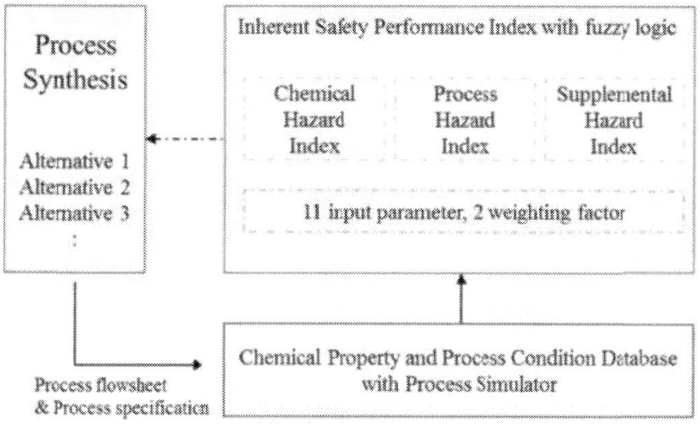


Figure 2. Procedure for calculation of ISPI.

## Case study

In this paper, the evaluation was done by using the six process routes for manufacturing methyl methacrylate (MMA) as a case study. To validate our proposed method, the experts ranking and several existed index are compared.

A remarkable feature of MMA is the variety in synthesis and production methods. It presents 17 different synthesis pathways of MMA reported [14, 15]. Among the developed synthesis pathway, several methods have industrialized and still studied on manufacturing process. Also, in order to improve efficiency, reduce waste and improve process safety, etc. various studies put forth a multilateral effort into solving mentioned problems including development of new catalytic, applicable operation procedures, and plant controllability, etc.

Traditional MMA production process (ACH process) and new developed and commercialized process are introduced such as Ethylene via Propionaldehyde based route (C2/PA), Ethylene via Methyl Propionate based route (C2/MP), Propylene based route (C3), Tertiary Butyl Alcohol based Route (TBA), Isobutylene based Route (i-C4) [11, 14]. Although several technologies have studied and developed, six routes are selected as case study to test proposed method in this study. Moreover, to validate proposed method in this work, selected routes have evaluated using ISI, PIIS, i-Safe and DOW F&EI, etc. compared with each other and experts knowledge.

The inherent safety performance index was evaluated and listed in Table 3. Based on the expert's opinions, the three different level of risk matrix was generated and performed to assess inherent safety performance index. As shown table 3, the expert ranking of MMA was same with based on the risk aversion (fuzzy-hard risk).

Table X. Results of ISPI with fuzzy rule on the six synthesis route of MMA

| | Fuzzy -easy risk | Fuzzy-standard risk | Fuzzy -hard risk | Ranking | | | | | | |
|---|---|---|---|---|---|---|---|---|---|---|
| | | | | ISPI (easy) | ISPI (standard) | ISPI (hard) | PIIS | ISI | i-SAFE | Expert |
| ACH | 11.692 | 16.966 | 32.345 | 6 | 6 | 6 | 6 | 6 | 6 | 6 |
| C2_MP | 7.26 | 9.89 | 20.653 | 3 | 3 | 3 | 3 | 3 | 2 | 3 |
| C2_PA | 9.350 | 13.01 | 24.89 | 4 | 4 | 4 | 5 | 4 | 5 | 4 |
| Propylene | 11.136 | 16.78 | 39.95 | 5 | 5 | 5 | 4 | 4 | 4 | 5 |
| iC4 | 5.067 | 8.74 | 14.92 | 1 | 2 | 1 | 2 | 1 | 3 | 2 |
| TBA | 5.874 | 6.767 | 16.541 | 2 | 1 | 2 | 1 | 1 | 1 | 1 |

## Conclusion

Sustainability study is mainly performed to assess sustainable development of process design in the chemical industries. In the sustainability study, economic, safety and social performance is mainly considered. Here, inherent safety index are need to evaluate safety dimension in the preliminary process design.

In this paper, inherent safety performance index (ISPI) was proposed, which is consists of chemical hazard index (flammability, explosiveness, toxicity and reactivity) and process hazard index by unit (temperature, pressure, inventory). Also, the complexity, process safety level index and operability of process are used as the weighting factors. In addition, the method is integrated with process simulator to get process operation date and chemical mixture properties such as flash point and boiling point. Also, fuzzy rule was adopted for reflection of expert's opinion and tendency of decision makers since existed and initial proposed index is hard to both reflect opinions and tendency and smooth boundary of category in the sub-indices.

The proposed methods are incorporated to computer aided tool using Matlab interface and process simulators. The evaluation was done by using the six process routes for manufacturing methyl methacrylate (MMA) as a case study. To validate our proposed method, the experts ranking and several existed index are compared. Also, the three level of rule was generated and performed to evaluate inherent safety design.

The proposed methods can be guideline to decision makers of chemical process design when the safety and economic have conflicts because of government aspects and for practical engineering reasons. It can make easy to affect the process configuration and inherent safety in the preliminary process design than in the later phase of process design.

## Acknowledgement

This paper was supported by Engineering Research Institute and ASRI (Automation and Systems Research Institute) in Seoul National University.

## References


1. Kim, Hosoo, I. K. Kim and E. S. Yoon (2010), "Multiobjective Design of Calorific Value Adjustment Process using Process Simulators", *Ind. Eng. Chem. Res.*, 49, pp. 2841–2848.
2. Azapagic, A., A. Millington, and A. Collett (2006), "A Methodology for Integrating Sustainability Considerations into Process Design", *Chemical Engineering Research and Design*, 84, pp. 439–452.
3. Cobb, C., D. Schuster, B. Beloff, and D. Tanzil, (2009), "The AIChE Sustainablility index: The Factors in Detail", *Chemical Engineering Progress*, January 2009, pp. 60-63.
4. Tanzil, D., G. Ma, and B. Beloff (2004), "Automating the Sustainability Metrics Approach", *2004 AIChE Spring Meeting*, New Orleans.
5. CCPS (2009), *Inherent Safer Chemical Process: A life cycle approach, 2nd edition*, New York: American Institute of Chemical Engineers, 2009.
6. Khan, Faisal I. and Paul R. Amyotte (2003), "How to Make Inherent Safety Practice a Reality", *The Canadian Journal of Chemical Engineering*, 81, pp. 2-16.
7. Gupta, J. P. and D. W. EDWARDS (2002), "Inherent safer desing-Present and future", *Trans IChemE*, 80, pp. 115–125.
8. Palaniappan, C., R. Srinivasan and R. Tan (2002), "Expert System for the Design of Inherently Safer Processes. 1. Route Selection Stage", *Ind. Eng. Chem. Res.*, 41, 6698-6710.
9. Palaniappan, C., R. Srinivasan and R. Tan (2002), "Expert System for the Design of Inherently Safer Processes. 2. Flowsheet Development Stage", *Ind. Eng. Chem. Res.*, 41, 6711-6722.
10. Khan, Faisal I. Khan and Paul R. Amyotte (2004), "Integrated Inherent Safety Index (I2SI): A Tool for Inherent Saferty Evaluation", *Process Safety Process*, 23 (2), pp. 136 – 148.
11. Lawrence, D. (1996), "Quantification Inherent Safety of Chemical Process Routes", a doctoral thesis, Loughborough University.
12. Heikkila , A. M. (1999), "Inherent Safety in Process Plant Design: an Index-based Approach", a doctoral thesis, Technical research center of finland.
13. Markowski, Adam S., M. Sam Mannan (2008), "Fuzzy Risk Matrix", *Journal of Hazardous Materials*, 159, 152–157.
14. Nagai, K. (2001), "New Developments in the Production of Methyl Methacrylate", *Applied Catalysis*, 221, 367-377.
15. Bizzari, Sebastian N. (2009), "CHE Marketing Research Report: Methyl Methacrylate", *SRI Consulting*.

## Simulation Study of Different Alternatives for Isoamyl Butyrate Production by Using Aspen Plus®

Jairo Duran, Karen Cañón, Iván Gil, Julio Vargas, Gerardo Rodríguez

*Grupo de Procesos Químicos y Bioquímicos. Department of Chemical and Environmental Engineering. Universidad Nacional de Colombia. Bogotá. Colombia*

**Keywords:** Acidic catalysts, coupled reactor/column, esterification, fixed costs, Isoamyl butyrate, liquid-vapor equilibrium, operational costs.

The Isoamyl butyrate is a value-added product widely used in fragrances and flavors; therefore its production can be interesting for this kind of industries and can be the basis for business approach. Otherwise, the production process may be included in the green chemical engineering field because the use of raw materials from renewable resources, for instance, isoamyl alcohol can be obtained from fusel oil, a well known by-product from ethanol production. This work presents the simulation and the analysis of the conventional scheme for ester production and the coupled reactor/column configurations. The total annual cost (TAC) has been used as a comparison and evaluation criterion to explore the alternatives of production and determine which one presents the best performance at a lower cost.

Due to the lack of experimental data on vapor-liquid equilibrium (VLE) of the isoamyl butyrate – isoamyl alcohol, it was necessary to obtain experimentally those data. Vapor-liquid equilibrium data were measured at 101.3 kPa. A dynamic equilibrium apparatus was used for the measurements (Fischer Labodest VLE-602). Equilibrium data were fitted with UNIFAC and NRTL models, obtaining a maximum absolute error of 2.5%. Herrington area test was performed using some criteria reported at literature. Also, Van Ness test was used as modelling capability test for NRTL and UNIFAC models [1]. Good thermodynamic consistency was found with both tests. VLE phase diagram is shown in figure 1.

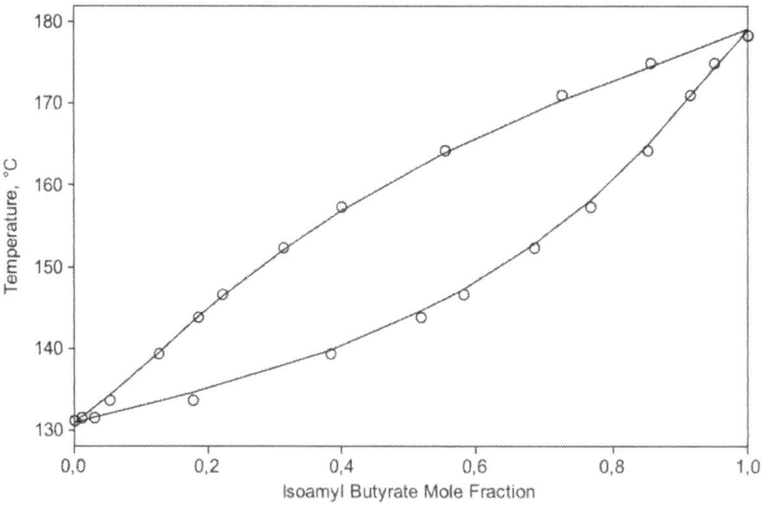


**Figure 1.** VLE diagram of the system isoamyl butyrate + isoamyl alcohol at 101.3 kPa. Continuous line represents NRTL model regressed.

After obtaining reliable thermodynamic data, it was possible to begin an analysis of each scheme of production. The process simulation has been performed by using Aspen Plus®. UNIFAC and Peng-Robinson with Wong Sandler mixing rules were set as property methods. It was preferred to use Wong Sandler mixing rules over others because the benefits of extrapolation on the model to other pressures as reported in the literature [2]. Operations where the isoamyl butyrate-isoamyl alcohol VLE is relevant for the design, such as distillation columns, were simulated by using parameters obtained from regressed model, however for the other operations, default UNIFAC parameters were used for modeling. Economic analysis includes the estimation of fixed and variable costs of the process by using equations and heuristics from literature. In this work, a homogeneous kinetics has been employed [3]. Those kinetic data were obtained by using p-toluenesulfonic acid as catalyst. Catalyst cost has been considered in the global economic analysis.

Conventional scheme consists on a continuous stirred reactor (CSTR) followed by a neutralization stage. Esterification temperature was fitted at 353.15 K, which is the maximum temperature studied by Páez *et al* [3]. Separation was simulated through two distillation columns where the isoamyl butyrate can be obtained at purity higher than 99% (mass basis). The first column is coupled with a decanter at the top, in order to obtain a water rich product and a reflux consisting mainly in isoamyl alcohol. Bottoms contains around 60% of isoamyl butyrate. Remainder consists in isoamyl alcohol and traces of butyric acid and water. Design specifications were tested in order to produce a high water removal and the maximum possible ester purity at the bottoms. The second column produces isoamyl butyrate at high purity (>99% mass basis) and a top stream containing mainly isoamyl alcohol available to recirculation to the esterification stage. In both columns, various sensibility analyses were done to fit the best operational parameters, such as number of stages, reflux ratio, feed stage and distillate to feed ratio. The conventional scheme is shown in figure 2.

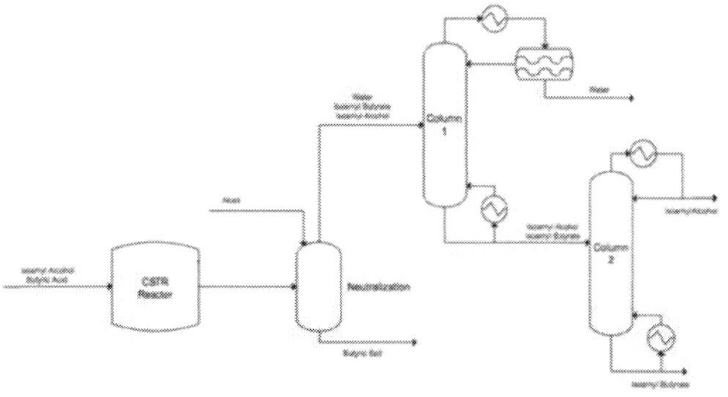


**Figure 2.** Conventional production scheme of isoamyl butyrate.

Further, a second separation alternative was studied, consisting on a single distillation stage. Figure 3 presents the scheme where a distillation column is tested with a side stream containing high purity isoamyl alcohol. Bottom product is composed by isoamyl butyrate (> 98% w/w). It was demonstrated that can be used a single column without greatly compromising the purity specification of product. Side product stream can be recycled to esterification stage. Conversion reached in the reactor is lowest than first studied scheme because the content of isoamyl butyrate present in the recycled stream. As it is expected, the single column requires considerably more stages to achieve the required separation degree.

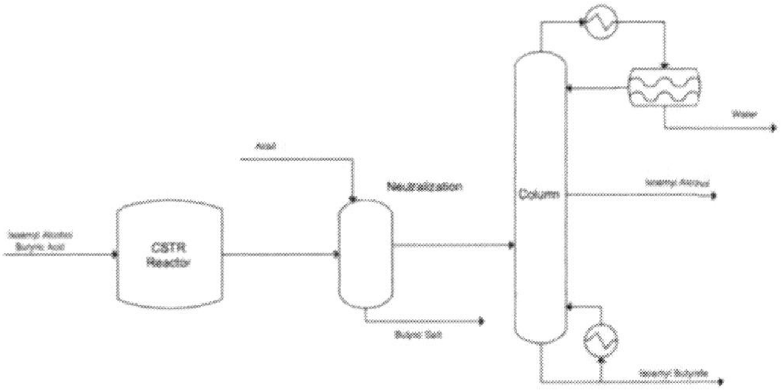


**Figure 3.** Conventional scheme with single distillation column.

Coupled reactor/column was also studied (figure 4). Stripping and rectifying sections are smaller than the columns in conventional scheme; thereupon it was obtained lower fixed costs in this stage of the process. At the top, it is obtained high purity water from the decanter, but at the bottom of stripping column, it was not possible to obtain the required specification for isoamyl butyrate. It is due the missing of neutralization stage in the scheme, and as it is expected, some impurity of isoamyl alcohol and butyric acid are drawn with isoamyl butyrate stream. Due to the previously exposed, it was necessary to use a side distillation column where the product specification is reached. This is the main difference with the work reported by Chiang *et al* [4], where the amyl acetate production is studied and the side column is not required because the high volatility of acetic acid in comparison with butyric acid volatility. In this study, the conversion achieved in the reactor is lower and the non-reacted acid tends to accumulate at the stripping column.

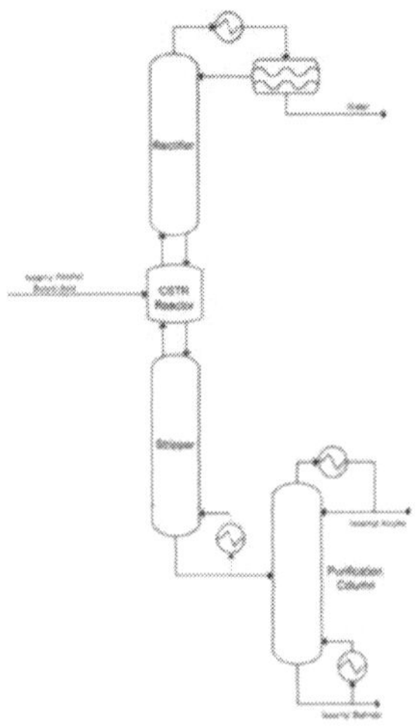

**Figure 4.** Coupled Reactor/Column scheme for isoamyl butyrate production.

All columns studied in this work were sized through ASPEN PLUS® 7.2, tray sizing utility. Sieve trays were preferred in the main calculation because their simplicity and wide use. Over-design factor was fitted in 50% and Fair method was used in the flooding approach calculations. Other parameters were kept as suggested by the simulator providers [5]. Economic evaluation was done with the approximations presented by Douglas [6] and Luyben [7,8]. Number of trays and column diameter were the cost evaluation parameters for distillation columns, whereas vessel diameter was used for reactor and decanters. Condenser and reboilers costs were calculated by using the heat exchange area. Steam and cooling water costs were determined with the relationships presented by Elliott and Luyben [9], as indicated in equations 1 and 2.

$$steam\ cost\ \left(\frac{\$}{year}\right) = \$\ 0.0306\ Q_R \qquad (1)$$

$$cooling\ water\ cost\ \left(\frac{\$}{year}\right) = \$\ 0.000977\ Q_C \qquad (2)$$

The annual equipment cost was assumed to be one-third of the capital cost (3-year payback). For the catalyst, it is supposed the impossibility of regeneration; therefore the cost can be calculated by mean the relationship exposed in equation 3.

$$catalyst\ cost\ (\$) = catalyst\ flow\ \left(\frac{kg}{hr}\right) \times 1.80\ \left(\frac{\$}{kg}\right) \times 8640\left(\frac{hr}{yr}\right) \qquad (3)$$

Finally, reactor/column resulted with the lower TAC using a base production of 200000 kg per year which is the average consumption of isoamyl butyrate in Colombia [10]. It is expected that the results presented in this work will be the comparison criteria of a base study on reactive distillation process for this system.

**Some Cited References**

**1.** Kang J.W., Diky W., Chirico R., Magee J.W., Muzny C.D., Abdulagatov I., Kazakov A, Frenkel M. *Quality Assessment Algorithm for Vapor-Liquid Equilibrium Data. J. Chem. Eng. Data*, 2010, *55* (9), pp 3631–3640.

**2.** Stanley I., Orbey H. *Mixing and Combining Rules.* Available on *Equations of State for Fluids and Fluid Mixtures* by Sengers J.V., Kayser R.F., Peters C.J., White Jr. H.J. (Editors). International Union of Pure and Applied Chemistry. 2000. pp 321-357.

**3.** Páez N., Duran J.A., Orjuela A., Gil I., Vargas J.C., Rodríguez G. *Homogeneously Catalyzed Esterification of n-Butyric Acid with Isoamyl Alcohol: A Kinetic Study*. Submitted paper to AIChE Annual Meeting 2011.

**4.** Chiang S.F., Kuo, C.L., Yu C.C., Wong D.S.H., *Design alternatives for the amyl acetate process: coupled reactor/column and reactive distillation. Ind. Eng. Chem. Res.* 2002, 41, pp. 3233-3246.

**5.** Aspen Technology, Inc. Aspen Plus 7.2. *User's Guide. Aspen Getting Started Building and Running a Process Model*, Burlington, 2010.

**6.** Douglas, J.M., *Conceptual process design*. McGraw-Hill, New York, 1998.

**7.** Luyben W., Yu C. *Reactive Distillation Design and Control.* AIChE and John Wiley & Sons, Inc. United States of America. 2008.

**8.** Luyben W., Distillation Design and Control Using ASPEN™ Simulation. AIChE and John Wiley & Sons, Inc. United States of America. 2006.

**9.** Elliott, T.R., Luyben W.L. *Quantitative assessment of controllability during the design of a ternary system with two recycle streams. Ind. Eng. Chem. Res.* 1996, 35, 3470.

**10.** BACEX, Foreign Trade Databank. Ministerio de Comercio, Industria y Turismo. Republic of Colombia. Consulted on December 16, 2010.

# (329f) Simulation and Economics of Biodiesel Production From Microbial Lipids

**Ramalingam Subramaniam**, Stephen Dufreche, Mark Zappi and Rakesh Bajpai, Chemical Engineering, University of Louisiana at Lafayette, Lafayette, LA

In this work, production of lipid from *Lypomyces starkeyi* on soluble sweet potato starch was simulated in Aspen Plus simulation. A kinetic model based on experimental results of medium optimization and high density cultivation was used for cell growth and lipid production. The simulation encompasses the entire production process, from raw feedstock to final product. Sensitivity analysis, heat integration and optimization were done to analyze the various cost factors on the production process. An Economic analysis will also be presented.

See more of this Session: Process and Product Development for Sustainability I
See more of this Group/Topical: Process Development Division

# (389a) A Suitable Disengagement: Adventures In a Multi-Phase Reaction System

**Cheryl Teich**, Core R&D/ Process Chemistry & Development, The Dow Chemical Company, Spring House, PA

ASuitable Disengagement: Adventures in a Multi-phase Reaction System

CherylTeich
The Dow Chemical Company, Spring House, PA
Core R&D/Process Chemistry and Development

In the early business development of a niche specialtyproduct, a small-scale CSTR provided adequate volumes. The product is generated as a gas that forms througha complex reaction pathway and must disengage from a mixture containing twodistinct solids and multi-component liquid and gas phases. Extended productholdup can lead to yield loss via byproduct and mer formation, and even to apotentially hazardous condition. Safe and successful operation of the processrelies on balancing the reaction pathway with adequate reactor mixing and gasdisengagement. This had been achieved with the existing process throughextensive operating experience.

The potential for step growth in the business led to anexamination of reactor options to supply the larger volumes. A concern wasthat a larger CSTR might not provide adequate gas disengagement and that theprocess would suffer as a result. The first step in generating the nextgeneration reactor design was to capture process understanding in the processscheme shown below.

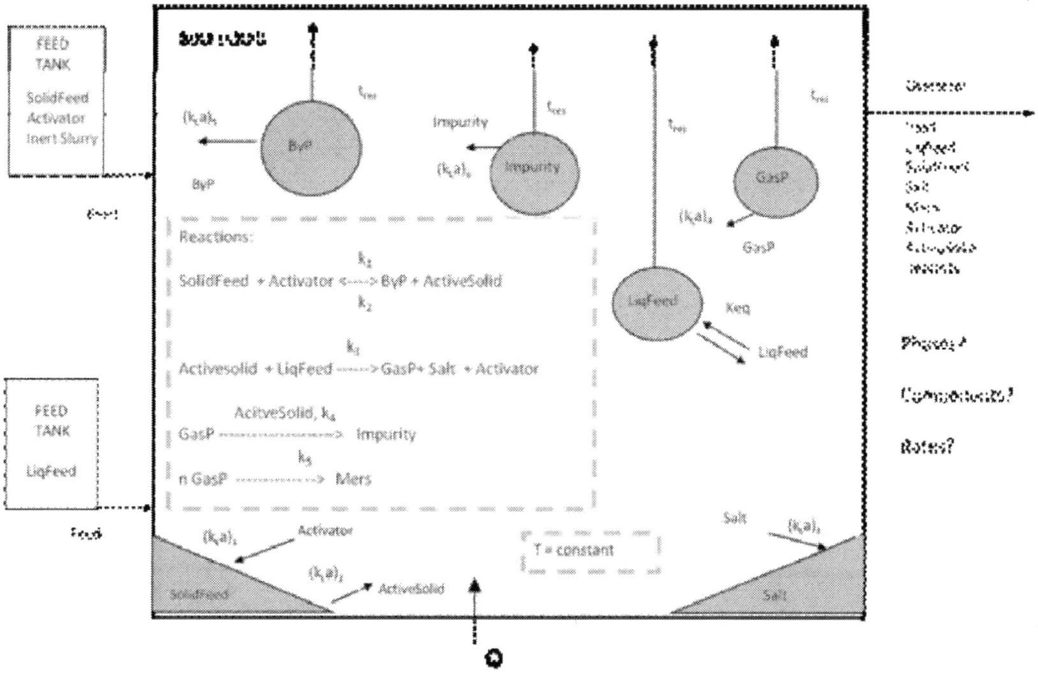


Development of the process scheme led to an examination of thescale-independent chemistry and its interaction with equipment-specificagitation configurations. The development program used various laboratoryreactors and modeling tools such as DynoChem and Aspen Plus to deliver volume-specificrecommendations on reactor type and designs to meet the growing business needs.

See more of this Session: Mixing In the Process Industries - How to Develop a Successful Process
See more of this Group/Topical: North American Mixing Forum

SABIC
Innovative
Plastics™

سابك
sabic

Mixing in the Process Industries | 2011 AIChE Annual Meeting

# Mixing Scale-Up for An Air Oxidation Process



Sharing our futures

## Introduction

- Scale-up of an air oxidation reaction to commercial scale was required for a plant expansion by Sabic Innovative Plastics.

- Heat and mass transfer considerations were critical to the success of the scale-up.

- The agitator design was a significant key to success:
  - Sufficient heat transfer to maintain the pilot plant cycle time
  - Mass transfer at least as good as seen at the pilot scale

- Agitation tests were performed in the vendor's facility to trial the proposed design and to optimize the performance.

## Air Oxidation Process - Reaction

organic substrate $+ O_2 + N_2 \xrightarrow[\text{diluent}]{\text{catalyst}}$ product $+ N_2$

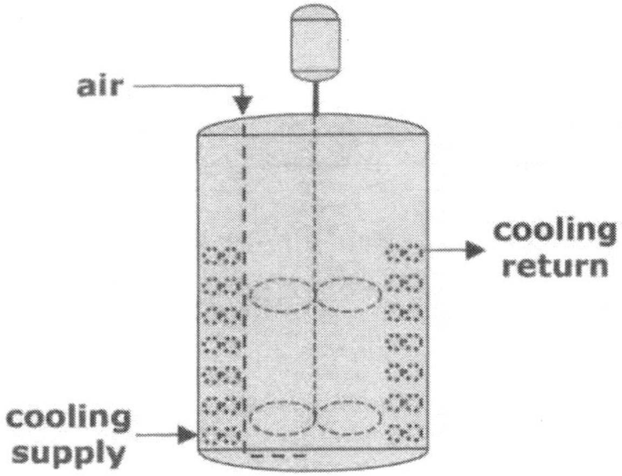


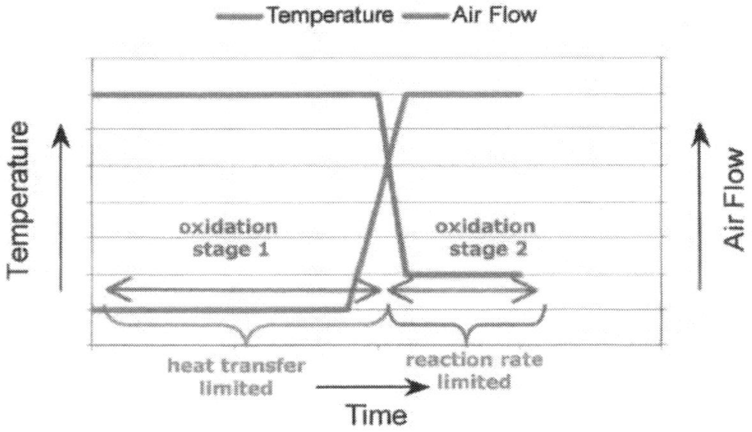


20X Scale-Up Required From Pilot Scale

## Scale-Up Considerations

Critical for Scale-Up

- Heat transfer (cooling)
  - rate limiting during oxidation reaction
  - jacket surface to Rx volume ratio significantly lower @ commercial scale vs. pilot scale
- Mass transfer
  - not rate limiting at the pilot scale
  - need at least equivalent mass transfer capability at the commercial scale
- Optimization of the proposed design
  - baffling required with the presence of cooling coils
  - location of impellers vs. point of air introduction
  - good distribution of air bubbles

## Agitation Hydrodynamic Study

A hydrodynamic study was commissioned in the agitator vendor's facility for:

- optimization of baffling, impeller placement, and air distribution
- mass transfer measurements
- heat transfer measurements
- mixing time analysis
- comparison with previous piloting configuration

- 1 meter ID transparent vessel
- Medium = water @ ambient T, atm. P
- Geometrically scaled from the proposed commercial design
- Internal coils with heating/cooling capability

SABIC Innovative Plastics™

# Design Optimization

**Baffle optimization:**

- Baffle configuration modified from initially proposed design.
- Vortex eliminated and full theoretical power input demonstrated

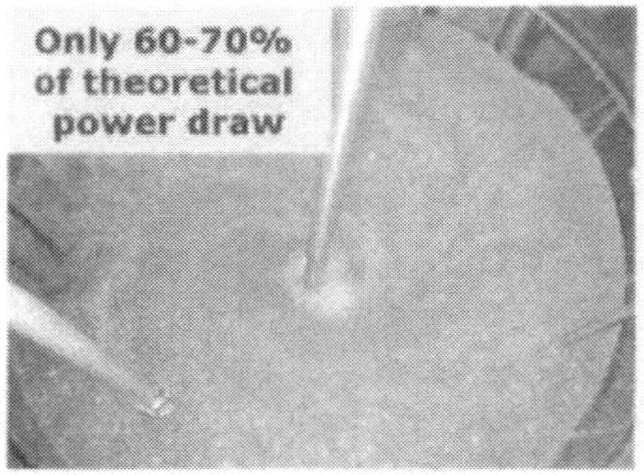


**Before baffle changes**

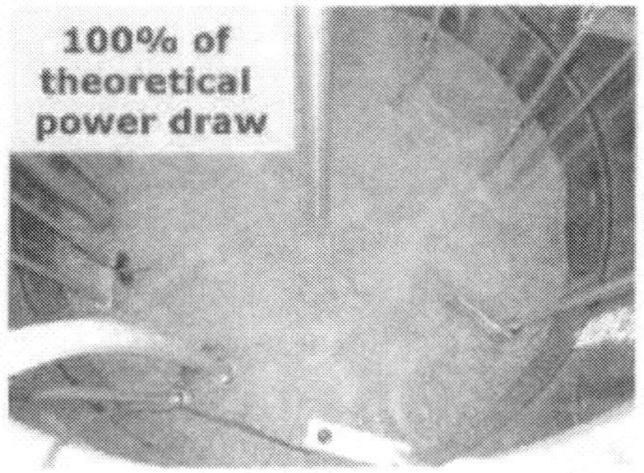


**After baffle changes**

**Lower impeller elevation:**

- Originally positioned to maximize heat transfer in the adjacent coils
- Lowered to improve overall performance
- Vertical fluid velocity between the coils increased from 0.18 to 0.37 m/sec.
- Gas hold-up increased from 5.6% to 6.4%

**"Spouting" Point:**

- Minimum rpm required to achieve good bubble distribution at a given gas flow
- Defined by observation as lack of large coalesced bubbles reaching the surface
- ~80 rpm @ 15 $Nm^3$/hr air flow
- Found to be ~70 rpm @ design air flow during water runs in the commercial reactor

SABIC
Innovative
Plastics™

sabic

## Mixing Time

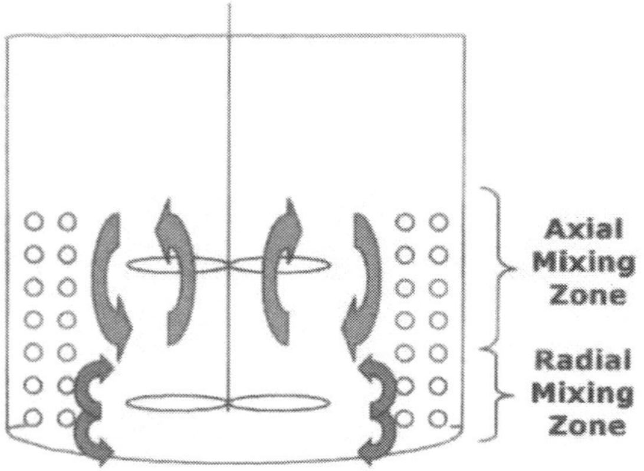


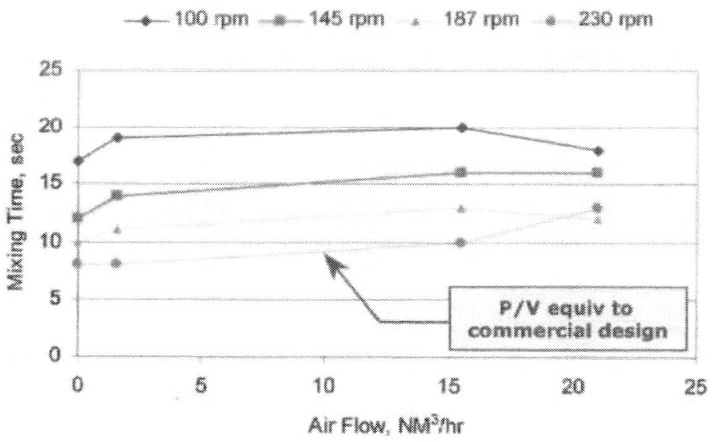


- Radial flow zone
  - intense gas-liquid contact for bubble formation
  - some finite residence time prior to the axial zone
  - minimizes headspace O2 – minimized gas by-passing
- Axial flow zone
  - enough energy input to maintain good bubble dispersion

**~ 7 seconds of bottom zone segregation at design conditions**

## Heat Transfer

Heat transfer scale-up critical - rate limiting during 1st stage oxidation

**Jacket area/volume of commercial reactor ~40% of the pilot scale**

$$Nu = C \cdot Re^{2/3} \cdot Pr^{1/3} \cdot (\eta/\eta_W)^{0.14}$$

**Generally for turbine type impellers:**
**C ~ 0.62 - for transfer to the jacket**
**up to 1.5 - for a single helical coil in the available literature**

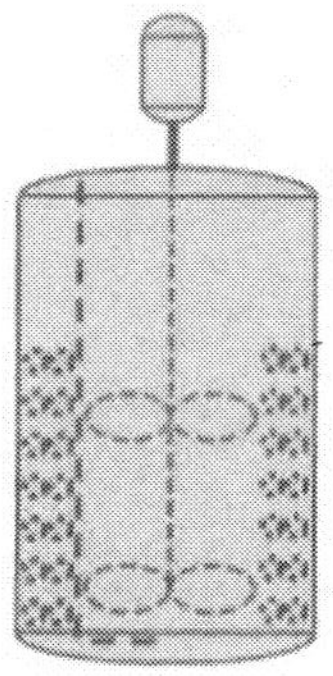

**Issues/Questions:**

- **Process side film coefficient = f(impeller type, impeller/coil geometry)**
- **Presence of coils reduce the film coefficient of the jacket**
- **Influence of multiple coil banks?**
- **Influence of air flow?**

Measurements with the proposed geometry were needed to reduce the risk of heat transfer scale-up

SABIC Innovative Plastics™

sabic

## Heat Transfer

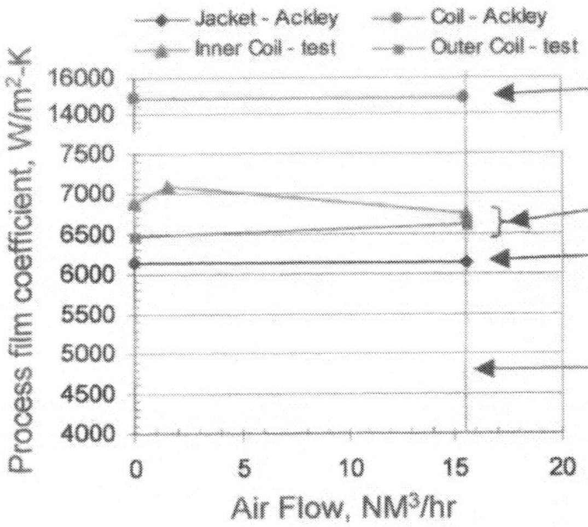


**Correlation for single coil (Ackley)**

**Test data for the inner and outer coils**

**Standard correlation for a jacket**

**15.5 $NM^3$/hr air flow – scaled volumetric equivalent to commercial reactor**

- Tests conducted in water - 25ºC to 35ºC
- C ~ 0.68 for the dual coil bank @ design air flow
- No significant effect of air flow seen
- Jacket area was ignored (used for design safety factor)
- Test data confirmed the adequacy of the dual coil design

Post Start-Up Plant Performance: +25% margin

SABIC Innovative Plastics™

## Mass Transfer

Reaction piloted with a single stage flat blade turbine (Rushton impeller)

- Need at least equivalent mass transfer ($k_La$) for the commercial reactors

$$k_La = A \cdot \varepsilon^B \cdot v_{sg}^C$$

where:
$k_La$ is liquid mass transfer coefficient, $hr^{-1}$
$\varepsilon$ is energy dissipation density, W/kg
$V_{sg}$ is superficial gas velocity, m/hr

- Agitation tests in water comparing $k_La$ for Rushton and proposed commercial impeller design

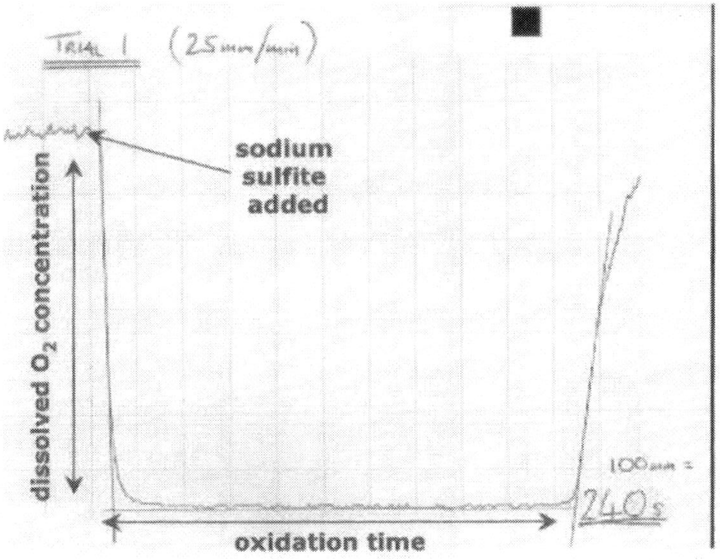


$$k_La = \frac{0.5 \cdot m \cdot MW_{O_2}}{t \cdot V \cdot c^* \cdot MW_{Na_2SO_3}}$$

where:
m mass of sodium nitrite added, g
MW molecular weight, g/gmole
T oxidation time, hours
V liquid volume, $m^3$
c* saturated O2 concentration, $g/m^3$

SABIC Innovative Plastics™

## Mass Transfer

Pilot scale reaction development:

- Single Rushton impeller
- P/V = 0.56 kW/m³

GOOD PERFORMANCE –
no sign of mass transfer limitation

⇩

Agitation trials @ 1m³ scale:

- Single Rushton vs. dual (proprietary) impellers at the same d/D
- In water at ambient T, P=atmospheric
- Scaled volumetric air flow

Higher $k_La$ with proposed design

⇩

**Proposed Design Shows Improved Mass Transfer vs. Pilot Reactor**

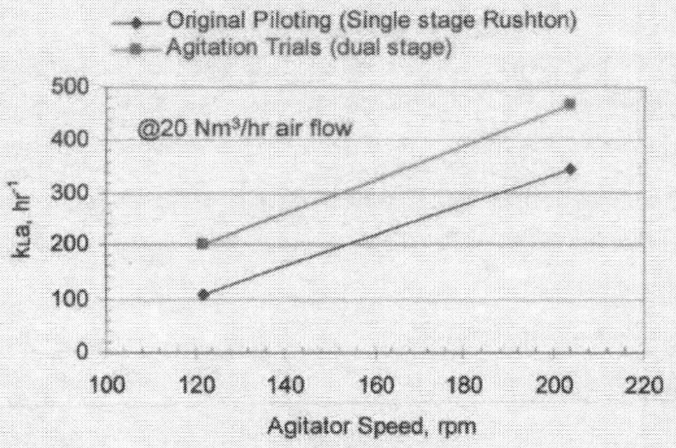


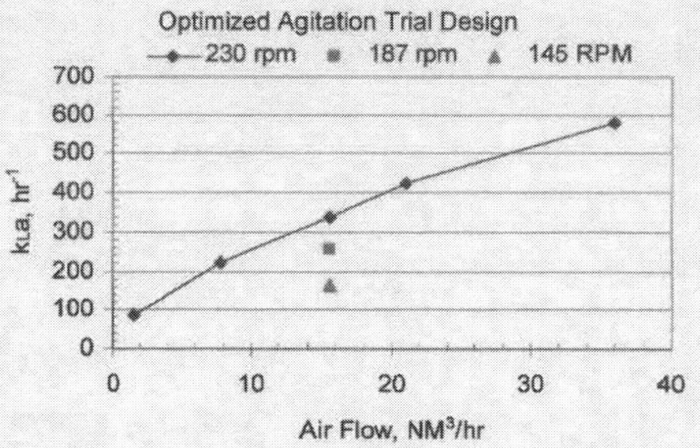


SABIC Innovative Plastics™

# Power Draw

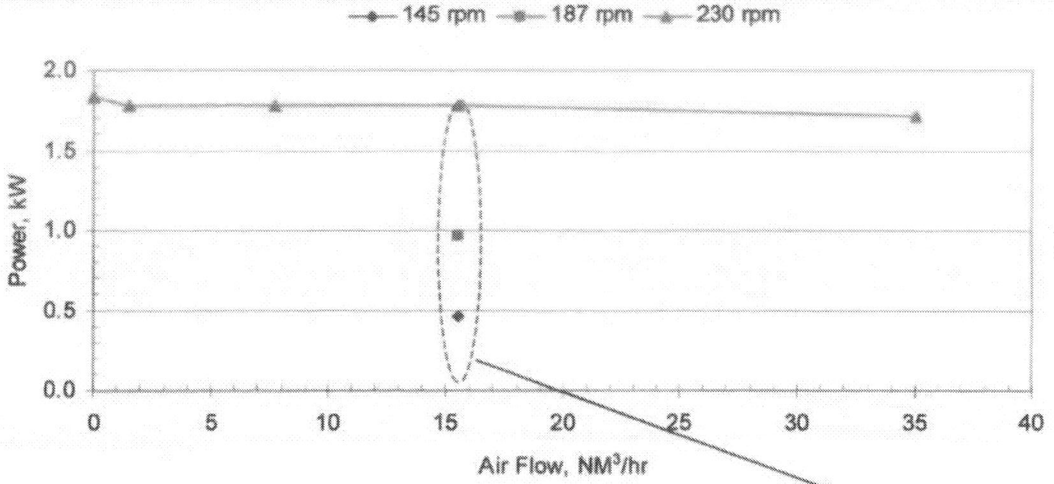


- *Power input ~ constant with changing air flow*

  Power/volume related to:
  - bubble size (original and coalesced)
  - bubble dispersion
  - gas hold-up

- Verified power proportional to $(rpm)^{2.9}$
  ~ same as standard $(rpm)^{3.0}$

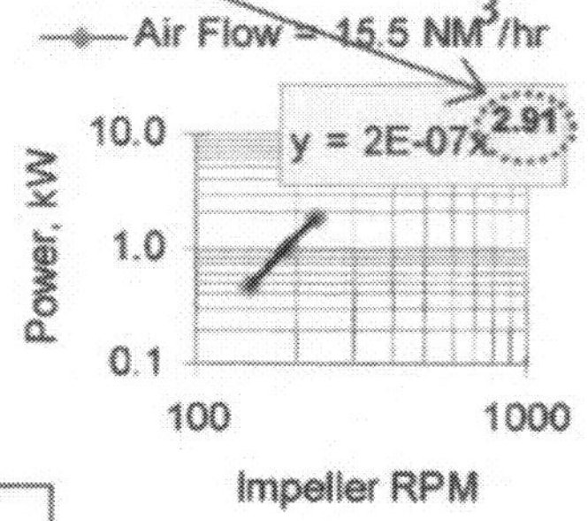


**Improvement vs. Conventional Impellers - Constant P/V with Changing Gas Flow**

SABIC Innovative Plastics™

سابك sabic

# Summary

<u>Goal Achieved – Successful Scale-Up</u>

- Design basis cycle time demonstrated
- 25% heat transfer margin realized – room for cycle time reduction
- Mass transfer is not limiting
- Low headspace concentration indicates efficient use of oxygen
- Product quality duplicates the pilot plant results

# (389c) Industrial Reactor Evaluation with Uncertainties In Mixing Performance and Reaction Kinetics

**Michael D. Cloeter**[1], Pramod Patil[1] and Aaron Sarafinas[2], (1)The Dow Chemical Company, Freeport, TX, (2)The Dow Chemical Company, Spring House, PA

Frequently the Fluid Mechanics & Mixing Discipline of The DowChemical Company (Dow) is asked to assess the mixing-sensitivity of industrial chemistries. The chemistries may be undergoing scale-up or are in some way experiencing processdifficulties. Sometimes the aim is to minimize impurities of any sort. Inother, perhaps more challenging, cases the aim is to maintain a constantdistribution of by-products to ensure that the product performance isunchanged. Most desirably, we have access to a detailed intrinsic kinetic ratemodel and detailed knowledge of the fluid dynamics, so that the pertinent timescales can be calculated and a Damkohler analysis completed for mixingsensitivity. Reality is frequently quite different.

Generally, we have the following scenarios with respect to kinetics:

- A kinetic model exists or can be derived from the correct type ofexisting lab or process data.

- A kinetic model does not exist, but a reactor mole balance & productdistribution is available or can be derived from process data. Reactorsampling may be impossible, meaning the engineer must resort to approximationsbased on plant feed and product flows, compositions, tank level changes, etc. Calorimetricdata may exist. If resources allow, a kinetic model can be derived from molecularmodeling and thermochemical kinetics, then validated with product data.

- No intrinsic kinetic data exist. Previous process data was taken with atransport limitation.

And generallywe have these scenarios for flow field:

- Vessel geometry is known precisely and the flow field iswell-characterized.

- Vessel geometry is known but flow characteristics are not known. If thereaction is localized, process instrumentation such as thermowells helpunderstand some spatial reaction characteristics.

- Studies in lab glassware, such as a round-bottomed flask, have givensatisfactory performance. Blending is inherently rapid. No significant changesto the flow field have been studied.

Furthermore,there can of course be large interactions between the turbulence and reactions.

The nextscale often must fit in to an already-existing reactor somewhere. In thiscase, the stakeholders need assurance that the existing reactor can work, and whetherany changes must be made to the agitation, feed introduction, etc. to ensuresuccess. Sometimes the reactor is yet to be built, so that many degrees offreedom for design are available.

This talkwill focus on real examples of recent industrial problems encountered and themethodologies used to address them.

See more of this Session: Mixing In the Process Industries - How to Develop a Successful Process
See more of this Group/Topical: North American Mixing Forum

# (389d) Designing In Reliability

**Todd M. Hutchinson**, Research and Development, Philadelphia Mixing Solutions, Ltd, Palmyra, PA

Most rotating equipment is selected as part of a Capitol Project govern by rules from Procurement with specifications written by in house engineers or outsources engineering partners. Because of Project spending limits the equipment supplier who is believed to meet the specification requirements with the lowest bid receives the contract. There are times when the equipment supplier who meets the project costs targets are awarded the orders regardless of the specification requirements. And then there are those times when the specifications do not adequately describe a process resulting in improperly designed equipment.

Regardless of this traditional process used unfortunately those responsible for Reliability inherit equipment and are tasked to operate the equipment at peak performance for continually longer periods of time. Of course this is nothing new. The question is how does someone in charge of predictable operating a plant impact the Project Process and selection of equipment. Specific applications will be studied showing failed agitation equipments and the efforts over the years to improve reliability. Unfortunately these efforts were focused on symptoms and not the True Root Cause. In some cases the plants were designed based on the perfect word and did not account for upset conditions. In other cases the basic data for the design was flawed. Either way those responsible for reliability face a seemingly never ending battle. The studied applications include staggering reduction in operating costs that have led to companies changing their Project Process.

See more of this Session: Mixing In the Process Industries - How to Develop a Successful Process
See more of this Group/Topical: North American Mixing Forum

# (389e) The Evolution of Process Understanding Through Micromixing and Mass Transfer Studies In the Successful Development of a Specialty Chemical Process

**Aaron Sarafinas**, Core R&D, Engineering Sciences, Fluid Mechanics & Mixing, The Dow Chemical Company, Spring House, PA

Researchers who develop new specialty chemical products are often faced with a variety of potentially conflicting rate processes which can influence reaction selectivity and final product properties. As the researchers' understanding of the process develops through laboratory studies and pilot plant trials, the original process paradigms must be critically examined in order to assure successful commercial implementation.

This paper examines the evolution of process understanding in the development of a specialty chemical product. The critical reaction step could produce the desired product (R) and two by-products (Q and S) by the reactions:

$$A + B \rightarrow Q$$

$$A + B \rightarrow R$$

$$R + B \rightarrow S$$

Researchers needed to minimize potential losses of the very expensive reagent B and the moderately expensive reagent A in the pilot plant. Recognizing the series-parallel nature of the R and S reactions, researchers used the Bourne reactions in the laboratory and found a strong correlation between the selectivity of the Bourne reactions and the production of R and S in the critical reaction step over a wide range of laboratory process conditions. The Bourne reactions were used in the pilot plant to establish operating conditions for the critical reaction step. The desired selectivity of the critical reaction step was achieved in the initial scale-up with high overall yield on the expensive reagents.

Though the initial results were rather satisfying, certain observations on the initial and subsequent scale-up campaigns indicated that the process understanding was not complete. This paper will further examine those observations and look in more detail at the phases, components, and rate processes controlling the process. Consideration of both the mass transfer effects and micromixing led to more complete process understanding and a highly successful commercial start-up.

See more of this Session: Mixing In the Process Industries - How to Develop a Successful Process
See more of this Group/Topical: North American Mixing Forum

# (389f) Further Studies of Micromixing: Scale-up, Baffling and Feed Pipe Backmixing

Melissa Assirelli, Huntsman Polyurethanes, Everberg, Belgium and **Alvin Nienow**, Birmingham University, Birmingham, UK

The work reported was undertaken at the Institute of Chemical and Engineering Sciences, Singapore. It consisted of an experimental study of micromixing in a semi-batch reactor agitated by a Rushton turbine using two different reaction schemes. With one, the iodide-iodate scheme, at the same mean specific energy dissipation rate and feed location, the selectivity was similar at the 3.6 L scale to that found earlier at the 25 L scale (Assirelli *et al.*, 2002). Also, at the 3.6 L scale, a study of back mixing at different feed locations showed that it was greatest when feeding into the trailing vortex, which is also the best feed location to improve micromixing. At that location, all pipes gave a similar selectivity at the same exit velocity. However, on closer inspection, it was found that even at the same velocity, in general, the bigger the pipe, the greater the level of backmixing. This finding raises certain questions about micromixing based on local specific energy dissipation rates. However, it is in agreement with some previous work (Bourne *et al.*, 1981). Comparison with the limited literature also showed that one of the two criteria for predicting backmixing fitted the present results quite well (Vicum *et al.*, 2004) but the other overpredicted it by two orders of magnitude (Jo *et al.*, 1994). The other reaction scheme involved two parallel reactions, consisting of an acid-base neutralization and the alkaline hydrolysis of ethyl monochloroacetate (Bourne and Yu, 1994). It was used to compare micromixing in a 3L vessel under baffled conditions with that without them. Such a comparison has only been made once before at a larger 25 L scale and with the iodide-iodate scheme (Assirelli *et al.*, 2008). The earlier surprising result was confirmed, namely that the mean specific energy dissipation rate required for equivalent selectivity without baffles was less than that required with them.

## REFERENCES

Assirelli, M., W. Bujalski, A. Eaglesham and A. W. Nienow, "Study of Micromixing in a Stirred Tank using a Rushton Turbine: Comparison of Feed Positions and Other Mixing Devices", *Chem. Eng. Res. Des.*, **80**, 855-863 (2002).

Assirelli, M., W. Bujalski, A. Eaglesham and A. W. Nienow, "Macro- and Micro-Mixing Studies in an Unbaffled Vessel Agitated by a Rushton Turbine", *Chem. Eng. Sci.*, **63**, 35– 46 (2008).

Bourne, J. R., F. Kozicki, U. Moergelit and P. Rys, Mixing and Fast Chemical Reaction - Model-Experiment Comparisons, *Chem. Eng. Sci*, **36**, 1655-1663 (1981).

Bourne, J. R. and S. Yu, Investigation of Micromixing in Stirred Tank Reactors using Parallel Reactions, *Ind. Eng. Chem. Res.*, **33**, 41-55 (1994).

Jo, M. C., W. R. Penney and J. B. Fasano, Backmixing into Reactor Feed Pipes Caused by Turbulence in an Agitated Vessel. *AIChE Symp. Ser.*, **90**, 41-49 (1994)

Vicum, L., S. Ottiger, M. Mazzotti, L. Makowski and J. Baldyga, Multi-Scale Modelling of a Reactive Mixing Process in a Semi-Batch Stirred Tank, *Chem. Eng. Sci.*, **59**, 1767-1781 (2004).

See more of this Session: Mixing In the Process Industries - How to Develop a Successful Process
See more of this Group/Topical: North American Mixing Forum

# (403a) Removal of Rhodamine B Dye with Fe/Mn Co-Doped Titania Nanotubes

Manohar Palla[1], **Daniel H. Chen**[2], Rakesh Sadu[3], R. Asapu[3] and George W. Spencer[3], (1)Dan F. Smith Dept. of Chemical Engineering, Lamar University, Beaumont, TX, (2)Dan F. Smith Department of Chemical Engineering, Lamar University, Beaumont, TX, (3)Chemical Engineering, Lamar University, Beaumont, TX

Fe/Mn co-doped titania nanotubes has been synthesized by solution combustion followed by the hydrothermal treatment. Combustion synthesis has emerged as a facile and economically viable technique for the preparation of advanced ceramics, catalysts and nanomaterials. Stoichiometric amounts of transition metal nitrates and fuel (Glycine) are mixed in a solution for the preparation of nanomaterials by using the single step solution combustion method at 350° C. The $Fe_xMn_yTi_{0.9}O_{(2-\delta)}$ (x,y ranges from 0.01-0.09, with x+y = 0.1) nanomaterials are made in the anatase phase by the solution combustion synthesis. The nanotubes are then fabricated using the hydrothermal method using 10 N NaOH at temperature of 130°C for 24 hours.

Characterization of the prepared co-doped titania nanotubes were performed using transmission and scanning electron microscopy (TEM and SEM), XRD analysis, and BET specific surface area analysis. The co-doped titania nanotubes thus prepared contain both monoclinic and anatase phases as evidenced by the XRD and EDX results. STEM analysis shows the formation of nanotubes. BET analyses show that the surface area for both the nanopowders and nanotubes increases with Mn doping and surface area increases when the nanostructure changes from a powder form to a tube form.

Degussa P-25, Titania nanotubes synthesized @ 180°C, co-doped Titania nanotubes ($Fe_{0.01}Mn_{0.09}Ti_{0.9}O_{(2-\delta)}$ , $Fe_{0.05}Mn_{0.05}Ti_{0.9}O_{(2-\delta)}$ & $Fe_{0.09}Mn_{0.01}Ti_{0.9}O_{(2-\delta)}$), co-doped titania nanopowders ($Fe_{0.01}Mn_{0.09}Ti_{0.9}O_{(2-\delta)}$ & $Fe_{0.09}Mn_{0.01}Ti_{0.9}O_{(2-\delta)}$), and Activated carbon were tested and compared for the adsorbtion of Rhodamine B. With adsorbent loading of 100 mg (0.06 wt%) in 150 ml beaker, the suspension was stirred using the magnetic stirrer. Adsorption of Rhodamine B was found to be the highest for the $Fe_{0.01}Mn_{0.09}Ti_{0.9}O_{(2-\delta)}$ nanotubes. The adsorption of Rhodamine B was found to increase with the adsorbent loading. Influence of pH was studied over a pH range of 2.5 – 10.5 and found that the removal of Rhodamine B is the highest at pH = 2.54. Adsorption of Rhodamine B was found to increase with temperature from 25 to 50° C.

See more of this Session: Process and Product Development for Sustainability II
See more of this Group/Topical: Process Development Division

# (403b) High Temperature Corrosion of Commercial Alloys At Various SO2/HCl Ratios

**Timothy Sharobem**, Department of Earth and Environmental Engineering, Henry Krumb School of Mines (HKSM), Columbia University, New York, NY

Waste-to-Energy (WTE) plants generate nearly 3.0 gigawatts of electricity in the United States annually via the combustion of about 28 million tons of municipal solid wastes (MSW). Several of the major operating problems facing these facilities (low efficiency, plant shutdowns and high capital costs) are a result of high temperature corrosion that occurs on boiler surfaces. With increasing steam temperature, unprotected metals that constitute the evaporator and superheater are subject to a series of reactions caused by the presence of hydrogen chloride (HCl) and sulfur dioxde ($SO_2$) in an oxidizing environment. The mechanisms thought to explain HCl and $SO_2$ corrosion can be complex, involving multiple gas phase and gas-metal surface reactions. Despite its corrosive nature, some studies have suggested that $SO_2$ can have a beneficial effect within certain concentrations. In the following presentation, the corrosion mechanisms of multiple commercial alloys are examined via lab scale corrosion furnace tests, simulating typical WTE conditions at various $SO_2$/HCl ratios.

See more of this Session: Process and Product Development for Sustainability II
See more of this Group/Topical: Process Development Division

# (403c) On Improvement of Solvent Quality In a Natural Gas Sweetening Plant

**Binay K. Dutta**, Chemical Engineering, The Petroleum Institute, Abu Dhabi, United Arab Emirates

**On Improvement of Solvent Quality in a Natural Gas Sweetening Plant**

**Binay K Dutta**

**Chemical Engineering Department**

**The Petroleum Institute, Abu Dhabi, UAE**

Sweetening of natural gas is mostly done by absorption in an aqueous amine, single or mixed. Mono-ethanolamine (MEA), diethanolamine (DEA), methyl-diethanolamine (MDEA), di-iso-propanolamine (DIPA) and 2-amino-2-methyl-1-propanol (AMP) are commonly used for this purpose depending on the composition of the raw gas and the process requirements. Solvent degradation and foaming are the most severe operating problems of the gas sweetening process. These two problems are, in fact, interrelated since the amine degradation products are the greatest contributors to the foaming problem. It is well known that foaming in an absorption column creates flooding as well as loss of separation efficiency. Corrosion by the dissolved acid gases at the high operating pressure in the absorption column and the moderately high temperature of the regeneration column adds substantially to the foaming problem. Another class of partially identified compounds, called heat-stable salts (HSS), are formed from the degradation products and the contaminants in the process water mixed with the amine.

A significant volume of literature exists on solvent degradation, identification of degradation products, foaming, formation heat-stable salts and corrosion in gas sweetening and solvent regeneration columns. The available literature may be classified into three categories – (i) identification of degradation products and mechanism of degradation; (ii) measurement of foaming tendency; and (iii) suppression of foaming. However, the majority of these literature pertain to the conditions of post-combustion carbon dioxide capture where the presence of oxygen in the flue gases creates an environment which is radically different from that of sour gas treatment where the environment is anoxic but characterized by the presence of a significant level of hydrogen sulfide. Chakma and Meisen (1988) reported one of the early studies on identification of the degradation products of MDEA using GC and GC-MS and listed a number of acids, alcohols and amines in the samples. Kadner and Rieder (1995) listed the anions (such as chloride, nitrate, sulfate, thiosulfate, formate, oxalate and acetate, phosphate and fluoride ) associated with HSS that appear mainly from the dilution water and keep on accumulating. Reza and Trejo (2006) conducted experimental studies on degradation of a few amines (DEA, MDEA and APM) up to 90 h at a maximum temperature of 200°C. Degradation was marginal in the absence of acid gases, particularly $H_2S$, confirming the need to run experiments at conditions nearing those in the actual sweetening plants in order to get realistic and representative results.

Foaming in amine solutions occur principally because of certain contaminants such as amine degradation products including heat-stable salts (HSS), corrosion inhibitors added to the solution, iron sulfide particles that get dislodged from the corroded pipe/vessel surface, carboxylic acids, alcohols and hydrocarbons (Pauley, 1991). Measurements have been reported on the foaming tendencies of a number of amines (single and mixtures) such as DEA and MDEA at pressure up to about 5 bar and temperature up to 85°C using $N_2$, $CH_4$ and $C_2H_6$ as the bubbling gas. The foam break time was also measured. Behaviors of DEA and MDEA were found to be similar in respect of foaming. Higher fatty acids ($C_5$ and above) positively contribute to foaming. Foaming behavior in the carbon dioxide absorption process has been reported in the literature. Surface tension of single and mixed amine solutions, that play a major role in foaming, as well as the foaming behavior were also measured by Aguila-Hernandez et al. (2007) in nitrogen atmosphere. Only a few reports are available on inhibition of degradation, particularly in the context of flue gas decarbonization in an oxygen environment (Bedell, 2010). However, degradation studies and remedial measures for foam suppression under actual industrial operating conditions such as $CO_2$ and $H_2S$ environment and modeling of the phenomena are practically more important.

The present work aims at addressing different aspects of the issues related to solvent degradation in a natural gas processing plant. The nature of the problems will be analyzed based on the experiences and data from a real life natural gas processing plant. Available plant data will be analyzed using the available literature information qualitatively as well as through modeling of the foaming and solvent degradation problems. The intensity of foaming will be related to the concentrations of heat stable corrosion and degradation products as well as dissolved hydrocarbons. The concentration of the hydrocarbon under actual plant operating conditions will be calculated by available thermodynamic models on solubility and its effect of surface tension of the solution will be analyzed in the context of foam generation.

**References**

Aguila-Hernandez, J., Trejo, A., and Garcia-Florec, B. E., Surface tension and foam behavior of aqueous solutions of blends of three alkanolamines as a function of temperature, *Colloids and Surfaces, Physico-chem. Engg Aspects, 308* (2007) 33-46.

Bedell, S.: Amine autoxidation in flye gas CO2 capture – mechanistic lessons learned from other gas treating processes, *Intern. J. Greenhouse Gas Control,* 2010, doi:10:1016/j.ijggt.2010.01.007.

Chakma, A., and Miesen, A: Identification of methyl-diethanol-amine degradation products by gas chromatography and gas chromatography-mass spectrometry, *J. Chromatography, 457* (1988) 287-297.

Kadnar, R., and Rieder, J.: Determination of anions in amine solutions for sour gas treatment, *J. Chromatography A, 706* (1995) 339-343.

Pauley, C. R.: Face the facts about amine foaming, *Chem Eng Progr.*, July 1991, 33-38.

Reza, J., and Trejo, A.: Degradation of aqueous solutions of alkanolamine blends at high temperature under pressure of $CO_2$ and $H_2S$, *Chem Eng Commun., 193* (2006) 129-138.

See more of this Session: Process and Product Development for Sustainability II
See more of this Group/Topical: Process Development Division

# (403d) Efficient Production of High Purity Phenolic Glycol Ethers

**Max Tirtowidjojo**, Hydrocrabons, Chemicals & Energy R&D, The Dow Chemical Company, Freeport, TX and Edward Daugs, Glycol Ether R&D, The Dow Chemical Company, Midland, MI

The manufacture of phenolic glycol ethers (also known as alkylene phenolic glycol ethers), e.g., propylene glycol phenyl ether (PPh) and/or ethylene glycol phenyl ether (EPh), is long known and practiced. These glycol ethers are typically produced in a batch process using excess ethylene oxide in presence of an acid or base catalyst. One common process for making phenolic glycol ethers, e.g., PPh and/or EPh, is a batch process in which propylene oxide (PO) and/or ethylene oxide (EO) is reacted with phenol in the presence of sodium hydroxide (NaOH) which serves as a catalyst. The oxides are added continuously into a mixture of phenol and NaOH catalyst until the amount of residual phenol in the reactor effluent is less than 100 parts per million (ppm). In order to achieve low oxide (less than (<) 15 ppm EO and <400 ppm PO) concentrations in the reactor effluent, a long residence time (e.g., greater than 10 hours) is necessary to complete the oxide conversion and this, in turn, imparts a low capacity (i.e., a poor production rate) to the process. Moreover, the long residence time and the oxide to phenol molar ratio of slightly (e.g., 5% excess oxide) greater than 1 employed to minimize unreacted phenol is such that a significant amount of higher homolog products, e.g., dipropylene glycol phenyl ether, and other impurities are produced. This, in turn, requires significant distillation effort to purify the EPh and PPh products, even if the reactor effluent is neutralized with an acid, e.g., phosphoric acid, to remove the NaOH catalyst in order to avoid further reaction and decomposition. In addition, filtration of the resulting salt, i.e., sodium phosphate, requires an intensive operation.

In those instances in which a significant amount of higher homolog is desired, e. g., diethylene or triethylene glycol phenyl ether, conventional practice is to recycle mono-product to react further with the oxides to produce the desired higher homolog products (in particular the di-product). However, this lowers the productivity of the process as a larger reactor volume is needed to accommodate the longer reaction time, or it requires a larger capital investment so as to allow recycling of mono-products. Moreover, while a simple phenol drying unit may be sufficient for mono-product production, usually a more intense operation, e.g., two or more phenol drying units coupled in series, is necessary to remove water from phenol recycle in order to obtain the desired glycol by-product purities.

In this work, an alternative process for the production of EPh and PPh as a continuous process will be presented that eliminates the need for catalyst neutralization, salt filtration, and long residence times for reaction completion while at the same time improving the selectivity, and hence quality, of the products

See more of this Session: Process and Product Development for Sustainability II
See more of this Group/Topical: Process Development Division

# EVALUATION OF DRY STRENGTH ADDITIVES FROM NATURAL RESOURCES OF INDIAN ORIGIN

**A.K. Ray, Ghosh U K, Sweta Gupta and Ravish Rajrana**

**Department of Paper Technology, Indian Institute of Technology Roorkee, Saharanpur Campus, Saharanpur**

ABSTRACT

There are many additives used during paper making, such as, in the beating or refining process and as strength additives. It is well known that additions of dry strength resin increase the mechanical strength properties of paper such as tensile and burst indices by increasing the interfiber bonding without affecting the tear strength. It is used to reduce the fuzz or lint on the paper surface, and to increase the internal ply-bond strength. The most common dry strength additives are: starch and its substitutes like oxidized starch, cationic starch and poly-acrylamide polymer. These are normally synthetic and some are semi synthetic in nature. Though additions of polyacrylamides can give superior properties but it is very expensive compared to starch based chemicals. In India gums like guar gum, locust bin gum and seed gum of Sesbania aculeate etc. are available in plenty which can be a good substitute of either cationic/oxidized starch or polyacrylamides. Earlier trials have been tried, but the data are not available in literature. An attempt has been made in the present investigation to explore the potential of guar gum as dry strength additives in paper industries.

Experiments were conducted with Guar Gum at various percentages (0.5 to 5.0%) on OD fibers. The freeness value (actually slowness), $^0$SR of the pulp were also varied between 35 to 45$^0$SR, and the results were evaluated. To compare its effects parallel runs were also tried with oxidized starch, maintaining the same doses of additive, $^0$SR value, mixing time and degree of agitation. For a given strength guar gum additive requires less magnitude of refining. It is expected that by decreasing refining, tearing strength, drainage aid requirements and permeability will decrease and consequently demand of refining energy will also decrease.

From the experimental data it is found that guar gum can be promising dry strength additives that give requisite dry strength in paper. A blank test without addition of dry strength additives has also been tried for comparison purposes. However all these cases that is blended with guar gum and with oxidized starch the internal sizing agent has been kept constant i.e. rosin and alum with a fixed ratio(1:4).

## INTRODUCTION

Papermakers are always attracted to a process that improves strength properties, reduces energy consumption and increases drainage. Additions of nonfibrous additives generally influence the appearance, resistance and strength properties of paper. An increase in the strength of paper theoretically be brought about by enhancing potential of fiber-fiber bonding (occurring at the fiber crossing points) by increasing hydrogen bonding between different fiber surfaces or increasing fibre-to-fibre contact area through beating/refining action. However, due to beating, the strength properties improve but optical properties deteriorate. It also increases the electrical energy consumption due to more refining action.

India has a large number of paper mills using waste paper or non-wood fiber. The nonwood fibers (except forest based e.g. bamboo) have comparatively small fiber size and papers made from them possess low strength. So the additions of dry strength additives for strength development with some impact on control of fine retention and water removal become more economically and ecologically important option.

## DRY STRENGTH ADDITIVIES

The dry strength additives are used to impart the strength after the dryness. These are mostly hydrophilic polymers. The magnitude of dry strength development depends on fiber strength and fiber-to-fiber bonds. This in fact, enhances the mechanical properties without affecting the optical properties of the paper. However, dry strength resin also effects some other properties such as permeability, ply bond and linting propensity which are important for heavy paper that are subjected to stresses, such as it might arise in a printing process where tacky ink are used. If ply bond strength is to be low the sheet will delaminate during printing. The influence of the dry strength additives on strength properties depends on the furnish pH, ionization of fiber carboxyl groups and many other parameters.

There are a large number of natural and synthetic polymers that are added to papermaking furnish(0.5-1%) in order to increase dry strength of paper at a given degree of refining or that require less refining action(consuming less refining energy) for a given target strength of paper. In some situations up to 2% of addition is required. These are detailed in various publications (1-12).

The most commonly used additives are: Starch and its derivatives, polyacrylamides and other synthetic resins, natural gums and their derivatives.

The dry strength resins are classified as under:

Naturally Occurring Additives

- Hemicelluloses
- Gums -(Mannogalactans)- Guar Gum, Locust Bean Gum , Tamarind seed gum , Sesbania Aculeata,
- Starches and is derivatives (Natural and Cationic)

Synthetic Additives

- Polyacrylamides (Non-ionic, Anionic, Cationic)
- Polyallylamine

- Poly acrylonitrile

Semi-Synthetic Additives

- Chitosan( up to 1.0% on O.D.pulp) , especially in acidic or neutral conditions

Other substances such as, unmodified and anionic starches, soluble cellulose derivatives, wet strength resins, polyvinyl alcohol and latex, account for approximately 1% of dry strength additives.

The most commonly used additives are however, starch and its modified forms, especially oxidized starch, cationic starch, and polyacrylamides, natural gums, and their derivatives. Though cationic starch has been favored by mills for a long time, compared to vegetable gums or synthetic and natural polymers, because of its lower cost and its biodegradable nature(8, 9) , oxidized starch has drawn attention by Indian mills due to its ease of processing or modification and lesser cost involved. However it has been observed recently that this creates runnability problems, is prone to hydrolysis leading to formation of glucose and consequently slime formation and odour problems. Another potential problem arises when treated with recycled papers, due to formation of foams and gases which disturbs paper formation in the wet end and loss of vacuum on the paper machine. The generated calcium ions could potentially form tacky metal soaps or calcium sulfate scale, a source of sticky problems. Synthetic resins are expensive and have many drawbacks, especially from environmentally unwanted chemicals. Natural gums and their derivatives may be alternatives for both starch as well as synthetic polymers if collection and processing are made easier and cheaper. In this present investigation therefore, effectiveness of gum (Guar gum) and oxidized starch as dry strength additives are investigated.

**STARCH AND ITS DERIVATIVES:**

The Source of starches is: Maize, Tapioca, Wheat, Potato, Sago ,Sorghum,Corn,Rice etc. in India, the major sources are Maize(80%) and Tapioca(20%). Commercial production of dry starch power from Maize and Tapioca are approximately 155,000 and 100,000 MTPA respectively in recent years..
Paper Industry consumes around 14% of the total starch produced; the largest is in Textile industry. The major areas of application of starch in Paper Industry are:

- Wet end for strength, retention, flocculent, drainage aid, size carrier
- Size press for surface sizing
- Coating as binder for fine pigments
- Effluent treatment as cationic polymer to control discharge of fibers etc.

World wide production of starch from various sources is shown in Table 1.

**Table 1: Worldwide source of production of starch**

| Raw material source | Production( Million TPA) | % |
|---|---|---|
| Maize | 27.38 | 74 |
| Tapioca | 3.70 | 10 |
| Wheat | 2.96 | 8 |

| | | |
|---|---|---|
| Potato | 2.59 | 7 |
| Others | 0.37 | 1 |

In India, the major sources are maize and tapioca. Commercial production of dry starch powder from maize and tapioca are app.155, 000 and 100,000 MPTA respectively.

**Consumption pattern**

Consumption pattern of starch in various industries vary to a great extent in India compared to developed countries. This is shown in Table 2

**Table 2: Consumption pattern of starches,%**

| Industry | India | USA |
|---|---|---|
| Textiles | 40 | 8 |
| Paper | 15 | 50 |
| Corrugation | 14 | 14 |
| Food | 9 | 13 |
| Others | 22 | 15 |

Comparison of general properties of starches from various sources is shown in Table 3.

**Table 3 General Properties of starches from various sources**

| Property | Corn | Tapioca | Potato |
|---|---|---|---|
| Anv. Particle Size, μm | 15 | 20 | 30 |
| Particle Size Range, μm | 5-25 | 5-35 | 15-100 |
| Bulk Density-loose (kg/m$^3$) | 575 | 590 | 735 |
| Bulk Density- packed (kg/m$^3$) | 800 | 830 | 900 |
| Avg. Amylose/Amylopectin (%)* | 28/72 | 17/83 | 21/79 |
| Avg. Degree of Polymerization | 800 | 3000 | 3000 |

Average molecular weights of potato and corn starch 3-3.5*10$^6$ and 0.8*10$^6$-1.8*10$^6$ respectively.

* Wheat starch contains slightly higher amount of Amylose (25%).

**Table 4 Average Composition of Commercial Starches on dry solid basis (%) after moisture**

| Components, % | Corn | Tapioca | Potato |
|---|---|---|---|
| Moisture | 12 | 12 | 15 |
| Starch | 98-99 | 95-99 | 99 |
| Protein | 0.3 | 0.8 | Traces |
| Fat | 0.5 | <0.1 | <0.1 |

| | | | |
|---|---|---|---|
| Ash | 0.2 | 0.1 | 0.4 |
| Fiber | <0.2 | 1.4 | 0.1 |

It is essential that most of the starches used in the paper industry are required to be modified to develop certain properties in the starch so that it will be more effective and adaptive in the papermaking media. The consumption of various unmodified and modified starch is shown in Table 5. Expected demand of modified starch in the year 2012 will be approximately more than 80,000 MTPA.

**Table 5. Various starches used in Indian Paper Industry:**

| Starch Type | Consumption, MTPA | % |
|---|---|---|
| Oxidized/**Acetylated** | 13,500 | 38 |
| Spray Starches (in paper machine) | 10,000 | 29 |
| Cationic | 5,000 | 14 |
| Unmodified | 3,500 | 10 |
| Thin boiling/Dextrin | 3,000 | 9 |

## Starch Modification Methods:

Unmodified starches differ in properties such as Amylopectin-to-amylose ratio, granule dimensions, shape, gelatinization temperature, and the molecular weight of the component fractions. Some of the methods are:

- Physical modification methods
- Fractionation into amylose and amylopectin components
- Thermo-mechanical conversion
- Acid Hydrolysis
- Chemical modification
- Oxidation
- Derivatization
- Enzyme conversion
- Chemically- Thermally Modification
- Genetic modification

Thermo-mechanical conversion, acid hydrolysis and enzyme conversion all lead to a decrease in the molecular weight of the starch and a corresponding decrease in the solution viscosity of starch solutions. This makes them more manageable and able to be pumped. The most important parameters for selecting modified starches are:

- Effect on Strength
- Filler Retention
- High retention by wood fibers
- Drainage
- High Degree of Dispersion
- Good Solubility

- Charge density and sign

The principle function of starch when applied at the beater or at wet end such as fan pump or head box to increase the strength of paper. When oxidized starch is used sizing is improved. The adhesive quality of starch are due to its contents of polar hydroxyl groups, these groups setup a fiber to starch to fiber bonding and as result the strength of the sheet is increased folding strength burst and tensile are greatly increased and tearing resistance is usually lowered.

**GUAR GUM**

The most common sources are vegetable gums, some of them are: seed gums like guar gum, tamarind gum, and locust gums (from locust bean or carob tree), Sesbania Aculeata, and tree exudate karaya gum (11-13). Guar & Tamarind are native and most widely used in India. Guar gum is extracted from the seed of Cyanopsis tetragonoloba, an annual leguminous plant originating from India , Pakitan, and US,. This gum is a hetero-polysacchrides , mainly mannogalactan(or galactomannan) , derived from the seed endosperm of the guar plant grown in sandy soils of West, and Northwest India. India is the largest producer of guar gum products used for many applications (13). Jodhpur city of Rajasthan in India is the most important processing center of guar gum and contributes approximately 40% of the World's guar gum supply/ consumption. There are considerable numbers of manufacturers and exporters of guar gum are available in India. Guar gum is sold as a white to yellowish odorless powder which is, available in different viscosities depending on the desired one.M/s Desmo Exports Gum Rosin WW grade for India.

Guar gum has some unique properties which make it useful in variety of applications. These are: easy solubility in hot water, fairly in cold water, film forming property, resistance to oils, greases and solvents, better thickening agent, water binding capacity, high viscosity and functionality at low temperatures.

This is used not only as beater or wet end additives to improve sheet formation and fiber bonding, primarily for improving strength properties as indicated earlier but also used as important chemicals in many applications including as a surface sizing agent, by providing denser surface to the paper used in printing.

Derivatives of guar gum include oxidized guar gum, cross-linked guar gum, guar gum amino ethyl gum, modified guar gum, cationic and anionic and non-ionic guar gum. Because of high costs, application guar gum has been slowed down world wide though it can develop better properties. These have been effectively employed at low addition rates especially for the manufacture of wrapping paper, kraft, photographic paper, filter etc.

A simple method to produce gum paste is take gum in some amount based on concentration and dissolve in cold water and water and then heat up to it became a paste form. The gum is used with the amount 0.5 to maximum 5% (based on dry fiber).

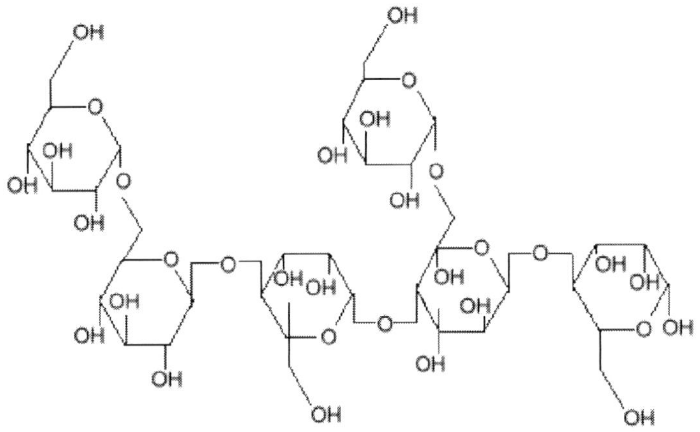


**Guar gum Structure**

Guar is more effective in dry strength improvement than starch from the same amount added. It is widely used for machine glazed bag and wrapping papers. It increases the strength, reduces the amount of mechanical treatment at low addition level and the filler retention also improves. But higher level of guar gum decreases the retention. Guar gum is generally used with 8-20 ibs /ton of pulp added ahead of refiner in the form of dry powder. The best temperature for applications of gum solutions was found to be higher than 50°C. Another property of guar gum useful in paper making is its ability to reduce friction in turbulent flow of fiber suspensions. This property extends pump life and lowers the requirement for pumping energy. Use of mill trials showed that filler losses were cut by 30 percent with additions not exceeding 0.25 percent (pulp basis) to the machine headbox.

The approximate costs of various wet end dry strength additives in Indian market are shown below.

**Approximate Costs of Dry strength additives in India in 2011**

| Components | Costs,Rs./kg | Components | Costs,Rs./kg |
|---|---|---|---|
| Native Starch/ spray Starch | 25-26 | Acetylated Starch | 45 |
| Oxidised Starch | 30-31 | Guar Gum | 110 |
| Amphoteric Starch | 36-37 | Cationic Guargum | 80-85 |
| Cationic Starch | 34-36 | Dextrin | 32 |
| Enzymetic Starch | 28-29 | Acrylamide | 175+VAT |

The cost data indicate that acrylamide are much more costlier than any modified starch, even more than guar gum or cationic guar gum. Cationic guar gum can compete with only oxidized , enzymetic or cationic starch provided doses of guar gum are within one third that of starch for the same degree of strength. Otherwise the strength gain and better formation ,higher retention,lower refining energy consumption in terms of cost must compensate the financial loss due to costs of guar gum.

In the present work therefore an attempt has been made to use natural polymers like guar gum as supplement with comparatively high cost synthetic polymer like acrylamide or lower costs for oxidized, cationic, anionic starch with an aim to enhance mechanical strength properties of the paper without affecting optical properties and to reduce high energy consumption during refining. The present investigation was carried out in the laboratory on the handmade sheets made from mainly from rice straw and wood based industrial pulps.

## EXPERIMENTAL

### Raw Material

Raw bleached pulps were procured from two Indian mills namely, Satia Paper Mill in the state of Punjab in Northern India and Star Paper Mills in Uttar Pradesh. Pulp from Satia paper mill was based on rice straw and imported waste paper and bleached pulp from star paper mill was made from 85% hard wood (Eucalyptus and Popular), 5% Bamboo and waste paper and negligible amount of soft wood. Pulps collected from the mill were dried and kept under ambient conditions. These raw pulps are mixed in 85:15 proportions and then subjected to refining action.

### Refining /Beating

The pulps were beaten in a valley beater up to get various freeness values using Shopper Reigler freeness tester namely, 35°SR, 40°SR and 45 °SR as per Tappi Procedure. The consistency and temperature were recorded and necessary corrections were made.

### Addition of Internal Sizing Agent

Alum and rosin were added in the ratio 1:4 (0.8% rosin and 3.2% alum) to the pulp and then agitated properly.

### Addition of Dry strength Additives

Guar gum and oxidized starch were added in different combinations. Both Guar gum and Oxidized starch were varied from 0.5 to 5% based on OD fibre. After adding the dry strength agents mixing is done properly with proper speed and the pulp were subjected to British sheet former. Hand sheet were made with sized pulp in a British sheet former as per Tappi procedure.

### Pressing And Drying

After making the sheet, the sheet had been kept in a hydraulic press under a pressure of 5 kg./$cm^2$ for 5 min. and then transferred to a rotary dryer for the dryness of the sheet.

### Evaluation of Mechanical Properties

The dried paper sheets were conditioned under standard temperature and humidity conditions in the laboratory as per Tappi procedure (14).

The tensile strength, tear strength, double fold and burst strength were measured by Alwetron tensile tester, Elmendorf tear tester, Kohler Mullen tester, Mullen tester respectively as per Tappi procedure. All data are expressed as under: Tensile Strength: kN /m X X 0.654 or $kg_f$/15 mm, Tear, $kg_f$ X 4;burst,$kg/cm^2$ respectively.

## RESULTS AND DISCUSSIONS:

The data obtained from various experiments evaluated in the laboratory were now subjected to analysis by drawing various plots. Some representative set of data are shown in the table 1 through table3. Y axes of all figures are depicted for strength properties whereas X axis is expressed either in terms of % guar gum or % starch. The natural polymers were expressed in percentage based on weight of pulp. The mechanical strength properties determined are double fold( DF), burst, tensile and tear. The Results are also shown in fig. 1 to fig 12.

Fig. 1 shows that when the percentage doses of the natural resin ( Guar gum) increase at 35°SR, then the strength properties (tensile, burst, double fold) increase. Without addition of additives (Guar gum), the tensile strength was found to be of the order of 1.798 and the percentage dose is 5 percent, the strength will be maximum around 3.54, the burst strength at 0 % resin 1.65 kg/sq.cm. and at the 5 % dose it increases upto 2.1 kg/sq. cm. Similarly the double fold will also be increase with increasing the doses of resin, at 0 % resin the value is 3.5, whereas at 5 % resin it will increase up to 7.5 %. According to Fig.2 one can see that the tear strength initially increases slightly and then decreases continuously. At the 0 % resin tear is 44 kgf , increase slightly at 0.5 % resin 44.5 $kg_f$ then continuously decreases. At the 5 % level the tear will decrease up to 32 kgf i.e. the decrease is affected up to 27%

In fig. 3 drawn for pulp of freeness 40° SR, one can see that as usual tensile , burst and double fold increase with the increase of dose of guar gum. The tensile at 0% resin is 2.87 and increases with the increase of percentage dose. At the 5% resin it will increase upto 3.29, burst at 0% resin is 1.8 kg/sq.cm. And at 5% resin it will increase up to 2.75 kg/sq.cm. The double fold increases with resin dose. At 0% resin the double fold is 3.66 and as resin percentage increases up to 5.0 then the double fold is obtained of the order of 6.5.

From fig. 4, it is evident that the tear strength increases with resin percentage. At the 0% resin it is 40, it slightly increases up to 3% and then decreases with the doses. At the 5% resin tear increases up to 40%. This trend is in contrast with the values obtained with other freeness values.

Fig. 5 is drawn for 45° SR, the tensile strength increases with percentage doses of resin. As the percentage doses increase up to **5.0**, the tensile strength increases with 35%. Burst strength at 0% resin is found of the order of 1.25 kg./sq.cm. And at 5% resin the burst is 2.35-kg/sq. cm. Double fold increases with percentage doses of resin i.e. at 5% resin the double fold is found on the order of **9.75** and at 0% the double fold is 6 numbers.

In fig. 6 it is clear that the tear will increase initially and then decreases rapidly. At 0% resin the tear is of the order of 32 kgf and at 5% it is 16. Therefore the tear decreases at 5% upto 56%.

Figs. 7 and 8 are drawn for 35°SR between strength as a function of percentage of starch added. It is seen that the tensile will increase with increasing the percentage doses of starch. At 0% the tensile strength is found 2.16 and at 5.0 the tensile is 3.07.Similarly the burst strength is also increased with percentage doses of starch. At 5% resin it increases up to 20 %. Double fold are also increased with increasing the percentage doses of starch and increase with 140% at 5.0% doses. Similarly as percentage of doses increases, the tear decreases. At 0% starch, tear is found of the order of 32 kgf and at 5% it is 14 kgf . So the decrease is affected of the order of 56%.

Figs. 9 and 10 represent the strength properties at 40°SR. With 0% dose of starch tensile is 3.76 and continuously increases. At 5.0% dose it increased upto 30%. Tear increases up to 2.0 % resin where it is found as 22.0 and then decreases to again 16.0. %. The tear is therefore decreased up to 27% after initial increase. Burst strength increases with 70% and double fold increases with 71.

Figs 11 and 12 are drawn for 45°SR for the strength properties as a function of percentage starch.It is evident from the figures that the the tensile strength increase with the increase of starch addition (up to 30% for the doses of 5%).As usual the burst strength, double fold also increase.
The figure 12 is drawn between tear strength and percentage starch doses. According to the data and plot the tear decreases with increasing 0 SR and additive level and decreases up to 60%.

Figures 13-16 indicate the comparison of the performances between strength properties attained by guar gum and starch for various doses. Figs. 17-18 show the tensile strength values as a function of freeness. These are self explanatory. On comparing the strength properties developed due to addition of guar gum and oxidized starch, the tensile values, generally ( beyond 1.0%) guar gum gave higher value than starch. But burst ,double fold and tear yielded higher value than those from starch.
Increasing the degree SR from 35 degree to 40 degree, starch gave higher values of tensile strength than guar gum but the latter scores over the former for burst, tear and double fold values. Much higher tensile values are noted for 45 degree SR when used starch compared to guar gum but burst, tear and double fold values are found to be inferior than those with guar gum.

It is found that with the increase of doses of both guar gum and starch, the tensile, burst and double fold increase continuously. Tear however, decreases after initial slight increase. But at 40 °SR the initial increase has been continued up to 3% dose in case of guar gum unlike up to 0.5% in other freeness values. Higher tensile values are always obtained in case of starch but tear, burst and double fold are found to be higher in case of guar gum irrespective of raw materials, rice straw or hardwood. The values of burst are found to be much lower in case of rice straw, even with addition of starch but the same for guar gum, higher values are obtained. At low refining (lesser degree of degree SR values),guar gum gives better results than those from oxidized starch

The tensile strength varying between 35°SR to $45^0$ SR slightly increases with the resin (guar gum) doses changing between 1%, to 5%. The maximum burst strength at $35^0$ SR is found 2.1 kg/sq.cm. With 5% resin but at $40^0$ SR it is 2.75 kg/sq.cm. And with $45^0$ SR it is 2.3 kg/sq.cm. The maximum double fold value with $35^0$ SR is 7.5 Nos., at $40^0$ SR is 6.0 Nos. and $45^0$ SR it is 9.2 Nos. Tear is also found maximum with $35^0$ SR with 0.5 to 1% resin addition (guar gum). Not much benefit is obtained by increasing the dose of guar gum beyond 1% in all such cases.

Therefore, according to above one can conclude that $35^0$ SR freeness is best for achieving the best results with 1.0%.of guar gum. This will also save a lot of refining energy.

The strength properties, viz, the tensile, burst, double fold and even tear values have maximum strength at $40^0$ SR with the starch resin. Therefore it can be concluded that the guar gum is beneficial than starch because the later consumes higher -energy compared to guar gum. Exact quantification is possible provided optimization is made through detailed cost estimates.

The plant of guar gum can be a good biomass for making good quality of pulp for writing and printing paper, the seeds out of the same plant which is a waste, can be utilized for the production of gum useful as wet end additive for paper industry. The husk and kernels obtained after removal of gum from guar gum seeds posses high protein value and can be utilized for cattle / poultry feed.

## CONCLUSIONS

Based on the analysis of data the following conclusions can be drawn.

With the increase of doses of both guar gum and starch, the tensile, burst and double fold increase continuously. Tear however, decreases after initial slight increase. But at 40 $^oSR$ the initial increase has been continued up to 3% dose in case of guar gum unlike up to 0.5% in other freeness values. Higher tensile values are always obtained in case of starch but tear, burst and double fold are found to be higher in case of guar gum irrespective of raw materials, rice straw or hardwood. The values of burst are found to be much lower in case of rice straw, even with addition of starch but the same for guar gum, higher values are obtained. At low refining (lesser degree of degree SR values), guar gum gives better results than those from oxidized starch.Not much benefit is obtained by increasing the dose of guar gum beyond 1% in all such cases. $35^0$ SR freeness is found best for achieving the best results with 1.0%.of guar gum for rice straw pulp. The strength properties, viz, the tensile, burst, double fold and even tear values have maximum strength at $40^0$ SR with the oxidized starch resin. Therefore it can be concluded that the guar gum is beneficial than starch because it consumes higher –energy. Refining at 5 degree lower SR values save a lot of refining energy.But detailed cost estimates are required taking care of all benefits of relative increase of strength properties by guar gum and comparatively three times higher costs of guar gum.

The biomass from the guar gum plant can be a good substitute for wood for making good quality of pulp for writing and printing paper, the waste seeds out of the same plant can be utilized for the production of gum useful as wet end additive for paper industry, especially as dry strength additives. The husk and kernels obtained after removal of gum from guar gum seeds posses high protein value and can be utilized for cattle / poultry feed or chief fertilizer.

## REFERENCES:


1. Kocurack M.J, Pulp and Paper Manufacture, Stock preparation Vol. 6, Joint Textbook Committee of the Paper Industry, Tappi, CPPA,1989

2. Thorn I., Wet end Chemistry ,Kluwer Academic Publishers Group,1994,
3. Casey J.P., Pulp and Paper Chemistry and Chemical Technology, Third Edition,Vol. 3 , John Wiley & Sons.,1981
4. Datta Tapas, Proceedings of IPPTA Workshop on Optimization of Wet-End technology in Paper & Boardmaking, IPPTA,2000,PP-1-10
5. Scott W.E. , and J.C.Abbott, Properties of Paper : An Introduction, Tappi Press, 1995
6. Scott W.E., Principles of Wet End Chemistry, Tappi Press, 1996
7. Rathi ,M.S. and Biermann,C.J.,Application polyallylamine as a dry strength agent for paper,Tappi J,,83(12):62(2000)
8. Lertsutthi,P,Stevens,W.F. and Chadrakrachang,S., Improved paper performance using chitosan, Proc.1st Thailand materials Science and Technology Conf.,Bangkok, Thailand, p.361,(2000).
9. Prannee Lertsutthiwong, S.Chandrkrachang, M.M. Nazhad and William F.Stevens,Chitosan as a dry strength agent for paper., Appita J,Vol.55(3),May(2002)PP.708-712.
10. Pelton,R,Zhang J.,Wagberg,L.,and Rundlof,M,The role of surface polymer compability in the formation of fiber/fiber bonds in paper, NordPulp Pap.Res.J. 15(5):400(2000)
11. Whistler r.L., Industrial gums, Second Edition,Academic press,1973
12. Ghosh,S.K., Saikia D.C.,and Saikia, C.N.,Dhainncha( Sesbania Aculeata) Seed A Source of Natural additives for Paper Industry,IPPTA,Vol.7(3), September,1995,PP.51-54.
13. http:/www.guargum.info/ tradeindia.com, 5/21/2011
14. Tappi Test methods-Tappi Press:1998-1999,Atlanta,,Georgia,USA,1999

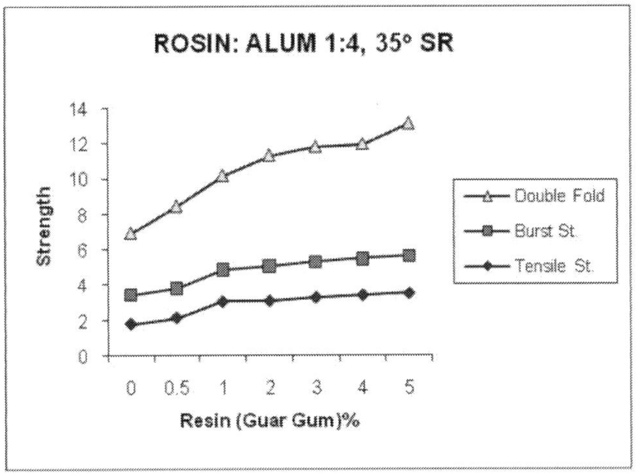


Figure 1: Strength properties as a function of Guar gum %

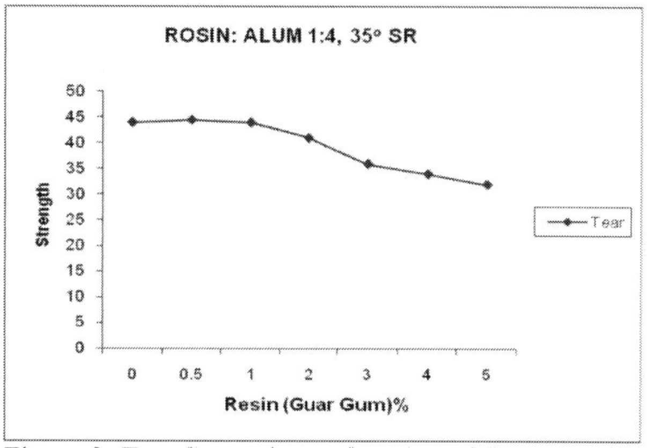

Figure 2: Tear Strength as a function of Guar gum %

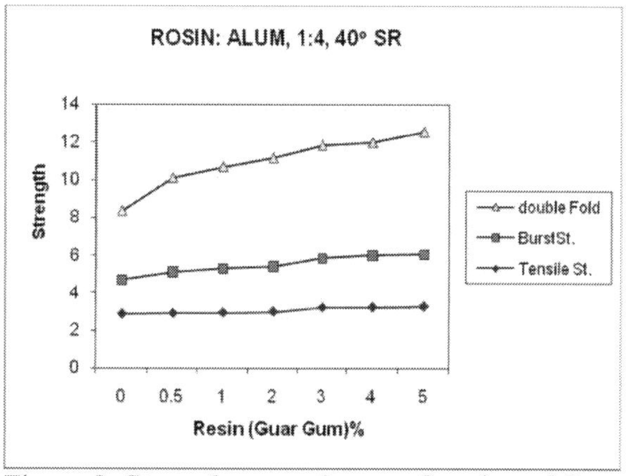

Figure 3: Strength properties as a function of Guar gum %

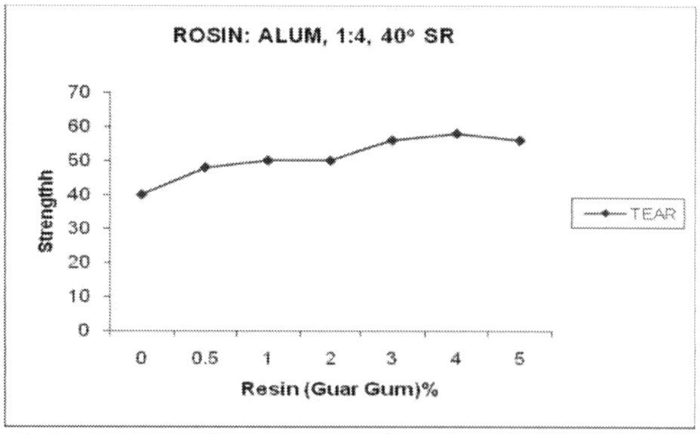

Figure 4: Tear Strength as a function of Guar gum %

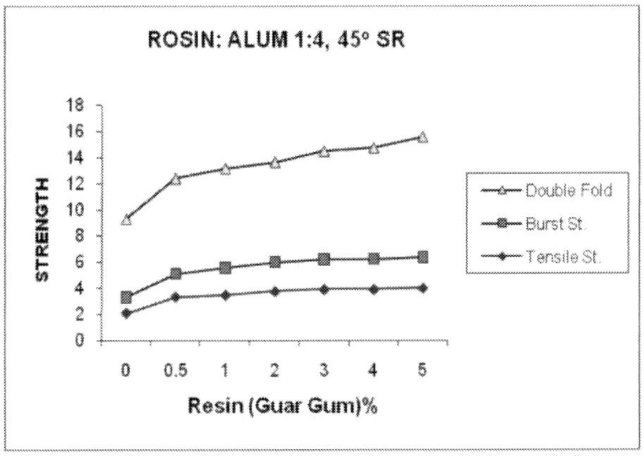


Figure 5: Strength properties as a function of Guar gum %

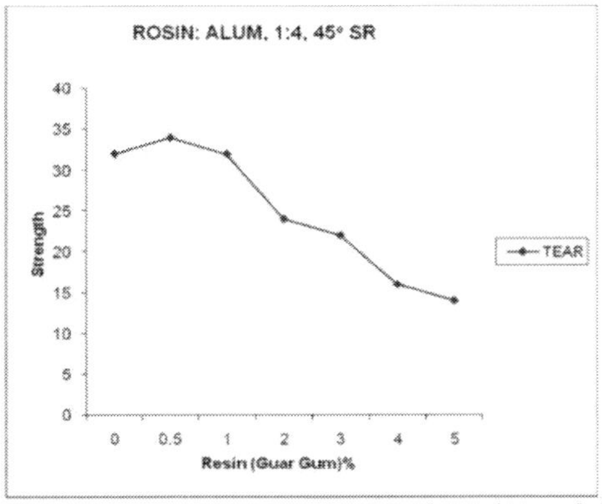


Figure 6: Tear Strength as a function of Guar gum %

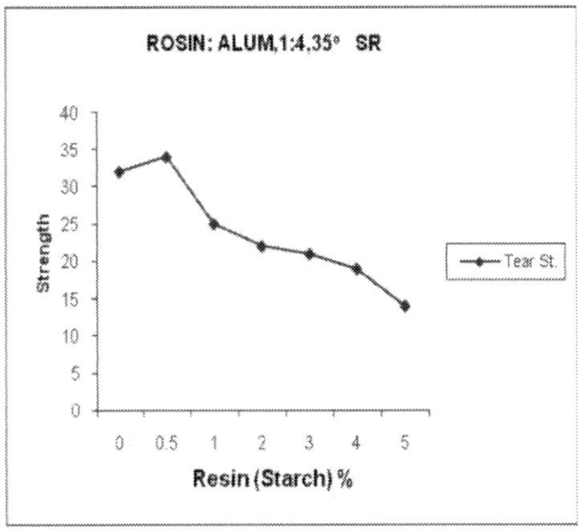


Figure 7: Tear Strength as a function of Starch %

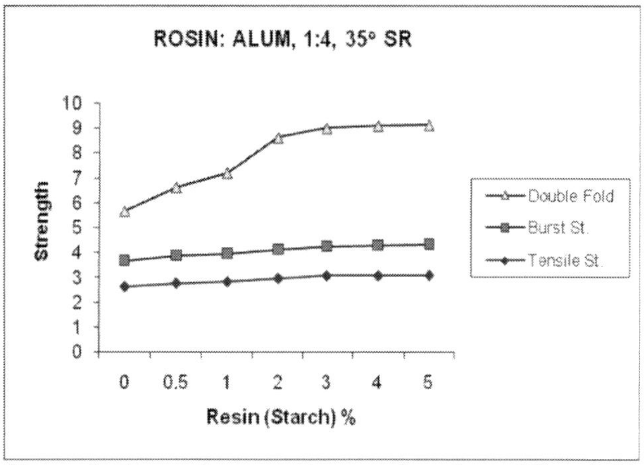

Figure 8: Strength properties as a function of Starch %

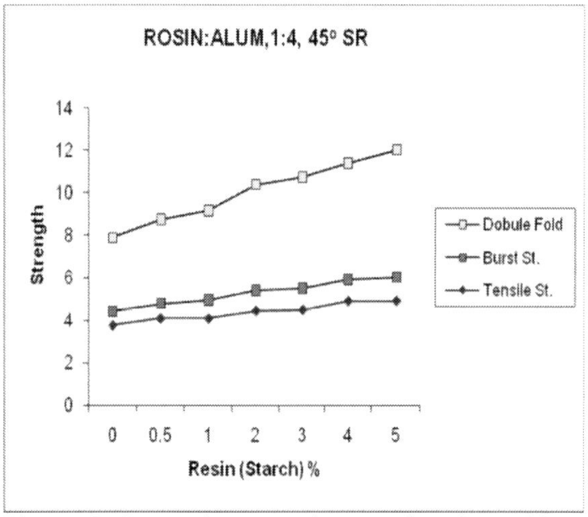

Figure 9: Strength properties as a function of Starch %

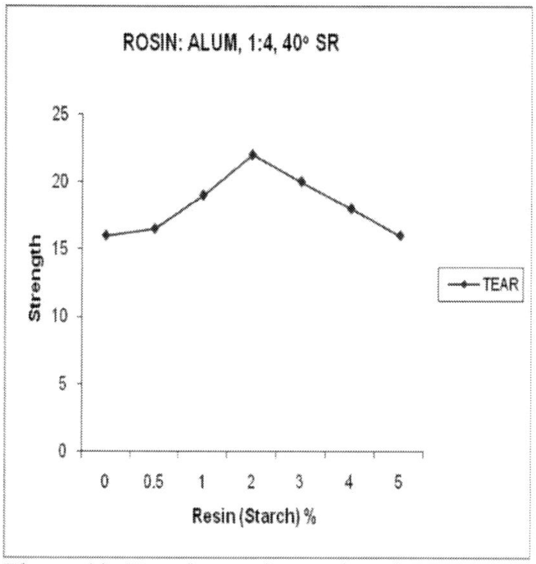

Figure 10: Tear Strength as a function of Starch %

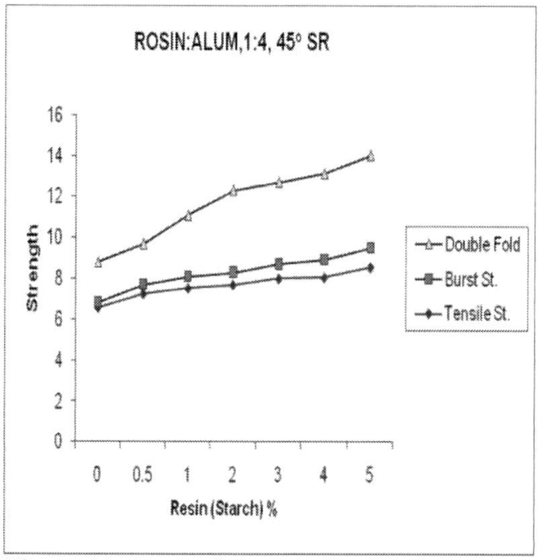

Figure 11: Strength properties as a function of Starch %

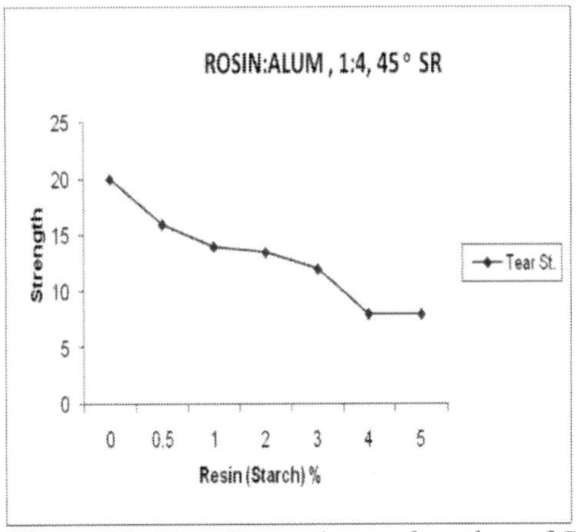

Figure 12: Tear Strength as a function of Starch %

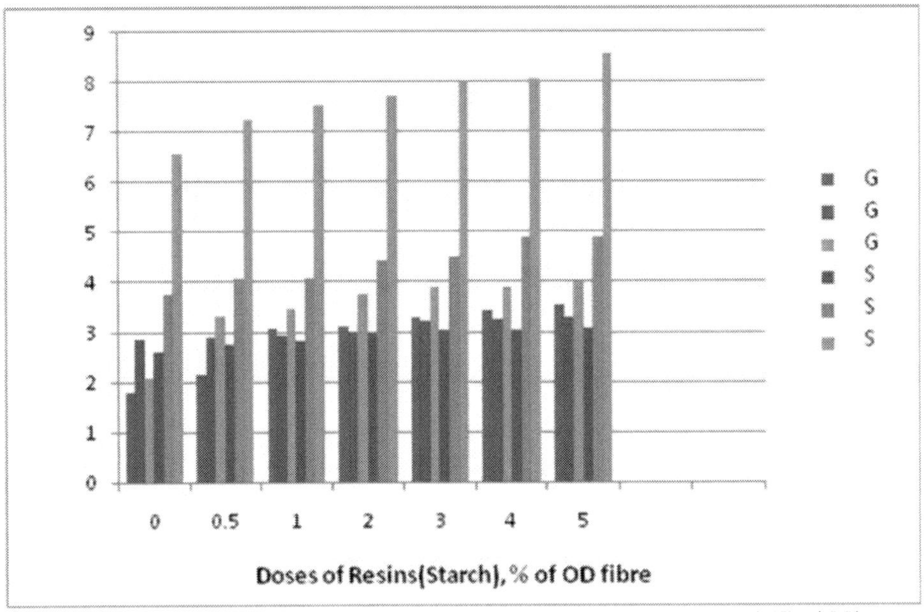

Figure 13: Tensile strength as a function of Guar gum (G) and starch(S), kN/m

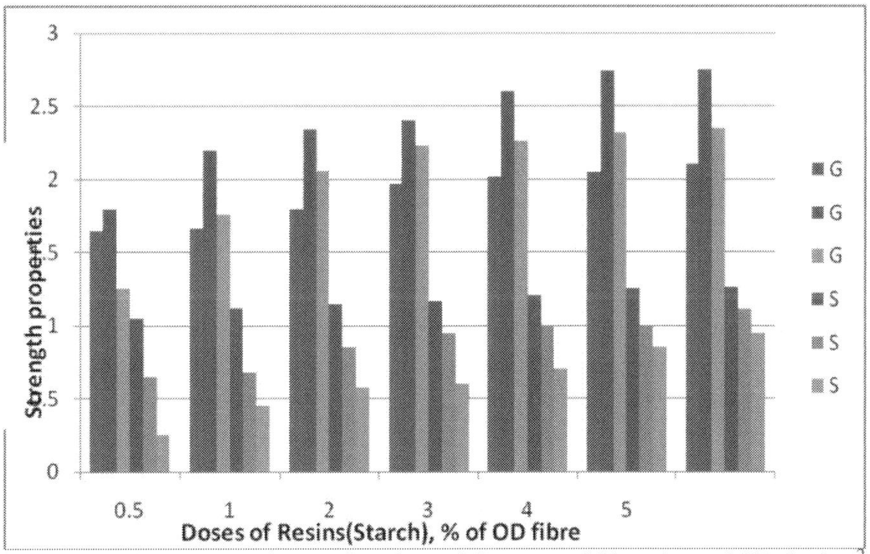

Figure 14: Burst strength as a function of Guar gum(G) and starch(S), kg/cm$^2$

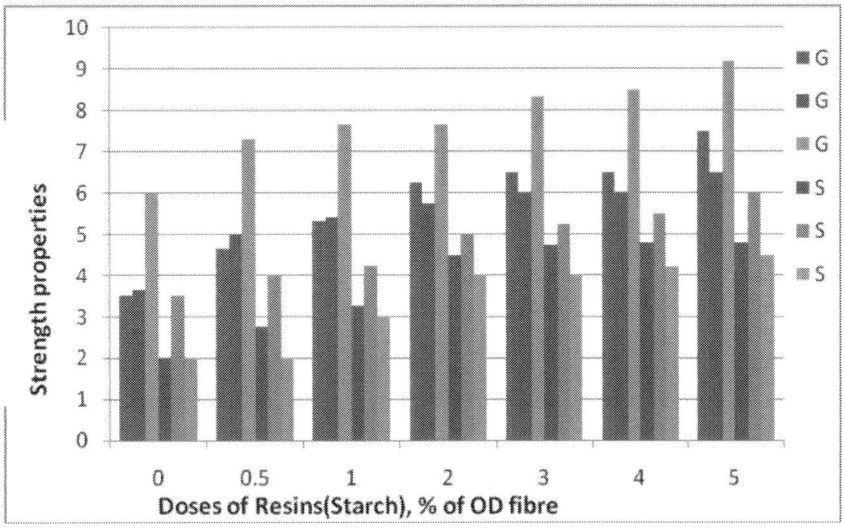

Figure 15: Double fold as a function of Guar gum(G) and starch(S).

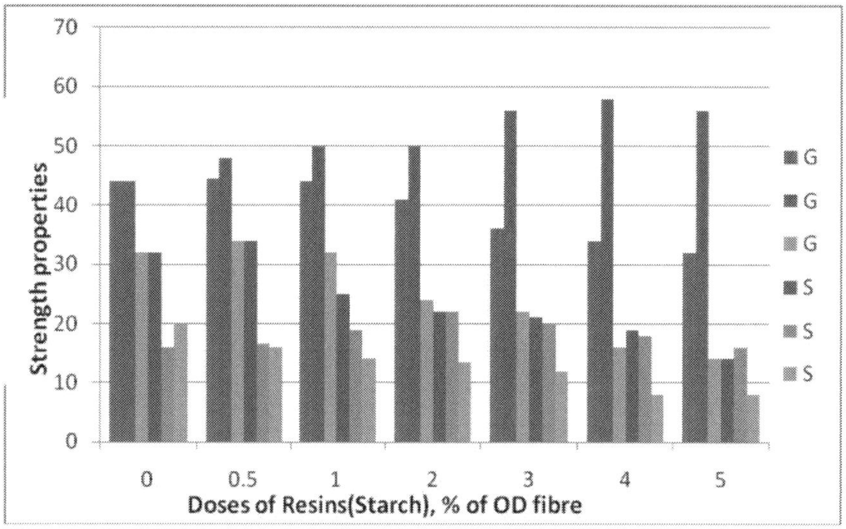

Figure 16: Tear strength as a function of Guar gum(G) and starch(S), kgf.

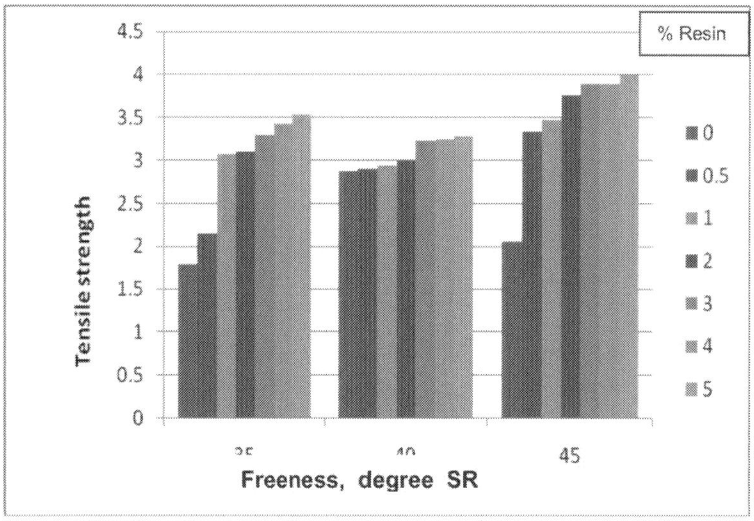

Figure 17: Tensile strength as a function of Guar gum at different freenesses.

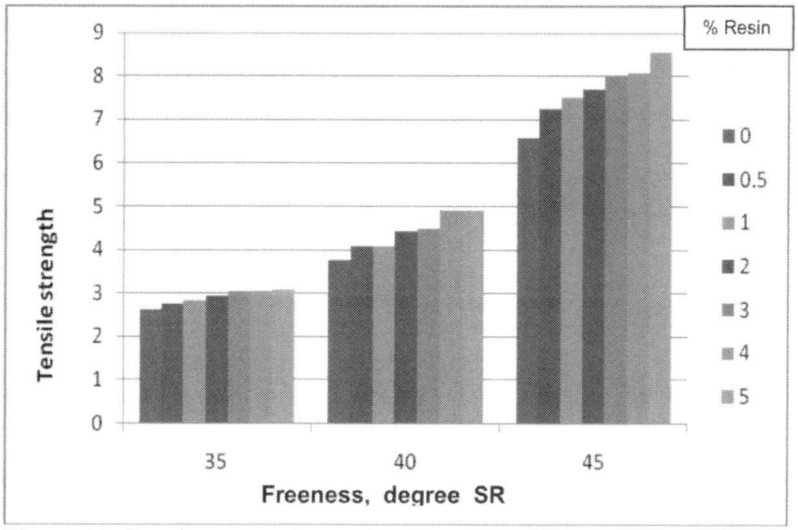

Figure 18: Tensile strength as a function of Starch at different freenesses.

# One-step Alkoxylation of Acetic Anhydride to Synthesize Diacetates of Ortho-Diols in High Yield Using a Novel Catalyst

*Jia Liu, Yun Fang*, Wei Tang*

*School of Chemical and Material Engineering, Jiangnan University, Wuxi, Jiangsu 214122, China*

## Abstract

Diacetates used to be synthesized in a two-step process with acetic acid or acetic anhydride as starting material. The disadvantage of the traditional route is low yield of di-ester with some undesirable by-products which would increase separation load and have to treat lots of waste acid solution. This paper reports a novel one-step synthesis of diacetates catalyzed by Mg-Al composite oxide, and achieved by alkoxylation of acetic anhydride with ethylene oxide or propylene oxide, respectively. The one-step process is a green chemistry approach with excellent atomic economy since neither solvent is needed nor lots of by-products are produced. More than 95% yield of the diacetates with high product selectivity(>99%) was obtained when the one-step reaction catalyzed by 1% Mg-Al composite oxide was performed in an autoclave in the molar ratio of acetic anhydride to epoxide 1:1 at 145 °C for 2.5 h.

***Keywords:*** one-step alkoxylation; epoxide; acetic anhydride

## Introduction

1, 2-propanediol diacetate and 1, 2-ethanediol diacetate are stable liquid compositions having properties of value as solvents and plasticizing agents in tobacco, packaging, cosmetic and pharmaceutical industries, as an alternative for triacetin [1-6]. One of the traditional synthetic routines is two-step esterification of glycol and acetic acid or acetic anhydride catalyzed by Lewis acid [3], acidic ion-exchange resin [7-8], or solid acid [9]. The other is the ring cleavage reaction of epoxide with acetic anhydride under a two-step mechanism, where monoester from the ring cleavage of epoxide with acetic anhydride is capped by acetic acid or acetic anhydride [10-11]. But these two-step methods often suffered from poor reaction selectivity, low yield of diacetate, the exhaustion of excess reagents, expensive and difficult catalyst separation, tedious work-up procedures, equipment corrosion, environmental toxicity and waste discharge. It is widely accepted that there is an urgent need to develop green and economical processes for diacetate synthesis.

In this paper, 1, 2-propanediol diacetate and 1, 2-ethanediol diacetate are obtained by one-step alkoxylation of acetic anhydride using Mg-Al composite oxide as catalyst. This new method helps in the effort to find a one-step synthesis of diacetates with high yield and excellent atomic economy, since neither solvent is needed nor by-product is produced. In a word, it is a green chemical synthesis process and has a prospected development in the near future.

* Corresponding author: yunfang@126.com.

## Experimental

### *Chemicals*

Acetic anhydride (AA) and propylene oxide (PO) were obtained from Sinopharm Chemical Reagent comp Ltd, China. Ethylene oxide (EO) was purchased from Sinopec Yangzi Petrochemical comp Ltd, Nanjing, China. Mg-Al composite oxide catalyst was prepared as described elsewhere [12-13]. All chemicals and reagents were purchased with highest purity and were used without further purification.

### *Synthesis*

The propoxylation was carried out in a 25 mL autoclave equipped with a magnetic stirrer. The mass of the acetic anhydride was 3 g. The mass of propylene oxide (PO) was 1.71 g. Two reactants were employed at the ratio of 1:1 without any organic solvent. After 2.5 h reaction at the desired temperature under stirring in the presence of 1 wt% Mg-Al composite oxide, the reaction mixture was cooled naturally to room temperature to obtain a liquid product. The crude product was centrifuged for 10 min at 10,000 rpm to remove the residual inorganic catalyst. An aliquot of 1 mL was taken from the supernatant for quantitative analysis of the product composition by GC analysis. The above procedure was repeated using ethylene oxide (EO). The main reactions were shown in scheme 1 and scheme 2.

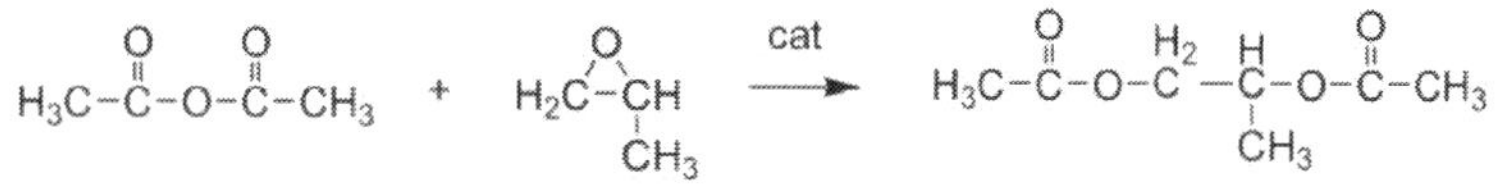


Scheme 1 one-step approach to synthesis of 1, 2-propanediol diacetate

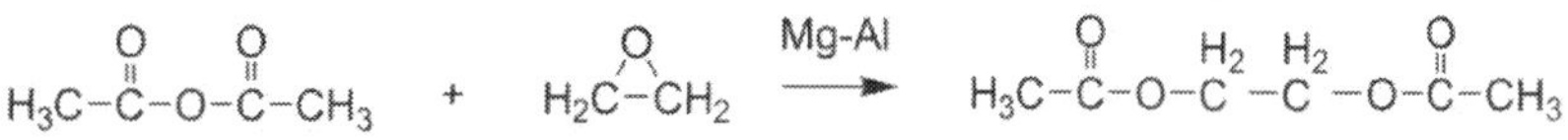


Scheme 2 one-step approach to synthesis of 1, 2-ethanediol diacetate

### *Characterization*

The progress of the reaction and the quantitative analysis were monitored using GC analysis. GC data was obtained on a GC970 gas chromatograph (Wenling Fuli Analysis Equipment comp Ltd, China) equipped with an OV-1701(3 m ×0.32 μm) column. A SrAdv chromatograph station was used for the signals produced by the hydrogen flame ionization detector to produce chromatograms and perform the quantitative analysis. The system allowed for the column temperature was 140 °C, when the detector temperature and the injector temperature were both 300 °C.

On completion, the products were extracted by distillation. The 1, 2-ethanediol diacetate and 1, 2-propanediol diacetate were characterized by FTIR spectra, GC-MS, $^1$H-NMR spectra and elemental analysis. The FTIR spectrum was recorded on a FTLA 2000(ABB, Canada) in a scanning range from 500-4000 cm$^{-1}$ for 7 scans. The GC-MS was equipped with a PEG-30 column. The GC temperature program was initially 50 °C for 2 min, increased to 200 °C at 10 °C /min, and then held at 200 °C for 15 min. $^1$H-NMR spectra was using a Avance III Digital NMR Spectrometer (Bruker, Switzerland) at a frequency of 400 MHz with $CDCl_3$ as solvent.

## Results and Discussion

### *Characterization of diacetates*

#### *FTIR characterization*

The FTIR spectra of 1, 2-propanediol diacetate, 1, 2-ethanediol diacetate and acetic anhydride are shown in Figure 1.

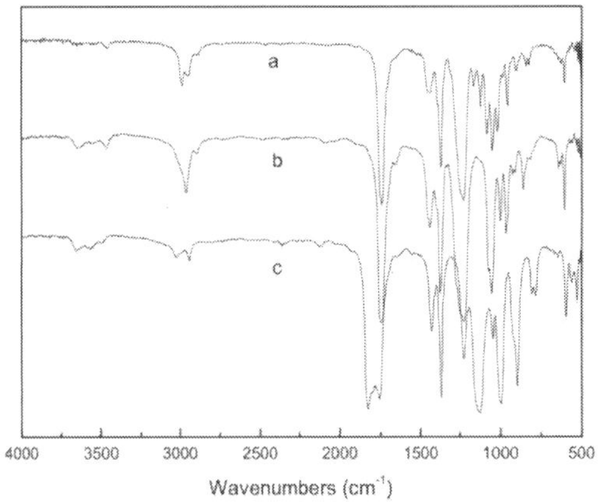


Fig. 1 FTIR spectra of 1, 2-propanediol diacetate (a), 1, 2-ethanediol diacetate (b) and acetic anhydride(c)

Compared the product diacetates (as shown in Figure 1a and 1b) with reactant acetic anhydride (as shown in Figure 1c), all these compounds have common peaks at 2893 $cm^{-1}$ and 2961 $cm^{-1}$ ($CH_3$ stretching vibration), 1465$cm^{-1}$ ($CH_2$ or $CH_3$ bending vibration), 1377 $cm^{-1}$ ($CH_3$ symmetrical bending vibration). It confirms the ester structures in the product that the characteristic band appeared at 1741 $cm^{-1}$ (C=O stretching vibration) in Figure 1a and 1b, instead of two strong absorption around 1826 $cm^{-1}$ (anhydride carbonyl symmetric stretching vibration) and 1754 $cm^{-1}$ (anhydride carbonyl symmetric stretching vibration) in Figure 1c. Another absorption appeared around 1240-1190 $cm^{-1}$ (Figure 1a and 1b), corresponding to the characteristic C-O-C group of ester due to introducing epoxide group into acetic anhydride.

#### *$^1$H-NMR characterization*

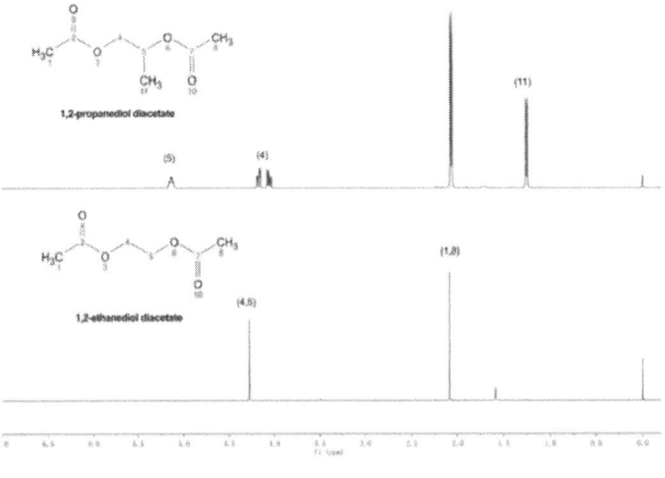

Fig. 2 $^1$H-NMR spectra of 1, 2-propanediol diacetate and 1, 2-ethanediol diacetate

In order to get the molecule structural information in detail, the $^1$H-NMR spectra of the diacetates were obtained. Figure 2 interprets the results of the $^1$H-NMR spectra using the signal of solvent $CDCl_3$ as a reference. The proton NMR peaks of 1, 2-propanediol diacetate are at 5.47-4.84 ppm for **CH** proton α to O, 4.11 ppm for $\mathbf{CH_2}$ protons α to O, 2.08 ppm for $\mathbf{CH_3}$ protons α to C=O, and 1.25 ppm for terminal $\mathbf{CH_3}$ protons α to CH . The proton NMR peaks of 1, 2-ethanediol diacetate are at 4.27 ppm for -$\mathbf{CH_2}$-$\mathbf{CH_2}$- protons α to O and 2.09 ppm for $\mathbf{CH_3}$ α to C=O. Besides, the peak of 1.56 ppm is due to the $H_2O$ in the $CDCl_3$. The number of protons is computed using a reference value of six protons for peaks at 2.08 ppm corresponding to two $\mathbf{CH_3}$ protons α to C=O. The calculated numbers of protons match well with the protons of the two molecules.

***GC-MS characterization***

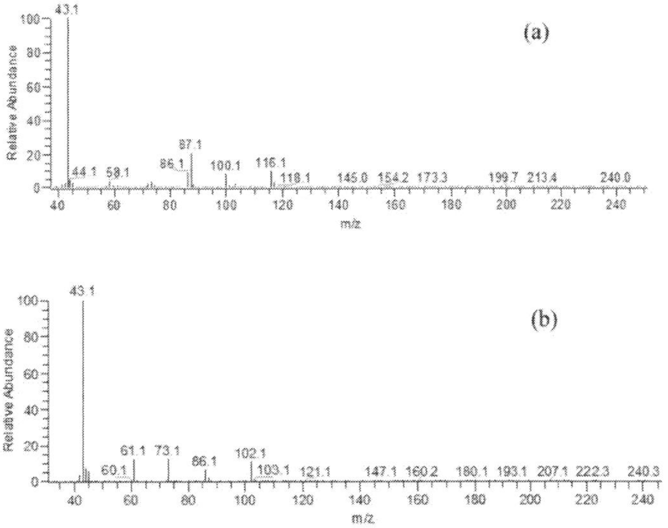


Fig. 3 The GC-MS of diacetates
(a) 1, 2-propanediol diacetate; (b) 1, 2-ethanediol diacetate

Further structural confirmations come through the use of the GC-MS fragmentation pattern in Figure 3. The GC-MS of two products show the typical m/z 43, which confirms the mass of the unit $[CH_3CO]^+$. In figure 3a, the m/z=87 for $[CH_3\text{-}CH\text{-}OC(OH)=CH_2]^+$, m/z=100 for $[CH_3COOCH_2C(CH_3)]^+$, and m/z=116 for $[CH_3COOCH_2C(CH_3)O]^+$ are providing a partial structure of 1, 2-propanediol diacetate. In figure 3b, a difference of 14 mass units with figure 3a is attributed to a reduced methyl group from propylene oxide to ethylene oxide.

***Elemental analysis characterization***

The found results of elemental analysis for 1, 2-propanediol diacetate are C 52.34 %, H 7.95 %, which matches well with the theoretical calculation compositions (C 52.49 %, H 7.55 %). The results of elemental analysis for 1, 2-ethanediol diacetate are C 49.32 %, H 6.97 %, which also matches with the theoretical calculation compositions (C 49.23 %, H 7.37 %).

All of the results from FTIR, $^1$H-NMR, GC-MS, and elemental analysis indicate that the diacetate present a sufficient purity.

***Effect of reaction parameters on the yield and selectivity of diacetates***

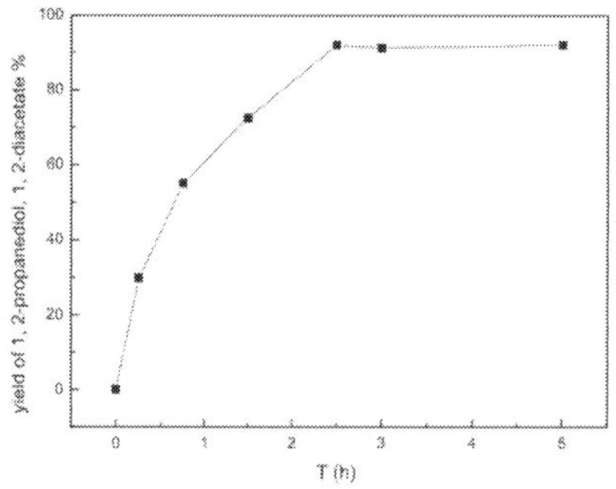


Fig. 4 Effect of reaction time on the yield of 1, 2-propanediol diacetate

Reaction conditions: acetic anhydride, 3 g; propylene oxide, 1.71 g; Mg-Al composite oxide, 0.047 g; T=145 °C

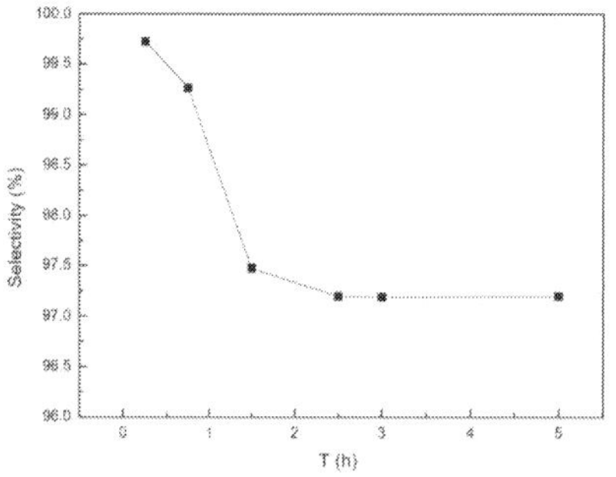


Fig. 5 Effect of reaction time on the selectivity of mono-adduct

Reaction conditions: acetic anhydride, 3 g; propylene oxide, 1.71 g; Mg-Al composite oxide, 0.047 g; T=145 °C

We selected acetic anhydride and propylene oxide as model substrates in order to get the optimum reaction parameters. Under the catalysis of Mg-Al composite oxide, two reactants were employed at the ratio of 1:1 without using any organic solvent. The kinetic curve obtained at various reaction times is presented in Figure 4. It shows that increasing the time increased the yield of 1, 2-propanediol diacetate. The highest yield of 1, 2-propanediol diacetate was above 92 % after 2.5 h. When the time was extended, the conversion remained almost unchanged. Surprisingly, the reaction showed a more "peaked" distribution and high selectivity (> 97 %) to mono-adduct (Figure 5). It is also observed that increasing the reaction time slightly decreased the selectivity of mono-adduct. In order to obtain the high yield of diacetates, suitable reaction parameters for one-step alkoxylation were Mg-Al composite oxide 1 wt %, reaction temperature 145 °C, reaction time 2.5 h, n(AA):n(PO) = 1:1. The ethoxylation of AA also have the similar results (94 % yield of 1, 2-ethanediol diacetate, 98 % selectivity to mono-adduct) under the condition.

## Conclusion

Mg-Al composite oxide has been used as a mild, non-hazardous and inexpensive solid catalyst for the complete and high-selective alkoxylation of acetic anhydride. The experimental results showed that the yield of diacetates could reach above 90 % and the selectivity to mono-adduct is > 97 % when the reaction in the molar ratio of acetic anhydride to epoxide 1:1 at 145 °C for 2.5 h. The reactions are green, operationally simple, high selectivity towards mono-adduct, excellent yielding, high atom economy and solvent-free. With these features, it should be considered as a green alternative for the production of 1, 2-propanediol diacetate and 1, 2-ethanediol diacetate.

***References***


1. H. Kato, T. Komai, K. Morita, S. Sakamaki (2006). "Eat-vaporizable air freshener compositions containing propylene glycol diacetate and/or phenoxyethanol and other solvents and method for controlling volatility of aroma," JP2006087512.
2. A. J. Daly, W. G. Lowe (1945). "Manufacture of mixed esters of glycols," US2454274.
3. X.-J. Zhang, C.-Y. Yan (2006)."Synthesis of plasticizer glycol diacetate catalyzed by $ZnCl_2$/ C," *Advances in fine petrochemicals*, December, pp. 25-27.
4. Y. Kubo, K. Takada, K. Kagosaki, Y. Imaizumi, H. Takahiro (2002). "Insecticide composition containing propione oxide derivatives," JP2002212010.
5. H. Seneca, S. Avakianm (1961). "Antibacterial and antifungal spectrum of 1, 2-propanediol diacetate," *Antimicrobial Agents and Chemotherapy*, pp.807-810.
6. J. W. Walker (1990). "Use of ethylene glycol diacetate in waterborne adhesives," US5100947.
7. B. Schmid, M. Dolker, J. Gmehling (2008). "Esterification of Ethylene Glycol with Acetic Acid Catalyzed by Amberlyst 36," *Ind. Eng. Chem. Res.* pp. 698-703.
8. S.-X. Yu, Y.-C. Tang, X.-W. Cai(1998). "Synthesis of ethylene glycol diacetate catalyzed with strong acidic cation exchange," *Chemical Reagents*, pp. 180-181.
9. S.-X. Yu, Q.-Y. Tang (2001). "Catalytic synthesis of ethylene glycol diacetate with p-toluene sulphonic acid," *Journal of shangrao normal college*, December, pp. 39-41.
10. I. B. Yanachkov, L. K.Yankov (1977). "Boric acid and boron fluoride-effective catalysts for preparation of glycol acetates from oxiranes and acetic anhydride," *Dokladi na B"lgarskata akademiâ na naukite*, pp. 29-32.
11. G. Fogassy, C. Pinel, G. Gelbard (2009). "Solvent-free ring opening reaction of epoxides using quaternary ammonium salts as catalyst," *Catalysis Communications*, pp. 557-560.
12. C.-Y. Yang, Y. Fang, J. Chen (2004). "Performances of ethoxylated fatty acid methyl esters and chemical modification with propylene oxide," *China Surfactant Detergent Cosmetics*, pp. 80-83.
13. Y. Fang, Y.-M. Xia, S.-R. Ji (2005). "One step catalytic preparation of polyoxyalkylene dicarboxylate from carboxylic anhydride and alkene oxide." CN 2005-10122794.

# (409a) Synapse: Software for Chemical Product Design

**Kevin G. Joback**, Molecular Knowledge Systems, Inc., Bedford, NH

The goal of chemical product design is to find molecular structures or formulations that possess a desired set of physical properties. Synapse, a commercial chemical product design software package, is capable of computationally generating and screening millions of candidate structures and formulations. Designers enter constraints on physical properties, molecular substructures, compositions, and chemical components. For each generated candidate Synapse performs molecular structure dissections and estimates the physical properties needed to evaluate each design constraint.

We will present an overview of the concepts behind Synapse's design algorithms, specifically detailing the enumeration of molecular structures and estimation of physical properties. We will then discuss some of the product designs investigated using Synapse, such as the design of replacements for volatile siloxane solvents and a new windshield washer fluid.

See more of this Session: Simulation Tools for Chemical Product Design
See more of this Group/Topical: Process Development Division

# (409b) Lipid Processing Technology: Shifting From Waste Streams to High-Value Commercial by-Products

**Carlos Axel Diaz-Tovar**[1], **Azizul Azri Mustaffa**[1], Georgios Kontogeorgis[2], Rafiqul Gani[1] and Bent Sarup[3], (1)Chemical and Biochemical Engineering (CAPEC), Technical University of Denmark, Kgs. Lyngby, Denmark, (2) Chemical and Biochemical Engineering (CERE), Technical University of Denmark, Kgs. Lyngby, Denmark, (3)Vegetable Oil Technology Business Unit, Alfa Laval Copenhagen A/S, Sborg, Denmark

In the last decades of the 20th century, a powerful international industry based largely on petroleum feedstock arose. This situation was driven mainly by the chemical industry that focused its production and research efforts on commodity chemicals. However, this situation has changed over the last few years. Nowadays, consumer needs and preferences for products based on performance have led the chemical industry to evolve beyond commodities toward specialty chemicals and 'consumer oriented chemicals based products'. Even though petroleum-based products are still ahead in world markets, the available market niche for these innovative products is not a constraint to them or to pharmaceutical products. Products obtained from renewable sources (e.g. vegetable oils and fats) have gained significance as the needs for healthier food products or the growing interest in biofuels, among others, have increased in the last decade. Regrettably, the oleochemical industry is far behind the chemical industry in terms of thermophysical property modeling and development of computational tools suitable for the design, analysis, and optimization of lipid-related processes.

The aim of this work is to present the development of computer-aided methods and tools for the systematic design, analysis, and simulation of processes/products employing lipid technologies. This includes, the identification and classification of the most representative lipid chemical species found in the edible oil and biodiesel industries and their representation in terms of molecular structure and the buildup of appropriate models for estimating pure component properties needed for model-based design and analysis of edible oil and biodiesel processes. In addition, for the modeling of phase behavior of relevant lipid mixtures, a combined group-contribution and atom connectivity approach (Group-Contribution$^{Plus}$ (GC$^{Plus}$)) that is able to extend the application range of property models has been developed. The main idea is the use of connectivity indices to describe the molecular fragmentation that relates properties which is the molecular interactions with the molecular structures.

For this purpose, a filled lipid-database (CAPEC_Lipids_Database) of collected experimental data from the open literature, confidential data from industry and generated data from validated predictive property models; as well as modeling tools for fast adoption-analysis of property prediction models to make it suitable for application with other computer-aided tools was developed. The CAPEC_Lipids_Database contains a total of 2603 data points collected for 12 pure component properties for a total of 243 lipid compounds. Among them 73 are triglycerides (C31-C63), 41 are diglycerides (C27-C43), 15 are monoglycerides (C11-C17), 29 are free fatty acids (C6:C24), 29 are methyl esters (C7-C25), 29 are ethyl esters (C8-C26), 4 are tocopherols (C27-C29), 4 are tocotrienols (C27-C29), 9 are terpenes (C30-C40), and 8 are sterols (C27-C53). For single value pure component properties and temperature dependent properties such as vapor pressure, enthalpy of vaporization, heat capacity, liquid viscosity, and surface tension GC-based models fine-tuned for the lipid chemicals were employed; in the case of the liquid density, the well-known modified Rackett equation and the Chemical Constituent Fragment approach were used. The regression of the adopted GC-based model parameters needed experimental data. PC-SAFT EoS was used to generate pseudo-experimental data when real data was not available. For the modeling of phase behavior of the relevant lipid mixtures, a master parameter table that contains all the needed binary interaction parameters fine tuned for lipid chemicals is made available. A user-added database (CAPEC_LIPIDS_x.x) that can be directly linked to PRO/II has been generated from the lipid-database. Through this database, PRO/II can be used to simulate and optimize edible oil process flowsheets.

Finally, the applicability of the developed methods and tools is shown through the analysis and simulation of the physical refining process of oils and fats. Within the traditional oleochemical industry this process is quite interesting because apart from obtaining the final oil product, it can now be employed to also produce (extract and refine) the high-value commercial by-products such as tocopherols and sterols.

See more of this Session: Simulation Tools for Chemical Product Design
See more of this Group/Topical: Process Development Division

# GENERATING LAMINAR, REPRODUCIBLE FLOW RATES BY MODELING TRANSPORT PROPERTIES OF THE HYDROFLUORIC ACID-SULFURIC ACID MIXTURE

*Tina Wang, Princeton University, Princeton, NJ, USA*

*Dr. Michael J. Kelley, Thomas Jefferson National Accelerator Facility, Newport News, VA, USA*

## Abstract

The Thomas Jefferson National Accelerator Facility (Jefferson Lab) aims to maximize the performance of particle accelerators by perfecting superconducting radiofrequency (SRF) technology. Smooth niobium interior cavities ensure that electron beams do not suffer from deflections. For this reason, electropolishing (EP) is used to rid the surface of existing defects. Previous studies have shown the dependence of EP performance on variables such as temperature. Yet, the effects of some other variables, such as those of surface flow rate, remain unmeasured. The purpose of this project was to design, build and test a controlled-flow apparatus to examine the effect of surface flow rates on the quality of EP. Proper apparatus dimensions were calculated using CFdesign simulation software. The built model demonstrated that simulated dimensions were accurate. The desired flow rates were successfully obtained. Because nine-cell niobium cavities cost over $50000, it is cost-prohibitive to conduct surface flow rate experiments directly on these cavities. The controlled-flow apparatus serves as a cost-efficient alternative. Additionally, it avoids many design issues that would have been encountered had the experiments taken place inside closed cavities. Hence, the successful construction of the controlled-flow apparatus is vital to future studies on EP.

## Introduction

Jefferson Lab research centers on the particle accelerator, which allows for examination of quarks and gluons by opening nuclei. To avoid deflections of the high energy electron beam within the niobium cavities, it is essential to maintain interior surface smoothness in the cavities. The high critical temperature and critical field of niobium makes the material ideal for SRF technology[1]. Electropolishing is currently the most effective method for removing surface detects on niobium. The EP solution, consisting of 1:9 volume ratio of hydrofluoric acid to sulfuric acid, causes a film, actually the reactant-depleted zone, to cover the niobium surface[2]. Film thickness varies according to topography. The film is thicker at microdepressions and thinner at microprojections[3]. The reaction mechanism is summarized in Eqn. 1[4].

$2Nb + 10HF + 2H_2O \rightarrow 2H_2NbOF_5 + 5H_2$ (Eqn. 1)

---

[1] H. Tian, S. G. Corcoran, C. E. Reece, and M. J. Kelley, "The Mechanism of Electropolishing of Niobium in Hydrofluoric-Sulfuric Acid Electrolyte," *Journal of the Electrochemical Society*, vol. 155, no. 9, pp. 563-568, July 2008.

[2] Ibid., 563-568.

[3] *Electropolishing: A User's Guide to Applications, Quality Standards, and Specifications.* 9th Ed. Jan. 2003. Delstar Metal Finishing, Inc.

[4] J. R. Delayen, J. Mammosser, L. Phillips, and A. T. Wu, "Alternative Electrolyte Composition for Electropolishing of Niobium Surfaces," Jefferson Lab publication 2001.

EP is widely studied in accelerator science because improvements in its effectiveness can prevent poor performance of the accelerator, a multi-million dollar project. For instance, researchers are working to find an alternative electrolyte solution due to the immense health risks associated with hydrofluoric acid[5]. Current tests suggest that methanol can serve as an adequate replacement for hydrofluoric acid. However, the method has yet to be fully characterized. Wide discrepancies in results were observed when the solution was tested at different concentrations or temperatures. Other EP studies geared towards evaluating different surface roughness measurement techniques for the electropolished surface[6].

Results from a computer simulation study suggested that controlling flow rates may be important[7]. In this study, the wall temperature became much higher than the bulk temperature of the electrolyte. Additionally, the surface flow rates changed drastically at different parts of the cavity. Simulations have suggested that flow rates can impact the quality of EP via changing the thermal conditions or through other unexamined mechanisms[8]. Full characterization of flow rate's impact on EP is necessary to provide consistent and efficient control on the polishing method. However, running full electrolyte flows through existing niobium cavities to obtain these measurements would cost up to tens of thousands per trial. Hence, a small-scale controlled-flow apparatus that supports niobium coupons can be best employed to achieve the task. The present study involves designing, building, and testing the first version of such an apparatus. The design needs to accommodate niobium coupons. Additionally, the set-up needs to ensure that the desired flow rates, ranging from 0 cm/s to 5 cm/s, can be adequately provided for liquids with viscosities similar to that of the EP electrolyte. A fully functioning apparatus can then be used to measure the effect of surface flow rates on the quality of EP.

## Methods

### *Flow Channel Design Using Simulation Software*

The dimensions of the flow channel were determined using simulations. The software CFdesign 2010 was used to produce simulated results. The apparatus needs to be long enough to accommodate the specimen, and, most importantly, ensure constant laminar flow across the specimen's surface. Due to the nature of piping, the apparatus is bound to see entrance effects in the flow. Initially, the flow would have to expand to the full cross-sectional area, resulting in a diagonal flow pattern. If the specimen is too close to the entrance, then the surface flow rate would not be at laminar steady-state. Such a design is undesirable, as flow patterns should be controlled in experiments testing the effect of flow rates on the quality of electropolishing. Hence, it is important that the specimen is placed at location that experiences laminar flow rates even at the highest tested volumetric flow rates.

---

[5] X. Zhao, S. G. Corcoran, M. J. Kelley, "Electropolishing of Niobium in Sulfuric Acid-Methanol Electrolytes for Radiofrequency Superconducting Accelerator Cavities Fabrication," J. Electrochemistry Society.

[6] H. Tian, G. Ribeill, C. Xu, C. Reece, M.J. Kelley, "A novel approach to characterizing the surface topography of niobium superconducting radio frequency (SRF) accelerator cavities," Applied Surface Science 257 (2011) 4781-4786.

[7] C. E. Reece, J. Ortega, and J. Mammosser, "Hydrodynamic Thermal Modeling of 9-Cell ILC Cavity Electropolishing and Implications for Improving the EP Process," Prof. 11th SRF Workshop, Beijing 2007.

[8] Ibid.

Although designing a very long apparatus would resolve the issue of entrance effects, such a design would restrict the choice of an effective pump. Hence, simulations are helpful at finding the proper balance in the trade-off between pump power and entrance effects. Models with varying dimensions were tested. Materials and boundary conditions were assigned to each design. Proper volumetric flow rates were calculated according to the corresponding cross-sectional area. These flow rates were set to the boundary conditions. After all the conditions have been set, iterations were run. Visual representations of the results were produced to analyze the flow paths of the simulant fluid. Various planes were created to observe the flow rates in different parts of the apparatus. Velocity vectors helped visualize the direction of the simulant fluids in the channel. Water and motor oil were chosen as simulant fluids.

### *Constructing the Controlled-flow Apparatus Model*

The overall design of the control-flow system is shown in Figure 1. The flow channel is situated below a reservoir and above a sump. The electrolyte is initially poured into the sump. A pump moves the electrolyte from the sump to the reservoir. The reservoir is connected to the flow channel. The electrolyte moves from the flow channel to the sump. Additionally, the reservoir has a drain to the sump for the remainder of the electrolyte. The chiller is attached to the sump. The flow channel models vertical electropolishing. The niobium coupon is attached to the side of the flow channel. One surface will be immersed in the electrolyte during electropolishing.

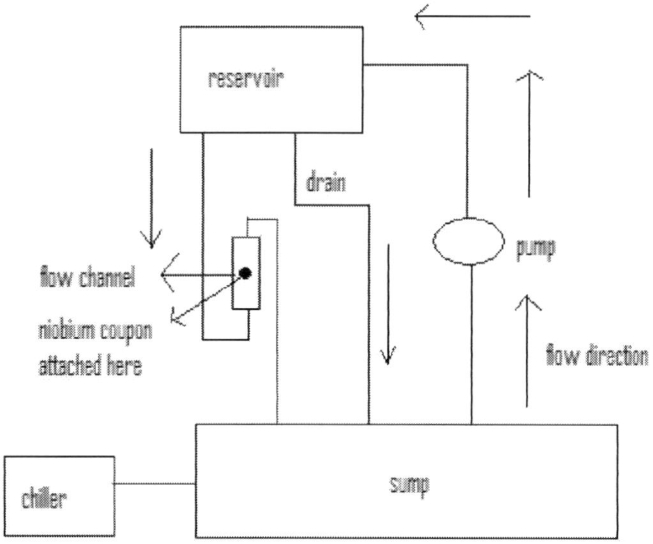


Figure 1: Simplified Flow Schematic

### *Testing the Apparatus*

The appropriate pressure heads needed to obtain certain flow rates with liquids with varying viscosities. Water was chosen as the first test fluid because it is the safest fluid. Due to the higher viscosity of the hydrofluoric acid-sulfuric acid mixture (typically ranging from 20 to 40 centipoises), further tests were completed using olive oil and antifreeze.

## Results and Discussion

### *CFdesign Simulation Results for Flow Channel: Water*

A selected set of simulation results are presented in order to demonstrate the design decisions of the apparatus. First, Figure 2a shows the velocity vector at the inlet and near the entrance. As can be seen in the image, the flow initially expands outwards to fill up the body of the apparatus. Eventually, the expansion will be complete and the velocity vectors would become vertical. Close observation of the velocity vectors at the surface of the specimen show the flow is constant (as the vectors are vertical). Figure 2b shows the velocity vectors approaching the location of the specimen. The velocity vectors are straight, indicating laminar flow.

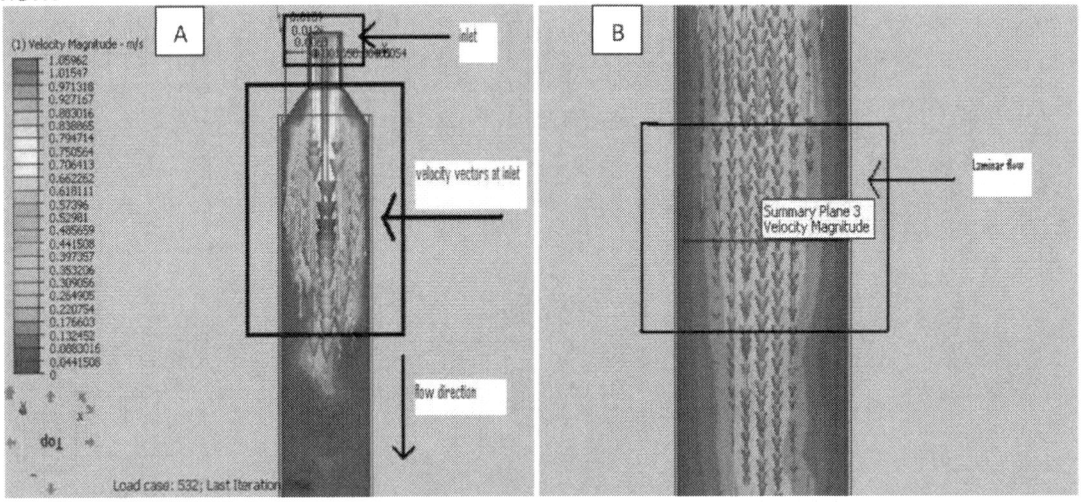


Figure 2: water velocity vectors inside the flow channel, (a) flow near entrance and (b) flow approaching niobium coupon

CFdesign shows that the flow across the specimen is laminar. In the simulation, a velocity flow rate resulting in a flow rate of 5 cm/s is chosen. Theoretically, a laminar flow regime has a Reynolds number between 1000 and 200000. The calculation of the Reynolds number is given in the Eqn. 2:

$$Re = \frac{\rho V^2 L}{\mu} \text{ (Eqn. 2)}$$

In Eqn. 2, V is the flow velocity of a single fluid particle, L is the length of the pipe, μ is the viscosity of the electrolyte, and $\rho$ is the density of the electrolyte. Since all the variables are fixed except for V, the velocities resulting in a laminar flow regime can be easily calculated.

The simulations show that the water achieves steady-state laminar flow in the flow channel with the dimensions: height 50 mm, width 5 mm, and depth 20 mm. The outlet pipe will have a diameter of 3 mm and the inlet pipe will have a diameter of 2 mm, see Figure 3:

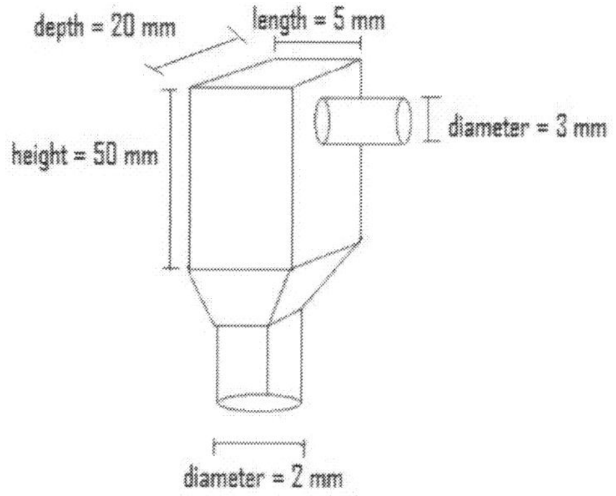

Figure 3: Design of Flow Apparatus

### *CFdesign Simulation Results for Flow Channel: Motor Oil*

Further simulations using motor oil as the simulant fluid, the 20 mm by 50 mm by 5 mm model has also proven to provide the desired steady-state, laminar surface flow.

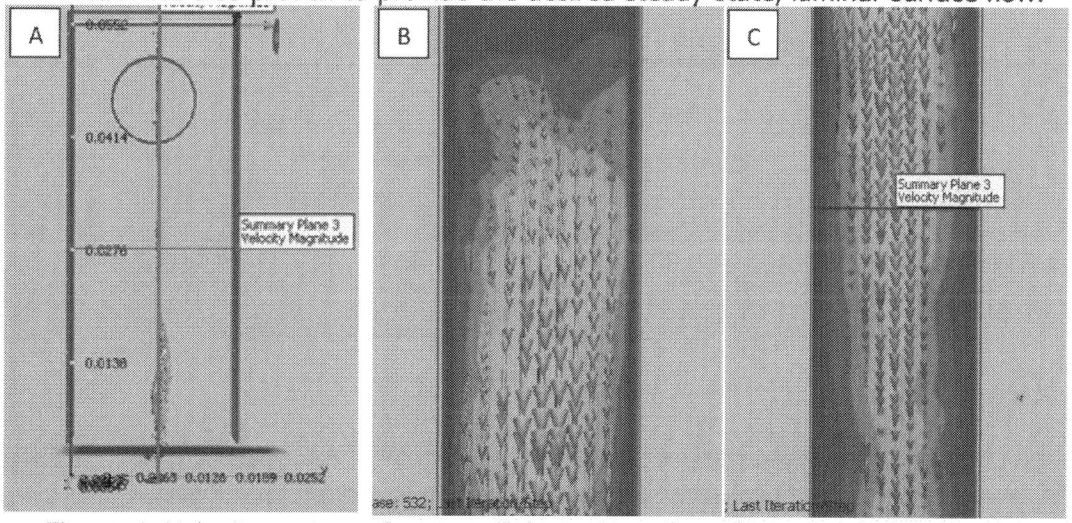

Figure 4: Velocity vectors of motor oil throughout flow channel, (a) side-view of velocity vectors at entrance, (b) frontal view of velocity vectors at entrance, and (c) velocity vectors approaching niobium specimen

While the initial flow vectors show expansion patterns, they reach steady-state far before reaching the specimen (Fig. 4). The side view of the flow vectors shows the fluid expanding outwards and then moving in a laminar fashion, hence moving out of the sight of the plane (Fig. 4a). This phenomenon is also evident in the front view, in which the velocity vectors are moving in a wavy, expanding pattern (Fig. 4b). Further down along the flow channel, the velocity vectors become straight, indicating laminar flow (Fig. 4c). The simulations using motor oil as the fluid confirmed that the determined dimensions (from the water simulations) for the flow channel were functional. Based on the design shown in Figure 1, a model apparatus was constructed. A pipe with the same length and cross-sectional area as the flow channel in the simulation acted as the channel in the model.

### ***Building the Controlled-flow Apparatus***

After a design was selected from our simulations and a preliminary version of the apparatus and its system was built accordingly, water and antifreeze were used as the simulant fluids. The two fluids were selected because they would test flow patterns at different fluid viscosities. Water has a viscosity of one centipoise at room temperature whereas antifreeze has a viscosity of fifteen centipoises, close to the viscosity of the hydrofluoric acid-sulfuric acid mixture. A pipe with the same length and cross-sectional area served as the flow channel in the test runs. The flow of stimulant fluids is shown in Figure 5.

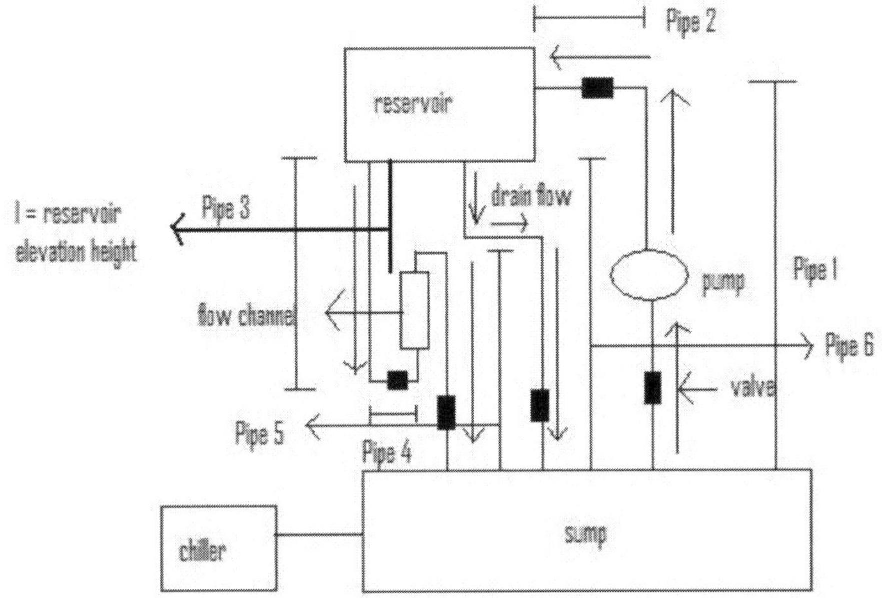


Figure 5: Design of Flow System

To obtain the desired surface flow rate across the specimen, the dimension *l*, the reservoir elevation height in Figure 3, was altered. Volumetric flow rates at each *l* were taken. The corresponding surface flow rate was calculated using the following equation:

Surface flow rate = Volumetric flow rate/cross-sectional area of the flow channel (Eqn. 2)

The relation between the reservoir elevation height, which is directly related to the pressure head, and surface flow rate is defined by the following equation:

$$\frac{v^2}{2g} = \varphi + z \quad \text{(Eqn. 3)}$$

In Eqn. 3, *v* is the electrolyte's flow velocity, *g* is the acceleration due to gravity, $\varphi$ is the pressure head in terms of the difference between the column elevation and the piezometer bottom, and *z* is the elevation at the piezometer bottom. According to Equation 3, it is expected that the fluid velocity would be inversely proportional to the reservoir elevation height. As the elevation of the reservoir decreases, less pressure head is available, hence reducing the flow velocity.

### ***Measuring Pressure Head Correlation with Flow Rates: Water***

Using the measurements, the complete dimensions for surface flow rates ranging from 0 cm/s to 5 cm/s can be found. The data presents a general correlation that relates $l$ to the desired flow rate. Ten trials were taken for various reservoir elevation heights. The surface flow rate showed a decreasing trend with decreasing elevation heights. The median, mean, and standard deviations of the flow rates at each elevation were calculated. Graphs depicting the dependence of flow rates on elevation heights are shown below:

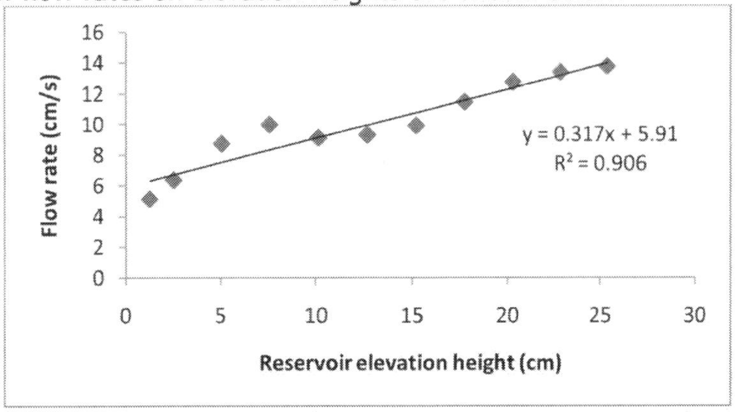


Figure 6: Median values of flow rates at various elevation heights

The linear trendlines serve as approximation tools for obtaining desired flow rates. Since the median is regarded to be the more accurate measure, the flow rate should generally be predicted with Eqn. 4:

$$v = 0.317 * l + 5.914 \text{ (Eqn. 4)}$$

Where v is the flow rate in cm/s and $l$ is reservoir elevation height in centimeters.

Based on Equation 3, it can be derived that:

$$v^2 = 2g(\varphi + z) \text{ (Eqn. 5)}$$

Here, it is shown that the squared value of the fluid velocity should be a linear function of the reservoir elevation height. The corresponding trendlines had higher $R^2$ values, indicating that this may be a more accurate prediction. The graphs depicting the relation between the squared values of velocity and reservoir elevation height are shown below:

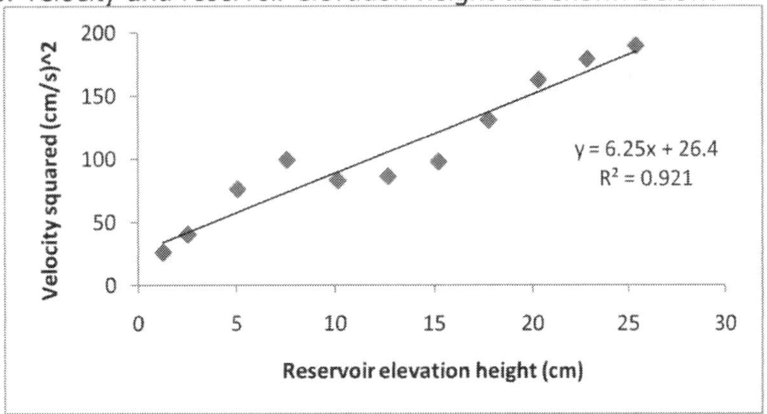


Figure 7: Median velocity squared values at various elevation heights

Using the median correlation, the flow rate can be predicted with Eqn. 6:

$$v = (6.25 * l + 26.4)^{1/2} \text{ (Eqn. 6)}$$

### ***Measuring Pressure Head Correlation with Flow Rates: Antifreeze***

Trials were conducted with antifreeze as the simulant fluid. With the higher viscosity of the fluid, the flow rate at each reservoir elevation height was reduced by the mean ratio of 1 to 7.30 (water flow rate to antifreeze flow rate). Antifreeze has a viscosity of 15 centipoises, close to the lower bound of the electrolyte's viscosity at 20 centipoises. The median flow rates at reservoir elevation heights 10 inches, 9 inches, and 8 inches were 2.10 cm/s, 1.68 cm/s, and 1.73 cm/s, respectively. Adjusting all length measurements into centimeters, the linear trendline for prediction flow rates of antifreeze (or any fluid with viscosity of 15 centipoises) is:

$$v = 0.0711 * l + 0.210 \text{ (Eqn. 7)}$$

As with the trials with water, Equation 3 predicts that the squared value of the flow rate correlates linearly with the reservoir elevation height. Adjusting the format of the equation accordingly, the following equation should be used to predict the flow rate:

$$v = (0.272 * l - 2.82)^{1/2} \text{ (Eqn. 8)}$$

### ***Measuring Pressure Head Correlation with Flow Rates: Olive Oil***

Trials were also conducted with olive oil as the simulant fluid. Olive oil has a viscosity of approximately 81 centipoises at room temperature. The flow rates were tested at reservoir elevation heights of ten inches, nine inches, and eight inches. The median flow rates at each height were 1.095 cm/s, 0.982 cm/s, and 0.943 cm/s, respectively. The mean flow rates at each height were 1.110 cm/s, 0.984 cm/s, and 0.918 cm/s, respectively. Using the medians and adjusting all length units to centimeters, the following linear trendline was produced to predict the flow rates of olive oil:

$$v = 0.0379 * l + 0.139 \text{ (Eqn. 9)}$$

Eqn. 9 is adjusted to fit the theory that Eqn. 3 predicts:

$$v = (0.0786 * l - 0.741)^{1/2} \text{ (Eqn. 10)}$$

### ***Error Sources***

Several conditions served as sources of error in the experiment. The changes in the diameter of the inlet stream resulted in occasional air bubbles in the flow channel. When the air bubble was present, the pipes needed to be bent and readjusted before the apparatus can function properly. In these cases, the times taken were generally shorter than the actual time due to the delay from rearranging the apparatus. Additionally, the beaker used to measure the volume of water for a given trial was not exact. Lastly, human reaction time affected the time recorded.

## Conclusions

A preliminary controlled-flow apparatus and a final model of the flow channel were built. Based on the CFDesign simulations, constructed model apparatus, and measured flow rates, a controlled-flow apparatus can be built out of fluoropolymer. The flow rate can be changed by adjusting the elevation height of the reservoir. Alternatively, the flow rate can be changed via changing the liquid level in the reservoir. This manipulation can be achieved with a valve downstream of the reservoir and at the discharge of the pump, both of which have already been incorporated in the flow schematic. Correlations between elevation height and flow rate were determined for water, antifreeze, and olive oil. The computational model, created using CFDesign, and the sample system provide an effective design for a controlled-flow apparatus. The fluoropolymer apparatus will be based on the constructed model and will test hydrofluoric acid-sulfuric acid and methanol-sulfuric acid solutions. The apparatus will determine the optimal flow rate for electropolishing. Operating the electropolishing process at the optimal flow rate would help obtain far more effective active layers in the niobium cavities, hence allowing the superconducting radiofrequency accelerator to operate efficiently. Because the fluoropolymer model would cost over $10,000 to build, it is critical that a preliminary apparatus is designed, built, and tested first. The overall saving achieved by this work is more significant as using an actual nine-cell cavity would cost over $50,000.

## Acknowledgments

The research was completed under the guidance of Dr. Michael Kelley. Resources and facilities were provided by the staff of Jefferson Lab. The Department of Energy and the College of William and Mary funded the primary researchers.

## References

[1] H. Tian, S. G. Corcoran, C. E. Reece, and M. J. Kelley, "The Mechanism of Electropolishing of Niobium in Hydrofluoric-Sulfuric Acid Electrolyte," *Journal of the Electrochemical Society*, vol. 155, no. 9, pp. 563-568, July 2008.

[2] *Electropolishing: A User's Guide to Applications, Quality Standards, and Specifications*. 9th Ed. Jan. 2003. Delstar Metal Finishing, Inc.

[3] J. R. Delayen, J. Mammosser, L. Phillips, and A. T. Wu, "Alternative Electrolyte Composition for Electropolishing of Niobium Surfaces," Jefferson Lab publication 2001.

[4] X. Zhao, S. G. Corcoran, M. J. Kelley, "Electropolishing of Niobium in Sulfuric Acid-Methanol Electrolytes for Radiofrequency Superconducting Accelerator Cavities Fabrication," J. Electrochemistry Society.

[5] H. Tian, G. Ribeill, C. Xu, C. Reece, M.J. Kelley, "A novel approach to characterizing the surface topography of niobium superconducting radio frequency (SRF) accelerator cavities," Applied Surface Science 257 (2011) 4781-4786.

[6] C. E. Reece, J. Ortega, and J. Mammosser, "Hydrodynamic Thermal Modeling of 9-Cell ILC Cavity Electropolishing and Implications for Improving the EP Process," Prof. 11th SRF Workshop, Beijing 2007.

# (409d) Application of Plant-Wide Dynamic Simulation for Safety Operation of An Ethylene Oxide Plant

**Xiongtao Yang**, Qiang Xu and Kuyen Li, Dan F. Smith Department of Chemical Engineering, Lamar University, Beaumont, TX

The manufacturing of Ethylene oxide (EO) involves critical exothermic reactions at high temperature and high pressure. EO is flammable and toxic gas In addition, the mixture feed of ethylene and oxygen has potential explosive dangers. It was reported a lot of fires and explosions occurred on EO Plants during the past decades, and most of those accidents were caused by the reactor entering explosive range, which were caused by equipment malfunctions, process failures, air/electricity/signal failure, etc. The EO manufacturing process is highly integrated process, where un-reacted reactants are recycled and reused in the reactors and lean water is recycled from downstream units to EO scrubber to knock down EO. Thus, the downstream process failures/upsets will propagate to the upstream reactions and affect the EO reactions. Therefore, EO plant operation needs sufficient dynamic investigation for its safety operation.

This work develops a plant-wide dynamic simulation model for EO manufacturing to study the safety impacts on the entire plant operations under various failure/ upset scenarios. The model includes two reaction subsystems and one recovery subsystem. It gives a comprehensive understanding of how the failures would affect the overall process. Decision-making information is provided so as to prevent dangerous accidents from occurrence. This information includes, but not limited to, the time duration between a failure/upset occurrence and resultant accident. The significance and benefits of this study are demonstrated through multiple case studies.

See more of this Session: Simulation Tools for Chemical Product Design
See more of this Group/Topical: Process Development Division

# (428a) Bioprocessing of Switchgrass to Ethanol Utilizing Thermostable Cellulases

**Lew P. Christopher**[1], Vasudeo Zambare[2], Archana Zambare[2] and Kasiviswanath Muthukumarappan[3], (1)Center for Bioprocessing Research and Development and Department of Chemical and Biological Engineering, South Dakota School of Mines & Technology, Rapid City, SD, (2)Center for Bioprocessing Research & Development, South Dakota School of Mines & Technology, Rapid City, SD, (3) Center for Bioprocessing Research and Development and Department of Agricultural and Biosystems Engineering, South Dakota State University, Brookings, SD

Switchgrass (*Panicum virgatum*), a perennial grass native to North America, is a promising energy crop for bioethanol production. The aim of this study was to optimize the enzymatic saccharification of thermo-mechanically pretreated switchgrass using a thermostable cellulase from *Geobacillus* sp. in a three-level, four-variable central composite design of response surface methodology (RSM). Different combinations of solids loadings (5 to 20%), enzyme loadings (5 to 20 FPU $g^{-1}$ DM), temperature (50 to 70 °C), and time (36 to 96 h) were investigated in a total of 30 experiments to model glucose release from switchgrass. All four factors had a significant impact on the cellulose conversion yields with a high coefficient of determination of 0.96. The use of higher solids loadings (20%) and temperatures (70 °C) during enzymatic hydrolysis proved beneficial for the significant reduction of hydrolysis times (2.67-times) and enzyme loadings (4-times), with important implications for reduced capital and operating costs of ethanol production. At 20% solids, the increase of temperature of enzymatic hydrolysis from 50 °C to 70 °C increased glucose concentrations by 34%. The attained maximum glucose concentration of 23.52 g $L^{-1}$ translates into a glucose recovery efficiency of 46% from the theoretical yield. Following red yeast fermentation, a maximum ethanol concentration of 11 g $L^{-1}$ was obtained, accounting for a high glucose to ethanol fermentation efficiency of 92%. The overall conversion efficiency of switchgrass to ethanol was 42%. High solids bioprocessing of switchgrass to ethanol at elevated temperatures could bring about significant savings of capital and operating costs, as it reduces the amount of enzyme and hydrolysis time needed, size of reaction vessels, water usage, and waste water treatment costs. The use of RSM for optimization of process parameters saves both time and costs, maximizing the amount of information that can be obtained while limiting the number of individual experiments. Furthermore, the RSM predicts the interaction between the independent variables, which results in improved accuracy and precision of the research data and facilitates their interpretation.

See more of this Session: Alternative Feedstocks for Energy and Chemicals
See more of this Group/Topical: Process Development Division

# (428b) Gas Fermentation of Industrial Waste Gases to Fuel and Chemicals

**Derek W. Griffin** and Michael A. Schultz, LanzaTech Limited, Roselle, IL

The food versus fuel debate is becoming increasingly controversial in the wake of the industrialization of eastern countries and the ever-growing population of the world. Sustainable and renewable feed stocks are being explored as energy sources in order to meet the increasing energy demand and decreasing fossil energy reserves. Current technologies to produce fuel or chemicals are limited to the use of sugar or cellulose feedstocks; these technologies are food-based, expensive, and capital intensive. As countries across the world continue to limit the amount of fuel/chemicals that can be produced from food sources, alternative feedstocks will have to be explored.

The LanzaTech process uses renewable resources to capture industrial waste gases and transform them into energy rich fuels and chemicals. Industrial waste gases or low value byproducts containing CO and/or $CO_2$ are fermented to chemicals/fuels such as ethanol, acetic acid, or 2,3-butanediol, among others. The LanzaTech technology can process a variety of industrial flue gases such as steel mill gas, coke oven gas, or syngas. These low value, non-food feedstocks are available at high volumes and can be utilized on-site, eliminating any feedstock handling or transportation costs. Initial estimates indicate that > 30 billion gallons per year of high value products can be produced from steel gases alone; this is a considerable contribution to the worldwide energy and chemical pool. The LanzaTech process offers superior carbon conversion, energy efficiency, and greenhouse gas emission performance compared to conventional and emerging routes to the same products.

Since 2005 LanzaTech has developed its gas fermentation technology from lab-scale batch experiments to a continuous pilot plant which has been in operation since 2008. The technology is described including the scale-up from lab reactors to pilot scale to the first demonstration plant that is currently under construction. The first LanzaTech technology to be commercialized is gas fermentation of raw steel mill waste gases to produce ethanol. This is the first process going to scale to produce ethanol from a non-food alternative feedstock. Ongoing technology development efforts include improving the first generation technology, modifying the technology to enable economic conversion of other gas sources, and producing next generation products.

# (428c) The Development and Commercialization of a Biodiesel Process Using Supercritical Fluid Methods

**Marc Privitera**, PreProcess, Inc., Walnut Creek, CA

Starting with high FFA waste feedstocks, a process was developed and scaled to commercial operations. This paper will survey the learnings of the effort by describing the chemical engineering techniques applied to solve the challenges presented to the team. The effort was successful in scaling the technology from bench (5 mls) to commercial. Many lessons were learned on the journey to producing and selling ASTM 6751 product for the market in truckload quantities.

See more of this Session: Alternative Feedstocks for Energy and Chemicals
See more of this Group/Topical: Process Development Division

# (428d) The Cultivation of Algae and Its Conversion to Biodiesel

**Spencer T. Glantz**[1], **Jasmin Imran Alsous**[1], **Daniel Choi**[1], Warren D. Seider[2] and Stuart W. Churchill[3], (1)Chemical and Biomolecular Engineering, University of Pennsylvania, Philadelphia, PA, (2)Chemical and Biomolecular Engineering, The University of Pennsylvania, Philadelphia, PA, (3)Chemical & Biomolecular Engineering, The University of Pennsylvania, Philadelphia, PA

**The Cultivation of Algae and itsConversion to Biodiesel**

DanielChoi, Spencer Glantz, JasminAlsous, Warren D. Seider, Stuart W. Churchill

Departmentof Chemical & Biomolecular Engineering,University of Pennsylvania

Apresentation at the 2010 Annual Meeting under the title "The Production ofAlkanes from Algae" drew a large audience, even though it was on Fridaymorning, and evoked considerable discussion. We have taken that discussion andthe report upon which that presentation was based as starting points for are-examination of the potential of cultivated algae as a source of biofuels.

Thepossibility of substitution of biofuels for petroleum-based ones, even in smallpart, has evoked considerable interest for economic, political, and environmentalreasons. There is a general recognition that the cultivation and processing ofalgae into biofuels have several advantages with respect to other agriculturalcrops in terms of the conversion of solar radiation into to usable forms ofenergy-containing liquids and solids. First, the cultivation of algae does notdirectly affect the food supply as does the diversionof corn or soybeans. Second, algae grow by absorbing $CO_2$, and if converted to a fuel and burned, are $CO_2$-neutral. Third, the production of lipids per acre far exceeds thatof any other agricultural commodity, and thereby lends itself to differentapplications than those of the low energy-density cellulosic ethanol derivedfrom corn and other such conventional crops. For example, the lipids in thealgae can be readily converted to biodiesel or conventional fuel-rangehydrocarbons. Fourth, it may be possible to grow algae with salty andcontaminated water and thereby on land unsuitable for traditional agriculture,as well as to capture phosphorous from runoff.

Theaforementioned prior investigation was based on the production of 17,500 BPD (225,000 lbs/hr) of biodiesel – a rate compatible in scale with those of petroleumrefining – on a site in flat, undeveloped land near the small town ofThompsons, Texas, which is about 25 miles southwest from Houston, 30, 60, and60 miles northward from three petroleum refineries on the Gulf Coast, and on abranch of the Atchison, Topeka, and Santa Fe Railway. This area receivesinsolation varying from the equivalent of 112 W/m$^2$ in January to 250W/ m$^2$ in June, so the plant can be operatedyear-round. A near-by coal-fired power-plantthat generates 2500 MW of electricity produces sufficient $CO_2$ in its stack gas to meet that need foralgae cultivation. The prior investigation utilized three modules: the Simgae$^{TM}$Algal Biomass Production System for the cultivation; the OriginOil$^{TM}$ single-step technology for theextraction of the algal lipids; and conventional catalytic hydrotreating for the conversion of the lipids inton-alkanes (green diesel). It was concluded that the process might beprofitable, but that a capital investment of the order of 2.8 billion dollarswould be required – far too great in consideration of the many technicaland fiscal uncertainties. The present analysis is based on that same locationand rate of production of biofuel but examines improved methods of cultivationand processing.

Ourobjective was to identify modifications in the processing that would reduce thecapital investment significantly while improving the technology and maintainingor decreasing the fixed and variable costs. Improved alternatives were soughtfor all three of the modules that characterized the previous investigation. Thefirst consideration was reduction of the principal factor in the capital cost, namely the acreage required for the cultivation of the algae. Two improvementswere identified in that respect. The first is the utilization of theautotrophic-heterotrophic process of cultivation proposed by Miao and Wu in2006 (*Bioresource Technology*), which involves aphotosynthetic stage but shifts most of the growth to a fermentationstage. In past decades, such a cultivation model has been economicallyunfeasible, given that the most widely employed source of organic carbon in algalfermentation has been glucose, a costly feedstock. However, in 2010, O'Grady (*Bioprocessand Biosystems Engineering)* found that crude glycerol could serve as alow-cost substitute for glucose for the cultivation of some heterotrophic culturesof algae. In a complementary study, Heredia-Arroyo (*Applied Biochemistry andBiotechnology)* found that, for a crude glycerol input of 83 poundsper 1,000 lbs of algae, the biomassproductivity and cellular lipid contents observed during fermentation weresimilar to those for algae grown in a glucose-rich media. At such aconcentration, roughly 35 million pounds of glycerol are required annually tomeet the 17,500 BPD biodiesel production rate. Given that 200 million pounds ofcrude glycerol are produced annually as a by-product of the transesterificationprocess by which biodiesel is chemically synthesized, crude glycerol willtherefore be a readily available feedstock and was proposed as the substratefor heterotrophic growth.

Thesecond modification is an increase of the diameter of the tubes through whichthe algae flows as a suspension from 6 inches to 12 inches. The improvedrate of cultivation results in a 7.5-fold decrease and the larger diameterin a roughly 1.5-fold decrease, for an overall 11-fold decrease in the requiredarea, and thereby in the cost of the land.

Theautotrophic-heterotrophic process of growth is dependent on the use of adifferent species of algae, namely *Chlorella protothecoides*, in place of the *Nannochloropsis*used in the prior study. This strain requires the use of fresh rather thansaline water and puts a premium on its recovery and reuse.

A betterunderstanding of the Single-Step Extraction process developed by OriginOil™for the separation of the lipids from the algae was recently made possible by thepublication of several patents that describe in detail the technologiesinvolved, particularly quantum fracturing and electromagnetic pulsation. Giventhis new information, the equipment required for Single-Step Extraction could bedesigned and analyzed in depth, permitting a more accurate economic estimationof the variable and fixed costs associated with the second module, namely of6/lb compared to 56/lb for conventional solvent- or mechanical-extraction.

As analternative to the catalytic hydrotreating processused in the prior analysis, we investigated a lipid-processing module based oncatalytic transesterification. In this process algallipids are converted to biodiesel, which consists of fatty acid methyl esters,as opposed to the n-alkanes that constitute "green" diesel. Abyproduct of this reaction is crude glycerol. A plant to carry out this latterprocess was designed. The production of pure biodiesel involves two reactors inseries with interstage glycerol removal, followed by aseparation train that requires vacuum distillation, neutralization and water-washing. This process was simulated using Aspen PLUS™software, andthe composition of the biodiesel-product stream was found to meet the ASTM D6751Standards for biodiesel fuel. The results were compared directly with thosepublished for a similarly sized catalytic hydrotreatingplant and found to be superior, especially when considering the potentialrecycle of glycerol for use in the fermentative phase of algal cultivation.

Thecombination of these three revised modules results in an overall process thatreduces both the costs and the technical uncertainties, although somesignificant uncertainties remain. In particular, pilot-scale tests areessential to be sure that some unacceptable aspect has not been overlooked orits impact under-estimated.

One majoruncertainty is the marketability and the price of the 1.9 pounds of dry biomassproduced as a byproduct for every pound of biodiesel. The overall processappears to be profitable even without selling the biomass, but if not sold itwould then pose a problem of disposal. The production of biomass from theoperation described herein is less than 2% of the national demand for animalfeed, and therefore its sale would not appear to disturb the market. Inaddition biomass from algae has a higher protein content than any of the commonagricultural sources of animal feed, and, on that basis, might evensupport a premium price. However, the proof of its acceptabilityby animals and the absence of any side-effectswill require tests and therefore some time. There are also other potentialapplications for the biomass, for example, in the pharmaceutical andpower-generation industries.

Aneconomic analysis based on the revised modules described above suggests that atthe current market price of $3.30 per gallon for pure biodiesel fuel, theprocess might be profitable, with net earnings of almost $340 million annually.Even more promising is the reduction in the capital investment to $1.2 billion,which is almost 60% less than that estimated for the prior process. The verytentative economics for the revised process are favorable and translate to aReturn on Investment (ROI) of 32% and an Investor's Rate of Return (IRR) of35%.

See more of this Session: Alternative Feedstocks for Energy and Chemicals
See more of this Group/Topical: Process Development Division

# Catalytic Technology for Converting Heavy Oils Derived from Alberta Oil Sands into Petrochemical Feedstock

*Duke DuPlessis, Katherine Wilson,*

Alberta Innovates: Energy and Environment Solutions

Suite 2540, 801 - 6 Avenue SW

Calgary, Alberta, Canada T2P 3W2

*Michel Berghmans, Chris Foy, Grace Kim and Vasily Simaznhenkov**

NOVA Chemicals Research and Technology Centre

2928 – 16 St. N.E.

Calgary, Alberta, Canada, T2E 7K7

*e-mail: simanzv@novachem.com

**ABSTRACT**

The North American petrochemicals market is currently undergoing significant changes which are caused by development of shale gas resources. This unconventional gas source recently became economical due to availability of new technologies. Eagleford and Marcellus shale gas resources, for instance, are proven to be very rich in high value gas liquids, which are a competitive feedstock for the petrochemicals industry. Ethane may be available in amounts, which ensure sustainable growth for the foreseeable future. However, this may bring new challenges in terms of a short supply of propylene monomer due to the use of lighter, gas-liquids feedstock. These produce less propylene than do heavier, crude-oil-based feedstock. New technologies which can economically produce propylene from sustainable hydrocarbon feed could fill that gap. On the other hand, oil sands operations in Alberta are poised to experience significant growth over the next number of years. The 2008 production of 1.31 million bbl/day of synthetic crude oil and bitumen blends from the oil sands is projected to increase to over 3 million bbl/day by 2018. This guarantees a large output of heavy oil fractions, such as vacuum gas oil, from oil sands upgraders. It reveals a new opportunity for production of petrochemical feedstock from an alternate source, provided these alternate feedstocks can be transformed into feed for petrochemical plants, which includes both fundamental building blocks of petrochemical industry: ethylene and propylene.

Over the past eight years NOVA Chemicals has systematically studied catalytic technologies that can be used to convert heavy oil sands derived fractions into olefins, aromatics-rich products and other high demand petrochemicals and refinery fuel products. Work was carried out in collaboration with external academic institutions and with the support of the Alberta government. This work has resulted in the development of two novel, breakthrough processes, which allow significant feedstock flexibility to petrochemicals producers and integration opportunities with heavy oil refining/upgrading operations. Both processes were specifically designed for treating heavy gas oil fractions, which are rich in aromatics and naphthenics, and which contain only small amounts of paraffins. This paper presents the process configuration and research results from studies undertaken, which show that it is feasible to produce petrochemical feedstock, petrochemicals and/or refinery fuel products from heavy oil fractions derived from oil sands at competitive prices via two different catalytic processes.

While originally NOVA Chemicals' goal was to develop technologies that were suitable to provide the olefins industry with an alternative feedstock source, the increase of NGL supply from shale gas sources would appear to make this less necessary. However, the market demand for propylene in North America may result in a renewed interest in the application of this process to fill the potential gap in propylene production.

Keywords: Oil Sands, Petrochemicals, FCC, Hydrotreatment, Gas oil

## 1. INTRODUCTION

Growing production of oil from oil sands in Alberta and the accompanying heavy oil fractions presents a new opportunity for feedstock diversification in the petrochemical industry and integration with refining and upgrading industries, provided that these heavy oils can be transformed economically into feed to petrochemical plants.

Most of the oil sands in Canada are located in three major deposits: the Athabasca-Wabasca oil sands of north northeastern Alberta, the Cold Lake deposits of east northeastern Alberta, and the Peace River deposits of northwestern Alberta. Oil sands are mined to extract the oil-like bitumen, which is then converted into synthetic crude oil or refined directly into petroleum products by specialized refineries. Current cost of production of synthetic crude from oil sands is approximately \$33-\$37 per barrel[1]. Bitumen from oil sands production is now at around one million bbl/day[2].

Today's modern petrochemical plants and oil refineries can process hydrocarbon feedstocks from conventional light and heavy crude oils in the following three technically advanced facilities to make substantial volumes of basic petrochemicals as long as the feeds are largely paraffinic.

1. Steam crackers for ethylene or ethylene & propylene or ethylene, propylene & benzene depending upon the type of feed used.
2. Catalytic crackers for propylene or ethylene & propylene depending upon the specific properties of the catalyst used.
3. Catalytic reformers for benzene.

The above processes are mature for which commercially proven technologies are available.

In contrast, the same facilities can produce only inadequate volumes of petrochemicals when the hydrocarbon feedstocks are derived from non-conventional sources such as oil sands bitumen. Most of the reduction arises because these hydrocarbon feeds are largely aromatic rather than largely paraffinic.

NOVA Chemicals has undertaken a project partly sponsored by the Alberta Innovates: Energy and Environment Solutions (AIEES) in the area of processing of heavy oil fractions from oil sands bitumen to petrochemical feedstock. In the framework of the project, NOVA Chemicals has developed two catalytic technologies: 1. NOVA Heavy Oil Cracking (NHC) technology, which converts heavy bitumen-derived gas oil into olefins, gasoline and cycle oils (aromatics rich diesel fraction); 2. Aromatic Ring Opening (ARO) technology, which converts low value cycle oils into valuable petrochemical feedstock containing mainly light paraffins and benzene, toluene and xylenes (BTX). NHC is a fluid catalytic cracking (FCC) type process, whereby, ARO is carried out in a trickle bed reactor. As a result, both NHC and ARO are based on mature industrial process platforms.

## 2. PROCESS SCHEME

A number of process schemes were considered in the development of business cases before arriving at few preferred schemes, one of which is reported in this paper. This flow scheme has an oil sands bitumen upgrader, with a NOVA Heavy Oil Cracking (NHC) unit plus an Aromatic Ring Opening (ARO) Unit and an Ethylene Plant.

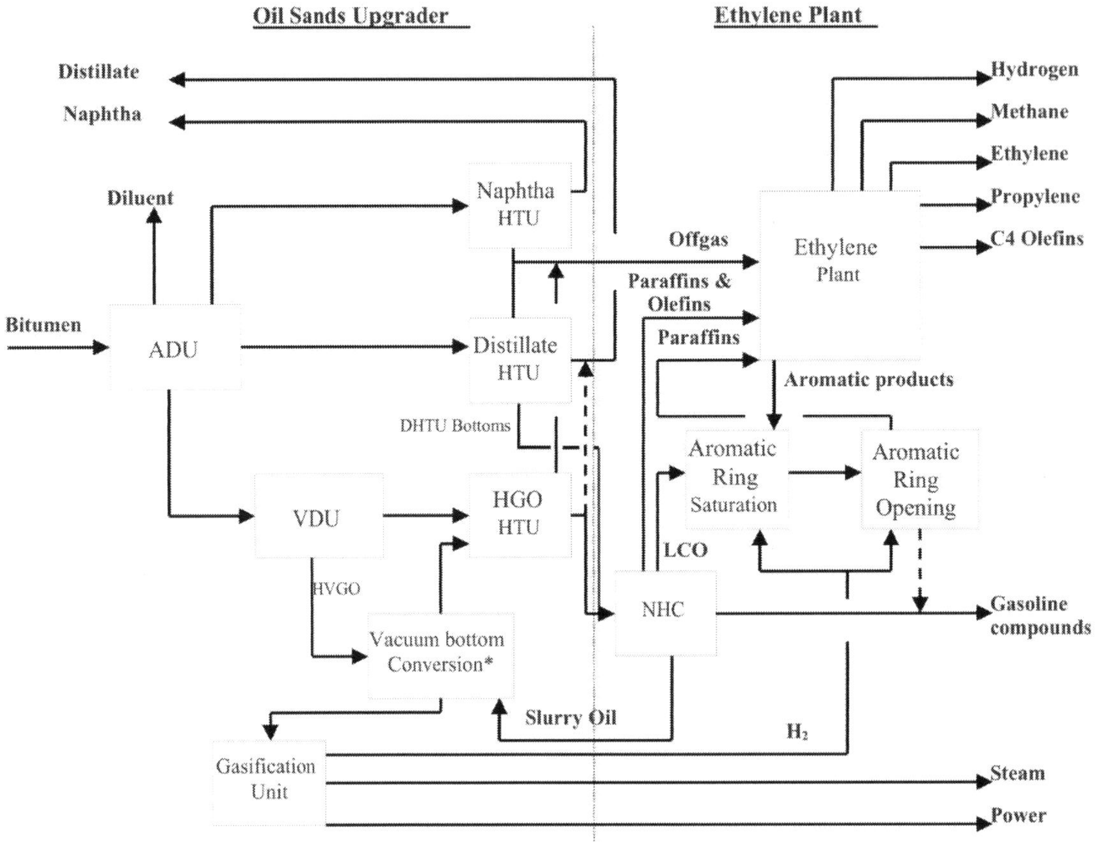


*Vacuum bottoms conversion options include coking, hydrocracking and asphaltene separation units.

**Figure 1.** Integration of Oil Sands Upgrader with a Petrochemicals Plant and the NHC and ARO Units.

Figure 1 shows an integration of oil sands upgrader with a petrochemicals plant and the NHC and ARO units. In Figure 1, bitumen is distilled in an atmospheric distillation unit (ADU) and the distillates are sent to the hydrotreating units (HTU). The atmospheric residue enters the vacuum distillation unit (VDU), where heavy gas oil is produced and the bottoms are sent into an asphaltene separation unit or an alternative vacuum bottoms processing unit. Ethane, ethylene, propane and propylene from the oil sands upgrader off gases are collected and passed to the separation section of the ethylene plant for processing. Separated ethane and propane are recycled to the heaters (furnaces) of the steam crackers for cracking into ethylene and propylene respectively. Heavy oils that have been hydrotreated in the upgrading section are passed on to the NHC unit to produce light olefins, gasoline, cycle oils, and slurry.

The cycle oils from the NHC Unit are transferred to the ARO Unit. This unit converts the aromatic compounds, (mainly two and three ring aromatic compounds) in the C5+ stream from the Ethylene unit as well as those in the cycle oil from the NHC unit into paraffins such as ethane, propane and butane. These paraffins can be used as feedstock for the Ethylene plant. As a result, the availability of feedstock for olefins production in an ethylene plant is higher with an ARO unit.

## 3. STEAM CRACKING

Steam cracking of largely paraffinic hydrocarbon feedstocks from conventional sources provides virtually all of the world's supply of ethylene along with substantial volumes of propylene, C4 olefins and BTX. The steam cracking method of producing olefins, especially for those who are not familiar with it, is shown below as Figure 2 in the form of a simplified flow diagram. This diagram is self-explanatory.

**Steam Cracking – Simplified Block Flow Diagram**

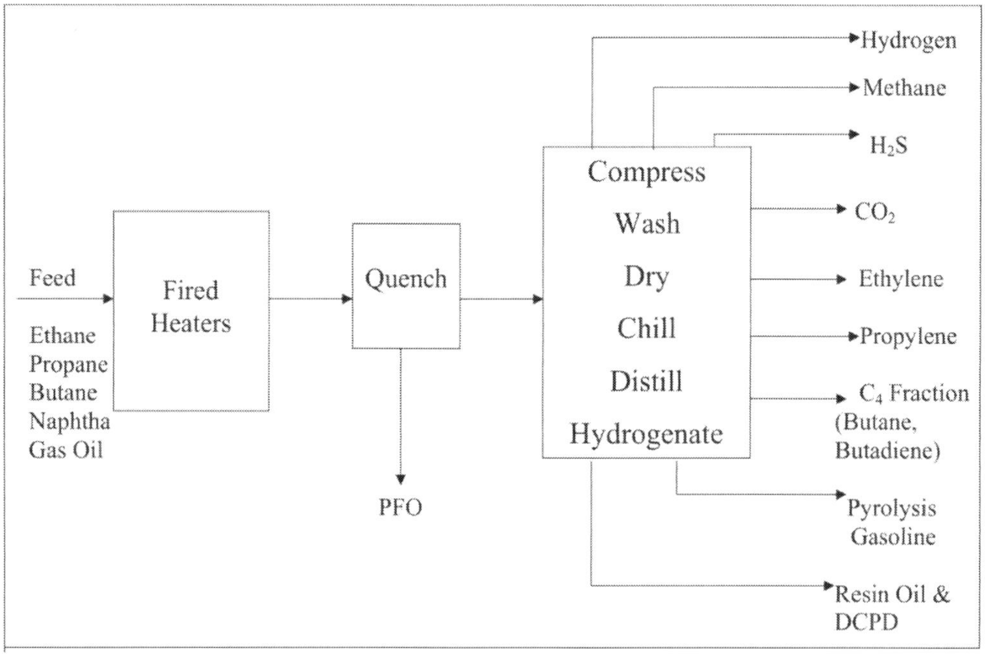


**Figure 2.** Simplified Block Flow Diagram of a Steam Cracking unit

A typical yield pattern of steam crackers in relation to the feed used as adapted from Wittcoff and Reuben[4] is shown in Table 1. One can see from the Table 1 that a shift from liquid to light feeds, particularly ethane, will result in significant reduction of propylene and BTX.

**Table 1.** Conventional Feeds for Steam Cracking – Dependence of Yields on Feedstock

| Prod. (wt. %) FEED | Ethylene | Propylene | Butadiene | BTX | Others |
|---|---|---|---|---|---|
| Ethane | 82.3 | 1.8 | 2.6 | 0.7 | 12.6 |
| Propane | 43.7 | 21.2 | 4.1 | 4.8 | 26.2 |
| Butane | 42.2 | 14.6 | 3.9 | 4.8 | 34.5 |
| Light Naphtha | 29.3 | 14.4 | 4.0 | 13.8 | 38.5 |
| Full Range Naphtha | 25.0 | 12.8 | 4.5 | 11.3 | 44.2 |
| Gas Oil | 25.2 | 8.3 | 4.8 | 11.2 | 46.6 |

## 4. FLUID CATALYTIC CRACKING TECHNOLOGIES FOR PETROCHEMICALS PRODUCTION

Fluid catalytic cracking (FCC) is a basic process used in the refining industry for converting heavy oil distillates into more valuable lighter components, notably motor gasoline. The first FCC unit was built by Standard Oil Company of New Jersey, (Esso), in 1941. The FCC process has continued to evolve, and its efficiency has greatly improved as a result of many technical innovations and the development of more active catalysts. Of particular importance was the introduction of zeolite catalysts in the late 1960s, which greatly increased the conversion to gasoline.

Fluid catalytic cracking of largely paraffinic hydrocarbon feedstocks from conventional sources in oil refineries provides about 35% of the world's supply of propylene. The proportion of the catalytic cracked feed in the form of propylene however is of the order of 5-9 volume percent. Numerous methods are now being employed or investigated for maximizing the yield of propylene or ethylene & propylene by catalytic cracking. Two examples of these are Deep

Catalytic Cracking (DCC) and the Catalytic Pyrolysis Process (CPP) developed by the Research Institute for Petroleum Processing (RIPP) of SINOPEC, Beijing, China[5-8].

The typical operating parameters for a DCC unit in comparison with a normal FCC and steam cracking (SC) units are shown in Table 2.

**Table 2.** Typical Operating Parameters for a DCC Unit[6]

| | DCC | FCC | SC |
|---|---|---|---|
| Res. Time, (sec.) | 10-16 | 1-30 | 0.1-0.2 |
| Cat/Oil, (w/w) | 9-15 | 5-10 | - |
| Steam, (wt.% of Feed) | 10-30 | 1-10 | 30-80 |
| Rx.Temperature (°C) | 549-593 | 510-549 | 760-871 |
| Pressure, (psig) | 10-20 | 15-30 | 15 |

Table 3 shows a comparison of CPP versus steam cracking in terms of yields. The CPP process results in high propylene yield and ethylene yields based on paraffinic feedstock.

**Table 3.** Yields from a CPP Unit versus Steam Cracking (SC)[9]

| Process | Steam Cracking | Steam Cracking | CPP |
|---|---|---|---|
| Feedstock | AGO<br>Atmospheric Gas Oil | LVGO<br>Light Vacuum Gas Oil | AR<br>Atmospheric Residuum |
| Reaction Temperature (°C) | 800 | 800 | 640 |
| Product Yields (wt.%) | | | |
| Ethylene | 26.60 | 23.00 | 20.37 |
| Acetylene | 0.20 | 0.20 | 0.002 |
| Propylene | 13.75 | 14.50 | 18.23 |
| Butadiene | 4.39 | 4.65 | 0.40 |
| Benzene | 5.46 | 5.12 | 2.05 |
| Toluene | 3.25 | 3.46 | 3.02 |
| $C_8$ Aromatics | 1.93 | 2.07 | 2.43 |

Others are developing or have developed FCC type technologies for cracking heavy, largely paraffinic feedstocks into light paraffins. An example is Universal Oil Product's (UOP's) "PetroFCC Unit". PetroFCC increases olefins while reducing fuel yields. Yields strongly depend on feed and catalyst type.

## 5. THE NHC PROCESS

In contrast to technologies, described above, the NHC technology starts with heavy, aromatic oil fractions, such as vacuum gas oil derived from oil sands bitumen, as feedstock. The process entails FCC technology, which is proven, with specially designed catalyst.

Table 4 summarizes some examples of catalytic processes now being used to convert hydrocarbon streams of different origin into petrochemical feedstocks or intermediates for petrochemicals production. This table is not complete with regards to all the available technologies and feed streams; however, it is a very good representation of industrial research trends in the field of converting petroleum fractions into petrochemical feedstock. One can see from the Table 5 that both NHC and ARO technologies are unique technologies, which allow converting heavy, aromatic oil sands derived fractions into valuable petrochemical feedstock.

It can be seen from Table 4 that only three technologies are available for conversion of largely aromatic crude oil fractions into petrochemical feed or intermediates. These are ARINO, ARO and NHC. It is important to emphasize the difference between ARINO and ARO technologies. ARINO technology is meant for treatment of highly aromatic, light fractions. This feed contains mainly single ring aromatics. Whereby, ARO technology was developed for heavy feeds, which contain poly-ring aromatics with up to five aromatic rings

**Table 4.** Feedstocks used for different catalytic processes to convert petroleum fractions into petrochemical feed.

| Crude nature / Crude source | *Olefinic* | *Paraffinic* | *Naphthenic / Aromatic* | *Aromatic* |
|---|---|---|---|---|
| ***Conventional crude derived light fractions*** | Superflex™ | ACO™ | MILOS | ARINO™ |
| ***Conventional crude derived heavy fractions*** | | HS-FCC<br>CPP<br>DCC<br>PetroFCC™<br>IndMax™ | Maxofin™<br>HS-FCC<br>NExCC™<br>High Olefins FCC<br>CPP<br>DCC<br>PetroFCC™<br>IndMax™ | |
| ***Conventional crude derived residuum*** | | MILOS | PetroRiser℠<br>High Olefins FCC<br>IndMax™ | |
| ***Oil Sands derived heavy fractions*** | | | **NHC** | **NHC**<br>**ARO** |

### *5.1 Equipment*

The reaction unit for the experiments was a confined fluidized bed reactor system, which had five sections – the oil and steam feed systems, the reaction zone, the temperature control system, and the product separation and collection system. For each experiment, a set amount of catalyst was loaded into the reactor with an effective volume of about 580 ml. A set amount of distilled water was pumped into the furnace to produce steam for fluidizing the catalyst. The steam was mixed with the feedstock, which was kept in an oven and fed simultaneously by a feed pump. The mixture was heated to approximately 500°C in a preheater, and then fed into the reactor. In the reactor, the feed was atomized and contacted with fluidized catalysts to cause catalytic pyrolysis to occur. The reaction product from the reactor was cooled and separated into the liquid and gas products, which were collected respectively. After the reaction, the spent catalyst was removed from the reactor by a vacuum pump.

The catalytic reaction product consisted of pyrolysis gas, liquid and coke. An Agilent 6980 gas chromatograph with Chemstation software was used to determine the volume percentages of pyrolysis gas components, which were converted to mass percentages using the ideal gas equation of state. The pyrolysis liquid was analyzed with a simulated distillation gas chromatograph to obtain the mass percentages of gasoline, diesel and heavy oil. Coke content on the catalyst was determined using a coke analyzer.

*5.2 Experimental Results*

The main reactions occurring during the NHC process are shown in Figure 3.

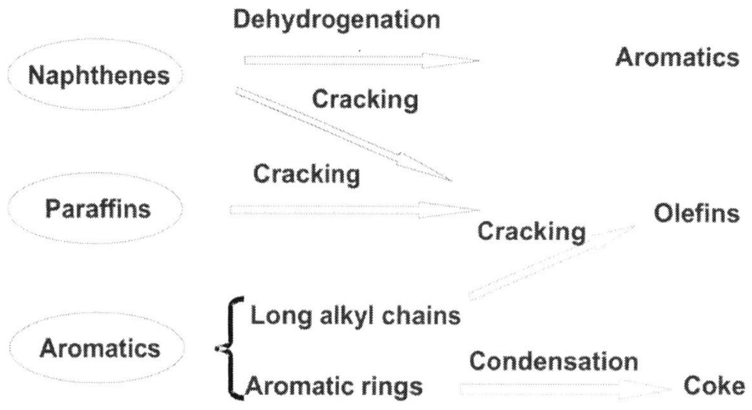


**Figure 3.** Main chemical reactions happening in NHC process.

The key to the success of the project is the ability to invent and produce catalysts capable of cracking the aforementioned feedstocks preferentially to light olefins. The significant difference between high severity FCC type catalysts and conventional FCC catalysts is higher acidity of high severity FCC catalysts. High acidity is typically reached by high content of zeolites in the catalyst. High content of zeolites in FCC catalysts is known to compromise an important property of the catalyst: attrition rate. This is set to be below 3% for majority FCC catalysts. Special additives to the NHC catalyst family allowed us to insure high catalyst acidity and very low (less than 2%) attrition rate of the catalyst. An extensive experimental program was conducted at eight different temperatures, five WHSV (Weight Hourly Space Velocity) points, five catalyst-to-oil weight ratio points and five steam-to-oil weight ratio points. Results show that NHC catalyst shows higher yield of light olefins, in particular propylene and ethylene compared to base high severity FCC catalyst. Table 5 shows yields of catalysts A and B compared to yields of a base high severity FCC type catalyst.

**Table 5.** Effect of Feedstock and Catalyst on NHC yields of light olefins.

| | Base Catalyst | Catalyst A | Base Catalyst | Catalyst B |
|---|---|---|---|---|
| Feed Type | HAGO* | HAGO* | HVGO** | HVGO** |
| Temp (°C) | 660 | 660 | 660 | 660 |
| Ethylene (wt.%) | 12.3 | 11.7 | 6.9 | 9.2 |
| Propylene (wt.%) | 19.4 | 22.3 | 10.7 | 16.1 |
| Butylenes (wt.%) | 9.0 | 12.0 | 5.9 | 9.5 |
| Total Light Olefins (wt.%) | 40.7 | 46.0 | 23.5 | 34.8 |

* Heavy Atmospheric Gas Oil

** Heavy Vacuum Gas Oil

Table 5 shows the yields of light olefins from largely aromatic gas oils when catalysts A and B are used in place of a base FCC catalyst, a technically advanced catalyst developed for cracking largely paraffinic gas oils to light olefins. Both catalyst A and catalyst B surpassed the base FCC catalyst in producing light olefins from the largely aromatic gas oils.

**Table 6.** NHC versus steam cracking yields of light olefins.

| | Steam Cracking | Catalyst C | Steam Cracking | Catalyst C |
|---|---|---|---|---|
| Feed Type | HAGO | HAGO | HVGO | HVGO |
| Temp (°C) | - | 660 | - | 660 |
| Ethylene | 18.8 | 11.9 | 15.6 | 8.9 |
| Propylene | 11.6 | 24.2 | 11.9 | 17.4 |
| Butylenes | 6.0 | 11.7 | 6.0 | 8.5 |
| Total Light Olefins (wt.%) | 36.4 | 47.8 | 33.5 | 34.8 |

Table 6 shows the advantage of NHC technology over steam cracking for the production of light olefins. It compares the yields of light olefins from steam cracking and from the NHC process. The catalytic process produced more light olefins in total from heavier, largely aromatic feedstocks although it produced less ethylene than steam cracking. Non-catalytic steam cracking is based on a free radical reaction mechanism favored by higher reaction temperatures. Catalytic cracking is based on the carbenium-ion mechanism, which preferentially produces more propylene than ethylene[8].

The results shown in Tables 5 and 6 indicate that new engineered catalyst variations, such as catalysts A, B and C together with a specially designed FCC unit allow substantial volumes of light olefins to be produced from largely aromatic gas oils derived from oil sands bitumen. Depending upon market conditions, heavy cycle oil (HCO) and light cycle oil (LCO) formed during catalytic cracking in the NHC process can be hydrotreated by the ARO process to transform multi-ring aromatics to paraffins for feed to a steam cracker.

A possibility is that alternate feedstocks for the NHC process could include a wider range of light and heavy gas oils, deasphalted oil or oil from waste plastics.

## 6. ARO TECHNOLOGY.

Today the refining industry is required to process increasingly heavy feedstocks and refiners have to look for routes to improve economic margins. One of those routes is increasing the processing of middle distillates, such as gas oils (GO) and light cycle oils (LCO) so that the processed products can be blended into various product streams as desired. LCO is one of the streams from the FCC type units that is normally disposed off as low-value heating oil. It is characterized by a high content of aromatics (> 70 wt.%) and relatively high amounts of heavy aromatic sulfur compounds and nitrogen compounds (3 and 0.08 wt.%, respectively)[10]. GO may contain more than 50 wt.% aromatics as well as relatively high amounts of sulfur and nitrogen (more than 2.5 and 0.08 wt.% respectively).

In the ARO process, LCO and GO upgrading may be achieved by a two-step process involving catalytic hydroprocessing, as a first step, followed by ring cleavage of naphthenic structures formed in the first step to produce light paraffins ($C_1 - C_4$) and a BTX rich stream. In the first step, hydrotreating removes heteroatoms such as sulfur and nitrogen and selectively hydrogenates poly-ring aromatics with little or no saturation of single ring aromatic compounds. This has been achieved by using a combination of commercial NiMo/Alumina and NiW/Alumina - based catalysts on staged beds. The selectivity of the saturation of poly-ring aromatics is governed by two major factors:

a) The kinetic factor - where the hydrogenation reactions of single ring aromatics are very slow compared to the hydrogenation of poly-ring aromatics[10].

b) The adsorptive hindrance factor – where poly-ring aromatics have higher adsorption strengths on the catalyst active centers compared to single ring aromatics, which makes active catalytic centers less accessible for single ring aromatics[10,11].

In the second step, the naphthenic ring cleavage is effected via a noble metal modified zeolite catalyst. This kind of catalyst has been shown to be effective for naphthenic and aromatic ring cleavage by some authors[12, 13]. The main reaction occurring at this step is hydrocracking of naphthenic and paraffinic compounds in the feed, resulting in the formation of light paraffins. The side chains of the single ring aromatics and condensed naphthenic rings on single aromatic rings are also hydrocracked during this step, resulting in the formation of light paraffins and C6 – C9 aromatics, which is the BTX rich stream. A diagram of the two-step process that we term the ARO Process is shown below as Figure 4.

Results from research on both steps of the technology as described above are shown in the next sections. The feed is VGO, (which is readily available) derived from Alberta's Athabasca Oil Sands Upgrader. However, LCO, pyrolysis gasoline, refinery fuel oil or waste plastics oil can be alternate feeds for this process.

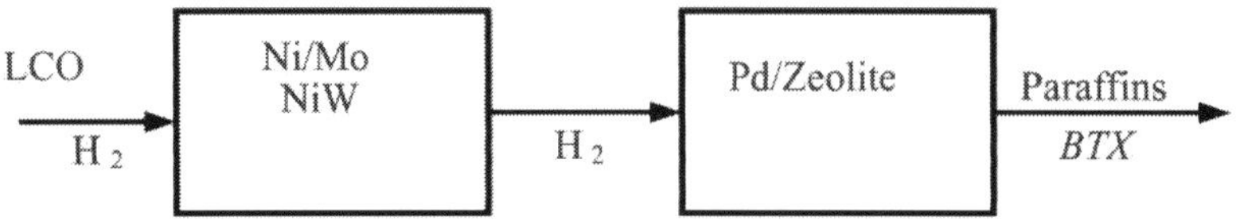


**Figure 4.** The two-step ARO Process

### *6.1 Experimental Equipment*

All the experimetal runs were carried out in a laboratory scale fixed bed-reactor in the up-flow mode. This type of reactor has been shown to closely represent the behaviour of an industrial gas oil hydrotreatment trickle-bed unit[14]. Because the unit contains only one reactor, all the experiments were carried out in such a way that the ring saturation step was carried out first. Thereafter, a zeolite-based noble metal catalyst was reloaded for the second step reaction to take place. The catalyst used for the first step was a stacked catalyst bed. The catalyst volume was chosen accordingly to meet the required liquid hourly space velocity (LHSV) for each stage. The runs for both the hydrogenation and the ring cleavage steps showed very good repeatability before the final runs were carried out. To minimize the effect of deactivation of the catalyst, the experimental runs were carried out during the day and the catalyst was regenerated during the night. This allowed catalyst activity to be the same for each of the runs.

### *6.2 Effect of Process Parameters on the Aromatic Rings Saturation Step*

This first stage of the ARO process is the heteroatom removal and ring saturation step. It is a conditioning stage for the next step, which is the ring cleavage step. In particular, poly- and di-aromatics are saturated to single ring aromatics. Only a small change of a total amount of aromatics has been observed in the first stage. Sulfur and nitrogen containing compounds in the feed have been reduced to levels that can be tolerated by the noble metal/zeolite catalyst in the next stage. Table 7 shows that the sulphur level was decreased from 2800 wppm in the original feed to only 50 wppm in the feed to the second stage reactor. Nitrogen level was also considerably reduced. Sulphur level in the feed to the second stage reactor should be kept to below 100wppm.

**Table 7.** Heteroatom reduction in the product of the first stage reactor

| Heteroatoms | Feed | Product |
|---|---|---|
| Sulfur [wppm] | 2800 | 50 |
| Nitrogen [wppm] | 867.1 | 13.6 |

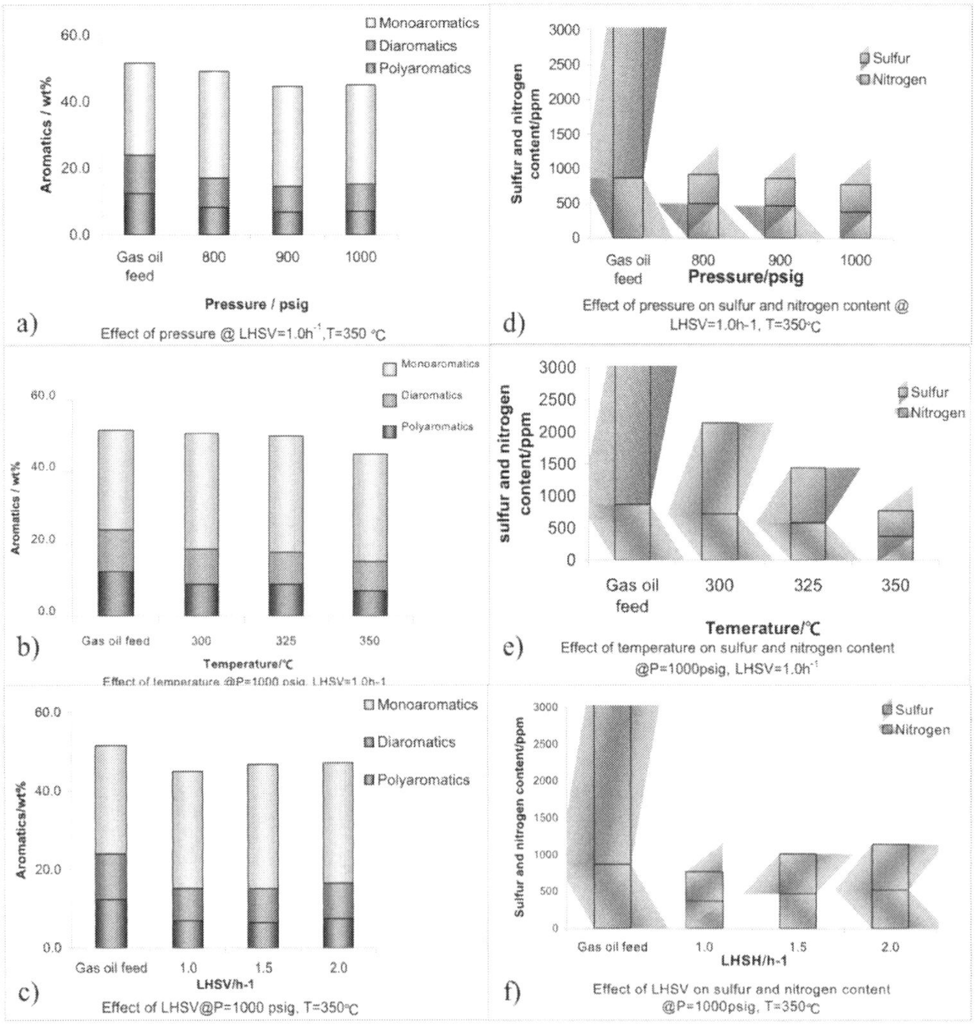


**Figure 5.** Effect of pressure a) & d), temperature b) & e) and LHSV c) & f) on the conversion of aromatic compounds and hetero-atom removal during GO hydrotreating.

It can be seen from Figure 5 that higher pressure, higher temperature and lower LHSV lead to both more extensive aromatics saturation and to higher heteroatom reduction.

*6.3 Effect of process parameters on the ring cleavage step*

An extensive experimental program was carried out to study the influence of process parameters on the ring cleavage step of hydrotreated GO (HTGO). This step of the process was revealed to be very flexible and easily adjustable via process parameters to the required product yields. Table 8 shows products yields obtained for the ARO technology being operated in three different modes. More detailed information on yields, selectivity and nature of ARO catalyst can be obtained from US7,513,988B2.

**Table 8.** Products yields from ARO process

| Units in (wt.%) | C2 to C4 Paraffins | C5 to C10 Paraffins | BTX + EB Aromatics |
|---|---|---|---|
| Max. Paraffin Mode | 69.23 | 6.71 | 4.6 |
| Moderate Mode | 51.2 | 9.61 | 18.4 |
| Max. Aromatics Mode | 14.93 | 14.87 | 43.77 |

Additional runs, were performed to test whether aged ARO catalyst can be regenerated. The regeneration of the deactivated catalyst was performed via in-situ coke combustion of the catalyst, followed by reduction of the catalyst in hydrogen flow. One can see from Figure 6 that regenerated catalyst shows the same activity as the fresh catalyst over test period of 400 hours.

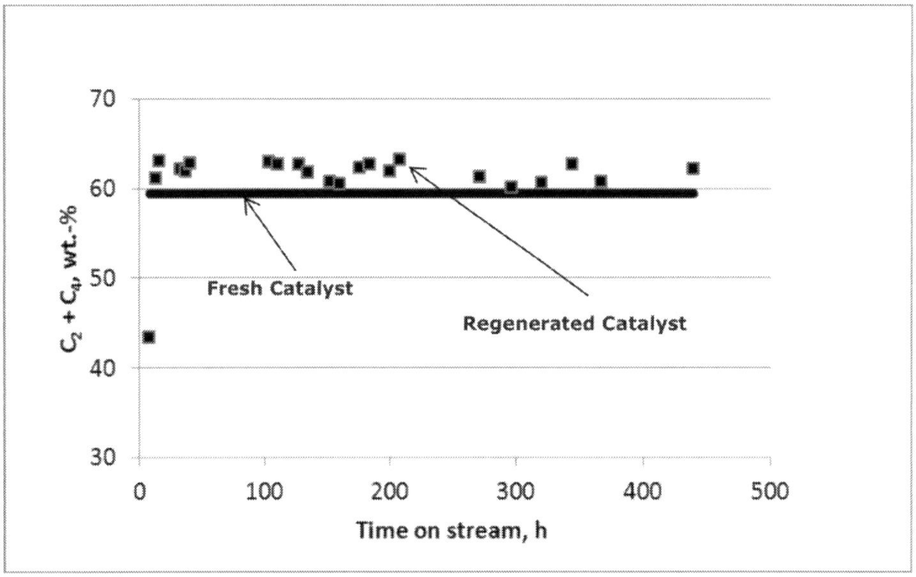


**Figure 6.** Performance of regenerated ARO catalyst in a long term run.

## 7. CONCLUSION

Technical advances are being made in linking Alberta's oil sands and petrochemical industries. Some olefins and aromatics by-products are currently being produced and used as petrochemical feedstocks and intermediates. However, little work has been done to date on using heavy bitumen-derived gas oil as a sustainable alternative, competitively priced feedstock for Alberta's petrochemical industry. The potential expansions of oil sands production and upgrading capacity create opportunities for phased development of new and improved technologies that can further improve economic returns.

The current shift of ethylene producers to lighter, shale gas based feedstocks will potentially result in shortening supply of propylene, unless new technologies and sources of propylene supply are identified.

This work provides a method for processing largely aromatic oil sands derived bitumen in Alberta to produce basic olefins (predominantly propylene) for the petrochemical industry. No commercially proven technologies are available on the market for converting this feedstock into petrochemicals.


## ACKNOWLEDGEMENT

This work is being carried out with the financial support and strategic input from the Alberta Innovates: Energy and Environment Solutions (AIEES). This program is aimed at developing and demonstrating commercially viable advanced technologies that convert bitumen-derived products into high value products and petrochemicals with reduced environmental footprint. We acknowledge the joint development effort with AIEES and the intellectual contributions of the University of Stuttgart, China University of Petroleum, together with members of our Research and Technology Centre at NOVA Chemicals who also work on this project.

## REFERENCES

[1] A Primer on the Canadian Oil Sands; An EPRINC Briefing Memorandum, November 2010

[2] Canadian Association of Petroleum Producers June report: Crude Oil Forecasts, Markets and Pipelines Expansions. 2007,

[3] Hou D., Wang X., Xie Ch., Shi Zh., Studies on the Reaction Mechanism of CPP and the Factors Affecting the Yields of Ethylene and Propylene. China and Petroleum Processing and Petrochemicals Technology 2002, 4, 51.

[4] Wittcoff, H. A., and Reuben, G.B., "Industrial Organic Chemicals", John Wiley & Sons Inc., New York, 1996, 58

[5] Zai-Ting L., Fu-Kang J, Chao-Gang X. You-Hao X. DCC technology and its Commercial Experience, 16th World Petroleum Congress, June 11 - 15, 2000 , Calgary, Canada

[6] Chapin L. and Zai-Ting L., "Deep Catalytic Cracking for light olefins", Paper at the 5th World Chemical Engineering Congress, San Diego, California USA; 1996 July 14-18, 4, 299.

[7] Shi,W., Xie,C., Huo,Y., Zhong, X. FCC Family Technology for Producing Light Olefins from Heavy Oils. China and Petroleum processing and Petrochemicals Technology, 2001, 2, 15.

[8] Hou D., Wang X., Xie Ch., Shi Zh., Studies on the Reaction Mechanism of CPP and the Factors affecting the Yields of Ethylene and Propylene, China Petroleum Processing and Petrochemical Technology, 2002, 4, 51.

[9] Laredo, G.C., S. Leyva, R. Alvarez, M.T. Mares, J. Castillo, J.L. Cano (2002) Nitrogen compounds characterization in atmospheric gas oil and light cycle oil from a blend of Mexican crudes. Fuel, (81), 1341-1350.

[10] Stanislaus A., Cooper B.H., Organic nitrogen compounds in gas oil blends, their hydrotreated products and the importance to hydrotreatment, Catal. Rev-Sci. Eng., 1994, 36/1, 75.

[11] Lylykangas M. S, Kinetic Modelling of liquid phase hydrogenation reactors, PhD Thesis, Helsinki University of Technology, 2004.

[12] Ringelhan, C., Burgfels, G., Neumayr, J.G., Seuffert, W., Klose, J., Kurth, V. Conversion of naphthenes into a valuable steamcracker feed using H-ZSM-5 based catalysts. DGMK- Tagungsbericht 2001, 4, 57.

[13] Ringelhan, C., Burgfels, G., Neumayr, J.G., Seuffert, W., Klose, J., Kurth, V. Conversion of naphthenes to a high value steamcracker feedstock using H-ZSM-5 based catalysts in the second step of the ARINO®-process, Catalysis Today, 2004, 97, 277.

[14] Myrstad, R., Rosvoll, J. S., Grande, K., Blekkan E. A. Hydrotreating of gas-oils: A comparison of trickle-bed and upflow fixed bed lab scale reactors. Stud. Surf. Sci. Catal. 1997, 106, 437.

# FUSEL OIL PROCESS SEPARATION

*Natalia Montoya*[1], *Fernando Córdoba*[1], *Carlos Trujillo*[1], *Iván Gil*[1], *Gerardo Rodríguez*[1]

[1]*Universidad Nacional de Colombia, Sede Bogotá.*

### *Abstract.*

Biofuels industry, especially fuel ethanol industry, has rapidly grown in the last twenty years. Consequently, by-product utilization has gained attention to make processes more efficient and profitable. Fusel oil is a by-product in ethanol production. It is mainly composed by isoamyl alcohol, which is an important raw material for the manufacturing of different products with a wide chemical market. In this work, a separation process to obtain high purity isoamyl alcohol from fusel oil was simulated using Aspen Plus. Results are satisfactory in terms of purity and recovery.

### *Introduction.*

Fuel ethanol production has become in one of the most studied processes in the last two decades. In countries where ethanol is produced in large scale, by-product utilization is an important issue for making the global process more profitable and less polluting. Fusel oil is a by-product in ethanol production. It is a mixture of higher alcohols, composed by isoamyl alcohol, as major constituent, and in minor proportion by ethanol, water and heavy components. Isoamyl alcohol is widely used in the manufacturing of different valuable products such as esters, solvents, lubricant oils, plasticizers, medicinal among others Güvenç *et al* (2007).

According to current regulations in Colombia it is expected an increase in ethanol production, as a consequence, the amount of fusel oil will increase too. In 2012, it is estimated that around 1400000 liters per day of fuel ethanol will be produced; as a result, around 2800 liter per day of fusel oil will be obtained. Taking into consideration, the mention before, the high content of isoamyl alcohol in fusel oil –its composition can vary from 60 to 80% [Patil *et al.*, 2002; Ikeda *et al.*, 1956; Schicktanz *et al.*, 1939)–, and its possibility to be converted in chemicals with a wide market, it becomes in an opportunity to develop a fusel oil separation process in order to obtain high purity isoamyl alcohol.

The aim of this work is to design and simulate a separation strategy to obtain isoamyl alcohol from fusel oil using Aspen Plus Simulator. As case base, a fuel ethanol production plant with capacity of 400000 liters per day was used. It was assumed a yield of 2 liters of fusel oil per each 1000 liters of absolute ethanol (Patil et al., 2002; Özgülsün, 2000; Küçük et al., 1998).

## *Thermodynamic Model*

According to the suggestions for selecting physical property methods given by Carlson (1996) and considering the polar and non-electrolyte nature of fusel oil and the moderate operating conditions of the process under study, NRTL thermodynamic package was chosen to calculate the activity coefficients in the liquid phase. Else, ideal behavior for the vapor phase was assumed.

In addition, some experimental data from VLE for ethanol – isoamyl alcohol, ethanol – water, and LLE for isoamyl alcohol – water systems were obtained using a Fischer Labodest VLE 602 apparatus and a LLE glass cell, respectively. These data were well fitted to NRTL thermodynamic model.

## *Process Description and Simulation*

The simulated process involves two main stages and is shown in Figure 1. The first one consists on separating alcohols and water from heavy components through distillation and the second one involves purifying of isoamyl alcohol.

Regarding to the first stage, Fusel oil (stream $F_1$) is fed into a distillation column (C-1) to separate the heavy components ($T_B$>124°C) at the bottom (stream $B_1$) and recover a mixture of the lighter components at the top (stream $D_1$), which is composed by water, ethanol and isoamyl alcohol. In the second stage, this mixture (stream $F_2$) is taken into a stripper (C-2) in order to draw off ethanol and water at the top (stream $D_2$) and to obtain high purity isoamyl alcohol at the bottom (stream $B_2$). Some isoamyl alcohol is dragged in $D_2$. Therefore, after cooling, this stream splits into two liquid phases in a decanter (D-1) and the organic phase ($O_1$) is mixed with the feed stream ($F_2$) to the stripper (C-2).

Preliminary specifications of the distillation columns were estimated using DSTWU short-cut method. Then, with the information obtained from this procedure, simulations were carried out using a rigorous method. The distillation column and the stripper were simulated using Aspen Plus RadFrac model and the liquid-liquid separator was simulated using Aspen Plus Decanter model. Sensitivity analyses were done by varying parameters such as number of stages, feed stage and reflux ratio, in order to settle the operating conditions. Also, design specifications were used to set up purity and recovery of the product in the process.

RadFrac Aspen Plus operation model was used to perform pressure drop calculations for packed columns. Columns diameters were estimated using PackSizing utility, considering Sulzer Nutter Ring No-0.7 as a packing. The distillation column and the stripper were also designed using a non-equilibrium approach in order to establish the effect of mass and heat transfer correlations in column performance.

## *Results*

The final results are satisfactory in terms of purity and recovery. They are summarized in figure 1. High purity and recovery of isoamyl alcohol, 99.7% and 99.3% respectively, are reached.

Sizing results show that a distillation column of 13.8 ft height and 4 in diameter and a stripper of 8.2 ft height and 3.5 in diameter, both, packed with Sulzer Nutter Ring 0.7 are able to manage a volumetric flow of 800 liters per day of fusel oil to obtain 792 liters per day of high purity isoamyl alcohol.

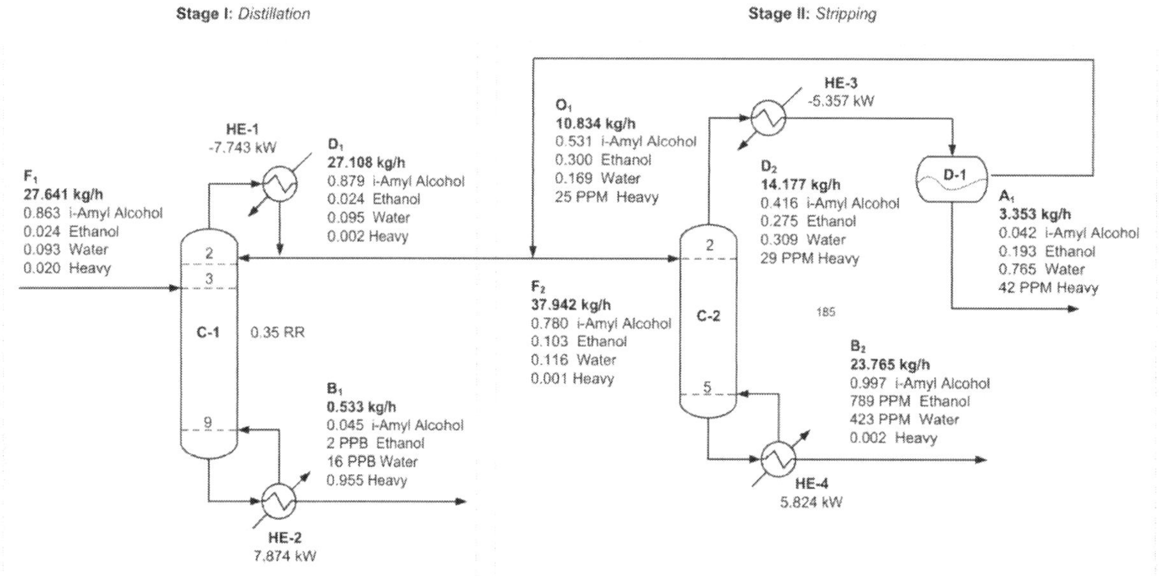


***Figure 1.*** *Mass and energy balances of the separation strategy*

In the case of non-equilibrium approach –Rate-based calculation type- liquid phase composition and temperature, as well as, mass and heat transfer rates profiles were obtained. They are shown in figures 2 and 3.

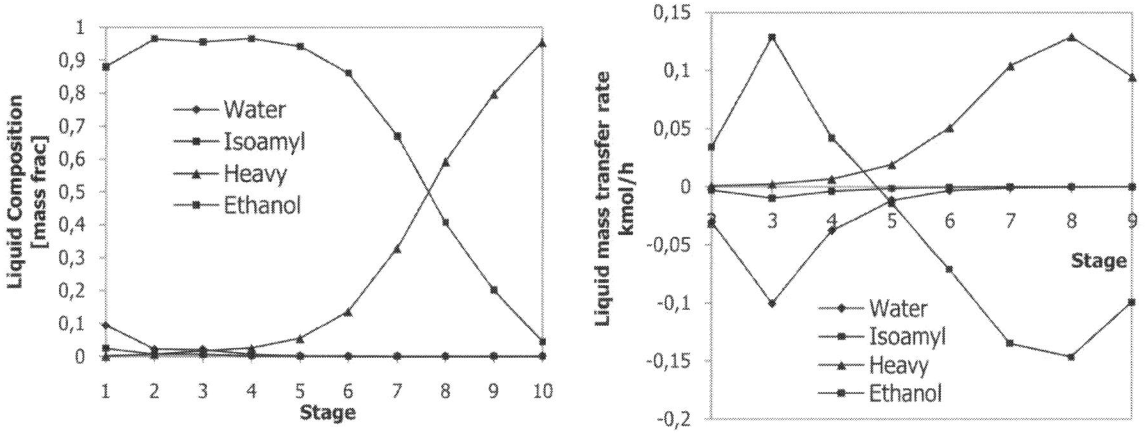


***Figure 2.*** *Liquid composition and mass transfer rate profiles for distillation column C-1*

There is a relation between the results from liquid composition and mass transfer rate profiles. Component mass transfer rates are positive for transfer from vapor to liquid phase. As can be seen in figure 2, at the top of the distillation column (stages 2 to 5) liquid phase profile is enriched in isoamyl alcohol, while at the bottom (stages 5 to 9) it is enriched in heavy components. These facts are caused by components transfer from vapor to liquid phase and involve a positive change in mass transfer rate profiles. For ethanol and water, mass transfer rate is low and almost constant; because of this their liquid composition profiles are flat.

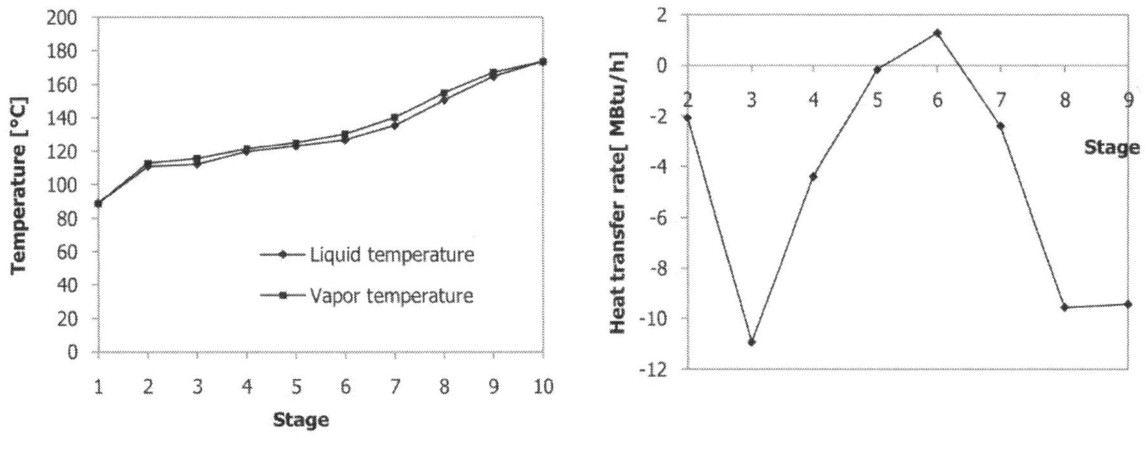


***Figure 3.*** *Temperature and heat transfer profiles for distillation column*

As can be seen in figure 3, there was not significant temperature difference between the two phases. Nevertheless, temperature of vapor phase was higher than temperature of liquid phase through the entire distillation column. This means that energy was transferred from liquid to vapor phase, which is evident in the negative values of the heat transfer rate profile (usual convention is heat transfer rate positive for transfer from vapor to liquid phase).

To develop equilibrium calculations a Murphree efficiency of 0.7 was assumed, according to average typical values reported on literature for distillation processes (Kister, 1992). From non-equilibrium approach HETP was calculated and is shown in figure 4. An increase in the HETP to the bottom of the column is observed and the most efficient zone is located between 6 and 10 segments.

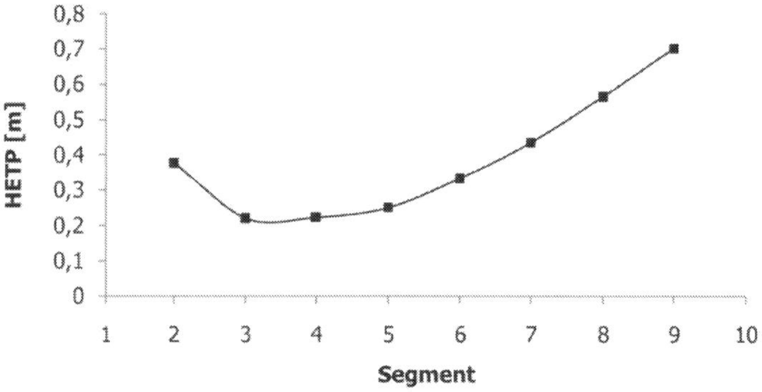


**Figure 4.** Height equivalent of a theoretical plate

In general terms, results from compositions and mass flows are nearly the same using both, equilibrium and non-equilibrium approaches. It can be explained, taking into account that the simulated process does not involve chemical reactions or formation of a second liquid phase inside the columns, which are the factors that most affected non-equilibrium calculations.

### *REFERENCES.*


1. Carlson, E.C., (1996). "Don't Gamble with Physical Properties for Simulations". *Chem. Eng. Prog.* 92 (10), 35-46.
2. Güvenç, A., Kapucu, N., Kapucu, H., Aydogan, Ö., Mehmetoglu, Ü., (2007). "Enzymatic esterification of isoamyl alcohol obtained from fusel oil: Optimization by response surface methodology". *Enzyme Microb. Technol.* 40 (4), 778-785.
3. Ikeda, R.M., Webb, A.D., Kepner, R.E., (1956). "Comparative analysis of fusel oils from Thompson Seedless, Emperor, and Muscat of Alexandria Wines". *J. Agric. Food Chem.* 4 (4), 355–363.
4. Kister, H., (1992). "Distillation Design". McGraw-Hill, pp 379.
5. Küçük, Z., Ceylan, K., (1998). "Potential utilization of fusel oil: A kinetic approach for production of fusel oil esters through Chemical Reaction". *Turk J. Chem.* 22, 289-300.

6. Özgülsün, A., Karaosmanôglu, F., Tüter, M., (2000). "Esterification reaction of oleic acid with a fusel oil fraction for production of lubricating oil". *JAOCS.* 77 (1), 105-109.

7. Patil, A.G., Koolwal S.M., Butala H.D., (2002). "Fusel Oil: Composition, removal and potential utilization". *Int Sugar J.* 104, 51- 58.

8. Schicktanz, S.T., Etienne, A.D., Steele, W.I., (1939). "Analysis of Fusel Oil by Azeotropic Distillation". *Ind. Eng. Chem. Anal. Ed.* 11 (8), 420-422.

# (428g) Simultaneous Saccharification and Fermentation of Switchgrass by Thermotolerant Kluyveromyces Marxianus IMB3: Effect of Enzyme Loading, Temperature and Operating Mode

Naveen Pessani[1], **Hasan K. Atiyeh**[1], Mark R. Wilkins[1], Danielle Bellmer[1] and Ibrahim M. Banat[2], (1)Biosystems and Agricultural Engineering, Oklahoma State University, Stillwater, OK, (2)School of Biomedical Sciences, University of Ulster, Coleraine, Northern Ireland, United Kingdom

Hydrothermolysis-pretreated Kanlow switchgrass was used in simultaneous saccharification and fermentation (SSF) with *Kluyveromyces marxianus* IMB3 at various enzyme loadings and temperatures. The pretreated switchgrass contained 57.7% glucan, 5.0% xylan and 35.1% lignin. Various loadings (0.3, 0.5 and 0.7 mL/g glucan) of the enzyme Accellerase 1500 (Genencor, Rochester, NY) were used in the SSF at 45°C (Accellerase 1500 activity was 82.2 FPU/mL). The highest ethanol concentration (22.3 g/L) was obtained in SSF with 8% solids and enzyme loading of 0.7 mL/g glucan. This enzyme loading was used with 8% solids in subsequent SSFs at 37, 41 and 45°C. The highest ethanol concentration (22.8 g/L) was obtained at 45°C after 96 h, corresponding to 83.9% of the maximum theoretical yield (MTY). However, ethanol yield was about 79% of MTY in SSF at both 37 and 41°C. No significant differences in the amount of ethanol produced were observed in the SSF operated in batch or fed-batch modes at 45°C and 12% solids ($p > 0.05$). However, the enzyme loading in fed-batch SSF was 33% lower than in batch SSF. About 32 g/L ethanol (81% MTY) was produced in all batch and fed-batch feeding strategies.

See more of this Session: Alternative Feedstocks for Energy and Chemicals
See more of this Group/Topical: Process Development Division

# (458a) Historical Perspective On Consumer Goods

**Tasha Desai**, Henkel Consumer Goods, Scottsdale, AZ

In the consumer products industry, convenience for the consumer motivates a lot of innovation. Over the years, detergents have become multifunctional, stain removers have become sprays or foams, and packaging for household cleaners has evolved to provide more convenience to the consumer. Historically, it is apparent how products have slowly changed as consumers began to prefer easier to use and multi-purpose goods.

See more of this Session: Innovation In Consumer Product Management and Design
See more of this Group/Topical: Management Division

## Using Other Peoples' Brains to Invent New Products

*Jack Hipple*
*Principal*
*Innovation-TRIZ*
*Tampa, FL*

### Extended Abstract

Inventing and commercialization of new products is a risky proposition. One of the ways of minimizing costs and time to market is to use technology approaches from parallel universes of technologies that are non-competitive in a commercial sense. "TRIZ" (Theory of Inventive Problem Solving) is a methodical way of identifying parallel functional universes where the same result and function is used, but possibly for a totally different commercial reason.

### What is the Function?

This seems like such a simple question to answer and it may be. However, we frequently answer it in highly specific technical terms unique to the industry or product area of concern. Let's look at an example. In the air traffic control business, FAA regulations are requiring more and more information to be displayed on air traffic control displays. This is done without thinking about the possible consequences of an air traffic controller trying to make quick decisions based on graphical display information which may cause information overload. If we were looking at a parallel universe which has a very similar problem, where would we look?

In a real problem solving situation with an air traffic control display manufacturer, the following technology areas were not even considered or reviewed in attempting to come up with new concepts relating to the air traffic control display challenge:

1. Chemical process control displays
2. Hospital operating room displays
3. Video games

If we frame the challenge as one of increasing the ability of an air traffic controller to respond to information change or emergency situations, our view of possible ideas is rather limited. For example, if we did a web search with the heading of "air traffic control displays, we would find about 4.7 million hits. However, if we thought for a minute about the function being performed, we might type in something like, "video displays", we would find almost a billion hits! We would industries such as chemical process control displays, heart surgeon operating room displays, video games, and many other products that have exactly the same problem.

Let's take another example in the auto industry. Though head rests in cars are mandatory, they are seldom neither comfortable nor easy to adjust. In a Wall Street Journal article highlighting the efforts of the Ford Motor Co. to address this problem, it was shown that the physics of baseball catchers' mitts have exactly the same problem. If we consider our web search again and type in the word "human headrests", we will find about 150,000 hits whereas if we type in "stop moving object", we will obtain 14,000,000 hits—nearly two orders of magnitude more. This article highlighted the fact that the physics of catchers' mitts (used since the late 1890's) had the same physics involved in their design and had never been considered a model for design of headrests.

So the first lesson when looking for parallel universes of knowledge is to use the most general description of the function being performed. Using language that a ten year old would understand is a good general guideline.

**TRIZ Thinking and Analysis**

"TRIZ" is a Russian acronym standing for "Theory of Inventive Problem Solving" and is an algorithm based problem solving technique based on the fact that when we study the global patent literature, we can only find 40 inventive principles that can be used to describe any patent, regardless of its field. Thinking about this fact in reverse means that the solution to most problems already exists. The exception to this would be the area of pure science such as new chemical or biological entities. It is very difficult for problem owners to want to admit that their problem is not special and unique. However, once this psychological barrier has been overcome, and the problem generalized, then it becomes easy to identify parallel universes of technology and markets where a problem has already been solved

**Application in Consumer Products**

Let's consider some applications in the consumer products area.

In the design of catsup bottles, Heinz finally discovered the force of gravity and redesigned their bottles to, in effect, flow upside down so that their high viscosity catsup formulation would more easily flow. Several years later, shampoo manufacturers inverted their bottle designs to take advantage of a free resource. Why was there a time delay?

In the paper towel industry, it was observed that many consumers do not need a normal sized paper towel for small spills and a second layer of perforations were incorporated into Bounty paper towels to allow the tearing of a towel into a smaller piece if desired. Sometime later, cheese manufacturers such as Kraft discovered that perforating cheese slices would allow consumers to tear off cheese squares in just the right size to fit on a cracker.

In another interesting parallel universe example, the same basic technology used in the biodegradable trash bag industry is now being used in dissolving heart stents under development in the pharmaceutical area. This did not happen until several years after the trash bags were on the market and their patents issues. There are two interesting questions here. Why did the trash bag formulators think about higher value added markets for their formulations? Second, why did the heart stent manufacturers not look in the garbage area for technology (possibly too low tech?). If we go through the same web search exercise, we would find the following results:

| | |
|---|---|
| Disappearing stents | 12,000 hits |
| Dissolving plastics, | 1,000,000 hits |
| Disappearing plastics | 2,200,000 hits |

The point in showing both "dissolving" and "disappearing" is that dissolving is only one way of making a material disappear! What might be some others?

**Summary and Conclusion**

When analyzing a new consumer product opportunity or a new product development opportunity more generally, it is critical to analyze the product or need from a functional, not a product, standpoint. This will allow a more general consideration of other approaches and technologies that may already exist to solve the problem or design the product.

# (458c) Facing the Top Five Innovation Barriers for Consumer Products: Systems and Strategies

**Jeff Lindsay**, Asia Pulp and Paper, China, Shanghai, China and Cheryl Perkins, Innovation Edge, Neenah, WI

In our experience with consumer product innovation, several common innovation fatigue factors repeatedly pose special challenges that can interfere with innovation success, especially in large companies. We explore these challenges and offer solutions that can strengthen culture, identify and eliminate bad behaviors that defeat innovation, and strengthen judgment in pursuing cost-effective innovation. Principles that must be understood by managers as well as researchers and product developers will be reviewed, with case studies drawn from global corporations as well as start-ups.

See more of this Session: Innovation In Consumer Product Management and Design
See more of this Group/Topical: Management Division

# (458d) The Science Behind the Innovative Laundry Sheet Purex Complete 3-In-1

**Matthew Petkus**, Laundry Care R&D, Henkel, Scottsdale, AZ

Consumers are king when it comes to the consumer goods industry. This is why products that are engineered for convenience take precedence in the category. In the fabric care category there was a need to develop a product that delivers cleaning, softening, and antistatic in an all in one laundry product. That is why Henkel designed and launched a novel laundry sheet that offers all of these benefits. The science and design behind the Purex Complete 3-in-1 laundry sheet was imperative to its successful launch.

See more of this Session: Innovation In Consumer Product Management and Design
See more of this Group/Topical: Management Division

# (458e) Rational Design of Pharmaceutical and Other Liquid Formulations

Elisa Conte[1], **Sascha Sansonetti**[1], Peter A. Crafts[2] and Rafiqul Gani[1], (1)CAPEC, Department of Chemical and Biochemical Engineering, Denmark Technical University (DTU), Kgs. Lyngby, Denmark, (2)Process R&D, AstraZeneca Pharmaceuticals Ltd., Cheshire, United Kingdom

Every year pharmaceutical, health-care and related industries invest a high percentage of their total time and resources on development of new products and improvement of existing (pharmaceutical) formulations. Product development is still mainly based on trial and error experimentation and laboratory procedures, indicating long research processes without any certainty of a definitive solution and far from achieving the *optimal formulation* of the product. Besides, (pharmaceutical) ingredients to be used for the experimentations are often very expensive and not available in the desired form or at the desired level of purity. Moreover, the experimental procedures are affected by unavoidable errors due to the complicated and often not-standardized procedures. Therefore, the need for a rational and systematic model-based approach to plan suitable experimental designs would constitute a fairly useful tool for pharmaceutical product design and product design in general. A model-based approach significantly reduces the number of experiments necessary for product design and at the same time gives a clear idea about the direction to take for further (reduced) experimentation. In such a way tedious and uncertain screening processes can be avoided with sensible reduction of both time and resources for the product launch on the market.

Identification of a pure solvent or anti-solvent for a specific Active (Pharmaceutical) Ingredient (API or AI) is a problem of major concern for the pharmaceutical and related R&D departments. Solvents, lipids and other compounds are commonly employed in product formulations as well as in API processing. In addition, the screening of several solvent mixtures that sometimes show improved characteristics of solubility toward a particular API is necessary. Besides, in liquid formulated products, API (or AI) needs to be mixed with additives to enhance the product quality and solvents to deliver the product.

This work presents a model-based method for solvent screening and solvent mixture design for pharmaceutical applications and API/AI based liquid formulated products. Two cases are investigated: a case where the solvent or solvent blend is designed for API processing (here the solvent is not part of the final API based product), and a case where the solvent or solvent blend is part of the API/AI based formulated product. In both cases, the solubility of the API/AI needs to be predicted in solvents (single solvent or blends). A general solubility modelling procedure having different options for the solubility calculation (such as UNIFAC, UNI-sac and PC-SAFT) is employed. A unique feature of this solubility calculation procedure is that it is truly predictive and does not need any new experimental data for the computer-aided design stage of the formulation design methodology (Conte *et al.* 2011).Once the domain search has been reduced to a few solvents or solvent mixtures, solubility can be computed by correlative models such as NRTL-SAC andPC-SAFT.

The newly developed method described above is to be integrated in a comprehensive framework for formulation design that was previously developed and tested for personal care products. Such a method was implemented in the software 'the virtual product-process design laboratory' (vPPD-lab) (Conte *et al.*, 2011).

Examples illustrating the application of the procedure for solvent, solvent-blend and/or AI-based formulation design will be presented. Also, a conceptual case study validating the models and the design steps involving the design of a solvent mixture that maximizes the API solubility as well as the identification of an optimal anti-solvent for a well-known API such as paracetamol will be presented.

**References**

Conte E, Gani R, Ng K M, Design of formulated products: A systematic methodology, AIChE J, 2011 (DOI 10.1002/aic.12458)

Conte E, Gani R, Cheng Y S, Ng K, M, Design of formulated products: Experimental component, AIChE J, 2011 (DOI 10.1002/aic.12582)

See more of this Session: Innovation In Consumer Product Management and Design
See more of this Group/Topical: Management Division

# (483a) The Final Step of a QbD Technical Transfer: Validation

**Kevin J. Bittorf**, CMC Process & Alliance Management, Strategic Development Networks, Vertex Pharmaceuticals, Cambridge, MA

The final step in a Quality by Design (QbD) pharmaceutical technical transfer process is Validation. Now that submitted QbD process what does it take to complete a QbD validation? The "Traditional" 3-Batch Validation has been a standard over the decades which implies a controlled product that meets release specifications; but how does one move away from a 3-Batch validation to continuous verification? In addition is there a requirement to move away from this type of validation? Finally once the process is validated, what's next? Presented will be an approach to answering these questions to finalize a successful technical transfer.

See more of this Session: Technology Transfer and R&D Portfolio Management
See more of this Group/Topical: Process Development Division

# Pilot Plant Funding from an R&D Perspective

*Dr. Merrill Galloway, SABIC Americas, Inc., Technology Center-H, Sugar Land, TX*
*Dr. Ashim Ghosh, SABIC Americas, Inc., Technology Center-H, Sugar Land, TX*
*Dr. Ramesh Rameswaran, SABIC Americas, Inc., Technology Center-H, Sugar Land, TX*
*Dr. Mohammad Shafiei, SABIC Americas, Inc., Technology Center-H, Sugar Land, TX*

## Extended Abstract

### *Introduction*

This paper considers issues related to funding for piloting a project from an R&D perspective. Most R&D projects in the CPI could be categorized as either process or product related. Process development is focused on defining a process in terms of material and energy balances, PFD's, front end equipment design, utility requirements, control and safety issues. Product development is focused on final product properties to satisfy customer needs and typically involves preparation of batch quantities for testing. Of course, there is a lot of overlap in most projects into both of these categories. Our particular discussion here comes more from the process side, but the fundamentals of funding for the piloting phase of a project are similar in both cases; i.e., who decides to go to the piloting phase, and what are the bases for that decision.

There are several sources of funding for piloting, including internal to the company that owns the project, funding from another producing company, funding from a venture capital source and funding from a government source. The latter two, while being very advantageous where applicable, are typically aimed at specific areas of development. We will consider only the first two funding sources in this paper.

### *Decision to pilot a process*

Typically R&D projects go through several stages and decision points before reaching the piloting stage. These are named differently in different organizations, but the concept is the same: Idea ➔ Business fit ➔ Technical/economic feasibility ➔ Lab/bench work ➔ Commercialization/piloting, where the arrows represent a review process (gate) before going to the next stage. The resources and financial commitment of the company increase at each stage. Many projects make it to the lab/bench phase, but relatively few make it to the piloting phase. Typically costs for the lab/bench phase are measured in \$100K's, while costs in the piloting phase are measured in \$MM's. Correspondingly, the decision to pilot must be made at a higher level in the organization. Several stake holders are usually involved. The central design/engineering group needs to approve. Manufacturing people may be involved if there would be modification of an existing plant. The business unit where the technology fits must approve, along with corporate planners. If licensing of the technology is planned, the licensing group must buy in. Typically, there must be at least one champion of the project at a high level who will keep it moving through the decision process.

Criteria for making the decision involve the strategic fit and economics for the project. Does the company plan to build a plant or license the technology? What do the projected economics look like in terms of IRR/NPV? This obviously involves market analysis and technical analysis, along with the uncertainty/risk involved in both of these. The plan for the technology affects the type of economic analysis that is performed. We will give an over-view of several alternatives: stand-alone plant; plant integrated into a complex of processes; technology licensed.

***Economic issues for piloting decision***

The economics for a stand-alone plant involve the cost of piloting (capital and operations), capital cost of the plant, operating costs of the plant and revenue generation. This analysis is relatively straight-forward. The economics for an integrated plant involve the cost of piloting, the capital cost of the integrated plant, the resulting incremental operating costs of the complex and the resulting incremental revenue generation from the complex. This involves how the plant is integrated and how it affects the rest of the complex, making the economic analysis much more complicated. The economics for technology licensing involve the cost of piloting, the cost of licensing/overheads, and revenue generation via licensing and catalyst fees. Careful market analysis is required to get a realistic estimate of the revenue potential for licensing the technology.

The cost of piloting may be a significant item in the overall project economics. This is particularly true if the process is to be licensed and it is the only major capital cost. Even if a plant is constructed, the piloting costs are up-front and thus get weighted more in a NPV calculation than plant construction costs. There are a number of factors that affect the cost of piloting:

Scale—A smaller pilot plant reduces capital and operation costs, but increases the risk of final scale-up to the plant.

Process complexity—Less complex reduces cost of equipment, controls, analytical effort, operator hours.

Duration of operation—Variable and operating costs ~ 50% of total cost.

Location—Located at the R&D site is less costly, but may impose unacceptable constraints.

Partnering—Formation of a partnership can greatly reduce the cost of piloting, as well as provide other benefits for commercialization. A partnership will allow the cost/risk of piloting to be shared. It also tends to confirm the value of the technology for internal stake holders. The partner may also bring additional technical know-how and market knowledge related to the process. There may be other synergies related to materials and/or technologies that the partner brings. The terms of the partnership will affect the economics of the project. If the partnership is for licensing the technology, then the agreement must include the plan for piloting cost sharing as well as licensing revenue sharing. If a JV is formed to build a plant, the agreement must include the plan for piloting/manufacturing cost sharing as well as product sales revenue sharing.

All of the above factors are inter-related, and also depend on the ultimate plan for the technology. They may make the difference between a go/no go decision for piloting of a project.

***Some specific piloting scenarios***

We now consider several common piloting scenarios to illustrate the ideas discussed above, particularly the funding issues.

Case 1

Funding: 100% internal
Location: R&D site
Pilot plant construction: Done internally or by a vendor
Commercialization goal: Build a plant internally (possibly license later)

Pros

Tight control of technology
Relatively low cost

Cons

Scale limited due to site constraints—increased scale-up risk
Lose outside technical/marketing knowledge related to process
Internal funding may be hard to get-stake holders question commercial viability

Case 2

Funding: 100% internal
Location: Company site with related operating commercial plant
Pilot plant construction: Done by piloting or EPC vendor
Commercialization goal: Build plant (possible subsequent licensing)

Pros

Larger scale (possibly a demonstration plant)
Makes use of some plant infrastructure and resources
Pilot plant and commercial plant products easily compared
Success at plant site makes commercialization easier

Cons

Relatively higher cost
Relatively longer time for piloting phase
Piloting operations limited by plant site constraints
Need to re-locate R&D people escalates cost
Need to obtain corporate funding and plant support

Case 3

Funding: 100% internal
Location: Piloting vendor site
Pilot plant construction: Likely done by external vendor
Commercialization goal: Build plant (possible subsequent licensing)

Pros

May be able to use existing equipment—reduce cost
Use existing facilities—utilities, material handling, analytical
Use vendor's operators—eliminate need to hire more people
Larger scale possible

Cons

Difficult to find vendor with suitable equipment
May need to re-locate R&D people for extended period

Internal funding may be hard to get-stake holders question commercial viability

Case 4

Funding: Cost shared with a partner

Location: Partner's site or 3rd party piloting vendor site

Pilot plant construction: Done by partner or 3rd party

Commercialization goal: Joint commercialization or licensing of technology

Pros

Lower financial risk

Partner familiarity with market facilitates technology licensing

Partner can help solve technical problems/optimize process

Plus all the pros from Case 3

Internal sponsors more convinced of commercial viability

Cons

Give up sole ownership of technology

More people involved in decisions slows progress

Cost of piloting not really cut in half

Difficult to find vendor with suitable equipment

May need to re-locate R&D people for extended period

***Summary***

In summary, the decision to fund piloting, or not, is based on total project economics starting from the piloting stage, and perceived technical and marketing risks. The project economics are affected by how the piloting is done, and how the technology is commercialized. 100% internal funding for piloting a project may not be appropriate in some cases due to cost and risk factors. Partnering the piloting/commercialization, with the right partner, may reduce the cost and risk, and improve the total project economics to where piloting can proceed. R&D can play an active role in this process in several ways: helping identify/screen potential partners and piloting service providers; planning a piloting program that minimizes cost while providing essential scale-up information; and, guiding economic analyses for the various project scenarios under consideration.

# (483c) Advancing Quality by Design Through Diverse Uses of Latent Variable Methods In Vaccines and Biologics Manufacturing

**Charles Miller**[1], Louis Obando[1], John Higgins[2], Gert Thurau[2] and Eric Ahuja[3], (1)Manufacturing Division, Process Analytical Technologies (PAT), Merck and Company, Inc., West Point, PA, (2)Merck and Company, Inc., West Point, PA, (3) Director Analytical Chemistry Development and Supply, Merck & Co, Inc, West Point, PA

It is generally understood that latent variable multivariate modeling methods have great value-add potential in vaccines and biologics enterprises. What is relatively less-known, however, is the diversity of tasks that these tools can support in such applications. Although the fundamental math and statistics behind these tools is fixed, each of these different tasks require unique considerations with respect to several attributes, including data alignment/integration, data preprocessing and scaling, model validation, computing hardware, and visualization of results.

This paper will use recent case studies to illustrate the various ways that latent variable methods support current vaccines and biologics efforts at Merck, which include QbD initiatives. These case studies will include retrospective analyses of existing operations for troubleshooting, exploratory modeling of processes for continuous improvement efforts, multi-scale models for scale-down model qualification, and consistency monitoring of manufacturing operations.

See more of this Session: Technology Transfer and R&D Portfolio Management
See more of this Group/Topical: Process Development Division

## Assessing Potential Value of R&D Opportunities via Explicit Assumption vs. Implicit Experience- or Learning-Curve Models

Joseph B. Powell and Juben N. Chheda
Shell Projects & Technology, Houston, TX

Experience or learning curves are often used to predict future cost reductions in known technologies, using a methodology where cost (C) is modeled as a function of cumulative production Q:

$$[1] \quad C = k\,Q^{-\alpha}$$

with "k" and "α" as positive constants. Cost of production is assumed to decrease with continued production via a process of "learning by doing" [1]. Decrease in construction or capex costs for replicate units, and improved operating efficiency are assumed responsible for progressive cost reductions. Unit costs of production are often not observable, whereas unit price P is. This can lead to other factors such as competition (4) which can lead to a decrease in price over time [2], which must be discerned from manufacturing "learnings" or experience.

While many different technologies are observed to follow a "learning trend" in cost reduction vs. cumulative production, in some instances this is not the case, which gives rise to analysis of reasons for "failure to learn" (3,6).

"Learning" is a psychological task. The labor cost of repetitive "hand assembly" jobs decreases with repetitive assembly, due to this phenomenon. Machines are more nearly fixed in performance, such that the "learning curve" associated with costs of production will in general be stronger for hand assembly vs. machine assembly jobs (5). In contrast to "Learning Curves", "Experience curves" reflect broader industry learnings beyond simply assembly and production.

Cost reductions in production of ethanol from sugar cane in Brazil are an example of three decades of governmental policy which has contributed to experience curve improvements. Detailed study reveals improvements in the efficiency of sugar cane production have been the dominant source of these improvements (7). This example highlights the use of government policy to subsidize early non-competitive technologies, until experience and learnings allow it compete with existing alternatives.

## Application of Experience Curves in Process Evaluation

Experience or Learning curves are derived from statistical improvements in commercial processes and manufacturing systems; i.e. from "processes that work". They are not well suited to projecting improvements in future cost at the R&D phase, where commercial production is initially zero. A majority, perhaps 90% of new concepts for R&D are never reduced in cost and rendered commercial, because they "do not work", or do not economically compete with existing technology. . Experience curves are thus best suited for expected improvements in manufacturing cost ***after*** construction and operation of the first commercial plant, not before all the learnings of scale up and commercialization have taken place.

Initial process development often appears as "unlearning" from a cost perspective, as costs generally increase over first projections from the research stage. Development is actually however a phase of rapid learning, despite losing ground relative to cost improvement. Extra purification steps, enhanced metallurgy requirements, constraints on catalyst life or unexpected costs, disposal and treatment of byproducts, energy costs, and other factors not consider in the research idea phase, generally drive up initial projected costs for a project entering the development phase.

Certain factors will essentially guarantee experience curve improvements for any successful launch of new technology which does grow in commercial volumes. Competition however will also tend to erode future margins, in compensation.

## Explicit Scenario Modeling

Rather than assuming a generic learning or experience curve cost improvement derived from general industry experience at some point after entering the process development phase, a more impactful way to assess potential future improvements in cost is to consider specific scenarios:

- Cost as a function of plant size is a no-brainer for explicit calculation. The well-known 0.7-power economics [8] for larger scale process units can be utilized to increasing advantage as market size grows, and to assess future unit capex charges as a function of capacity. The scale advantage is however less for modular systems or expansion by number of trains, and hence warrants explicit consideration of relevant scale-improvement factor.
- Cost improvements if material yields are increased to 100%, or energy usage reduced as low as practical, can also be explicitly calculated. The probability of achieving these improvements can be assessed by the presence of screening data showing potential to improve yields or reduce byproducts via further development of novel catalysts or reaction conditions.
- Potential for capex reduction via process intensification can be assessed by screening experiments and modeling, taking into account the explicit potential for integration, and ignoring integration that is obviously mismatched or incompatible.

Thus, a process that is already operating at greater than 90% yields is less attractive for future improvements than one where yields are 60% for a base case catalyst system, especially if scouting experiments exist, showing potential for catalyst or conditions capable of significantly greater yields. Similarly, grossly dissimilar process steps are not likely to be combined into intensified process steps at lower capex. Consideration of these explicit scenarios is inherently more accurate in predicting the range of allowed future costs, than applying generic learning or experience curve factors to estimate future costs at a particular but potentially non-representative point in process development.

## Summary

In summary, experience and learning curves are useful for commercial technologies in predicting rates of future cost reductions and improvements. They are not useful at the R&D or project "select" phase as a generic "it will improve in the future" concept, which sometimes is erroneously used for investment decisions. If technology improvements such as higher mass and energy yields, lower capex due to larger- scale plants are already considered explicitly in future economic scenarios, arbitrary inclusion of additional future generic "learning curve" improvements of unknown origin, can be redundant, and can lead to incorrect, overly optimistic assessment. Costs will come down due to manufacturing experience and learning (especially upon expansion of scale), but only for viable commercial processes. Margins will also likely be eroded due to competition, such that assumption of generic future learnings improvements of unknown origin are unlikely to be a valid premise for investment in what appears to be an borderline uneconomic project.

## References


1) Arrow, K., 1962. The economic implications of learning by doing. *Review of Economic Studies,* 29, 155–173.
2) M. Greaker, E. Lund Sagen, *Energy Economics, 30* (2008) 2899–2911.
3) K. Schoots, F. Ferioli, G.J. Kramer, B.C.C. van der Zwaan, Learning curves for hydrogen production technology: An assessment of observed cost reductions, *International Journal of Hydrogen Energy, 33* ( 2008 ) 2630 – 2645.

4) M. B. Lieberman, The Learning Curve, Diffusion, and Economic Strategy, *Strategic Management Journal, 8* (1987), 441-452.

5) http://cost.jsc.nasa.gov/learn.html

6) Martin Junginger, Erika de Visser, Kurt Hjort-Gregersen, Joris Koornneef. Rob Raven, Andre´ Faaij, Wim Turkenburg, Technological learning in bioenergy systems, *Energy Policy 34* (2006) 4024–4041.

7) J.D. van den Wall Bake, M. Junginger, A. Faaij, T. Poot, A. Walter, Explaining the experience curve: Cost reductions of Brazilian ethanol from sugarcane, *Biomass and Bioenergy, 33 (2009), 644-65.*

8) Peters, M., Timmerhaus , K. 1991. *Plant design and economics for chemical engineers*. 4th edition. McGraw-Hill, New York.

# (490a) Scale-up Methodologies When Handling Extremely Dangerous Intermediates and Products

**Jerry S. Salan**, Shannon Lenahan and Matt Jorgensen, Nalas Engineering Services Inc., Salem, CT

The Department of Defense is currently investigating 'green' primary explosives. As a result, dozens of novel primary explosives have been synthesized and some scaled up. Nalas Engineering Services, Inc. has assisted in the process development of these extremely dangerous materials. Traditional methods for process development were avoided due to safety concerns of isolating large quantities of the intermediates and products. Non-traditional sampling methods, analytical methods and kinetic model development were used to gain process understanding for these materials. In-situ tools included in-situ infrared, focused beam reflectance measurement, and particle vision and microscopy. Off-line tools included atomic absorption, high performance liquid chromatography, karl fischer, differential scanning calorimetry and NMR. The combination of these tools assisted the chemical engineers in solving some very intersting process challenges.

See more of this Session: Advanced On-Line Analytical and Optimization Tools In Pilot Plants
See more of this Group/Topical: Particle Technology Forum

# (490b) Automated Pilot Plant System Producing 3,3-Diamino-4,4-Azoxyfurazan

**Kimberly Hanson**[1], Jerry S. Salan[2], Andrew G. Pearsall[1], Matt Jorgensen[2], Patrick B. Skahan[1] and Tod F. Ricks[1], (1)Chemical Development, Naval Surface Warfare Center, Indian Head, MD, (2)Nalas Engineering Services Inc., Salem, CT

ABSTRACT:

The United States Navy developed a novel continuous process to produce kilogram quanitites of 3,3-diamino-4,4-azoxyfurazan (DAAF) for inclusion in explosive formulations. The process required extremely tight control of pH to maintain the desired impurity profile and resulting thermal stability of the final product. The process required the development of offline analytical techniques to monitor starting materials, intermediates, product and impurities in reaction mixtures and isolated solids. This data was incorporated into a kinetic model which was utilized to design a continuous process to DAAF. The details of the process development, kinetic model and final pilot plant controls will be discussed.

See more of this Session: Advanced On-Line Analytical and Optimization Tools In Pilot Plants
See more of this Group/Topical: Particle Technology Forum

# (490c) Process Monitoring of An Energetic Material Pilot Plant

Jerry S. Salan, Nalas Engineering Services Inc., Salem, CT and **Thomas Palathra**, Indian Head Division, Naval Surface Warfare Center, Indian Head, MD

The Naval Surface Warfare Center, Indian Head Division develops processes for energetic material ingredients and their formulations. The energetic material ingredients are developed using similar protocols utilized by pharma and fine chemical companies. However, the Navy is focused on removing operators from the hazardous environment. The Navy has purchased and currently installing novel reactors and chemical reaction monitoring equipment in order to mitigate hazards to operators while maintaining process control.

See more of this Session: Advanced On-Line Analytical and Optimization Tools In Pilot Plants
See more of this Group/Topical: Particle Technology Forum

# (490d) Using Process Analytical Technology (PAT) Tools to Support Flow Chemistry Development and Production

**Yanqiao Xiang**[1], John Lucas[2], John G. Van Alsten[2], Bryan Li[3] and Nandell F. Keene[3], (1)Research Analytical & Technology, Pfizer Inc., Groton, CT, (2) Chemical Research & Development, Pfizer Inc., Groton, CT, (3)Research API, Pfizer Inc., Groton, CT

This work reports applications of process analytical technology (PAT) to flow chemistry development and manufacturing. Case studies demonstrate that the PAT tools provide insight into chemical mechanisms and generate kinetic data to facilitate flow process scale-up. The case studies also present the utility of the PAT tools for on-line monitoring flow production process as it provides real-time data on flow stream dispersion, impurity profile, reaction progress and flow system cleaning verification, proving that PAT are convenient and versatile tools for rapid optimization and synchronized control of continuous flow chemistry processing.

See more of this Session: Advanced On-Line Analytical and Optimization Tools In Pilot Plants
See more of this Group/Topical: Particle Technology Forum

**Process Analytics – Intensive Pilot Plants: Learn Early, Save Later**

Steven F. Wright, Technology Fellow, Process & Environmental Analytics Group, Eastman Chemical Co.

It is well established that appropriate piloting efforts tend to produce new plants that start up on-time, on budget and running at expected capacity. This is also true for analytical instrumentation used throughout the development and piloting process. Correct inclusion of at-line and on-line analyzers in labs and pilot plants can support rapid, early process knowledge learning while it is still possible to change design features at lower cost, to strengthen models and reduce unpleasant surprises. While good goals, many piloting efforts are hindered because they fail to dedicate sufficient resources to analysis, generally breaking the knowledge link somewhere between the development lab and the full-scale plant.

The author will review some suggested approaches that can be used to identify and justify situations where at-line and on-line analytical will help speed your piloting process, when the normal analytical lab turnaround times just won't suffice for cost, process and/or safety reasons. Examples include highly dynamic reactions, complex samples and intense 24x7 analytical support bills that can wreck the budget while not really giving you the information you need.

Instrumentation is no better than its sampling strategy. The author will review sampling strategies that minimize process disruptions, while insuring representative sampling, especially important for smaller volume units found in bench and small piloting operations.

On-line Analyzers, such as mass spectrometers and Raman spectrometers, while valuable, are expensive to purchase and own. Approaches to instrument selection for maximum flexibility, process sharing, calibration and long-term maintenance in an ever-changing piloting environment will be discussed.

# (512a) Academics and Open Innovation

**Richard L. Long**, New Mexico State University, Las Cruces, NM

Open Innovation generally refers to property rights issues associated with orgainzations such as nine sigma. In a broader sense, it refers to the issue of "credit" of which cash awards are one type. The issue of "control" and the financial and other rewards have to do with the relationship between buyer and seller. Transactions must have a degree of confidence in the integrity of the parties and establishing that confidence is a key requirement.

See more of this Session: Challenges In Open Innovation and Growth
See more of this Group/Topical: Management Division

# Effective Quantitative Decision-Making based on Lowering Cost of Capital and ROI thru Safety Eng Studies

**Gutiérrez Capriles Glenda J**
***Risk Project Analysis Leader Engineer Advisor**
**Avenida Circunvalación del Sol, Urbanización Santa Paula, Torre Inelectra, piso 7, Municipio Baruta, Estado Miranda, Venezuela**
**glenda.gutierrez@inelectra.com**

Prepared for Presentation at
AIChE 2011 Annual Meeting
Minneapolis, MN, United States
October 16-21, 2011

UNPUBLISHED

# Effective Quantitative Decision-Making based on Lowering Cost of Capital and ROI thru Safety Eng Studies

**Gutiérrez Capriles Glenda J**
***Risk Project Analysis Leader Engineer Advisor**
**Avenida Circunvalación del Sol, Urbanización Santa Paula, Torre Inelectra, piso 7, Municipio Baruta, Estado Miranda, Venezuela**
**glenda.gutierrez@inelectra.com**

**Keywords:** Management of projects, processes of management, areas of management, phases or service life of the project, Decision-making methodology, instrumented protective systems (IPS), spills, return on investment (ROI), hazardous events, fire zones

## Abstract

The intention of this article is to reflect on the main aspects that affect the performance of the projects, for such effect took place a diagnosis of the schemes of management used by the company of the consulting sector. For the study one approached the areas of generation business, petroleum, refining, petrochemical, technology, etc of inelectra and the operational area of a production facility.

The process industry has adopted automation to improve product quality and production rates, to reduce the potential for operator error, and to decrease resource requirements. Process automation includes many different instrumented systems, such as process controls, alarms, instrumented protective systems (IPS), and emergency shutdown systems. These systems range from simple hardwired systems to complex programmable electronic systems.

Good design leads to safe operation by anticipating potential errors and failures and implementing means to detect their presence, so that they can be corrected once implemented, the increased complexity may require a significantly greater level of attention during assessment, design, inspection, testing, and maintenance.. Good design ensures that the core attributes of independence, functionality, integrity, auditability, reliability, access security, and management of change are addressed consistent with owner/operator requirements. Poorly implemented design can lead to significant hazardous events, involving impact to people, the environment, and plant assets.

Various valuable information scattered all over CPI (Chemical Process Industry) needs to be integrated and retrieved efficiently to manage technique, facility and personnel. This paper focuses on a new decision-making methodology of safety-related investment and management of main causes for accidents and failures. The methodology integrates safety technology, process data and business information in a petroleum and oil-refining industry. This methodology suggests a quantitative investment procedure to reduce safety accidents and plant failure cases within limited budget efficiently. Major factors are considered as accident history, human error, facility, service, operation, maintenance, emergency plan, and priority of investment, effectiveness of improvement, environmental concerns, loss control and cost.

# 1. Introduction

Any industry, principally oil-refining industry, have to keep safety management methods that ensure risk of fire, explosion and toxic material release which is very important in saving life, assets and environment. The right decision-making for improving safety in chemical plants requires more quantitative and systematic methods of analyzing previous accident records, process data and business information. This safety engineering studies are necessary to be assure to the whole decision-making procedure for project risk analysis evaluation, to prevent major accidents and to reduce the risk of a plant.

The initial design effort includes the prediction of potential hazards. Numerous types of hazard analysis methods have been proposed and used. Hazard analysis methods are classified by qualitatively and quantitatively. The qualitative methods include Check-lists, What-If Analysis, Failure Modes and Effects Analysis, Preliminary Hazards Analysis or HAZID and HAZOP. The quantitative methods include Event Tree Analysis, Fault Tree Analysis, Quantitative Risk Analysis, Reliability Analysis and Safety Integrity Level (SIL) / Safety Instrumented System (SIS).

New technology is of value only if it provides a suitable financial return. Many articles have been published on the potential economic benefits and return on investment (ROI) of facility and advanced automation technology in the process industries. Unfortunately the benefits claimed are often unrealistic and unsubstantiated; this leads to significant credibility issues when the forecast benefits are not achieved and to a lack of confidence on the part of management on proposals in this area. In this article a short review of the proper way to perform financial analysis as part of project risk analysis that was affected by improper safety engineering studies execution.

# 2. Review of Plant Economics

The first step is to examine the plant as a financial asset. We may think of plants as a collection of equipment and personnel that convert raw materials into products. From a financial point of view, a plant is an asset that consumes money and produces money, hopefully producing more than is consumed. The major monetary components are shown in the figure following:

Figure 1: The Plant as a Financial Asset

By convention, financial inputs are classified as either capital or expense. Expenses include all the ongoing costs of producing the products such as raw materials, net utilities (used less produced), operating costs, maintenance expenses, and other miscellaneous costs. Capital has two components – working and investment.
Investment capital refers to the cost of major equipment or system additions that will last for several years and can be depreciated for tax purposes. Working capital is the value of inventory and required net short term financial funding.

## 3. Return on Investment Capital taking into account also, Procurement and Safety

To properly prepare an investment evaluation we must understand the language and objectives of the plant's financial management. In this section a quick overview of the subject is given. For more background and a more comprehensive review.
Ultimately the management of a company has an obligation to increase the long term financial value of the corporation to its owners, who are normally the shareholders. This value can be measured in many ways but stock market valuation is one of the more important and ultimately is a function of financial performance. There are equally many ways of evaluating financial performance but the primary measure used here will be Return on Invested Capital (ROIC). This is not only a good internal measure but also correlates with long term stock market performance. The consulting company McKinsey (2003), recently published results of a study by of 130 publicly traded chemical companies in the US and Europe and their financial and market performance over 40 years. Their conclusion: "only returns on invested capital drive market-to-book valuations".

Figure 2: Return on Invested Capital

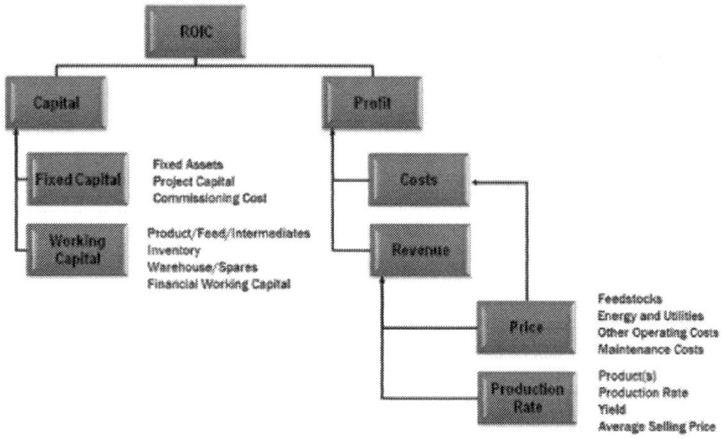


To increase ROIC, capital must be reduced or profit increased or preferably both at the same time. The primary areas where automation and advanced automation savings are normally found are illustrated in the boxes at the bottom of the figure. When a project is considered, all possible savings areas should be evaluated.

Potential capital savings include both fixed and working capital components. Potential project capital cost reductions due to automation choices can include savings in:

- Engineering
- Procurement Costs
- Purchase Price
- Installation, Configuration, Calibration & Commissioning
- Project Execution

Working capital can be reduced by reducing raw material, intermediates, and product inventories and also by reducing owned spares for equipment stocked in the warehouse. Capital deferred also results in savings due to the time value of money.

Deferred capital can result from longer equipment life due to better control or from more production from the same equipment which results in a postponed plant expansion.

Expenses can be reduced by lowering energy /utilities usage and decreasing raw material costs through increasing yields of desirable products. In special cases it may be possible to substitute a lower valued raw material.

There are several types of maintenance costs including:

- Scheduled Maintenance
- Unscheduled Call Out
- Major Shutdown

Reductions can occur in all types of maintenance costs due to improved automation performance and enhanced monitoring of process equipment. Specific savings can include reductions in:

- Unscheduled maintenance
- Number of routine checks
- Time to perform required maintenance tasks
- Maintenance materials purchase
- Number and costs of required tasks during scheduled shutdowns

Reduced off specification material can lower reprocessing costs. Demurrage is the cost charged by ships waiting to load. Waits beyond contracted amounts incur penalties. In selected cases it may also be possible to reduce average staffing levels.

Abnormal event prevention deserves some special comments. Reducing Health, Safety, and Environmental (HS&E) issues is at the top of every plant manger's agenda.

Improved automation is often a key to reducing these types of events. If your proposed investment will result in safer operation or reduced emissions it should definitely be included in the justification description even if no quantitative economic value can be attached to these improvements.

Next is the area of increased average selling price for products. Automation can contribute to this goal by increasing the yield of more valuable products. Reduced amounts of lower valued byproducts will increase average revenue per unit feed. Off specification material, which can be reduced through improved control, usually brings a discounted selling price. Occasionally an

increase in product quality due to improved control will actually permit selling the existing product(s) at a higher price.
With regard to increased production, the key question is whether or not the additional product(s) can be sold. Increases are only financially valuable for production limited plants, i.e. those manufacturing products where the market can absorb the increases. If not, no benefits can be claimed for the improvements. Increased production can result from the capability, with better control, to operate closer to production limits at constant product quality. Increased production can also come from reduced unscheduled downtime for equipment due to better reliability, shorter batch cycle times, lower grade transition time, less product reblending, and decreases in scheduled shutdown duration and frequency. The shutdown frequency can be reduced, for example, by increasing a furnace run time between required decoking or not cleaning a heat exchanger prematurely.
To estimate the potential magnitude of the savings, examine the normal expenditures and look for specific quantifiable areas where savings are possible. Often historical data will provide a basis for this analysis. External consultants with experience in this area can also help insure that potential savings are not overlooked and that there is proper documentation for the estimated savings used for the justification.

### *3.1. The Cost of Tendering*

Major equipment manufacturers have calculated that it costs them up to thousand dollars to bid on a complex tender. These costs are eventually passed on to the customer although be it indirectly.
Among the major contributing factors, are:
**Long Lead-Time of the Process -** There are many time consuming tasks that contribute to the long lead-time of tendering including: internal collaboration within a multi-disciplinary group of people within a single organization (but often geographically distributed), obtaining approvals (both financial and technical), preparing the tender specifications and documents, interfacing between customers and suppliers, interfacing with other companies, third party consultants, partners in a consortium and sub-contractors, assembling accurate bidding documents and calculating prices and time estimations.
**The Quality of Tendering and Bidding Documents -** This is a crucial element for successful and efficient tendering and involves criteria such as the clarity of requirements and specifications, compliance with quality certification standards, clarity of the bid, exhaustiveness, adequate structure, nice presentation, compliance with specific requirements set-out in the tender, completeness of the information provided, etc.
**Inability to Quickly and Reliably Assess the Bids -** Differences and inconsistencies in the bid formats make it very difficult to assess the bids and compare apples with apples. A large amount of time is often spent collating data from the various bidders into a format that allows for proper comparison and evaluation.
**Reliability of the Cost and Time Estimate -** Ultimately companies who are consistently accurate in their pricing and delivery estimates will win in the long run. The ability to provide reliable estimates depends both upon the capability of the bidders and the clarity and unambiguousness of the tenders. Clear, complete and non-ambiguous tenders will also naturally

lead to lower bidding prices. As uncertainty diminishes so does the "safety price" or padding added to the actual estimate.

**Monitoring Delivery of the Works -** The ability of the customer to continuously and transparently monitor the current status of the works is a definite advantage in that it reduces delays and ensures delivery in compliance with the contract.

### *3.2 The ROI of Safety*

Although it is difficult to prove that incidents will happen in the future, management and financial representatives of the company need to be aware of the consequences of an ineffective safety solution. By understanding the elements of risk management and the methodology of a safety investment analysis, safety engineers or risk management professionals can present the financial justification to help make operations safer and reduce both workplace injuries and insurance costs.

Integrating a safety solution to manage risk is a process that is best employed via up-front teamwork between operations and risk management, such as risk managers, safety professionals and claim analysts. In addition, it is important to understand risk and the different types of risk that exist on a plant floor. Risk awareness often leads to better decisions that can reduce soft costs. By providing an easy-to-be-safe environment for operators with safety solutions that are designed into the machine and focus on complete machine life-cycle requirements, companies are able to reduce indirect costs. In this paper, we will highlight key elements of effective risk management and the risks that affect the engineering, procurement, construction and operation of Production facility, discuss the implications on workers and provide ideas to measure potential savings that can be found in proactive safety investments.

Risk management is a two-step process that identifies and analyzes uncertainty in decision-making. It consists of determining risks that exist in an equipment investment and managing those risks in a way best-suited to achieve investment objectives. For managers and financial professionals investing in safety, proving the benefit of the safety investment is difficult since it could prevent injury and avoid costs for a company. Using the risk management process, professionals can determine the level of safety investment that meets their needs and would provide the best result for their situation.

An unplanned interruption can threaten a company's supply chain because of the decreasing amounts of raw materials readily available on the plant floor. These limited resources affect productivity levels and operation costs and threaten the overall health of a company as lower levels of inventory put manufacturers at increased risk for losing revenue each time a production line goes down. Effective safety programs can be implemented to help minimize this risk so manufacturers can better protect their supply chains.

Once the risks are fully defined and clearly understood, they must be designed out or mitigated to the greatest extent possible. Risk mitigation measures are necessary in order to physically improve the machine to prevent personal injury or property damage. Risk mitigation can be accomplished through a variety of activities, as demonstrated in Figure 2 below. One effective mean is to use safeguarding equipment such as installing physical guards or safeguarding control products, such as light curtains, safety relays and cable pull switches to reduce risk to employees.

Figure 3 – Hierarchy of Protective Measures

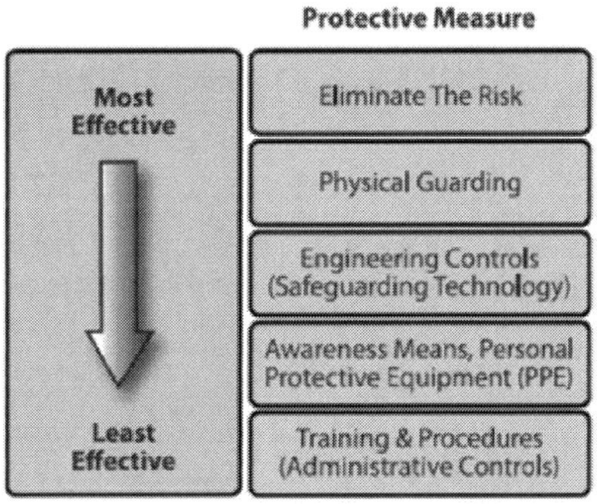


Using a formal risk assessment process also provides the benefit of documenting how the risks have been reduced or designed out via safeguarding solutions or other protective measures. The use and documentation of the risk assessment process is a critical tool for companies to show their due diligence and good engineering practices in providing a safe work environment. By effectively implementing and documenting the process, the resulting protective measures, and training programs, a company can potentially lower risks of litigation should an incident occur. After implementing and documenting the process, it is important to provide the appropriate training and supervision. It is crucial to that operators understand the safety measures, including usage of personal protective equipment. Operators must be trained.

Safety is not a luxury reserved for developed countries. One topic of discussion is this idea that safety and health programs benefit a company's bottom line. For many analysts, the bottom line is simple: "The CEOs 'who get it' recognize that a corporate commitment to safety has its costs, but, more important, it pays tremendous dividends in business and human capital."

So there is a clear connection between safety, quality, production, and profitability."

## 4. Investment Costs

Next investment costs must be estimated. When performing investment analysis, it is important to consider the full life cycle costs, not simply the initial purchase price, of new equipment and new software systems. Many studies have shown that more than two thirds of their life cycle costs occur after installation.

In the project initial, or capital costs area, include the following:

- Installation costs
- Required infrastructure upgrades
- Software licenses
- Application specific services – design, configuration, installation, integration and commissioning
- Project expenses – project management, purchasing
- Training – System support and end user

Also, ongoing maintenance and support cost estimates. Don't forget the last item. Make sure that you budget for ongoing training. The best long term job security today is maintaining you and your staff's skill levels and expertise current. It is easy to fall into the trap of being so busy with day to day problems that you lose sight of the longer term requirements.

## 5. How Can ROI be Useful?

Profits happen when a company operates effectively. We can tell that the management team is doing its job well if the company prospers, obtains funding, and rewards the suppliers of its funds.ROI is the principal tool used to evaluate how well (or poorly) management performs.

## 6. Safety Decision-Making Procedure

A framework for viewing management information systems is essential if an organization is to plan effectively and make sensible allocation of resources to information systems tasks. A framework is needed to provide a more efficient allocation of resource in safety information. This study suggests the procedure, development and information management of safety decision-making methodology in two different information areas as shown in Table 6.1. A framework is suggested for Decision-Making by using model structure and implementation procedure which differs sharply between the structured and no-structured information area.

**Table 6.1. Framework of Safety Information Management System**

| | Safety | Process | Business |
|---|---|---|---|
| **Structured** | MSDS<br>Regulations<br>Standards<br>Company Policy | P&ID<br>PFD<br>Plot Plan | Department<br>Personal<br>Facility<br>Warehouse<br>Buildings |
| **Non-structured** | Accidents<br>Measures<br>Risk<br>Accident History<br>Causes Analysis | Technical Reports<br>Technical Audits<br>Design Execution with Quality<br>MOC History | Cost<br>Strategy<br>Planning<br>Investment Priority<br>Project Risk Analysis<br>Benefit-Cost Analysis |

CCPS (Center for Chemical Process Safety, 1998) in AIChE presents effective taxonomy of safety data for reliability analysis as shown in Figure 4.

The basic building blocks for the reliability analysis fall into two main classes. One is the data regarding specific events occurring in a piece of equipment. The other is the data describing the sample space of equipment. Event data in the first class is used as a numerator in the reliability calculation and inventory data in the latter class is used as the denominator.

Classification of safety accident is important while collecting safety information, so we classified it as accident results, direct and indirect causes of accident, primary causes, process status, utilities, departments etc.

Significance and operating variables also affect the business strategy in plant site also of Project Risk Analysis.

Figure 4. Database Taxonomy and Structure

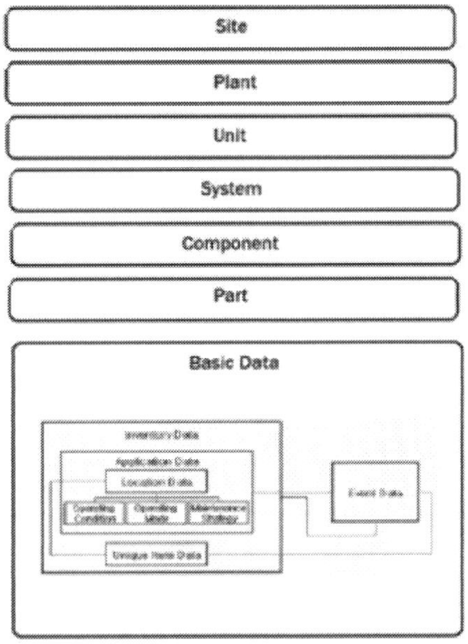


This study describes an overall systematic decision making procedure of preventing or minimizing safety accident under the given budget as shown in Figure 5. To decide the investment of safety related facility, two inputs are required: 1) the priority of investment and 2) the criteria of investment. To list up the priority of investment requires analyzing plant accident data and process data with the general safety information by using matrix computation method that we propose. To set up the criteria of investment requires surveying business-level executives regarding acceptable safety damage such as casualty, cost loss and business interruption.

Figure 5. Safety related Business Decision-Making Procedure.

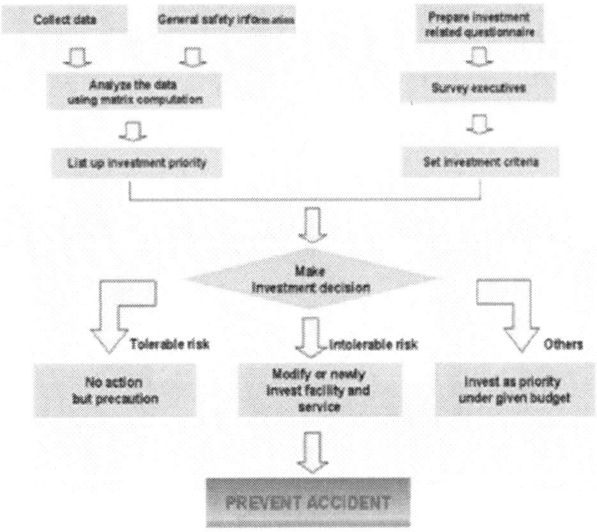

Based on the above two preparations such as the priority and the criteria, the decision is made according to the three risk probability regions: 1) definitely tolerable risk region, 2) definitely intolerable risk region and 3) other region in between the two. If the risk of a hazard is tolerable, the company does not need to invest any money into the relevant event and, in many cases, just careful attention is necessary. If the risk is definitely intolerable, the company must invest money to modify the process or purchase new facility and related service regardless of the budget if the plant needs to be run continuously. If the risk is between definitely tolerable and intolerable region, the investment is negotiable with the given budget that is, the investment to each event is selected by the priority under the condition that the given budget allows. As following this decision procedure, the company invests money only to necessary facility and service for preventing safety accident under the given budget.

## 7. Study Case: Surface Upstream Project

### *7.1 Scope*

The whole Station Upgrade Project comprises the installation of new facilities, and revamp of existing ones needed to increase the produced water treatment capacity during production of Extra Heavy Oil (EHO) of 100 KSBPD to 200 KSBPD.
In more detail (Figure 6), MSUP includes:

- A new Diluted Crude Oil (DCO) Treating Train (Train C), capable of treating 95 KSBPD of EHO diluted with 34.4 KSBPD of Diluent, with 77 KSBPD of associated produced water.
- A new Produced Water Treating Plant of 200 KSBPD.
- Utilities revamp, including a new 25 MW Turbogenerator and 100 KBBL Diluent tank.
- Minor upgrades in existing DCO Treating Trains.
- New Flare
- New Fuel Gas Skid
- New Chemical Injection Packages
- New Desanding Systems
- New Open and Closed Drain System
- New Control Room (MSCR)

### *7.2 Location*

Main Station facilities are located to 7,5 km South-Eastern from San Diego de Cabrutica, near from region Pariaguán, to South Anzoátegui State, in Venezuela.

### *7.3 Facilities Description*

Main Station facilities are part of the Upstream Section of an overall heavy oil production and upgrading concept that will lead to the production of Synthetic Crude. Development of this concept is under way through two different Engineering Projects. The project that deals with the production and initial conditioning of the heavy oil is referred to as the Upstream Surface

Facilities Project, while the one that deals with the transformation or upgrading of the heavy oil to Synthetic Crude is referred to as the Downstream Facilities Project.
The major elements of the Upstream Section are:

- Oil Wells, grouped in Clusters, to produce the heavy crude.
- Diluent System, to dilute the heavy crude at the well head and improve its pumpability.
- Gathering Network, to collect and send the diluted heavy crude for initial conditioning.
- Main Station, to carry out the initial conditioning.

The facilities in the Main Station Upgrade Project are divided into Third Oil Dehydration Train called "Train C", a Production Water Treatment Plant (PWTP), and some utilities revamp (fuel gas, Diluent, open and closed drains).
The existing MS Oil Processing Trains (A and B) are not capable to handle the new production target 246 KSBPD EHO, 89 KSBPD Diluent and 200 KSBPD Water due to the increasing water cut. In order to handle the fluids coming from clusters a Third train is required.

- Export Pipeline, to send the conditioned heavy crude to an intermediate Pump Station, operated by another party that will finally get the oil to the Upgrader Section.

Main Station's primary objective is to process the incoming fluids from the gathering network, i.e. oil, associated gas and water, so that oil can be consistently exported to the Upgrader within the required specs and at proper pumping conditions.
Oil to be exported from Main Station shall have a maximum water cut of 2%, a Reid Vapor Pressure no greater than 0.76 $bar_a$ (11 $psi_a$), a specific gravity of 0.953 (17° API) and shall be pumped at approximately 89 °C and 25 barg.
The crude oil is separated from the gas and is stabilized to less than 0.76 $bar_a$ (11 $psi_a$) RVP and the water content is reduced to less than 2% vol. The API gravity of the crude oil is adjusted by adding diluents to 17 °API and is then pumped to the export pipeline. The facilities at the Main Station are continuously manned.
The fluids coming from the gathering network enter Main Station and split into two identical trains at the upstream of the Slug Catchers (SC) where most of the associated gas is separated and sent to the fuel gas system.
The liquid from the SC is heated through heat exchangers and direct fired heaters so it gets to the right temperature to release enough light ends (gas) to meet the RVP spec in the High Temperature Separators (HTS). The vapor released in the HTS is cooled down so the diluents fraction that evaporated in the HTS is condensed, separated from water and returned to the export pump suction. The remaining gas is sent to the fuel gas system and the water is sent to treatment.
The new PWTP is located at Main Station and it is designed to process 1325 $Sm^3/h$ (200 KSBPD) of production water. Treated water has oil content below 30 ppm and total solids content below 20 ppm making it suitable for re-injection.
The PWTP consists of the following water clarifying equipment:

- Skimmer Tanks T-3900 C/D and associated Gas Boots D-3900 C/D (working in parallel), with associated pumps: Skimmer Tank Transfer Pumps, P-3901 A/B/S (P-3901 A for T-3900 C and P-3901 B for T-3900 D; P-3901 S is a common spare) and Oil Recovery Pumps, P-3951 A/B/S (P-3951 A for T-3900 C and P-3951 B for T-3900 D; P-3951 S is a common spare).
- Flotation Units X-3900 B/C with associated Flotation Transfer Pumps, P-3921 A/B/S (P-3921 A for X-3900 B and P-3921 B for X-3900 C; P-3921 S is a common spare).

- A Flotation Overflow Separation Tank T-3930 B, with associated pumps: Flotation Recycling pumps P-3926 A/S and Flotation Sludge Recovery Pumps P-3961 A/S.
- A Clean Water Tank T-3920 B to store filtered water and,
- High - pressure Water Re-injection Pumps P-3932 A/B/S.

Utilities such as Fuel gas, Firewater, Instrument Air, Plant Air, Drinking and Utility Water, Diluents required by Train C and PWTP are provided by Main Station.
The Fuel Gas used in Main Station it is provide by the incoming fluids from clusters through Trains A/B/C (Gas Oil Ratio between 50 - 250 SCF/BBL, about 12,3 MMSCFD – 61,5 MMSCFD), and the one coming from the Gas Gathering Network, for a maximum gas flow rate of 120 MMSCFD, according to gas production forecast by MSUP Study.
In order to handle this peak flow rate new Fuel Gas facilities and LP Flare System will be enhanced.
All this gas is sent to the Fuel Gas KO Drums, D-3201 A/B/C, and Fuel Gas Filters F-3201 A/B/S, and then it is distributed to the users, namely:

- Oil Heaters, H-3001 A/B/C,
- Power Generation Units, GT-3201 A/B/C/D,
- Off-spec Storage Tank, T-3002 and the Production Water Treatment Plants for blanketing purposes.
- Flare system (LP and LLP), for continuous purge.

Gas for turbogenerators is compressed in the Fuel Gas Compressor Packages, K-3201 A/B/C/S while the heaters and relief system can use it at the pressure level supplied by the KO drums.
Firewater, Instrument Air and Utility Air, Utility Water and Drinking Water connection points to Main Station are listed in Tie-in Point List. Electrical power is supplied by Turbo Generators, GT-3201 A/B/C, for MSUP a new Power Generation Unit, GT-3102D, will be installed, with its associated equipment (new generator substation G/S-02 plus an electrical substation T/S-04 dedicated to New Oil and Water Processing Trains). Additionally MS has two Emergency Diesel Generators, EDGS-1/2.
The PWTP normal operation is manual but it can be monitored from the Main Station Control Room and by sample testing at different points throughout the Plant.

### *7.4 Risk Analysis - HAZOP*

The key to the Process Hazard Analysis (PHA) is a simple concept offers ways to answer the following four (04) questions, also a greater risk reduction, beyond the goals above, may be appropriate where an analysis cost-effectiveness of permits:

a. What can go wrong?
b. What are the causes?
c. What are the consequences?
d. What is the probability of occurrence?

The implementation of the entire set of techniques PHA comprises a quantitative review risks (Train C, New Water Treatment Production Water (NWTPW), Generation System and Common Services including here: Fuel Gas System, Low Low Pressure and Low Pressure Relief System, Diluent System, Open and Closed Drain Systems, Instrument and Plant Air System, Potable and Service Water System, Chemical Injection Systems) that may be in an installation, ranging from frequent incidents low-stakes to major incidents rare and, using a methodology uniform and consistent.

Once identified risk processes, the techniques can PHA help focus control studies risk. The PHA must address the following:

- The hazards of the process
- Identification of any previous incidents with had a likely potential for catastrophic consequences
- Engineering and administrative controls applicable to hazard, their interrelationships and consequences of failure of these controls, especially those affecting employees and environment
- Facility siting
- Human factors
- Evaluation of effect of failure of controls
- The requirement to promptly resolve PHA findings and recommendations

The largest contributors to risk can be identified and can be recommendations and take decisions, as corrective measures, on the basis objective and consistent.

The business value of the hazards analysis is the use of report to train people on how deviations from normal operation propagate to process hazards. The risk analysis defines the risk reduction strategy selected to address unacceptable risk. Plant management, engineering, operations, and maintenance must understand this information if they have responsibility for decisions affecting the hazards or risk.

Operator and maintenance procedures should include a description of the process hazards where their actions potentially increase risk. The hazards analysis documents should be discussed at Operations' safety meetings.

Residual risk should be addressed with compensating measures and action plans to reduce risk as required.

In this case, don't was identified the hazard about:

- Wrong specification for suction and discharge valves at Fired-direct Furnace, H-3001C, (Figure 8)
- Lack of control for valves at inlet pipelines
- Among others

**Figure 8. Fired direct Furnace H–3001C**

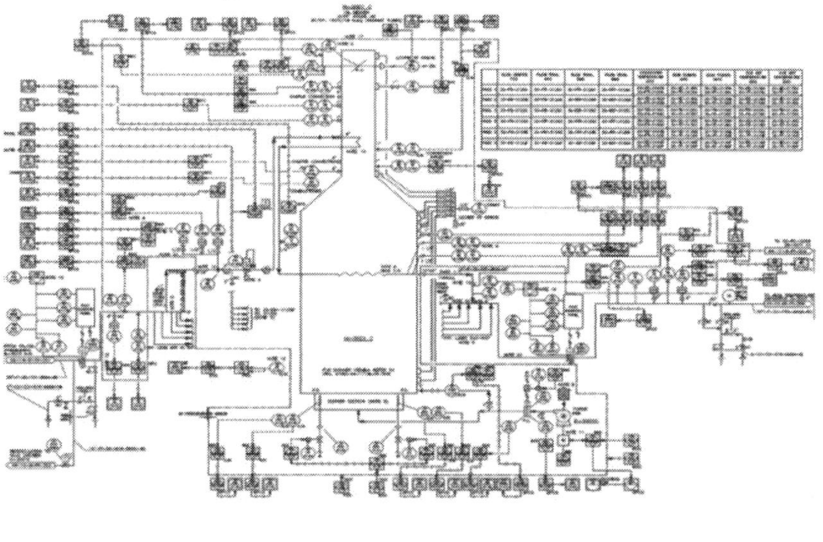

### 7.5 *Allocation of Requirements of Safety Integrity Level (SIL) of Safety Instrumented System (SIS)*

Due to the worldwide major accidents, most industrialized nations established process hazards control regulations. Germany passed the Hazardous Incident Ordinance (1980) and the European Community created the Major Accident Hazards Directive (1982). The United States passed legislation creating the Emergency Planning and Community Right-To-Know Act (EPCRA, 1986). In 1990, the Clean Air Act Amendments mandated that the Occupational Safety and Health Administration (OSHA) and the Environmental Protection Agency (EPA) develop accident prevention regulations, leading eventually to the 1992 OSHA Process Safety Management (PSM) and the 1996 EPA Risk Management Program (RMP) regulations.
None of the regulations were prescriptive. Most were intended to be flexible, simply requiring industry to examine the risk posed by their processes and to work to minimize these risks. Only a few European regulations provided quantitative risk targets. Many European regulations, as well as those of the U.S., relied on industry to develop its own tools to identify risk and to make appropriate efforts to achieve lower risk. As an incentive, industry was required to communicate the identified process risks to the public and emergency responders through public forums and documentation.
The industrial community responded to the regulations by issuing industry standards and guidelines concerning the evaluation of risk and the design of safety systems. The American Petroleum Institute (API), the American Society of Mechanical Engineers (ASME), and the National Fire Protection Agency (NFPA) created codes and practices for specific applications. In 1988, the International Society for Measurement and Control (ISA) began an eight year odyssey to develop a standard for safety instrumented system design (SIS) for the process industries, ANSI/ISA 84.01-1996; late In the mid-1980s, the International Electrotechnical Commission (IEC) began the development of an international standard for the design of all safety-related systems, covering transportation, medical, manufacturing, and process industries as IEC 61508. And then at 2003, ANSI/ISA.S.84.00.01-2004 and IEC 61511.
Consequently, it is regulations the mandate that industry identify and address risk. Industry standards are simply providing the tools for assessing the adequacy of industry's efforts. The difficult task of creating one standard for all of the process industry made the creation of a prescriptive standard impossible. After all, could one prescriptive standard cover refinery furnaces, chemical reactors, pulp and paper digesters, and utility boilers and reduce risk appropriately at a reasonable cost? Thus, the standards had to be performance based. Both standards chose to rely on the establishment of a design process called the SIS lifecycle, throughout which the performance of the instrumented systems must be maintained.
The lifecycle is intended to address the primary causes of control system incidents. Compliance with the industrial standards requires four essential elements:

a. Identification of safety functions required for safe shutdown;
b. Assignment of a safety integrity level (SIL) for each safety function;
c. Use of the safety lifecycle for the SIS design; and
d. Verification of the SIL achieved for each safety function.

These elements result in a major paradigm shift for SIS design. While it has been accepted practice for many years to mitigate potential incidents with instrumented systems, there has generally been no assessment of what type of SIS provides the appropriate risk reduction. These standards have now established requirements that industry document the rationale behind the use of a particular design, evaluate that design for its integrity, and demonstrate that the integrity can be maintained. This is perhaps the most significant impact of ANSI/ISA 84.01 and IEC 61508 on process safety, in general.

As performance based standards, the race for compliance is marked with the broadly defined finish line of good engineering practice and highly recommended practices. As industry races toward compliance, it must work hard to prevent the creation and acceptance of bad engineering practices, which threaten the economics of plant operation and erode the effectiveness of SIS designs, addressing seven bad engineering practices:

1. Believing that, if something is not specifically stated, either "shall do" or "shall not do," in the standards; you do not have to worry about it.
2. Thinking that meeting the minimum requirements means the process is safe and the SIS is compliant with the standard.
3. Ignoring the importance of good engineering practice.
4. Designing systems that meet safety requirements but not economic protection requirements.
5. Focusing only on SIL and not on preventing nuisance trips.
6. Neglecting the human factors.
7. Focusing on capital cost and not lifecycle costs.

A special attention should be paid to the possible common mode causes which may induce a failure of several (or all) barriers of the protection system. They include namely:

- The power supplies of the system.
- The logic control system(s)
- The valves activation system(s)
- The corrosion/erosion/plugging effects of the crude fluid.
- The operation of the system.

The Active Protection System is to be based on different independent barriers acting at different levels in case of an escalation of any deviation from the normal operation:

- Alarm: provides a warning to the operators in order to take a corrective action, given the deviation is not expected to become hazardous during the performance of the action.
- Production Shutdown System (PSDS): avoids the occurrence of an accidental event upon detection of a hazardous situation. The related action induces generally the shutdown of the related equipment through the implementation of one or several of the following barriers:
    - Primary barrier: aims at suppressing the cause of the deviation.
    - Secondary barrier: aims at limiting the effect of the deviation onto the equipment.

The PSDS may include several sub-levels, given there is General production Shutdown level initiating the shutdown of the whole production equipment.

Emergency Shutdown System (ESDS) triggers the safety shutdown of the installation upon detection of, either an **accidental event** or of a hazardous situation which may escalate rapidly. The ultimate ESD level causes the isolation of the installation at its limits through ESDV's and SCSSSV's (wells). It initiates the General production Shutdown and the shutdown of the related utilities. It keeps the necessary resources for emergency response. The ESDS may include several sub-levels, given the ultimate level, as defined here above, has priority on the other levels.

The Industry Practice regarding the control of the overpressure risk is to consider, by order of priority, the following possible options:

1. Design the equipment and/or the pipeline system to sustain the maximum possible pressure which may occur in the equipment (safety by design).
2. Design a protection system which, in accordance with API RP 14C, is made of:
    - A primary barrier aiming at isolating the heated equipment from the source of overpressure (suppressing the cause).
    - A secondary barrier aiming at relieving the pressure through a suitably sized venting or flaring system (limiting the effects).
3. Design an instrumented protection system made of two independent barriers acting at different levels to isolate the heated equipment from the source of overpressure before the pressure exceeds the maximum allowable pressure. This option named IPPS is considered when it is shown the options (1) and (2) induce disproportionate costs, technical or environmental problems. When a rapid escalation of the overpressure is expected, the IPPS may trigger the closure of the SCSSV's and/or ESDV's through the ESDS, in that case, the ESDS provides an additional and independent barrier.

The system should be designed fail-safe which means that the system will revert to a predetermined safe state in the event of failure of its components or of its power supplies. These include:

- Loss of the actuation energy of the valves.
- Loss of the control signal of the valves.
- Loss of the overpressure detection sub-system.
- Loss of the logic sub-system.

The overpressure protection system is to be based on the concept of two independent active barriers automatically initiated, at different levels, by the early detection of the pressure built-up into the protected equipment.

The system should survive any single failure of its components without jeopardizing its protection function. This is to be implemented through the research and elimination of the possible common mode failures, implementation of redundancies or by fail-safe action. Consideration should be given in the design of the redundancies to the using of different technologies (electrical/hydraulic valve activation as example).

The system has to be designed in order to reduce the probability of unrevealed failures of its components. This is achieved by implementation, where applicable, of monitoring and/or automatic testing of the components triggering an alarm in case of detected failure.

The system has to be designed in order to facilitate the periodic testing of the system. Consideration should be given to the possibility of performing a partial closure and re-opening of the valves while producing when this improve the availability of the protection system.

The overpressure protection system including the dedicated IPPS components should be considered as a whole system regarding its dedicated target and general objective. The reason of this approach is to facilitate the control of the overpressure hazard by considering all possible causes and their effects and namely to tackle the possible common mode failures which might involve a failure of the whole system, for example, for fired-direct furnace H-3001C it is considered an overpressure protection system that don't operate because the shutdown valves (input and output; continuing service) are specified as "Fail Last Position" and then, occurring an event will not protect the equipment, operators, assets and environment.

7.5.1 Evaluated Systems as New Train C, New Water Treatment Production Water (NWTPW), Generation System and Common Services including here: Fuel Gas System, Low Low Pressure and Low Pressure Relief System, Diluent System, Open and Closed Drain Systems, Instrument and Plant Air System, Potable and Service Water System, Chemical Injection Systems

In general, SIL3 is configured on all systems which is not sufficient safety to conduct a risk that has negative consequences and high exposure to high toxic concentrations.

Provided the single safety function (SIF) as HAZOP results, you're talking about lies within the scope of IEC 61511/62061 (i.e., a SRECS for instruments and machinery), and – as long as it functions properly – is sufficient to reduce the residual risk to an acceptable level, a reliability level of SIL3 for that single SIF is sufficient. By design that single SF will have to include sufficient redundancy/diversity/monitoring to be able to achieve SIL3.

With I mean that a if the single SF concerns e.g. the guarding of a shutdown valve to a danger zone, no other means of access to that same danger zone should exist, unless these are also guarded by SIL3 safety functions.

Remember that according to the principle of integrated safety for equipments, a SIS should only be considered "after" the possibilities of design changes and of physical protection measures have been exhausted. If the residual risk at the end is still too high, should consider additional constraints on the intended use of the equipments.

If use a simple pressure switch (diaphragm to lever-arm to electrical switch) in a safety interlock circuit where $PL_a$ is called for, is a standard off-the-shelf pressure switch appropriate?, and If $PL_c$ is called for (due to higher pressure), is still; possible to use a high reliability pressure switch, or would one use a pressure transmitter in a SIL1 safety function.

Can use a commercially off the shelf pressure switch provided it is a well tried component with known failure modes for a $PL_a$ safety related control function, however to achieve $PL_c$, you will need to use redundant pressure switches. Using a single pressure switch may not give you the desired reliabilities due to HFT = 0 using IEC62061 or even with the use of ISO13849-1 due to a single channel circuit. However, it is important to bear in mind that the safety related control function extends from the sensing switch right up to the actuating device; and the lowest PL or SIL in the chain will influence the final result for the entire safety loop. A pressure vessel would fall under the pressure equipment directive 97/23/EC and the IEC 61511 functional safety standard for such systems would take over. Principles are similar using SIL too.

$PL_c$ capability can be achieved in a single channel subsystem providing that the components have high $MTTF_d$. For a SIL1 according to Table 5 in IEC 62061 it is possible to achieve SIL1 with HFT = 0 providing that SFF is >60%. Single channel sensor subsystems can also achieve SIL1 according to Table 6 in IEC 61511.

The common cause and systematic failures that affect even a SIL Logic Solver will affect all protection layers that the logic solver is part of, and these will as a consequence of this not be independent. In fact this is subject to discussion, where some are of the opinion that a SIL3 logic solver is able to cover both the 2 IPLs required by ISO 10418.

These SIFs would not qualify to be IPLs. However, I would think that the SIFs should reinforce the protection towards the same risk, especially if the Logic Solver has very low failure rate compared to the field devices.

The independent rule would be hurt. According Layer of Protection Analysis, CCPS 2001: "In order to be considered an IPL, a device, system, or action must be independent of the initiating event and the components of any other IPL already claimed for the same scenario..." However, one of them could be considered as a safeguard, none SIL claimed.

Depends on having independent paths through the SIL3 Logic Solver. It is possible to have two separate SIF's with different initiators and different final elements to cater for the same risk but same logic solver (SIL3 rated) are these IPL's.

Adding SIFs for the same risk don't provide you with more IPL because:

- They aren't independent, because even they use different instruments; they share the same logic solver.
- They aren't different IPL, because both of them are SIF. An extra PL could be a mechanical relief valve, and plus, it will be independent from the logic solver, so you will have another IPL. That way if the inner previous IPL doesn't work (you can consider your SIF as this inner IPL) you will have another extra IPL.
- The common cause and systematic failures of the SIL3 Logic Solver will affect both protection layers. Therefore the LS IPL load is limited to 3, one could load a 2 SIL1 loops or 1 SIL1 + 1 SIl2 on this system, but one could NOT load 2 SIL2 loops on this system. This according IEC 61511-part 1, clause 9.5.1, part 2 Clauses 9.5.3.
- As you know that SIS consists of SIF's and you can use single or redundant logic solver for those SIF's (in SIS). So my answer for using same logic solver for many SIF is yes you can, single/redundant LS will be depend to your verification that you will count many things (diagnostic covered, beta factor for common cause, etc) to make sure that your SIF equipment reliability will meet level target (SIL target / PFD).

A Common Cause Failure (CCF) occurs when a single fault results in the corresponding failure of multiple components, such as a miscalibration error on a bank of redundant transmitters. The frequency of common cause faults is difficult to estimate. The modeling techniques and available failure rate data make the predictive calculations of these failures cumbersome and the results obtained questionable. Therefore, a more meaningful way for most SIS designers is to eliminate the potential source of CCF in the SIS design, installation, operation and maintenance.

The SIS should be designed to meet the required safety integrity level as defined in the safety allocation requirement specification. Moreover, the SIS design should be performed in a way that minimizes the potential for common mode or common cause failures. A CCF occurs when a single fault results in the corresponding failure of multiple components. Thus, CCF's can result in the SIS falling to function when there is a process demand. Consequently, CCF's must be identified during the design process and the potential impact on the SIS functionality must be understood.

Unfortunately, there is a great deal of disagreement among the experts on how to define CCF and what specific events comprise a CCF. The following are often cited as examples of common cause faults:

- Miscalibration or no calibration of sensors
- Pluggage of common process taps for redundant sensors
- Incorrect maintenance or no maintenance
- Improper bypassing
- Environmental stress on the field device
- Process fluid or contaminant plugs valve

But the examination of these faults, in light of any SIS design, will indicate that any of these six examples can disable single I/O systems, as well as redundant I/O systems. However, many of the proposed methodologies for assessing CCF ignore this fact and only penalize redundant sensors and final elements.

The ultimate failure of all is that the safety requirement specification (SRS) is incorrect at the beginning of the design process and the transmitter cannot detect the potential incident. This is the most disastrous common cause failure that directly leads to the hazardous incident that the designer is seeking to prevent. A bad SRS is the ultimate "trump card" for the entire SIS and is one failure that most of the proposed methodologies ignore.

The choice of the evaluation technique is typically dependent on experience of the User with the particular SIS design. This would include documented historical performance of instruments, installation details, and design engineering teams. Experience with the specific application environment is required, because a device or installation detail that works well in one type of application may not work well in another. For example, standard taps into a vessel for mounting a transmitter may work extremely well in clean service, but may plug very quickly in a service where solids can deposit. Three qualitative techniques are often used to assess SIS designs:

- Industrial Standards
- Engineering Guidelines and Standards
- Qualitative Assessment as Checklist

## 8. Conclusions and Recommendations

These new rules require a change of focus for everyone associated with these systems. The change of focus will result in a significant learning curve for many engineers. In order to make this transition as smooth as possible, rigorous qualitative techniques should be employed to provide a method for assessing these systems for adequacy. Clear communication of the requirements for effective SIS design is essential. All procedures, internal guidelines, and checklists must be evaluated by all personnel involved to ensure a compliant and cost effective SIS.

One suggestion that is obvious but difficult due to human nature is – help management (refinery/corporate) to truly see the OSHA inspection as a positive learning opportunity.

An efficient, accurate OSHA inspection can provide direction to upgrading a refineries PSM program into one that supports sustainability. Long term maintenance of a rigorous refinery PSM system will prove its business case in terms of:

- Corporate Responsibility
- Business Flexibility
- Risk Reduction
- Sustained Value

An effective PSM system upgrade resulting from the inspection can actually increase a refinery's reliability and ability to sustain long term success.

The refining industry will have a directed opportunity to improve overall performance in process safety due to the OSHA refinery process safety management national emphasis program. Leveraging this opportunity requires a culture that views personnel and process safety as second nature – a conditioned first response to address for any issue related to refinery operations or maintenance.

High reliability organizations (such as refineries, petrochemical complexes, chemical plants and power plants) whose upper management values long term growth and more predictable physical plant reliability will embrace the chance to use the effort involved in a PSM program implementation based upgrade to bolster these benefits.

We now have benefits and costs. How are these converted into a proper overall project financial analysis? How can we demonstrate that the proposed investment will increase the long term corporate ROIC more than other potential investments?

The cultural level of operational discipline at the corporation, team, and individual level can better support a high level of operational excellence – doing those critical measurable things that support success – when a program is detailed and monitored. Business excellence can be attained when a company values a holistic sense of what comprises quality.

The Figure 9 represents this view. OSHA inspectors and management have gained a new level of process safety management expertise based upon their roles in the Industry. This can only enhance OSHA's ability to help industry protect the workforce from preventable risk. If production facility management adopts this view, the industry's level of catastrophic risk prevention will increase.

Figure 9. The Holistic View of Quality

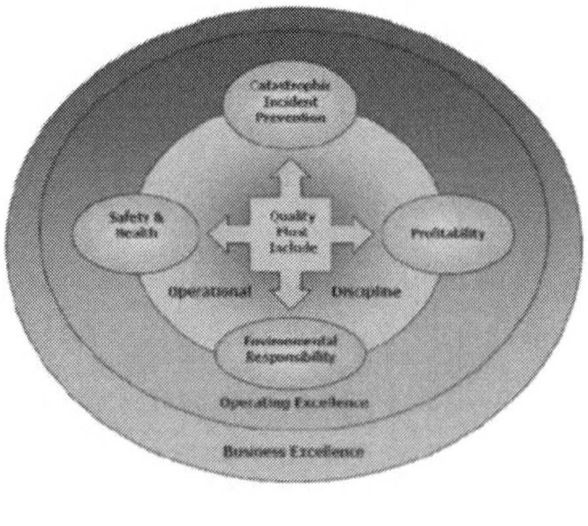

Although the specific investment evaluation protocols vary from company to company and you need to follow your company's standards, they all have at their center the calculation of the net present value of the incremental after tax benefits generated by the plant due to the equipment or technology and comparing it with the net present value of the required investment. In financial terms, all assets, be they stocks, bonds or plants, are valued the same, i.e. as the net present value of the future sequence of after tax cash flows. If the value is greater than the investment the project is considered for funding. Net present value implies discounting a sequence of net cash flows (receipts less expenditures) back to the current time. But there are many subtleties.

The Operation and Maintenance staff of today must justify how the SIS is operated, when it is bypassed, when it is tested, and when maintenance is performed.

## 9. Acknowledgement

The author would like to thank our Meeting assistant colleagues for many discussions, which assisted in the evolution of the ideas presented herein to:

AIChE's 2011 Annual Meeting , United States of America, October 16-21, 2011, Minneapolis, MN, and

***inelectra*** Operations VicePresidency (Eng. Milagros Sulbarán) and Engineering Management (Eng. Victoria Granados), for their cooperation and support.

## 11. Author: Glenda J. Gutiérrez Capriles

She was born in Valencia, Carabobo, Venezuela on January, 1974. She obtained her B.Sc. from Los Andes University (ULA) in Merida - Venezuela, in 1996. Then in 1999-2000 obtained her title as Magisterium Scientarium Specialist in Environmental, Health and Safety, ULA.

Also she holds certification in PMP and TUV FS.

Between 2000 – 2008 she served from Process Specialist, Instrumentation Engineer, Risk Engineer, Maintenance Centered in Reliability Supervisor, Operational Engineering Superintendent for PDVSA E&P, PDVSA GAS and PEQUIVEN.

Gutiérrez Capriles is currently worked as Project Risk Engineer for inelectra Venezuela from beginning of 2009.

Gutiérrez Capriles has over fourteen (14) years of experience as Engineer, Consultant, Operator, highlighting those regarding Risk Studies as Inherently Safer Designs, HAZID, HAZOP, SIL/SIS, FTA, ETA, FMEA, Reliability Analysis as RCM, RBI, OEE, RAM, RAMOS or MORAS, Project Risk Analysis as LCC, DSD, PEP, among others.

She is currently worked as Project Engineer for inelectra Venezuela from the beginning of 2009. She has over fifteen (15) years of experience as engineer, consulter, operator, highlighting those regarding risk studies, reliability analysis, among others.

Her experiences in Process Safety Knowledge are cited on many national and International Journal and Seminars. She has published and presented papers at several events organized by relevant institutions at national and international levels.

- 2010 Chemical Engineers Argentine Congress, Mar del Plata, Argentina.
- 3rd CCPS Latin America Conference and Expo, Buenos Aires, Argentina.
- 2011 Annual Meeting AIChE, Minnesota, USA.
- AIChE Journal, March 2001 Performance Assessment of Constrained Model Predictive Control Systems.
- Gutiérrez-Capriles G. "I Jornadas de Protección Integral". Universidad de Los Andes. Facultad de Ingeniería. Mayo 1996.
- Gutiérrez-Capriles G. Ergonomía en los diferentes Ambientes de Trabajo (Tesis de Magister). Mención Especial de Honor por la Sociedad Venezolana de Medicina Ocupacional, Escuela de Salud Ocupacional de la Facultad de Medicina y Escuela de Protección Integral de la Universidad de Los Andes. Universidad de Los Andes. Mérida, Venezuela (1999)
- Gutiérrez-Capriles G. Anteproyecto del Diseño de una Planta para la Obtención de Lecitina Blanqueada de Soya (Diseño de Plantas). Memorias del II Congreso Nacional de Estudiantes de Ingeniería Química. Universidad de Los Andes. Mérida, Venezuela (1999)
- Gutiérrez-Capriles G. Peña A. BATCH: Programa de Computación simulando una Torre de Destilación por Lotes Batch. Memorias de las IV Jornadas Científico-Técnicas de la Facultad de Ingeniería. Universidad de Los Andes. Mérida, Venezuela (1998)
- Gutiérrez-Capriles G. "III Congreso Internacional de Educación Superior en materia de Drogas". Universidad de Los Andes. Abril 1996.
- Gutiérrez-Capriles G. "I Congreso Internacional de Educación Integral en materia de Seguridad sobre Riesgos Sísmicos". Vicerrectorado Académico y Brigada de Rescate de la Universidad de Los Andes. Abril 1996.
- Gutiérrez-Capriles G. Optimización de Mezclas de Policloruro de Vinilo utilizadas para la Elaboración de Envases Plásticos (Tesis de Grado). Mención Especial de Honor por la Sociedad Venezolana de Ingenieros de Polímeros, Departamento de Química de la Facultad de Ciencias y Departamento de Ingeniería Química de la Universidad de Los Andes. Universidad de Los Andes. Mérida, Venezuela (1996)
- Gutiérrez-Capriles G, Belandria J. Graficador Termodinámico de Compuestos Químicos Puros en Fase Fluida por 4 diferentes Métodos Termodinámicos. II Congreso Interamericano de Computación Aplicada a la Industria de Procesos. Santiago de Chile, Chile (1994)

# (512c) Exogenous Uncertainty: Choosing Innovation Projects

**Diana Matonis**, Flucomp Enterprises, Hickory Hills, IL

What is the difference between risk and uncertainty?

What are some quantitative methods that aid decision making under uncertainty?

What are some qualitative methods that aid decision making under uncertainty?

How can decision makers effectively combine qualitative and quantitative methods?

See more of this Session: Challenges In Open Innovation and Growth
See more of this Group/Topical: Management Division

# (512d) Identifying & Transforming Local Best Practices to a Global Competence

**Marcus A. Tennant**, Yokogawa Corp. of America, Pleasant Prairie, WI

Often in global corporations, best practices emerge that enable a company to improve business performance at alocal facility or region. In this presentation we will look at how best practices emerge locally but are often ignoredin a global perspective due to their local or regional origin. Many of these best practices are developed "below theradar" and work very well for a local market; however small changes and the proper strategy for adaptation canturn the best practices into company-wide global competence. We will we will discuss methods that a companycan "dig out" these practices and find the critical success factors, discuss strategies of implementing these bestpractices in other regions and how they could be used to improve business performance globally. Finally, we willalso look at potential pitfalls and how to overcome obstacles in a global implementation effort.

See more of this Session: Challenges In Open Innovation and Growth
See more of this Group/Topical: Management Division

# SYNTHESIS OF SUSTAINABLE PROPERTY- BASED WATER NETWORKS

Luis Fernando Lira-Barragán,[a] José María Ortega-Ponce,[a] Medardo Serna-González,[a] Mahmoud M. El-Halwagi[c]

[a] *Universidad Michoacana de San Nicolás de Hidalgo, Morelia, Mich., México*

[b] *Chemical Engineering Department, Texas A&M University, College Station, TX, USA*

## Introduction

Water is one of the most important natural resources that support the life in the world. The chemical and process industry is one of the biggest consumers of the available clean water, and at the same time the industry is one of the highest producers of polluted wastewater. In this regard, recently there have been proposed a set of techniques to economically reduce the clean water consumption in the process industry considering the possibility to recycle, reuse and regenerate the wastewater produced in the industrial facilities. The proposed techniques have addressed the problem from different perspectives (i.e., minimizing the fresh water consumption, minimizing the wastewater discharge, minimizing the regeneration costs, etc) and using diverse techniques (i.e., graphical, algorithmic and mathematical programming based approaches). The heuristic approaches allow to easily identify targets ahead of the design and to obtain a graphical representation of the problem (see for example Wang and Smith, 1994; El-Halwagi and Spriggs, 1998). All previous methodologies only have considered the processes that happen inside the industrial facilities, without taking into account the processes that occur outside the plant for the wastewater streams discharged to the environment.

Therefore, the goal of the proposed work is to overcome the drawbacks of the previously reported methods for the synthesis of water integration networks, considering the sustainability of the surrounding watershed that is based on properties for the cases that multiple pollutants are presented and that make difficult to characterize the streams in terms of the composition. The optimal quantity and quality for the wastewater discharged from the new plant depends on the location selected to install the plant because the impact in the watershed depends strongly on the location selected (considering the integration with other streams and the chemical and biochemical reactions that occur in the watershed system); and therefore, the recycle and reuse network is highly influenced by the location of the new plant.

## Model Formulation

The main stream of the river is fed by several streams or tributaries ( $FT_{r,t}$ ). As it can be seen in **Figure 1**, the watershed system exchanges water during its trajectory (i.e., water for agricultural use, wastewater discharged to the river with and without treatment, industrial and residential effluents, etc.) and finally the watershed discharges to the final disposal. Besides, natural phenomena like precipitation, filtration and vaporization are taken into account in the model. These processes modify significantly the characteristics of the river; therefore, to track adequately the mean properties, the river is sectioned in parts where the properties can be considered constant (these sections are represented in **Figure 1** with ovals). These parts o sections are called reaches and these identify zones where no big intermediate effluents are discharged and/or extracted. In this sense, the model also indentifies the main effluents discharged in each reach; these effluents are called tributaries. The tributaries can be channels or branches of the river, which may contain discharges with/without treatment, industrial discharges, etc., and their flowrates ( $FT_{r,t}$ ) and properties ( $P^{Trib}_{p,r,t}$ ) affect the reaches where they are discharged. The discharges to the watershed contain compounds that interact chemically and biochemically with the system thought the flora and fauna established in the watershed.

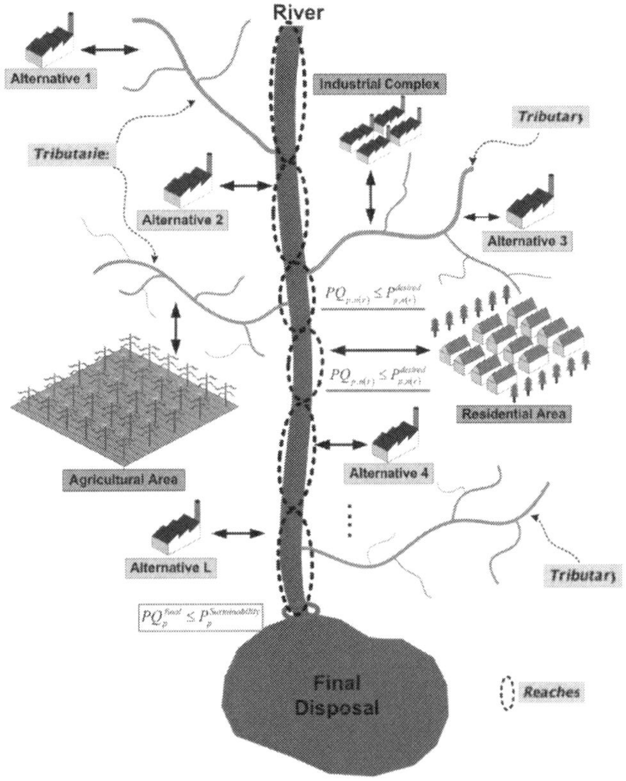


**Figure. 1** Alternatives to locate the new industrial plant in a watershed system.

To ensure the sustainability of the surrounding watershed system, it is required to carry out an analysis prior to the optimization respect to the capacity of the final disposal to decompose chemically and biochemically the pollutants, and as consequence it is necessary to quantify these processes in terms of properties. The analysis yields the sustainability constraints which ensure that the pollutants will not be accumulated in the system and to maintain the properties of the watershed under reasonable limits.

In this sense, to consider the performance function of the interceptors, it is used a correction factor ($\gamma_p^{int^1,\ldots,int^N}$) that can be calculated prior to the optimization process, obtaining a linear function to model the interceptors. The correction factor must consider several aspects like the value of the property at the inlet conditions of the interceptor, the flowrate and design and operating parameters.

Thus, the correction factor is a very useful tool in this work, where the available technologies considered in the property interception network can be simulated (i.e., using ASPEN) to get the efficiency factors. It is worth of notice that the model includes a set of fictitious interceptors for each property to be intercepted with conversion factors and costs equal to zero. Besides, the property operators values for the inlet condition of the process sources are also known prior to the optimization process, $\psi_p\left(P_{i,p}^{InSource}\right)$, then the property values at the exit conditions of the property interception network can be calculated prior to the optimization process as follows:

$$\psi_p\left(P_{i,p}^{OutPIN,\text{int}^1,\ldots,\text{int}^N}\right)=\gamma_p^{\text{int}^1,\ldots,\text{int}^N}\psi_p\left(P_{i,p}^{InSource}\right),\quad i\in I, p\in P, \text{int}^1\in INT^1,\ldots,\text{int}^N\in INT^N$$

**MFA Model based on properties**

$$Q_r=Q_{r-1}+P_r+D_r+H_r+\sum_{t=1}^{N_{t(r)}}FT_{r,t}+QPNEW_{r(l)}-L_r-U_r,\quad \forall r\in R \tag{1}$$

$$\begin{aligned}\psi_p\left(P_{p,r}^{Reach}\right)Q_r &= \psi_p\left(P_{p,r-1}^{InReach}\right)Q_{r-1}+\psi_p\left(P_{p,r}^{Prec}\right)P_r+\psi_p\left(P_{p,r}^{Ind}\right)D_r+\psi_p\left(P_{p,r}^{Resid}\right)H_r+\sum_{t=1}^{N_{t(r)}}\psi_p\left(P_{p,r,t}^{Trib}\right)FT_{r,t}\\ &+\psi_p\left(P_{p,r(l)}^{PNEW}\right)QPNEW_{r(l)}-\psi_p\left(P_{p,r}^{Loses}\right)L_r-\psi_p\left(P_{p,r}^{Uses}\right)U_r-\int_{V=0}^{V_r}r_{p,r}dV_r,\ \forall p\in P, r\in R\end{aligned} \tag{2}$$

$$FT_{r,t} = S_{r,t}^{Untreat} + S_{r,t}^{Treat} + I_{r,t} + P_{r,t} + D_{r,t} + QPNEW_{r,t(l)} - L_{r,t} - U_{r,t}, \qquad \forall r \in R, t \in T \tag{3}$$

$$\begin{aligned} \psi_p\left(P_{p,r,t}^{Trib}\right)FT_{r,t} = {} & \psi_p\left(P_{p,r,t}^{Untreat}\right)S_{r,t}^{Untreat} + \psi_p\left(P_{p,r,t}^{Treat}\right)S_{r,t}^{Treat} + \psi_p\left(P_{p,r,t}^{Ind}\right)I_{r,t} + \psi_p\left(P_{p,r,t}^{Prec}\right)P_{r,t} + \psi_p\left(P_{p,r,t}^{Direct}\right)D_{r,t} \\ & + \psi_p\left(P_{p,r,t(l)}^{PNEW}\right)QPNEW_{r,t(l)} - \psi_p\left(P_{p,r,t}^{Loses}\right)L_{r,t} - \psi_p\left(P_{p,r,t}^{Uses}\right)U_{r,t} - \int_{V=0}^{V_{r,t}} r_{p,r,t}\,dV_{r,t}, \qquad \forall p \in P, r \in R, t \in T \end{aligned} \tag{4}$$

$$\int_{V=0}^{V_r} r_{p,r}\,dV_r = k_p\left[\psi_p\left(P_{p,r}^{Reach}\right)\right]^{\sigma_p} V_r, \qquad \forall p \in P, r \in R \tag{5}$$

$$\int_{V=0}^{V_{r,t}} r_{p,r,t}\,dV_{r,t} = k_p\left[\psi_p\left(P_{p,r,t}^{Trib}\right)\right]^{\sigma_p} V_{r,t}, \qquad \forall p \in P, r \in R, t \in T \tag{6}$$

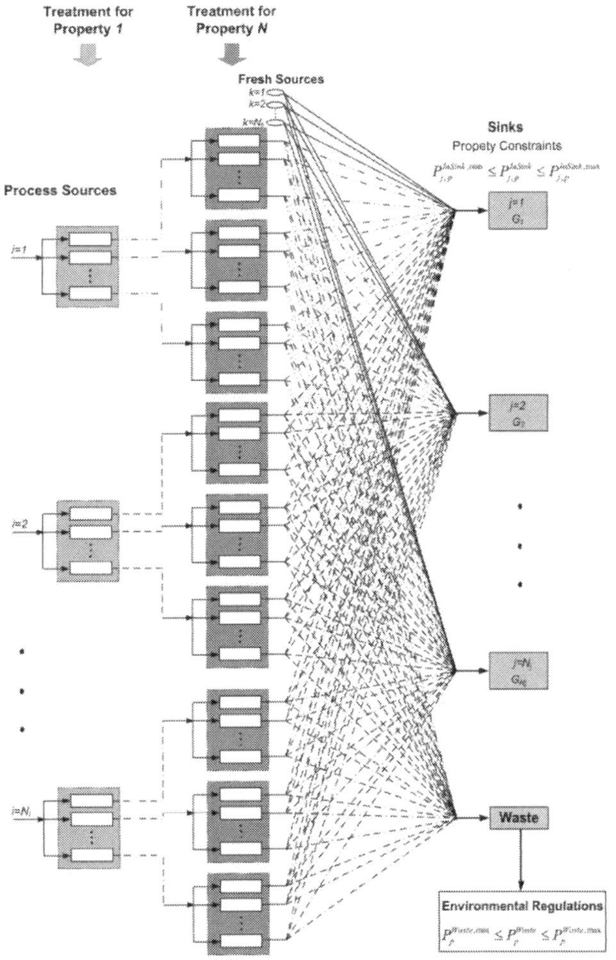


**Figure 2.** Superstructure for the recycle and reuse network

## Location of the new plant

This work considers a set of alternatives to locate the new industrial facility ($L$), these options are generated prior to the optimization process. Then, for each alternative, it is necessary to consider the possible discharge of the new wastewater flowrate ($QPNEW_l$) and their property operators ($\psi_p\left(P_{p,l}^{PNEW}\right)$), thus, if the new industrial plant is located in the alternative 1, the wastewater flowrate and their property operators associated to this site must be bigger than zero; otherwise, if the new plant is not located in the position 1, the wastewater flowrate and their property operators discharged would be zero. This is included in the following disjunction:

$$\begin{bmatrix} Y_1 \\ \psi_p\left(P_{p,1}^{PNEW}\right)\geq 0, QPNEW_1 \geq 0 \\ \psi_p\left(P_{p,2}^{PNEW}\right)= 0, QPNEW_2 = 0 \\ \vdots \\ \psi_p\left(P_{p,l}^{PNEW}\right)= 0, QPNEW_l = 0 \end{bmatrix} \vee \begin{bmatrix} Y_2 \\ \psi_p\left(P_{p,1}^{PNEW}\right)= 0, QPNEW_1 = 0 \\ \psi_p\left(P_{p,2}^{PNEW}\right)\geq 0, QPNEW_2 \geq 0 \\ \vdots \\ \psi_p\left(P_{p,l}^{PNEW}\right)= 0, QPNEW_l = 0 \end{bmatrix} \vee \cdots \vee \begin{bmatrix} Y_l \\ \psi_p\left(P_{p,1}^{PNEW}\right)= 0, QPNEW_1 = 0 \\ \psi_p\left(P_{p,2}^{PNEW}\right)= 0, QPNEW_2 = 0 \\ \vdots \\ \psi_p\left(P_{p,l}^{PNEW}\right)\geq 0, QPNEW_l \geq 0 \end{bmatrix}, \quad \forall p \in P$$

The reformulation includes the transformation of the Boolean variables ($Y_l$) into a set of binary variables ($y_l$), when the Boolean variables are true, the associated binary variables must be set as one; otherwise, when the Boolean variables are false, the associated binary variables are zero. Therefore, the following relationship is used to select only one location to install the plant:

$$\sum_{l\in L} y_l = 1 \tag{7}$$

$$QPNEW_l \leq QPNEW^{Up}\, y_l, \qquad \forall l \in L \tag{8}$$

$$\psi_p\left(P_{p,l}^{PNEW}\right) \leq \psi_p\left(P^{PNEW}\right)^{Up} y_l, \qquad \forall p \in P, l \in L \tag{9}$$

where $QPNEW^{Up}$ and $\psi_p\left(P_{p,l}^{PNEW}\right)^{Up}$ are the upper limits for $QPNEW_l$ and $\psi_p\left(P_{p,l}^{PNEW}\right)$, respectively. Notice that $\psi_p\left(P_{p,l}^{PNEW}\right)^{Up}$ is an upper bound for all properties $p$. These constraints are required to ensure that $QPNEW_l$ and $\psi_p\left(P_{p,l}^{PNEW}\right)$ take values bigger than zero only when the new plant is located in a position $l$ in particular.

$$\sum_{l\in L} QPNEW_l = Waste \tag{10}$$

$$\sum_{l\in L} \psi_p\left(P_{p,l}^{PNEW}\right) = \psi_p\left(P_p^{disc}\right), \qquad \forall l \in L \tag{11}$$

**Superstructure for the water network**

$$F_k = \sum_{j\in J} f_{k,j}, \quad \forall k \in K \tag{12}$$

$$W_i = \sum_{int^1 \in INT^1} w_i^{int^1}, \quad \forall i \in I \tag{13}$$

$$w_i^{int^1} = \sum_{int^2 \in INT^2} w_i^{int^1,int^2}, \quad \forall i \in I, int^1 \in INT^1 \tag{14}$$

$$w_i^{int^1,int^2,\ldots,int^{N-1}} = \sum_{int^N \in INT^N} w_i^{int^1,int^2,\ldots,int^N}, \quad \forall i \in I, int^1 \in INT^1,\ldots,int^{N-1} \in INT^{N-1} \tag{15}$$

$$w_i^{int^1,\ldots,int^N} = \sum_{j\in J} g_{i,j}^{int^1,\ldots,int^N} + g_{i,Waste}^{int^1,\ldots,int^N}, \quad \forall i \in I, int^1 \in INT^1,\ldots,int^N \in INT^N \tag{16}$$

$$G_j = \sum_{i \in I} \sum_{int^1 \in INT^1} \cdots \sum_{int^N \in INT^N} g_{i,j}^{int^1,\dots,int^N} + \sum_{k \in K} f_{k,j}, \quad \forall j \in J \tag{17}$$

$$\psi_p\left(P_{j,p}^{InSink}\right) G_j = \sum_{i \in I} \sum_{int^1 \in INT^1} \cdots \sum_{int^N \in INT^N} \left[\psi_p\left(P_{i,p}^{OutPIN,int^1,\dots,int^N}\right) g_{i,j}^{int^1,\dots,int^N}\right] + \sum_{k \in K}\left[\psi_p\left(P_{k,p}^{Fresh}\right) f_{k,j}\right], \quad \forall j \in J, p \in P \tag{18}$$

$$Waste = \sum_{i \in I} \sum_{int^1 \in INT^1} \cdots \sum_{int^N \in INT^N} g_{i,Waste}^{int^1,\dots,int^N} \tag{19}$$

$$\psi_p\left(P_p^{Waste}\right) Waste = \sum_{i \in I} \sum_{int^1 \in INT^1} \cdots \sum_{int^N \in INT^N} \left(\psi_p\left(P_{i,p}^{OutPIN,int^1,\dots,int^N}\right) g_{i,Waste}^{int^1,\dots,int^N}\right), \quad \forall p \in P \tag{20}$$

$$\psi_p\left(P_{j,p}^{InSink,\min}\right) \le \psi_p\left(P_{j,p}^{InSink}\right) \le \psi_p\left(P_{j,p}^{InSink,\max}\right), \quad j \in J, P \in P \tag{21}$$

**Environmental constraints**

$$\psi_p\left(P_p^{Waste,\min}\right) \le \psi_p\left(P_p^{Waste}\right) \le \psi_p\left(P_p^{Waste,\max}\right), \quad \forall p \in P \tag{22}$$

**Constraints for the quality of specific reaches**

$$\psi_p\left(P_{p,m(r)}^{DesiredMIN}\right) \le \psi_p\left(P_{p,m(r)}^{Reach}\right) \le \psi_p\left(P_{p,m(r)}^{DesiredMAX}\right), \quad \forall p \in P, m(r) \in M(R) \tag{23}$$

**Sustainability constraints**

$$\psi_p\left(P_p^{SustainableMIN}\right) \le \psi_p\left(P_{p,final}^{Reach}\right) \le \psi_p\left(P_p^{SustainableMAX}\right), \quad \forall p \in P \tag{24}$$

**Objective Function**

The objective function consists in the minimization of the total annual cost ($TAC$), which is constituted by the installation costs for the new plant, the fresh sources costs, the treatment cost (including operational cost, variable and fixed capital cost) and the piping costs. The installation cost ($Cland_l$) takes into account the land cost, the transportation costs for raw materials, products and services. Fresh sources cost ($FreC_k$) is caused by the required flowrate of fresh sources to satisfy the inlet flowrate in the process sinks as well as the properties constraints imposed on them. Also, process sources need to be treated prior to be sent to the process sinks, as well as prior to be discharged to the environment, in this sense; a set of technologies must be installed to treat the process sources yielding capital and operational costs for each one of the interceptors required. The operational cost ($VarC_{op}^{int^N}$) for the interceptors depends on the treated flowrate, while the capital cost contains a variable cost ($VarC_{cap}^{int^N}$), which also depends on the treated flowrate and a fixed component ($FixC_{cap}^{int^N}$). Finally, the piping costs for the four sections showed in the superstructure are considered. The first section are the pipes requires when the process sources are sent to treatment ($TrPipC$), once the process sources are treated they can be sent to wastewater stream ($WaPipC$) or process sinks ($EqPipC$), the next part covers the pipes to connect the fresh sources with process sinks ($FrPipC$).

$$
\begin{aligned}
Min\,TAC = & \sum_{l \in L} Cland_l y_l + H_Y \sum_{k \in K} FreC_k F_k \\
& + H_Y \sum_{i \in I} \left[ \sum_{int^1 \in INT^1} VarC_{op}^{int^1} w_i^{int^1} + \cdots \sum_{\text{int}^N \in INT^N} \left( VarC_{op}^{\text{int}^N} w_i^{\text{int}^1,\ldots,\text{int}^N} \right) \right] \\
& + H_Y \sum_{i \in I} \left[ \sum_{int^1 \in INT^1} VarC_{cap}^{int^1} w_i^{int^1} + \cdots \sum_{\text{int}^N \in INT^N} \left( VarC_{cap}^{\text{int}^N} w_i^{\text{int}^1,\ldots,\text{int}^N} \right) \right] \\
& + k_f \sum_{i \in I} \left[ \sum_{int^1 \in INT^1} FixC_{cap}^{int^1} z_i^{int^1} + \cdots \sum_{\text{int}^N \in INT^N} \left( FixC_{cap}^{\text{int}^N} z_i^{\text{int}^1,\ldots,\text{int}^N} \right) \right] \\
+ H_Y & \left[ \begin{array}{l} TrPipC \sum_{i \in I} \left( \sum_{\text{int} \in INT} w_i^{int^1} + \cdots \sum_{\text{int}^N \in INT^N} w_i^{\text{int}^1,\ldots,\text{int}^N} \right) + WaPipC \sum_{i \in I} \sum_{int^1 \in INT^1} \cdots \sum_{\text{int}^N \in INT^N} g_{i,Waste}^{\text{int}^1,\ldots,\text{int}^N} \\ + EqPipC \sum_{i \in I} \sum_{int^1 \in INT^1} \cdots \sum_{\text{int}^N \in INT^N} \sum_{j \in J} g_{i,j}^{\text{int}^1,\ldots,\text{int}^N} + FrPipC \sum_{k \in K} \sum_{j \in J} f_{k,j} \end{array} \right]
\end{aligned}
\qquad (25)
$$

## Results and discussion

### Case of study

This case considers the Balsas watershed that is one of the biggest systems in Mexico. This system transports several types of discharges until the Pacific Ocean. CONAGUA 20010a, 2010b reported the information for the system sowed on **Figure 6**. In this case, the properties considered were the composition, toxicity, density, COD, odor and color; however, there is a set of properties restricted in the environment and other one restricted in the process sinks. Thus, the properties restricted for the stream discharged to the environment are composition, toxicity, COD, odor and color, while the process sinks limit the composition and density. The system offers twenty possible locations to install the new industrial facility and their installation costs are presented in **Table 6**

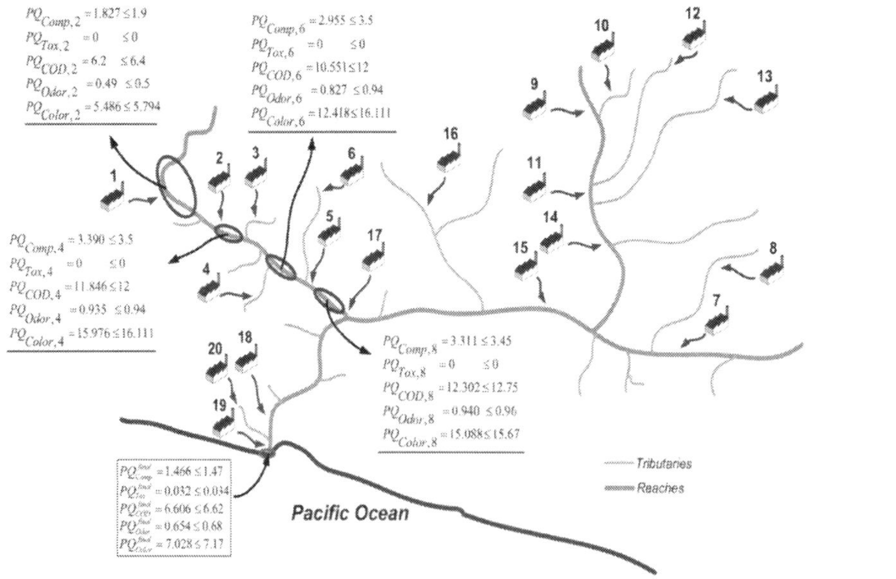


**Figure 6.** Balsas watershed system (possible locations and the constraints on the river)

**Table 6.** Installation costs for Example 2

| Site | Annualized installation cost, $/year |
|---|---|
| 1 | 8,000,000 |
| 2 | 8,500,000 |
| 3 | 9,000,000 |
| 4 | 10,500,000 |
| 5 | 9,500,000 |
| 6 | 9,000,000 |
| 7 | 15,500,000 |
| 8 | 14,500,000 |
| 9 | 17,700,000 |
| 10 | 22,000,000 |
| 11 | 18,000,000 |
| 12 | 17,500,000 |
| 13 | 24,000,000 |
| 14 | 28,000,000 |
| 15 | 23,500,000 |
| 16 | 17,000,000 |
| 17 | 20,000,000 |
| 18 | 19,000,000 |
| 19 | 16,000,000 |
| 20 | 17,000,000 |

**Table 7**. Fresh and process sources for Example 2

| *Sources* | *Flowrate, $m^3/s$* | *Composition, ppm* | *Toxicity, %* | *COD, $mgO_2/lt$* | *Odor* | *Color* | *Density, g/ml* |
|---|---|---|---|---|---|---|---|
| | | | *Process* | | | | |
| 1 | 0.363 | 64.6 | 0.11 | 149 | 6.3 | 250 | 0.94 |
| 2 | 0.241 | 58 | 0.04 | 88 | 7.7 | 234 | 2.41 |
| 3 | 0.046 | 45.3 | 0 | 108 | 5.1 | 298 | 1.14 |
| 4 | 0.447 | 60 | 0.13 | 185 | 4.8 | 202 | 1.20 |
| 5 | 0.196 | 52 | 0.21 | 120 | 7.6 | 194 | 2.70 |
| 6 | 0.204 | 62 | 0.25 | 76 | 8.2 | 164 | 1.64 |
| 7 | 0.281 | 54.5 | 0 | 243 | 3.7 | 179 | 1.85 |
| 8 | 0.329 | 47.8 | 0.13 | 164 | 7.4 | 346 | 1.75 |
| 9 | 0.282 | 46 | 0.09 | 297 | 4.8 | 264 | 2.45 |
| 10 | 0.266 | 82.1 | 0.7 | 326 | 8.5 | 394 | 0.89 |
| 11 | 0.215 | 26.4 | 0.3 | 146 | 5.7 | 195 | 2.15 |
| | | | *Fresh* | | | | |
| 1 | - | 0 | --- | --- | --- | --- | 2.3 |
| 2 | - | 1 | --- | --- | --- | --- | 2.45 |
| 3 | - | 2 | --- | --- | --- | --- | 1.8 |

**Table 7** shows the characteristics for the fresh and process sources. **Table 8** shows the requirements for the flowrate, maximum composition as well as the minimum and maximum density for each process sink. The cost for fresh source 1 is $0.128/m$^3$, for fresh source 2 is $0.113/m$^3$, for fresh source 3 is $0.097/m$^3$. **Table 9** shows the characteristics for the property interceptors. For this case, the variables costs are included in just one term.

**Table 8**. Requirements for process sinks for Example 2.

| *Process Sink* | *Flowrate, $m^3/s$* | *$Comp^{max}$* | *$Density^{min}$* | *$Density^{max}$* |
|---|---|---|---|---|
| 1 | 0.348 | 16 | 2.2 | 2.4 |
| 2 | 0.428 | 19.4 | 1.95 | 2.1 |
| 3 | 0.406 | 23.3 | 2.35 | 2.55 |
| 4 | 0.312 | 28.5 | 1.7 | 1.9 |
| 5 | 0.336 | 18 | 2.2 | 2.3 |
| 6 | 0.294 | 14 | 1.8 | 2 |
| 7 | 0.196 | 22.5 | 2.45 | 2.55 |

**Table 9**. Available interceptors for Example 2.

| Interceptor | Efficiency Factor | Variable Cost, $/m$^3$ |
|---|---|---|
| | ***Composition*** | |
| $Rec^1$ | 0.89 | 0.026 |
| $Rec^2$ | 0.77 | 0.021 |
| | ***Toxicity*** | |
| $Tox^1$ | 1 | 0.054 |
| $Tox^2$ | 0.92 | 0.049 |
| | ***COD*** | |
| $Aer^1$ | 0.82 | 0.022 |
| $Aer^2$ | 0.57 | 0.018 |
| | ***Odor*** | |
| $Ads^1$ | 0.86 | 0.013 |
| $Ads^2$ | 0.61 | 0.011 |
| | ***Color*** | |
| $Dec^1$ | 0.90 | 0.019 |
| $Dec^2$ | 0.80 | 0.015 |

On the other hand, the environmental legislation imposes the next constraints for the wastewater stream: $P_{Comp}^{Sustainable} = 50$, $P_{Tox}^{Sustainable} = 0.1$, $P_{COD}^{Sustainable} = 75$, $P_{Odor}^{Sustainable} = 2.7$, and $P_{Comp}^{Sustainable} = 100$. While the piping cost for each section in the superstructure are: $TrPipC = 0.001\,\$/m^3$, $WaPipC = 0.002\,\$/m^3$, $EqPipC = 0.005\,\$/m^3$, and $FrPipC = 0.003\,\$/m^3$.

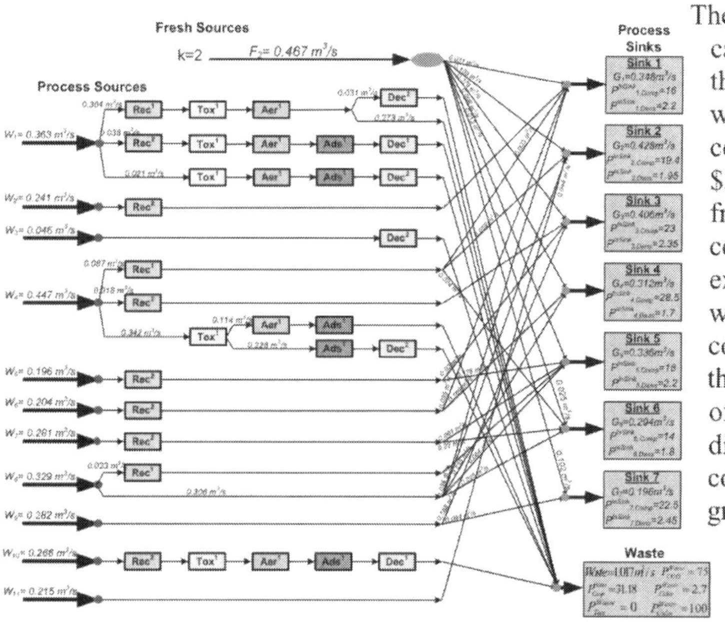


Figure 7. Mass integration for Example 2

Then, the proposed model was applied for this case of study, giving an optimal solution where the new facility must be placed in alternative 6 with a total annual cost of \$14.9x10$^6$/yr, composed by \$9x10$^6$/yr of installation cost, \$3.6x10$^6$/yr of treatment cost, \$1.5203x10$^6$/yr of fresh source cost and \$779,040/yr of piping cost. **Figure 7** shows the water network this example, notice that the value of toxicity in the wastewater stream is zero, satisfying the constraint in the reach 8, where is not allowed the existence of this property. Notice that most of the properties in the wastewater streams are discharged in the limit of the environmental constraints; however, the composition requires a greater reduction to satisfy the constraints imposed over the reach 8 ensuring the quality of the water extracted for human activities.

## Conclusions

This paper proposes a new water integration approach to consider simultaneously the integration inside the plant and the sustainability of the surrounded watershed. The model is based on the properties of the streams, which is suited for the wastewater streams that are constituted by several pollutants. The proposed approach considers the integration of the wastewater discharged by the industrial facility to the other discharges and uses through the optimal selection of the location of the industrial facility. In addition, the model considers the natural degradation of the pollutants through the watershed and the effect to the specific properties that are constrained in the system. The proposed model is applied to one case from Mexico. The results show that the traditional water integration schemes that do not consider the surrounding watershed yield inappropriate solutions, because the environmental regulations are not enough to satisfy the sustainability of the system.

## References


CONAGUA (2009a). Mexican National Water Commission. Water Statistics 2008. http://www.conagua.gob.mx/OCB07/Temas/EstadisticasBALSAS-FINALago09.pdf

CONAGUA (2009b). Mexican National Water Commission. Water Atlas in Mexico 2009.

El-Halwagi, M. M., and Spriggs, H. D. (1998). Solve design puzzles with mass integration. *Chemical Engineering Progress*, *94* (8), 25-44.

Wang, Y. P., and Smith, R. (1994). Wastewater minimization. *Chemical Engineering Science*, *49* (7), 981-1006.

# (521b) Comparing Continuous and Batch Process Design Under Uncertainty for Biodiesel Production

**Pahola T. Benavides**[1], Urmila Diwekar[2] and Juan M. Salazar[2], (1)VRI- Custom, Chicago, IL, (2)Vishwamitra Research Institute, Center for Uncertain Systems: Tools for Optimization and Management, Clarendon Hills, IL

**Abstract:**

The continuing depletion of fossil fuel reserves and the increasing environmental concerns has encouraged engineers and scientists to look for an alternative, clean and renewable fuel that can reduce the negative environmental impact. Biodiesel has been considered as one of the best candidate of one of these renewable fuels. One of the pathways to biodiesel production is the transesterification reaction of triglycerides from vegetable oils and short-chain alcohols. A batch reactor is employed for the production of biodiesel. However, most of the other unit operations like decanters and distillation columns are operated continuously. The success of biofuel introduction in economy depends on the economics of these fuels. One of the factors which are problematic in this is the seasonal variation of feedstock, and transportation and storage costs of material for continuous processing. In this work, we compare batch and continuous processing options for biodiesel production under feedstock variation. These models will allow us to choose an optimal process configuration, depending on the size of the market, distance to market, and feedstock availability,

***Key words***: *batch process, continuous process, biodiesel, stochastic simulation, economical analysis.*

See more of this Session: Design of Sustainable Processes
See more of this Group/Topical: Computing and Systems Technology Division

# (521c) Simultaneous Optimization of Flowsheet, Heat Recovery, and Water Network

**Linlin Yang** and Ignacio Grossmann, Chemical Engineering, Carnegie Mellon University, Pittsburgh, PA

Significant research work has been done in the area of heat integration in the past thirty years due to the incentive of improving energy recovery. Also, stricter environmental regulations have caused an increasing demand for minimization of water consumption and effluent generation for industrial processes. Various techniques have been developed in response to this need for process integration and they are presented in these reviews[1][3][4]. However, most of this research has been focused on minimization of energy and water separately. Furthermore, traditionally, process synthesis design is performed sequentially - the flowsheet is optimized first, then the resulting operating parameters are used in heat exchanger network synthesis (HENS) and water network (WN).

This presentation describes a novel and effective approach for the simultaneous optimization of process flowsheet subject to heat and water integration. Through simultaneous optimization approach, the idea is to represent the HENS and the WN through design targets such as energy consumption and water consumption to avoid solving a very large MINLP problem. The advantage of performing simultaneous optimization is that proper trade-offs of operation cost and investment cost can be taken into account.

While effective NLP targeting formulation for simultaneous process optimization with heat integration has already been presented by Duran & Grossmann[2], this is not the case for targeting minimum water use. Typical WN problems are formulated using NLP or MINLP models using a detailed superstructure as reported in many references[5]. Freshwater is first used for process units, and then wastewater generated from these processes is treated in various treatment units. Process units are defined by their allowable contaminant concentration level, whereas wastewater treatment units are defined based on their contaminant removal ratio. The superstructures consider systematic alternatives for water recycle and water reuse with the objective of minimizing the freshwater consumption, or more generally the total cost. In general, single-contaminant systems can be modeled using LP formulations, while multi-contaminant systems take the form of NLP models.

In order to incorporate water integration in the simultaneous optimization framework, a new LP formulation is presented to determine the target for a multi-contaminant water network with process units only. In particular, the bilinear equations of the mixer are replaced by a set of inequalities that are evaluated at the limits of inlets and outlet concentrations of the units. This relaxation, which follows the optimality conditions proved by Bagajewicz & Savelski[6], yields an LP that can be shown to predict the exact minimum amount of freshwater that is required by the process units. The LP targeting model is incorporated in the simultaneous optimization of a methanol synthesis process flowsheet with heat and water integration and resulted in an improvement of the economical objective function. The important observation here is that in the simultaneous approach, heating and cooling are integrated as part of the optimization so that improved operating conditions can be achieved. It is shown for a specific instance that the resulting flowsheet from the simultaneous method improves the profit by 17%.

For the case of multi-contaminant water network problems including wastewater treatment units with process units, the proposed LP targeting formulation is shown to predict a rigorous minimum freshwater requirement for special cases of the WN problem. For the general case, the lack of exact upper bound of contaminant concentration in the treatment units complicates the interactions in the mixer units. To circumvent this problem, a number of linear relaxations are used to obtain approximate but tight water targets for general WN problems. The proposed targeting model yields an MILP problem. The application of this model is also applied in the simultaneous optimization, heat and water integration of the methanol process yielding further improvements in the profit.

**References:**

[1] Bagajewicz, M., (2000). A review of recent design procedures for water networks in re0neries and process plants. *Computers & Chemical Engineering,* 24 (9), 2093-2115.

[2] Duran, M.A. & Grossmann, I.E. (1986). Simultaneous optimization and heat integration of chemical processes. *AIChE Journal*, 32, 123-138.

[3] Foo, D.C.Y. (2009). State-of-the-art review of pinch analysis techniques for water network synthesis. *Ind. Eng. Chem. Res.* 48, 5125–5159.

[4] Furman, K.C. & Sahinidis, N.V. (2002). A critical review and annotated

bibliography for heat exchanger network synthesis in the 20th Century. *Ind. Eng. Chem. Res.*, 41, 2335-2370.

[5] Karuppiah, R. & Grossmann, I. E. (2008). Global optimization of multi-scenario mixed integer nonlinear programming models arising in the synthesis of integrated water networks under uncertainty. *Computers & Chemical Engineering*, 32(1-2), 145-160.

[6] Savelski, M. & Bagajewicz, M. (2003). On the necessary conditions of optimality of water utilization systems in process plants with multiple contaminants. *Chemical Engineering Science*, 58(23-24), 5349-5362.

See more of this Session: Design of Sustainable Processes
See more of this Group/Topical: Computing and Systems Technology Division

# (521d) Incorporating Safety and Efficiency Considerations In Sustainability Assessment of Poly-Generation Systems

**Preeti Gangadharan**[1], Kailiang Zheng[2], Abhishek V. Jayswal[3], Anand Zanwar[1] and Helen Lou[1], (1)Dan F. Smith Department of Chemical Engineering, Lamar University, Beaumont, TX, (2)Department of Chemical Engineering, Lamar University, Beaumont, TX, (3)Dan F. smith Department of Chemical Engineering, Lamar University, Beaumont, TX

Sustainability assessment determines the most sustainable pathway, taking into consideration the economic, environmental, efficiency and societal aspects. Embedding exergy analysis and novel safety evaluation approach to complement the widely-accepted economic assessment and environmental impacts assessment in a decision tree evaluation framework provides a more comprehensive, yet faster methodology to compare related processes in terms of sustainability. Considering their promise to provide high material and energy conversion from natural resources, poly-generation systems are recognized as promising technologies for future chemical and power industries.

The commonly used inherent safety approach considers chemical and process safety, but does not give due emphasis to complexity of the process flow sheet. This becomes important in the case of poly-generation systems that accept multiple feeds and produce multiple products through many pathways for chemical and energy conversion. Since the simultaneous production of chemicals and electricity offers both economical viability and thermodynamic efficiency, exergy efficiency analysis is employed for thermodynamic comparison of proposed processes.

In this paper three different poly-generation systems, which use coal and natural gas as feed to produce DME and power are compared using a comprehensive sustainability assessment methodology, which takes into account the economic, environmental, societal and efficiency factors.

See more of this Session: Design of Sustainable Processes
See more of this Group/Topical: Computing and Systems Technology Division

# (521e) Technology Evaluation and Decision Making for Sustainability Enhancement of Industrial Systems Under Uncertainty

**Zheng Liu**, Department of Chemical Engineering and Materials Science, Wayne State University, Detroit, MI and Yinlun Huang, Dept. of Chemical Engineering & Material Science, Wayne State University, Detroit, MI

Depletion of natural resources, environmental pressure, economic globalization, etc., demand seriously industrial organizations to ensure that their development is sustainable. Technology based sustainability enhancement is one of key approaches for industrial sustainability realization. However, identification of effective technologies for a given industrial system could be a combinatorial problem; this is especially true when multiple technologies are considered for adoption together. On the other hand, if the available data and information about the industrial system and the known technologies are incomplete, imprecise, and uncertain, the technology identification could be even more difficult.

In this paper, we introduce a simple approach for systematic sustainability assessment of industrial systems and technologies, and effective system sustainability enhancement under uncertainty. An industrial system can be an industrial zone, corporate, company, or plant of any complexity. We assume that the industrial organization has set a strategic plan that includes detailed development goals of economic, environmental, and social sustainability with a budget set aside for goal achievement. It is known that a number of technologies are seemingly promising for sustainability enhancement. A challenge is how to derive efficiently the most suitable solutions for identification of superior sustainability technologies.

The technology-assisted sustainability enhancement methodology to be introduced will consist of a sustainability assessment method for evaluating the performance of industrial systems as well as technologies, and a sustainability decision making and analysis method for identifying superior technology combinations for achieving the sustainability goals of industrial systems. The interval parameter based information processing and decision making permit the acceptance of uncertain information that could be available at the best. The logically designed solution identification procedure can make the combinatorial problem solved efficiently through reducing the solution space stage-wisely using different criteria set by the strategic plan. The identified solutions are also sufficiently detailed which can greatly facilitate the industrial organization in final decision making. The methodology is general that can be applied to the sustainability enhancement problems of any size and scope, and the efficacy of the approach is illustrated through a sustainability development problem about regional biodiesel manufacturing improvement.

See more of this Session: Design of Sustainable Processes
See more of this Group/Topical: Computing and Systems Technology Division

# (521f) Integration of Production Processes, Waste Management and Energy Utility Generation In Chemical Batch Plants

**Stavros Papadokonstantakis**, Claude Rerat and Konrad Hungerbuhler, Chemistry and Applied Biosciences, Institute for Chemical- and Bioengineering, Swiss Federal Institute of Technology, Zurich (ETHZ), Zurich, Switzerland

The production of chemicals on industrial scale heavily relies on the availability of utilities such as energy in different forms e.g., steam, electricity, cooling water, brine and others. These utilities represent a significant share of the production cost of chemicals and define to a major degree the environmental impact resulting from chemical production. The efficiency of utility usage in chemical industry is primarily defined by two major factors, the efficiency of the utility generation (e. g. steam production) and the efficiency in using the generated utilities in achieving a productivity of chemicals as high as possible.This description evokes the idea of two separated problems and unfortunately nowadays the two fields are often treated like this and consequently substantial integration potentials are overlooked resulting in increased cost and environmental impact. However, the generation and the usage of utilities in chemical production are highly interlinked and integration potentials must be identified during early design stages of both utility generation systems and chemical production processes. Major aspects in this context are waste treatment in general and solvent management in particular. Increased process integration, e.g., recycling of material flows, reduces the amount of waste to be treated and therefore the additional utility input required by many waste treatment technologies (e.g., stripping, sewage treatment). However, waste can also be a source of utility generation, e.g., by generating steam through waste incineration. This is particularly an issue for the huge amount of waste solvents that are generated in pharmaceutical and fine chemical production typically operating in batch mode.

Chemical batch plants are characterized by frequent changes in production portfolio which implies dynamic patterns in energy demand and waste generation. Since waste treatment plays an important role in the energy system of a chemical plant, these dynamic patterns can increase energy demand and simultaneously offer opportunities for primary energy savings. In constrast, utility generation systems cannot be changed frequently because high investments are associated and high efficiency can only be achieved for a well-defined utility mix (e.g. the proportion of electricity to steam in Combined Heat and Power devices).

In order to study the potential of an integrated design in this three-component system (i.e., production processes, waste management and energy utility generation) suitable models are needed. In the present study the energy demand of batch chemical production processes was based on bottom-up unit operation models adjusted to real process measurements using calibrated control valve openings. The difference between theoretical and real utility consumption was used to develop empirical parametric equations describing thermal losses due to inefficient heating/cooling system operations. Validation of these bottom-up models was performed at different aggregation levels, including equipment specific unit operations, production lines and overall batch plant energy consumption. The results indicated that for all types of energy utility, equipment and unit operation the consumption could be estimated with a relative error between 5% and 35% depending on the aggregation level.

Typical batch plant waste treatment facilities were considered like incineration, wastewater treatment and wet air oxidation, while solvent recovery was treated with continuous and batch distillation units. This waste management system can make full use of the dual properties of waste solvents which can be an energy source as combustible for incineration or energy sink as material for recycling in distillation units. In order to include life cycle analysis aspects in the investigation, multi-input allocation models were used for the waste treatment units to include energy requirements and pollutant emissions of waste handling. These models assess emissions and utility consumption with few waste stream data using linear transfer coefficients derived by extensive industrial studies of the respective operations.

The energy utility generation system includes steam boilers and the associated steam network, gas turbines, a refrigeration cycle producing brine and of course the waste incineration plant. The heat load provided by utilities corresponds to the heat demand of the process taking into account an average efficiency for heat exchangers. Concerning the process streams temperatures, it is possible to assume that they are $\Delta T_{min}$ lower (or higher, if the utility is a cold stream) than the utility lower (higher) temperature. The size of the corresponding heat exchanger network and the amount of necessary primary energy sources can be optimized using Pinch analysis.

For the optimization of the integrated system in a modular form, three nested optimization problems were solved: the production planning of the batch processes (MILP), the mixing problem of the generated waste streams (MINLP), and the heat integration problem for energy utilities geneartion (LP). The operational costs and diverse life-cycle impact oriented environmental indicators were used as objective functions. A synthetic case study plant consisting of a mono-product and a multi-product building was used for investigating the efficiency of the integrated design. The case study plant reflects the complexity of a real batch plant that provided the data for the development of the bottom-up energy consumption models, the desired production capacities, the waste management constraints and the utility generation system. In this way, it was possible to define a base case corresponding to the current plant operation, to which the optimization results were compared. It is shown that an optimal waste mixing together with a suitable production planning (i.e., more homogeneously distributed production over the time horizon) can result in a significant reduction of ancillaries in waste treatment operations and primary sources for energy utility generation (approx. up to 50% less energy consumption related operational costs and 20% less environmental impacts expressed in Eco-Indicator-99 points).

See more of this Session: Design of Sustainable Processes
See more of this Group/Topical: Computing and Systems Technology Division

# (521g) Economic Analysis of Municipal Wastewater Utilization for Thermoelectric Power Production

**Iman Safari**[1], Michael E. Walker[1], Javad Abbasian[1], Hamid Arastoopour[2], Ming-Kai Hsieh[3], Ranjani B. Theregowda[3], David A. Dzombak[3] and David C. Miller[4], (1)Chemical and Biological Engineering, Illinois Institute of Technology, Chicago, IL, (2)Department of Chemical and Biological Engineering, Wanger Institute for Sustainable Energy Research (WISER), Illinois Institute of Technology, Chicago, IL, (3)Department of Civil and Environmental Engineering, Carnegie Mellon University, Pittsburgh, PA, (4)U.S. Department Of Energy, National Energy Technology Laboratory, Morgantown, WV

The thermoelectric power industry in the U.S. uses a large amount of freshwater. The large water demand is increasingly a problem, especially for new power plant development, as availability of freshwater for new uses diminishes in the United States. Reusing non-traditional water sources, such as treated municipal wastewater, provides one option to mitigate freshwater usage in the thermoelectric power industry. The amount of freshwater withdrawal that can be displaced with non-traditional water sources at a particular location requires evaluation of the water management and treatment requirements, considering the quality and abundance of the non-traditional water sources.

This paper presents the development of an integrated costing model to assess the impact of degraded water treatment, as well as the implications of increased tube scaling in the main condenser. The model developed herein is used to perform case studies of various treatment, condenser cleaning and condenser configurations to provide insight into the ramifications of degraded water use in the cooling loops of thermoelectric power plants. Further, this paper lays the groundwork for the integration of relationships between degraded water quality, scaling characteristics and volatile emission within a recirculating cooling loop model.

See more of this Session: Design of Sustainable Processes
See more of this Group/Topical: Computing and Systems Technology Division

## Considerations for Research, Development and Commercialization/Scale-up of Biomass Pretreatment Processes

***Dale Monceaux***[1], *Richard C. Agar*[1] *and Charles E. Wyman*[2], *(1)AdvanceBio, LLC, Milford, OH, (2)College of Engineering-Center for Environmental Research and Technology, University of California, Riverside, Riverside, CA*

Global efforts to cost effectively convert biomass-based polysaccharides to sugars via enzymatic hydrolysis and subsequently to fuels and chemicals have been ongoing since the early discoveries by the US Army Natick Research & Development Command. These, and other early developments related to enzymatic hydrolysis of cellulose were followed by advancements in microbial genetic engineering. Genetically modified microorganisms capable of utilizing xylose-rich hemicellulose fractions of biomass feedstocks in addition to the cellulose derived glucose, promised improved ethanol yields and process economics. The past fifteen years of research has led to other major advancements in cellulase protein chemistry as well as an improved understanding of the microstructure of lignocellulose, moving the world ever closer to commercially viable biomass-based fuels and chemicals.

Early research ultimately revealed the sensitivity of the feedstock, pretreatment, enzyme and organism regime. This, in turn has exposed the need for consistency in methodology as research proceeds from concept to commercial-scale operations.

Much current activities focusing on the commercialization of biomass based fuels and chemicals hinge on the development of cost-effective biomass pretreatment processes. This "first step" in the process of converting biomass to end-products is critical to successful commercialization efforts for a number of reasons. First, it provides reactive cellulose that can be efficiently enzymatically hydrolyzed to oligomeric or monomeric glucose. Second, it efficiently solubilizes hemicellulose fractions of the biomass, producing oligomeric and or monomeric xylose, mannose, galactose, arabinose, etc. Third, it produces minimal reaction by-products that inhibit enzymatic hydrolysis and fermentation processes. Last, it is mechanically and economically scalable.

Many pretreatment technologies employ thermo-chemical reactions that reduce the recalcitrance of ligno-cellulosic matrices, solubilize fractions of the biomass and expose reactive sites to subsequent reactions. Various pretreatment technologies are under development, each with a range of design basis conditions. **Table 1** illustrates general conditions for six common pretreatment chemistries.

| Pretreatment | Temp (°C) | Residence Time (min) | Water Loading* | Catalyst | Catalyst Loading** | Catalyst Recyclable |
|---|---|---|---|---|---|---|
| AFEX | 150 | 30 | 200 | Anhydrous $NH_3$ | 152 | Yes (95-95% of loaded) |
| Dilute Acid | 140 | 40 | 891 | 1% $H_2SO_4$ | 9 | No |
| Lime | 120 | 240 | 1500 | $CaOH/O_2$ | 100/100 psi | Maybe/Ne |
| Liquid Hot Water | 200 | 10 | 663 | Water | N/A | N/A |
| SAA | 160 | 60 | 765 | 15% Aqueous $NH_3$ | 135 | Yes, ecess |
| $SO_2$ | 180 | 10 | 900 | 5% $SO_2$/Steam | 5/? | Yes, excess |

* Water Loading = L/100 kg of dry biomass input to pretreatment

** Catalyist Loading = kg/100 kg of dry biomass input to pretreatment

Table 1

Typical Thermochemical Pretreatment Chemistries

Reaction kinetics varies significantly with feedstock type, mechanical pretreatment, catalyst type, catalyst loading, water loading, residence time and temperature. Tight control of key parameters is required to maximize conversion efficiency. Elevated catalyst concentration and residence times increase reaction rate but also xylose destruction as illustrated in **Table 2**. Consequently, optimum reaction conditions require mechanical designs that deliver controlled residence times under near-plug flow conditions.

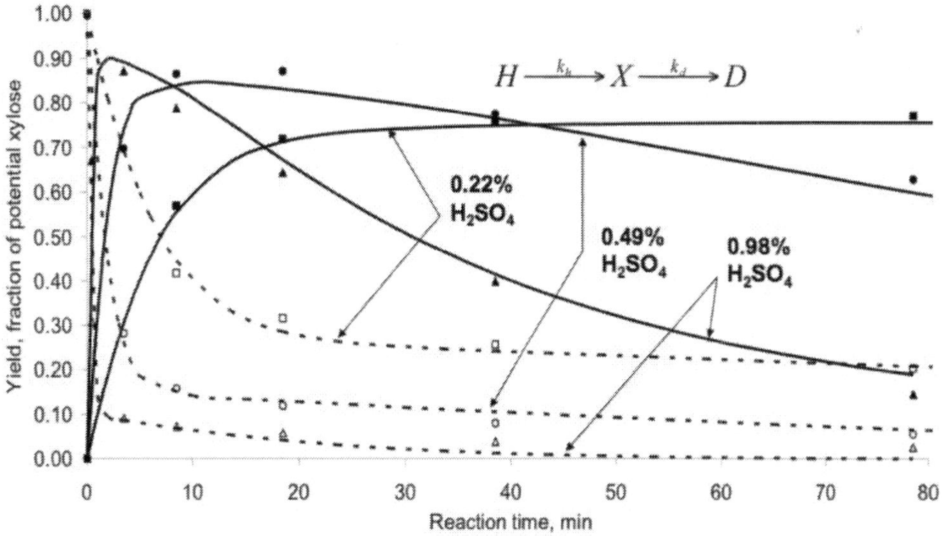


Table 2

Xylose Yield in Pretreatment vs Time at 160°C

Resource limitations, including feedstock supply, workspace, staffing and funding constrain the research and development efforts at each stage. Real-world feedstock properties such as foreign matter content, chemical composition variability, recalcitrance, bulk density, etc. introduce constraints that further complicate research and commercialization efforts. Feedstock selection dramatically impacts capital investment and production cost. Feedstock recalcitrance in conjunction with chemical composition effectively determines yield, and therefore the quantity of feedstock required per unit of fuel or chemical produced. Feedstock chemical composition and bulk density, in conjunction with reactor residence time establishes the size of the major components of the capital and energy intensive pretreatment reactor unit operation, as illustrated in **Table 3**.

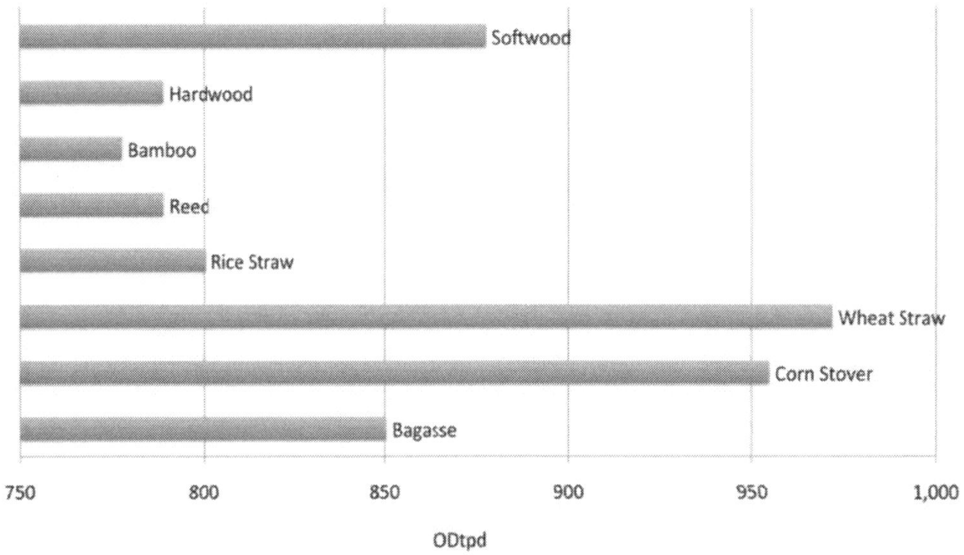


Table 3

Estimated Feedstock Tonnage per 100 Megaliters of Ethanol

High yields are critical to distribute both operating and capital costs. Low operating costs, including chemicals, energy and labor, are essential to providing adequate margins for return on capital. General operating cost categories must be lower for a new process than for established technologies making conventional products since low capital costs are essential to minimize financial exposure. The driving forces to contain capital investment include general technologies and processes that minimize equipment sizes, operate at lower pressures, operate at neutral or higher pHs as well as thermal processes that operate at lower temperatures and biochemical processes that operate at elevated temperatures.

Technical compromises frequently lead to approaches that ultimately expand the research and development program effort, schedule and budget. The impact of decisions made early in the technology development phase of a project may not fully materialize until late in the project after significant design and engineering expenditures have been made. The time period between research decision-making and engineering-based capital cost estimating delays the commercialization or leads to technology dead-ends. Visualization of the expected commercial scale process should occur very early in the technology research and development effort. A process "scale-down" approach to optimize research and technology development, scale-up and commercialization efforts has proven to be and effective strategy to accelerate project development and finance.

# (522b) Improving Corn Stover Reactivity In Low Severity Pretreatment Through Deacetylation and Disc-Refining

**Xiaowen Chen**[1], Joseph Shekiro[1], mary Ann Franden[1], Min Zhang[1], Bonwook Koo[2], Sunkyu Park[2] and Melvin Tucker[1], (1)National Bioenergy Center, National Renewable Energy Laboratory, Golden, CO, (2)Department of Forest Biomaterials, North Carolina state university, Raleigh, NC

The current "state-of-art" dilute acid pretreatment to produce bio-ethanol from corn stover requires severe conditions to overcome biomass recalcitrance, which, in turn, offers 70~80% xylose monomer yield and 5~7% furfural yield. However, high acid loadings and high reaction temperatures harm the economic feasibility of the process. In addition, the amount of acetate and furfural produced during pretreatment is in a range that strongly inhibits cell growth and impedes ethanol fermentation. To decrease the impact of these issues, deacetylation, pretreatment with lower acid concentrations, lower temperatures, and integration of disc-refining was carried out in this work. The results show that deacetylation with dilute NaOH (0.1M) combined with low severitydilute-acid pretreatment increased corn stover reactivity by increasing the xylose monomer yield by 20% in pretreatment while degradation of xylan to furfural remained under 2%. Glucose yield increased by 10% and xylose yield by 20% during dilute solids enzymatic hydrolysis as a result of deacetylation of the feedstock prior to pretreatment. Particle size reduction by mechanical refining after pretreatment is shown to assist achieving> 90% cellulose yield in high solids (20 wt% total solids) enzymatic hydrolysis. Bioscreen C growth assays and mini fermentation at high solids (20wt %) showed less toxicity of deacetylated pretreated corn stover and enhanced xylose utilization by 20%.

See more of this Session: Developments In Biobased Alternative Fuels I
See more of this Group/Topical: Sustainable Engineering Forum

## Reducing Process Complexity for Cellulosic Ethanol

Lonnie O. Ingram, Department of Microbiology and Cell Science
University of Florida, Box 110700, Gainesville, FL 32611 Ingram@UFL.edu

Lignocellulosic biomass represents a potential source of carbohydrates (cellulose and hemicellulose) for microbial fermentation into fuels and chemicals. Steam pretreatment with dilute mineral acids serves as an efficient approach to depolymerize hemicellulose and enhance cellulose digestion with enzymes. However, side products (furfural, 5-hydroxymethyl furfural, acetate, phenolics,and others) formed during pretreatment inhibit microbial growth and retard fermentation. Potential engineering solutions to this problem (separation of hemicellulose hydrolysate, counter-current-washing of fibers, and mitigation of toxins) add complexity and increase cost. The level of inhibitors produced during acid pretreatment is affected by the severity of conditions and the choice of acid. Weaker acids such as phosphoric acid produce lower levels of toxic side products and reduce the need for exotic metal alloys. Since mineral acids are not consumed by the pretreatment process, the higher cost of phosphoric acid as compared to sulfuric acid could be offset in part by recovery and reuse as a dilute fertilizer when crops are nearby. Recent progress has been made in the development of more robust *E. coli*-based biocatalyst with increased resistance to dilute acid hydrolysates. Using this biocatalyst and phosphoric acid pretreatment, we have been able to co-ferment sugarcane bagasse hemicellulose hydrolysate and enzyme-hydrolyzed cellulose in a single vessel using a variation of the simultaneous saccharification and co-fermentation process (SScF; Fig. 1). Hexose and pentose sugars from phosphoric acid pretreated sugarcane bagasse were fermented in a single vessel, eliminating process steps for solid–liquid separation and sugar cleanup. After pretreatment, a liquefaction step (L) with cellulase was included to improve mixing and saccharification (L + SScF), analogous to a corn ethanol process. Bagasse slurries containing 10% and 14% dry weight (fiber plus solubles) were tested using pretreatment temperatures of 160–190 C (1% phosphoric acid, 10 min). Enzymatic saccharification and inhibitor production both increased with pretreatment temperature. The highest titer (30 g/L ethanol) was obtained after incubation for 122 h using 14% dry weight slurries of pretreated bagasse (180 C). The highest yield (0.27 g ethanol/g dry bagasse) was 82 gal ethanol per dry ton of bagasse using this L+SScF process and 10% dry weight slurries.

# Comparison of conversion processes

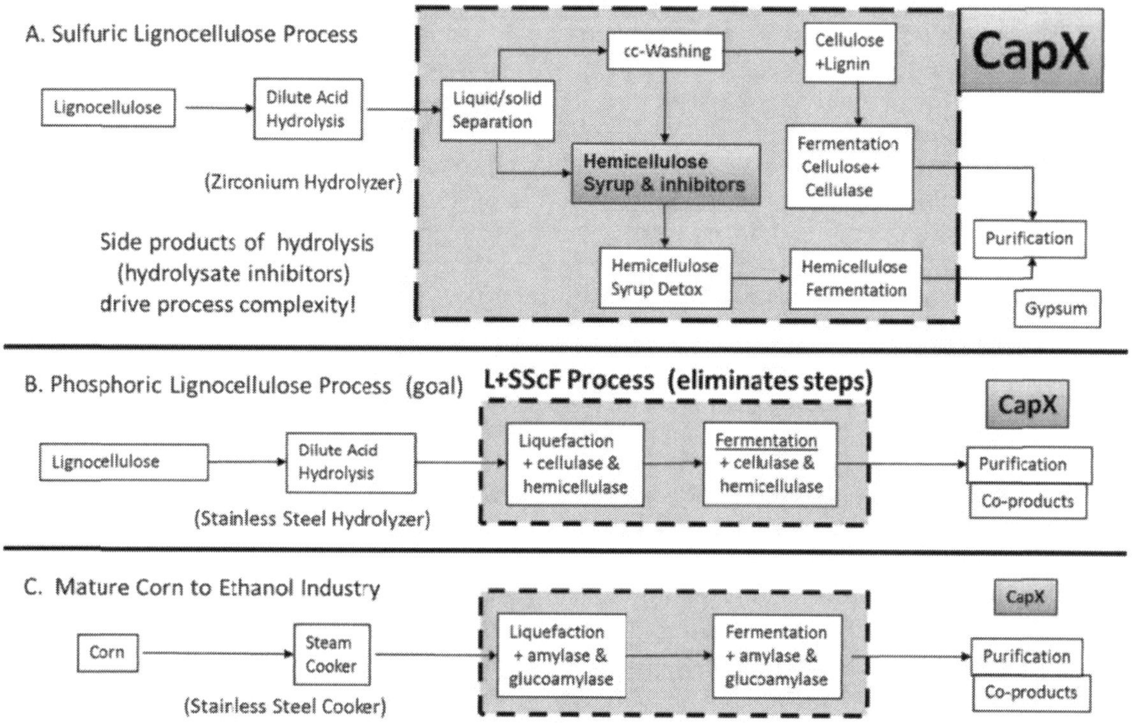

# (522d) Co-Utilization of Methane In Steam-Biomass Gasification Using Concentrated Solar Energy

**Aaron W. Palumbo**, Erica L. Jorgensen and Alan W. Weimer, Chemical & Biological Engineering, University of Colorado at Boulder, Boulder, CO

Presented is a novel gasification application where biomass is steam-gasified and natural gas is steam-reformed simultaneously in a solar reactor. Solar-driven gasification is a proven technology, and can augment the biofuel industry as a domestic fuel source. Gasification performed in an indirectly heated gasifier can achieve temperatures greater than 1150 °C [1], mitigating or eliminating the production of $CO_2$, tar, char, and light hydrocarbon gases in the product gas. This reduces the need for gas clean-up technology and will not require a downstream water-gas shift reactor to adjust $H_2$:CO. Thermodynamics favor the formation of only hydrogen and carbon monoxide in the product synthesis gas. Therefore, a predictable $H_2$:CO can be controlled and tuned for downstream fuel conversion. Since the majority of biomass carbon is converted to CO, more than twice as much biomass is converted to a useable liquid fuel than in a traditional gasification process. Preliminary experiments show that reactor temperatures must exceed 1350 °C in order to reduce the effects of the water-gas shift reaction and reform practically all light hydrocarbon gases, including $CH_4$. This operating temperature is consistent with realizable concentrated solar energy systems.

1. Piatkowski, N., et al., *Solar-driven gasification of carbonaceous feedstock-a review.* Energy & Environmental Science, 2011. **4**(1): p. 73-82.

See more of this Session: Developments In Biobased Alternative Fuels I
See more of this Group/Topical: Sustainable Engineering Forum

# (522e) Poly(oxymethylene) Dimethyl Ethers As Components of Tailored Diesel Fuel - Properties, Synthesis and Purification Concepts

**Jakob Burger**, Laboratory of Engineering Thermodynamics, Technische Universitt Kaiserslautern, Kaiserslautern, Germany

Poly(oxymethylene) Dimethyl Ethers as components of tailored diesel fuel - properties, synthesis and purification concepts

**Jakob Burger [1*], Markus Siegert [2], Eckhard Ströfer [2], Michael Nilles [2], Hans Hasse [1]**

[1] Laboratory of Engineering Thermodynamics, University of Kaiserslautern, Germany;

[2] BASF SE Company, Ludwigshafen, Germany
*Corresponding author: Jakob Burger, Laboratory of Engineering Thermodynamics, University of Kaiserslautern, Germany. Email: jakob. burger@mv.uni-kl.de

**Keywords:** Fuel additive, Soot reduction, Process development

Poly(oxymethylene) Dimethyl Ethers (POMDMEs), oligomers of the general structure $CH_3$-O-($CH_2$-O)$_n$-$CH_3$, open a new route for tailoring diesel fuels.

POMDMEs belong to the group of oxygenates which reduce soot formation in the combustion when added to diesel fuels. They can be produced on a large scale based on Methanol which can be produced from renewable resources. This contribution presents the process chain, possible synthesis routes and some fuel related properties of POMDMEs. Comparisons to other oxygenates, e.g. Dimethyl Ether, are given.

A particularly favourable route for the large scale production of POMDMEs in which they are formed from Methylal and Trioxane is proposed. The chemistry behind this route is given in detail. Measurements of the reaction kinetics and the reaction equilibrium were carried out in a batch reactor using acidic ion exchange resin as heterogeneous catalyst. The parameters of a pseudo-homogeneous model were fitted to the results of these experiments. The formation of side-products was studied experimentally and concepts to minimize them are discussed. Based on the reaction model a process concept was developed. It consists of a reactor and a downstream distillation sequence to yield pure POMDMEs of the most favourable chain length n=3-5.

The use of distillation technology allows an easy scale-up to huge single train plants which can potentially fulfil the need for millions of tonnes of the innovative oxygen carrier POMDME needed for the supply of clean diesel fuel in the future.

See more of this Session: Developments In Biobased Alternative Fuels I
See more of this Group/Topical: Sustainable Engineering Forum

# (522f) Synthesis of Green Diesel Fuel From Fatty Acid Feedstocks Via Electrochemical Hofer-Moest Decarboxylation

**Cody Wagner**, Charles Coronella and Rainer Busch, Chemical Engineering, University of Nevada, Reno, Reno, NV

Production of biodiesel by esterification or transesterification of long-chain fatty acids is a well known process, but the fuel product has characteristics making it not fully compatible with petroleum diesel engines. A better option is to convert fatty acid molecules to hydrocarbons by removing their carboxyl group. Decarboxylation of fatty acid feedstocks was performed by electrolysis using graphite electrodes, which selectively causes a Hofer-Moest (or non-Kolbe)-type reaction. Methanol was used as a solvent for a robust selection of fatty acid feedstocks.

The liquid product created by the electrochemical reaction was composed of mainly alkenes with at least one double bond, ethers, and methyl esters. This fuel product has advantageous cold-flow properties, no sulfur content, heating values close to those of diesel fuels, and is capable of running in current diesel engines as a drop-in replacement for petroleum diesel, unlike biodiesel. The temperature, pH, concentrations of ions, and type of base used were all varied to optimized the current density, which limits the production rate. A separation method was tested for removal of the fuel product from the electrolyte. An efficiency was then calculated for several varieties of fatty acid feedstocks, comparing the electrical energy input to the product generated from the electrolysis. Overall first-law thermodynamics and economics were also evaluated, examining production rates and the large-scale viability of the process.

See more of this Session: Developments In Biobased Alternative Fuels I
See more of this Group/Topical: Sustainable Engineering Forum

# (544a) Lignin Depolymerization and Deoxygenation In Ionic Liquids

**Blair J. Cox**, Thong Q. Ngo and John G. Ekerdt, Chemical Engineering, University of Texas at Austin, Austin, TX

Lignin, an amorphous aromatic biopolymer that comprises up to 30% of plant matter, was dissolved and depolymerized in ionic liquids under acidic conditions. The acidic ionic liquid 1-methylimidazolium chloride (HMIMCl) acts as a combination solvent and catalyst that facilitates degradation of lignin into smaller fragments. This degradation occurs through the catalytic hydrolysis of the β-O-4 ether linkage, the most prominent linkage in the lignin polymer. Studies with guaiacylglycerol-β-guaiacyl ether and veratrylgycerol-β-guaiacyl ether as phenolic and non-phenolic lignin model compounds demonstrated that acid catalyzed degradation of lignin in HMIMCl proceeded efficiently at 150 °C, primarily by eliminating water to produce an enol ether intermediate. This intermediate then undergos hydrolysis of the β-O-4 ether linkage, which is analogous to degradation of the β-O-4 linkage in the lignin polymer. These process conditions were next applied to lignin that was extracted from pine wood biomass through dissolution in the HMIMCl ionic liquid and subsequent precipitation. Analysis with gel permeation chromatography indicates significant decrease in size after treatment in HMIMCl at 150°C for 40 min. After depolymerization, the lignin fragments will be subjected to hydrodeoxygenation (HDO), while in the 1-butyl-3-methyl imidazolium chloride ionic liquid, over supported metal catalysts. Preliminary HDO studies using phenol with ceria- or zirconia-supported copper and nickel catalysts show complete conversion and suggest a viable path to HDO of lignin.

See more of this Session: Novel Catalytic and Separation Process Based on Ionic Liquids
See more of this Group/Topical: International Congress on Energy 2011

# (544b) Preparation of Cellulose-Rich Materials From Cornstalks Using Biodegradable Ionic Liquids

**Xiaofeng Zhang**, Institute of Nuclear and New Energy Technology, Tsinghua University, Beijing, China

In our work, we obtained cellulose-rich materials via utilizing a kind of economic biodegradable IL, choline acetate ( [ChO][OAc] ), demonstrating the possibility of the partial dissolution and separation of cornstalks after mild grinding. The selected [ChO][OAc] IL presents very low toxicity, due to the benign nature of the cation and its high biodegradation potential which was evaluated under aerobic conditions. Delignification of cornstalks using the [ChO][OAc] IL was successfully achieved that lignin content decreased from 14% of cornstalk raw materials to 2% of undissolved residue under mild conditions. Thus, cellulose-rich materials accounting for 98% could be obtained. Furthermore, the cellulose fibre from [ChO][OAc] solutions can be regenerated to form cellulose membrane on a glass plate. In this way, partial separation of the cornstalk components becomes available with relative efficient, nontoxic and biodegradable ionic liquids.

See more of this Session: Novel Catalytic and Separation Process Based on Ionic Liquids
See more of this Group/Topical: International Congress on Energy 2011

# (544c) Preparation of Cellulose-Rich Materials From Cornstalks In Biodegradable Ionic Liquids

Shan Li[1], Minli Zhu[1], Xingmei Lu[2], Zengxi Li[1] and **Suojiang Zhang**[2], (1)College of Chemistry and Chemical Engineering, Graduate University of Chinese Academy of Sciences, Beijing, China, (2)Institute of Process Engineering, Chinese Academy of Sciences, Beijing, China

Agricultural wastes such as bagasse, cornstalks, are relatively cheap sources of lignocellulose consisting of lignin, cellulose and hemicellulose. However, most of them are burned as waste without rational use, even causing severe environment contamination. Lignocellulosic materials have aroused wide public concern due to their potential as starting materials for biofuels and other commodity chemicals.[1]

The unique physicochemical properties of ionic liquids (ILs), such as non-flammability, negligible vapour pressure, high thermal and chemical stability, have attracted increasing interests in their potential applications in various areas. It is showed that lignocellulosic materials can be dissolved in ILs which are almost imidazolium-based ionic liquids (IBILs)[2]. The differences of solubility among the components of lignocellulose make it possible to separate them from each other. However, the scope of possible applications of imidazolium-based ionic liquids is restricted due to their high cost, toxicity and the non-biodegradability of the imidazolium ring[3].

In our work, we obtained cellulose-rich materials via utilizing a kind of economic biodegradable IL, choline acetate ( [ChO][OAc] ), demonstrating the possibility of the partial dissolution and separation of cornstalks after mild grinding. The selected [ChO][OAc] IL presents very low toxicity, due to the benign nature of the cation and its high biodegradation potential which was evaluated under aerobic conditions. Delignification of cornstalks using the [ChO][OAc] IL was successfully achieved that lignin content decreased from 14% of cornstalk raw materials to 2% of undissolved residue under mild conditions. Thus, cellulose-rich materials accounting for 98% could be obtained. Furthermore, the cellulose fibre from [ChO][OAc] solutions can be regenerated to form cellulose membrane on a glass plate. In this way, partial separation of the cornstalk components becomes available with relative efficient, nontoxic and biodegradable ionic liquids.

**Reference**

1 Suzie S. Y. Tan, Douglas R. MacFarlane, Jonathan Upfal, Leslie A. Edye, William O. S. Doherty, Antonio F. Patti, Jennifer M. Pringle and Janet L. Scott, Green Chem., 2009, 11, 339-345

2 Ning Sun, Mustafizur Rahman, Ying Qin, Mirela L. Maxim, H°Ctor Rodríguez and Robin D.Rogers, Green Chem., 2009, 11, 646-655

3 Helga Garcia, Rui Ferreira, Marija Petkovic, Jamie L. Ferguson, Maria C. Leitão, H. Q. Nimal Gunaratne, Kenneth R. Seddon, Luís Paulo N. Rebelo and Cristina Silva Pereira, Green Chem., 2010, 12, 367-369

See more of this Session: Novel Catalytic and Separation Process Based on Ionic Liquids
See more of this Group/Topical: International Congress on Energy 2011

# (544d) Preparation and Testing of Adsorbents for the Separation of Ionic Liquids From HMF

**Yuxiang Rao**[1], Alan Cooper[2], Heather Brown[2] and Wei Liu[3], (1)Energy and Environment Directorate, Pacific Northwest National Laboratory, Richland, WA, (2)Pacific Northwest National Lab, Richland, WA, (3)Pacific Northwest National Laboratory, Richland, WA

HMF is an important chemical product since it has the potential to be used as a feedstock to replace petroleum-derived aromatics for production of a range of specialty chemicals and plastics.In this work, we will demonstrate an economically-viable, ionic liquid catalytic process for conversion of biomass-derived raw sugar to 5-hydroxymethylfurfural(HMF). The experimental work includes preparation of adsorbents materials by membrane coating, catalytic reactions in ionic liquid fluids and separation of reacted products by extraction and adsorption, which includes (i) catalytic reaction tests in both a batch and flow reactor system to develop systematic understandings to critical process operation parameters and establish material balances, and (ii) screening of adsorbents and identification of an effective adsorption process for separation of HMF product from reaction mixtures and for complete recovery of the catalyst and ionic liquid. The adsorbent materials tested for this project were characterized by XRD, SEM/EDS and Nitrogen adsorbtion/desorption technology.

See more of this Session: Novel Catalytic and Separation Process Based on Ionic Liquids
See more of this Group/Topical: International Congress on Energy 2011

# (544e) Isobutane/1-Butene Alkylation Catalyzed by Ionic Liquid-Superacid Coupled Catalysts

Xueqi Xing, Guoying Zhao, **Suojiang Zhang**, Peng Cui and Hailing Ren, Institute of Process Engineering, Chinese Academy of Sciences, Beijing, China

Abstract-FThe alkylation of isobutane with 1-butene was synergeticly catalyzed by trifluoromethanesulfonic acid and ionic liquids in a batch autoclave. A high-quality alkylates with up to 80% desired trimethylpentane and 95 research octane number were obtained using the coupled catalyst with 33% trifluoromethanesulfonic acid. The coupled catalyst was reused 10 recycles without significant decrease in catalytic performance, compared to the today commercial H2SO4 catalyst. The middle- length cation of ionic liquids, for example C6, was shown to be good for both the quality and yield of alkylates.

See more of this Session: Novel Catalytic and Separation Process Based on Ionic Liquids
See more of this Group/Topical: International Congress on Energy 2011

# (544f) Degradation of Poly(ethylene terephthalate) Wastes Catalyzed by Mn-Containing Ionic Liquid

Qian Wang[1], Xingmei Lu[1], Xiangping Zhang[1], **Suojiang Zhang**[1], Chunshan Li[1], Zengxi Li[2], Xueyuan Zhou[1] and Minli Zhu[2], (1)Institute of Process Engineering, Chinese Academy of Sciences, Beijing, China, (2)Graduate University of Chinese Academy of Sciences, Beijing, China

With the increasing of poly(ethylene terephthalate)(PET) consumption in recent years, the effective recycling of PET wastes becomes an urgent affair focused on by many governments and scientific researchers. Ionic liquids (ILs) have been reported as catalysts to degrade PET and most of them have showed some advantages of mild reaction condition and being easily separated from the main product. But the conversion of PET and the selectivity of the monomer bis(2-hydroxyethyl terephthalate)(BHET) are low. Therefore, the exploitation of novel ionic liquids that can efficiently degrade PET is significant.

In this study, Mn-containing ionic liquid was synthesised, characterized by Raman spectroscopy and Electrospray Ionization-Mass Spectrum and used to the catalytic degradation of PET in ethylene glycol. The results shows that the IL exhibits higher catalytic activity, compared with conventional ionic liquids,such as [bmim]Cl, [bmim]$H_2PO_4$, [bmim]$HSO_4$. The conversion of PET and the selectivity of the BHET reach up to 100% and 78.9% respectively under the conditions of reaction pressure 1atm, reaction temperature 170°C and reaction time 2.5h. They are both higher than 88.6% and 62.8% catalyzed by 1-butyl-3-methylimidazolium tetrachloroferrate ([bmim]$FeCl_4$) and the reaction time were shortened greatly from 4h to 2.5h(Wang et al, 2010). In addition, the method has advantages that the main product is easily separated from catalyst and has good quality. Finally, the mechanism is proposed and shows that the higher catalytic activity of the IL is owing to the synergistic effect of its cation and anion.

**Reference**

Wang, H., Yan, R.Y., Li, Z.X., Zhang, X. P., Zhang, S.J.(2010). Fe-containing magnetic ionic liquid as an effective catalyst for the glycolysis of poly(ethylene terephthalate). *Catal. Commun.*, 11, 763-767.

# (548a) Estimation and Modeling of Crystal Size and Shape Evolution Using In Situ Tools

**Mo Jiang**[1], Mark Molaro[1], Michael L. Rasche[2], Haitao Zhang[1], Keith Chadwick[1], Lifang Zhou[1], Minhao Wong[2], Zhilong Zhu[2], Dominique Hebrault[3], Des O'Grady[3], John Tedesco[3] and Richard D. Braatz[1], (1)Chemical Engineering, Massachusetts Institute of Technology, Cambridge, MA, (2)Department of Chemical and Biomolecular Engineering, University of Illinois at Urbana-Champaign, Urbana, IL, (3)METTLER TOLEDO AutoChem, Columbia, MD

A large proportion of pharmaceutical crystallizations produce crystals with high aspect ratio, which can cause problems in operations downstream from the crystallizer, such as washing and filtration (e.g., see Refs. 1-24). This presentation describes the application of the Focused Beam Reflectance Measurement (FBRM) for the in-process estimation of the crystal characteristics and ReactIR Attenuated Total Reflection Fourier Transform Infrared Spectroscopy for solution concentration estimation during the nucleation, growth, and dissolution of crystals of varying shapes (e.g., see Refs. 3, 8, 12-13, 15, 18 and citations therein). The particle characterization includes the estimation of the mean crystal length and mean crystal width for rod-like crystals, which are commonplace in the pharmaceutical industry. Chord length distributions and their derived statistics are compared for the S- and G-Series FBRM probes for various types of systems, with additional particle characterization using Process Vision Measurement (PVM) and off-line optical microscopy. Experiments that cycle the temperature from conditions of positive and negative supersaturation are used to assess reproducibility of the estimated particle and solution properties while providing a wide range of changes in the aspect ratio in a single well-characterized experiment for assessing the different relative dependen-cies of growth and dissolution on supersaturation along the different crystal axes. The crystal size and solute concentration estimates are used to estimate kinetics in a multidimensional population balance model for prediction of the crystal size and shape distribution. In contrast to previous studies that have estimated kinetics along multiple crystal axes in mixed-tank crystallizers (e.g., Refs. 14 and 24), this study employs standardized commercial instrumentation that is readily available in pharmaceutical laboratories (FBRM) rather than using specialized instrumentation or excessive sampling. In contrast to Ref. 24, the size characterization in this work is of much higher quality. In contrast to past studies, this work characterizes the different dissolution kinetics along each crystal axis, and is able to collect enough experimental design for kinetics estimation in a single experiment (instead of multiple experiments). The subsequent model predictions are validated with experimental data.

References:

[1] Yang, G.; Kubota, N.; Sha, Z.; Louhi-Kultanen, M.; Wang, J.; Crystal shape control by manipulating supersaturation in batch cooling crystallization, *Crystal Growth & Design* **2006**, *6*, 2799–2803.

[2] Zhang, Y.; Sizemore, J.; Doherty, M.F.; Shape evolution of 3-dimensional faceted crystals, *AIChE J.* **2006**, *52*, 1906–1915.

[3] Gunawan, R.; Ma, D.L.; Fujiwara, M.; Braatz, R.D.; Identification of kinetic parameters in a multidimensional crystallization process. Special Issue on Crystallization and Interfacial Processes, *Int. J. of Modern Physics B* **2002**, *16*, 367-374.

[4] Lee, K.; Lee, J.H.; Fujiwara, M.; Ma, D.L.; Braatz R.D.; Run-to-run control of multidimensional crystal size distribution in a batch crystallizer, in *Proceedings of the American Control Conference*, Anchorage, AK, pages 1013–1018, **2002**.

[5] Wan, J.; Wang, X.Z.; Ma C.Y.; Particle shape manipulation and optimization in cooling crystallization involving multiple crystal morphological forms, *AIChE J.* **2009**, *55*, 2049–2061.

[6] Briesen, H.; Simulation of crystal size and shape by means of a reduced two-dimensional population balance model, *Chem. Eng. Sci.* **2006**, *61*, 104–112, 2006.

[7] Ma, C.Y.; Wang, X.Z.; Roberts, K.J.; Morphological population balance for modeling crystal growth in face directions, *AIChE J.* **2008**, *54*, 209–222.

[8] Patience, D.B.; Rawlings, J.B.; Particle-shape monitoring and control in crystallization processes, *AIChE J.* **2001**, *47*, 2125–2130.

[9] Puel, F.; Fevotte, G.; Klein, J.P.; Simulation and analysis of industrial crystallization processes through multidimensional population balance equations. Part 1: A resolution algorithm based on the method of classes, *Chem. Eng. Sci.* **2003**, *58*, 3715–3727.

[10] Calderon De Anda, J.; Wang, X.Z.; Lai, X.; Roberts, K.J.; Classifying organic crystals via in-process image analysis and the use of monitoring charts to follow polymorphic and morphological changes, *J. Process Control* **2005**, 15, 785–797.

[11] Larsen, P.A.; Rawlings, J.B.; Ferrier, N.J.; An algorithm for analyzing noisy, in situ images of high-aspect-ratio crystals to monitor particle size distribution, *Chem. Eng. Sci.* **2006**, *61*, 5236–5248

[12] Ma, Z.H.; Merkus, H.G.; Scarlett, B.; Extending laser diffraction for particle shape characterization: Technical aspects and application, *Powder Technology* **2001**, *118*, 180–187.

[13] Pollanen, K.; Hakkinen, A.W.; Reinikainen, S.P.; Louhi-Kultanen, A.; Nystrom, L.; A study on batch cooling crystallization of sulphathiazole-Process monitoring using ATR-FTIR and product characterization by automated image analysis, *Chem. Eng. Res. Des.* **2006**, *84*, 47–59.

[14] Oullion, M.; Puel, F.; Fvotte, G.; Righini, S.; Carvin, P.; Industrial batch crystallization of a plate-like organic product. In situ monitoring and 2D-CSD modelling, Part 1: Experimental study, *Chem. Eng. Sci.* **2007**, *62*, 820–832.

[15] Eggers, J.; Kempkes, M.; Mazzotti, M.; Measurement of size and shape distributions of particles through image analysis, *Chem. Eng. Sci.* **2008**, *63*, 5513–5521.

[16] Togkalidou, T.; Tung, H.-H.; Sun, Y.; Andrews, A.; Braatz, R.D.; Parameter estimation and optimization of a loosely-bound aggregating pharmaceutical crystallization using in-situ infrared and laser backscattering measurements, *Ind. Eng. Chem. Res.* **2004**, *43*, 6168–6181.

[17] Grön, H.; Mougin, P.; Thomas, A.; White, G.; Wilkinson, D.; Hammond, R. B.; Lai, X.J.; Roberts, K.J.; Dynamic in-process examination of particle size and crystallographic form under defined conditions of reactant supersaturation as associated with the batch crystallization of monosodium glutamate from aqueous solution. *Ind. Eng. Chem. Res.* **2003**, *42*, 4888–4898.

[18] Togkalidou, T.; Tung, H.-H.; Sun, Y.; Andrews, A.; Braatz, R.D.; Solution concentration prediction for pharmaceutical crystallization processes using robust chemometrics and ATR FTIR spectroscopy, *Org. Process Res. Dev.* **2002**, *6*, 317-

322.

[19] Hulburt, H.M.; Katz, S.; Some problems in particle technology, *Chem. Eng. Sci.* **1964**, 19, 555–574.

[20] Qamar, S.; Noor, S.; ul Ain, Q.; Seidel-Morgenstern, A.; Bivariate extension of the quadrature method of moments for batch crystallization models, *Ind. Eng. Chem. Res.* **2010**, *49*, 11633–11644.

[21] Gunawan, R.; Fusman, I.; Braatz, R.D.; High resolution finite volume methods for simulating multidimensional population balance equations with nucleation and size-dependent growth, *AIChE J.*, **2004**, 50, 2738-2749.

[22] Ma, D.L.; Tafti, D.K.; Braatz, R.D.; High resolution simulation of multidimensional crystal growth, *Ind. Eng. Chem. Res.* **2002**, 41, 6217-6223.

[23] Ma, D.L.; Tafti, D.K.; Braatz, R.D.; Compartmental modeling of multidimensional crystallization, *Int. J. of Modern Physics B* **2002**, 16:383-390.

[24] Gunawan, R.; Ma, D.L.; Fujiwara, M.; Braatz, R.D.; Identification of kinetic parameters in a multidimensional crystallization process, *Int. J. of Modern Physics B* **2002**, 16, 367-374.

See more of this Session: PAT for Crystallization Development and Manufacturing
See more of this Group/Topical: Separations Division

# (548b) Enhanced Resolution for More Accurate Inline Particle Measurements In Crystallization

**Anjan P. Pandey**, Des O'Grady and Ben Smith, METTLER TOLEDO AutoChem, Columbia, MD

The use of Process Analytical Technology (PAT) tools for real-time crystal size distribution and supersaturation measurements have been established for crystallization optimization and control. Yet up to now, interpreting measurement artifacts has been a challenge.

Paradigm shifts in signal processing have dramatically improved the resolution and accuracy of *in situ* particle size (chord length) measurements and advanced optics has dramatically enhanced the count linearity with increasing concentrations.

Examples will be presented where the Next Generation FBRM® technology is used to

- Enhance sensitivity to secondary nucleation

- Improve nucleation and growth kinetics calculations

- Detect probe fouling and measure crystal distributions in an inherently fouling processing

- Improve correlations with downstream process efficiency and product quality

See more of this Session: PAT for Crystallization Development and Manufacturing
See more of this Group/Topical: Separations Division

# (548c) PAT Based In Situ Monitoring and Control of the Polymorphic Purity of Pharmaceuticals In Batch Cooling Crystallization

**Zoltan K. Nagy**, Chemical Engineering Department, Loughborough University, Loughborough, United Kingdom, M. R. Abu Bakar, Department of Pharmaceutical Technology, International Islamic University Malaysia, Pahang, Malaysia and Ali N. Saleemi, Chemical Engineering, Loughborough University, Loughborough, United Kingdom

Sulfathiazole, an antibiotic compound, has been reported to have five polymorphs [1]. Although its polymorphism has been extensively and repeatedly investigated by various researchers, it remains difficult to produce a pure polymorph; most of the time, the desired polymorph contains impurities from at least one other form [2]. In this paper the application of Process Analytical Technology (PAT) tools such as focused beam reflectance measurement (FBRM), and attenuated total reflectance ultraviolet/visible (ATR-UV/Vis) spectroscopy in combination with chemometrics will be corroborated for the in-situ monitoring and control of the polymorphic purity of sulfathiazole. The approach will use a feedback control strategy, which will drive the process in the phase diagram by manipulating the operating temperature profile to achieve a desired polymorphic form. An experimental case study demonstrates that the proposed approach is able to selectively control the formation of form I and form IV of the sulfathiazole crystals. In this case, form IV is the desirable polymorph and in order to eliminate the presence of form I, the operating profile is driven in such a way that after the nucleation of an unknown polymorph or a mixture of the polymorphs, it is controlled to operate below the solubility curve of form I but in the supersaturation region of form II. An additional case study for the polymorphic control of o-amino benzoic acid is also presented. Characterization of the polymorphs obtained will be performed using x-ray diffraction, thermal and microscopy methods to validate the success of the control approach.

[1] D. S. Hughes, M. B. Hursthouse, T. Threlfall and S. Tavener, (1999), Acta Cryst. C, 55, 1831.

[2] A. Hakkinen, K. Pollanen, M. Karjalainen, J. Rantanen, M. Louhi-Kultanen

and L. Nystrom, (2005), Biotechnol. Appl. Biochem., 41, 17.

See more of this Session: PAT for Crystallization Development and Manufacturing
See more of this Group/Topical: Separations Division

# (548d) Experimental Investigation of Crystal Shape Evolution During Growth and Dissolution

**Meenesh R. Singh**[1], Stephan X. M. Boerrigter[2], Christian Borchert[3], Kai Sundmacher[3] and Dorwaiswami Ramkrishna[1], (1)School of Chemical Engineering, Purdue University, West Lafayette, IN, (2)Industrial and Physical Pharmacy, Purdue University, IN, (3)Max Planck Institute for Dynamics of Complex Technical Systems, Magdeburg, Germany

Controlling crystal morphology is a hot topic in current research on crystalline materials for diverse industrial applications. In this connection, experimental detection of polymorphs through on-line measurement of crystal morphology in a crystallizer is a problem of crucial interest. XRD and Raman Spectroscopy are generally accepted as a primary method to identify polymorphs. We present evidence of a new, strikingly simple and inexpensive method of identifying polymorphs through computer-aided analysis of images produced with a Confocal microscope. This methodology has the following features.

- A software with an automated program for polymorph detection integrated with a Confocal microscope which can be used for on-line monitoring and control of particle shape, size and polymorph content.
- Different polymorphs can be identified by recognizing inter-planar angular patterns specific to a given polymorph. It yields the absolute amounts of different polymorphs present in a population of crystals.
- It has the potential for on-line measurement of 2D and 3D crystals of complex shapes.

Consequently, this methodology can:

- Replace currently used FBRM (focused-beam reflectance measurement), digital video imaging and microscopy based sensors that have limitations with real shapes.
- Eliminate the need for multiple sensors currently used in industry.

See more of this Session: PAT for Crystallization Development and Manufacturing
See more of this Group/Topical: Separations Division

# (548e) Interpretation of FBRM Data At High Crystal Density

**Huayu Li**[1], Yoshiaki Kawajiri[1], Martha Grover[2] and Ronald W. Rousseau[1], (1) School of Chemical & Biomolecular Engineering, Georgia Institute of Technology, Atlanta, GA, (2)Chemical & Biomolecular Engineering, Georgia Institute of Technology, Atlanta, GA

Focus Beam Reflectance Measurement (FBRM) can be used in situ to monitor crystallization processes. If such measurements are sufficiently accurate and robust, it should be possible to utilize the FBRM in a scheme to control crystallizer operation. To fulfill this goal, both models for crystallization kinetics and the sensor are needed: the former is part of a population balance or method of moments, while the latter converts the measured chord-length distribution (CLD) to the crystal size distribution (CSD). To identify the sensor model, Yu et al [1] found linearity between chord count and the number of polystyrene particles in an aqueous slurry having low particle density. However, this has not been validated for a broader range of particle density, as is the case in the example system involving batch cooling crystallization of paracetamol from ethanolic solutions.

We have begun expanding the range of particle density coverage with experiments and analysis of paracetamol crystals suspended in a non-solvent, toluene. As crystals at certain sizes are added into toluene continuously, the increase on total count number and chord length histogram are recorded and analyzed to modify the linear relationship between number of chord and number of crystals. Our investigations include the following:

-A regime where linearity holds is identified.

-The linear coefficients are found to be different for different sizes of crystals in the linear regime.

- Linear additivity among different sizes of crystals is studied.

And we also perform Monte Carlo simulation to obtain the linear operator for normalized CSD to normalized CLD [2, 3]. The shape of crystals in the simulation is octahedral, the aspect ratio of which is defined by microscope images and adjusted to fit the measured CLD. The linear coefficients of chord count and the matrix for normalized CLD are combined to formulate a sensor model for FBRM.

Reference:

1. Yu, Z.Q., P.S. Chow, and R.B.H. Tan, *Interpretation of focused beam reflectance measurement (FBRM) data via simulated crystallization.* Organic Process Research & Development, 2008. **12**(4): p. 646-654.

2. Ruf, A., J. Worlitschek, and M. Mazzotti, *Modeling and experimental analysis of PSD measurements through FBRM.* Particle & Particle Systems Characterization, 2000. **17**(4): p. 167-179.

3. Kail, N., W. Marquardt, and H. Briesen, *Estimation of particle size distributions from focused beam reflectance measurements based on an optical model.* Chemical Engineering Science, 2009. **64**(5): p. 984-1000.

See more of this Session: PAT for Crystallization Development and Manufacturing
See more of this Group/Topical: Separations Division

# (548f) Calibration Free Reactive Dissolution Kinetic Modeling with In-Situ FBRM, ATR-IR, and Calorimetry

**Sebastien Cap**, Institute for Chemical and Bioengineering, Swiss Federal Institute of Technology (ETH) Zrich, Zurich, Switzerland, Levente L. Simon, Institute for Chemical and Bioengineering, Safety and Environmental Technology Group, ETH Zurich, Zurich, Switzerland and Konrad Hungerbuehler, Institute for Chemical and Bioengineering, ETH Zurich, 8093 Zurich, Switzerland

Modeling efficiently the kinetic of dissolving particulate systems, is a key step in several fields, such as for the controlled released of any active pharmaceutical ingredient. Since a decade the Focused Beam Reflectance Measurement (FBRM) technique became the most common method for the direct in-situ study of particulate systems. The FBRM modeling has been addressed by several authors [1]. A majority of these models aim to recover the Particle Size Distribution (PSD) from the Chord Length Distribution (CLD), i.e., the FBRM raw signal. The matrix problem formulation and its inversion has shown to be ill-posed, and several regularization techniques have been proposed. These techniques are often numerically complex and not suitable to be nested in within a calibration free kinetic analysis framework.

In this contribution, a solvent free reactive dissolution is performed in a custom developed dissolution calorimeter and simultaneously monitored by in-situ ATR-IR, calorimetry and FBRM. The kinetic modeling is based on the population balance framework and combined with a calibration free use of the spectroscopic and calorimetric data. Furthermore, we will discuss two quantitative chord length distribution (CLD) to particle size distribution (PSD) transformation options suitable to be nested within the calibration free dissolution modeling strategy.

We acknowledge financial support by the SNF (grant no 200021-113473).

[1] Kail, N., et al. Chem. Eng. Sci., 2009, 69, 984-1000.

[2] Maeder M. and Neuhold Y.-M., Practical Data Analysis in Chemistry. Elsevier, Amsterdam 2007

# (550a) Design for Six Sigma (DFSS) Approach to Copolymer Process Development

**Mohan Khadilkar**, SABIC Innovative Plastics, Mt. Vernon, IN and William Hollar, Sabic Innovative Plastics, Mount Vernon, IN

The SABIC Innovative Plastics LEXAN* polycarbonate business has introduced several new copolymer platforms into existing resin manufacturing assets in the past decade. These new copolymers expand the application space of traditional polycarbonates and allow business growth into new applications. Commercialization of new copolymers involves developing processes that address challenges due to copolymer chemistry, new product requirements, and fit with existing assets and investment constraints. Having a systematic and proactive methodology that provides "designed in" process capability and product quality is critical to successful process and product scale-up. The Design for Six Sigma (DFSS) methodology provides such a framework and set of tools to ensure success by design. This paper will survey critical tools for managing scale-up and product performance requirements. An example of development and scale-up of one of these copolymers will be reviewed.

* LEXAN is a registered trademark of SABIC Innovative Plastics LLC

See more of this Session: Process and Properties In Polymers
See more of this Group/Topical: Process Development Division

# (550b) Thermal Degradation of Nylon 66: Modeling and Parameter Estimation

**Kim B. McAuley**, Hadis Karimi and Mark A. Schaffer, Chemical Engineering, Queen's University, Kingston, ON, Canada

A fundamental dynamic model is developed to predict the influence of reactor operating conditions on the rates of undesirable reactions that can occur in the low-moisture, high-temperature finishing stage of commercial nylon 66 production. Kinetic parameters and initial values for unknown concentrations are estimated using experimental data. The proposed model is an advance over previous models because it accounts for the effect of melt-phase water concentration on thermal degradation. Estimability analysis and a mean-square error (MSE) criterion are used to determine which parameters can and should be estimated using the available data, and which parameters should remain at their nominal values. Parameters are ranked from most estimable to least estimable based on their influence on the model predictions, correlation with other parameters and uncertainty in initial guesses. Subsequently, the MSE criterion is used to determine that 48 of the 56 parameters should be estimated. The remaining 8 parameters are fixed at their initial guesses. The fitted model matches the data well, with typical errors of 1.6% for amine ends, 1.5% for carboxyl ends, 7.6% for gas evolutions rate of $CO_2$, 15.3% gas evolutions rate of $NH_3$ and 18.1% for gas evolutions rate of cyclopenatone. This model will help engineers to select temperature and moisture level setpoints that lead to high polymerization rates while ensuring good product quality.

See more of this Session: Process and Properties In Polymers
See more of this Group/Topical: Process Development Division

# FLOW ACCELERATES THE REACTION KINETICS OF MULTIPHASE POLYMER SYSTEMS

*Jie Song[1], Randy H. Ewoldt[2], Christopher W. Macosko[1*]*

*[1]Department of Chemical Engineering and Materials Science, University of Minnesota, Minneapolis, Minnesota, 55455*

*[2]The Institute for Mathematics and its Applications, University of Minnesota, Minneapolis, Minnesota, 55455*

Coupling reactions between functional group grafted macromolecules have been proven to be a powerful way to facilitate compatibilization and adhesion. Yet their reaction kinetics requires more fundamental understanding. The common wisdom for reactions between *small molecules* is that external mass transfer increases the reaction rate by increasing concentration of reactive species without affecting reaction mechanism and transition states at a certain temperature. However, for reactions between *macromolecules*, the kinetics is more complicated because of their relatively slow diffusion and complex configurations. It is generally difficult to study the effects of flow on macromolecules reactions during mixing because interfacial area generation is always convoluted and hard to isolate.

In this paper, a reactive bilayer system was created to explore this issue. In an effort to study the reaction between polyethylene (PE) and thermoplastic polyurethane (TPU), maleic anhydride (MA), hydroxyl (OH) and secondary amine (NHR) functionalized PEs were synthesized via reactive extrusion and blended into non-modified PE. We bonded these PEs to TPU via lamination and coextrusion. In order to compare reaction rates of the two processes we determined the interfacial copolymer density $\Sigma$ considering both advection and interfacial area generation. We found that the reaction rate in coextrusion was much faster in comparison with lamination at the same temperature. This difference was attributed to the extensional and compressive flow in coextrusion overcoming the diffusion barrier at the interface and forcing reactive species to penetrate into interface. The effects of functional group reactivity and processing variables on adhesion were correlated with interfacial copolymer coverage. Amine functionalized PE showed dramatic adhesion improvement even at 1 wt%. We also determined that flow accelerates the reaction between polyamide 6 and maleic anhydride high density polyethylene. New die design is investigated to isolate the effect of extensional and compressive flow.

# (550d) Continuum Lumping Modelling for Polymerisation Kinetics

Mustafa Adam[1], **Raffaella Ocone**[1] and Valeria Arrighi[2], (1)Chemical Engineering, Heriot-Watt University, Edinburgh, United Kingdom, (2)Chemistry, Heriot-Watt University, Edinburgh, United Kingdom

Polymerisation occurs when two "units", called monomers, join by either chain addition or step-growth process leading to a product with a higher molecular weight characterised by repeating chemical units. A model based on the continuum lumping methodology is proposed and implemented to predict the molecular weight distribution during polymerisation. The work is aimed at assessing the benefit of the continuum lumping approach to obtain information about some specific features of the reaction. The model is tested for two selected case studies furnishing the product distribution at various times. Since many properties of a polymer depend on the molecular weight distribution (MWD) and on the average of such distribution, the results will pave the way for future work assessing the predictive capability of the lumping methodology as applied to polymerisation. The aim is twofold: to develop a model which could predict the polymer MWD at various times, and to assess the capability of the continuum lumping methodology as applied to polymerisation. The work focuses on step growth polymerisation only, since we think that the underlying kinetic mechanism is amenable to be described through a continuum lumping procedure. We show that the polymerisation yields distribution can be obtained through the methodology. The model is based on the assumption that the MWD evolves in time as described by an integro-differential equation. A Skewed Gaussian distribution function for determining product yield distribution of the polymerisation reaction is introduced. The model is based on the assumption that the kinetics is liner and first order, and the reaction is irreversible. The model captures successfully the effect of the reaction time and reaction rate constant on the MWD.

See more of this Session: Process and Properties In Polymers
See more of this Group/Topical: Process Development Division

# (550e) Stress Cracking of Poly (ethylene terephthalate) Beverage Bottles: Mechanism, Structure/ Property Relationships, and Methods of Prevention

**Eric Morrison**[1], Megan W. Malvey[1], Richard D. Johnson[1], Jeffrey S. Hutchison[1], Jessica L. Anacker[1] and Keith A. Brown[2], (1)Ecolab, Eagan, MN, (2)Plastic Technologies, Inc, Holland, OH

Starting in 1978 with the introduction of one piece poly(ethylene terephthalate) (PET) soft drink bottles with molded-in feet, there has been the problem of environmental stress cracking in bottles used for packaging carbonated beverages. Environmental stress cracking is delayed failure in which a large crack forms in the amorphous base portion of the bottle through which all of the liquid contents are lost. Using a full package test method first described in 2008, the mechanism of stress cracking was determined to be alkaline hydrolytic cleavage of ester bonds in the PET backbone and not a Volynskii crazing mechanism as was previously thought. Hydrolysis leading to failure does not happen in the case that naturally occurring water alkalinity is precipitated as carbonate minerals during evaporation. For this reason, eliminating contact of filled PET bottles with softened water has proven to be a highly effective means of stress crack prevention. Physical aging greatly increases amorphous PET susceptibility to hydrolytic stress crack failure and minimizing exposure of empty bottles to environmental conditions which promote physical aging provides another important strategy for the elimination of stress crack failure.

See more of this Session: Process and Properties In Polymers
See more of this Group/Topical: Process Development Division

# (550f) A Polymer Isolation Process for Powders with Designed Properties

**William Hollar**, Sabic Innovative Plastics, Mount Vernon, IN and Viswanathan Kalyanaraman, SABIC Innovative Plastics, Mount Vernon, IN

Most polymer process development programs for new products focus on molecular structure to achieve desired product properties. This relationship tends to lead to a reaction focus for polymer process development. There are cases where the final properties of the polymer powder also determine its functionality in a final product. This is the case for polymer additives used for advanced composites. A new process for producing fine, porous engineering thermoplastic powders has been developed for a range of these applications. The powder formation process determines the functionality and desirability of the powder and leads to powders with unique properties.

This paper will focus on some of the key chemical engineering aspects of the process conditions that can be used to tailor properties such as porosity and particle size distribution. Such changes can enable or enhance product performance. Process alternatives for powder production will be presented and contrasted. A two-fluid atomization process and its applicability to tailored polymer powders will be described. A review of applications and desired product characteristics will be discussed.

See more of this Session: Process and Properties In Polymers
See more of this Group/Topical: Process Development Division

## From Multi-Scale Methods to Virtual Process Engineering --- Efficient and Accurate Simulation of Multi-Phase Complex Systems

**Ge Wei**, State Key Laboratory of Multi-Phase Complex Systems, Institute of Processing Engineering, Chinese Academy of sciences, Beijing, China

Dynamic simulation of the equipments and reactors in process engineering typically requires very long time and enormous computational cost. The major challenge lies in the multi-scale nature of the structures and behaviors of these systems. The structural consistency between the system simulated, the algorithm employed and the computing hardware used may provide an effective way to speedup the computation and to improve the accuracy, which can be well implemented in a multi-scale discrete computing mode with stability constraints. This presentation will give an overview of the work at our Institute in this direction, focusing on its development and extension in a variety of complex multi-phase systems, and establishment of the corresponding supercomputing software and hardware. The possibility of achieving real-time simulation, that is, virtual process engineering in this computing mode is prospected finally.

# (589b) Point of Use Recyclable Bioreactor System for Bio-Agricultural Applications

**Matthew A. Robinson** and Henry Y. Wang, Department of Chemical Engineering, University of Michigan, Ann Arbor, MI

Farmers have long been respected for their self-sufficiency, acting as primary producers for society. Yet as time progresses, productivity demands drive farmers to require increasing levels of technology in terms of seed strains, fertilizers and pesticides in order to remain competitive. In turn farmers have grown dependent upon manufactured products with rising price tags. Current market prices leave biologically derived fermentation based active ingredients and cultures, like reduced risk certified natural insecticides, uncompetitive against traditional synthetic insecticides placing them out of reach of small farmers, particularly in the developing world. In order to reduce the basic manufacturing cost, we have designed a point of use recyclable bioreactor system capable of fermenting off patent compounds for direct field application eliminating the non-value added concentration step. With recyclable single use inserts, the system can produce an array of bioproducts including nitrogen fixing cultures, phosphate solubilizing cultures, and various natural pesticides without the need for extensive sterilization. The system empowers farmers to do what they do best: cultivate, returning the power of self-sufficiency to the hands of the farmer.

See more of this Session: High-Throughput and Compact Chemical Processing Technologies
See more of this Group/Topical: International Congress on Energy 2011

# (589c) The Study of Ultrasonic Irradiation Facilitating Fat Hydrolysis Reaction

**Guangming Du**, College of Chemical Engineering, Xinjiang Agricultural University, Urumqi, China

See more of this Session: High-Throughput and Compact Chemical Processing Technologies

See more of this Group/Topical: International Congress on Energy 2011

# (589d) Autothermal Hydrogen Production From Isobutanol for Fuel Cells

**Reetam Chakrabarti**, Jacob S. Kruger and Lanny Schmidt, Department of Chemical Engineering and Materials Science, University of Minnesota, Minneapolis, MN

We demonstrate a process which converts isobutanol, a fermentation product of lignocellulosic biomass, to a high selectivity hydrogen stream for fuel cells by autothermal reforming. We use staged reactor configuration, with a Rh-Ce/$Al_2O_3$ catalyst used to produce a high selectivity syngas stream by partial oxidation and reforming, and an additional Pt-Ce/$Al_2O_3$ catalyst to consume CO by the water gas shift reaction (WGS). The Rh-Ce/$Al_2O_3$ catalyst showed high conversion of isobutanol resulting in a product stream consisting mostly of CO, $CO_2$, $H_2$ and $H_2O$. These products react further on the Pt-Ce/$Al_2O_3$ catalyst. The effect of steam addition to further lower the concentration of CO by promoting WGS has also been investigated. Thermodynamic calculations were performed to optimize the feed compositions and the amount of steam added to obtain a product stream containing minimal amounts of CO for use in a fuel cell.

See more of this Session: High-Throughput and Compact Chemical Processing Technologies
See more of this Group/Topical: International Congress on Energy 2011

# (589e) Sulfuric Acid Catalyzed Isobutane-Butene Alkylation Promoted with Ionic Liquids or Trifluoroethanol

Hailing Ren, Guoying Zhao and **Suojiang Zhang**, Institute of Process Engineering, Chinese Academy of Sciences, Beijing, China

**Sulfuric acidcatalyzed isobutane-butene alkylation promoted with ionic liquids ortrifluoroethanol**

**HailingRen[ab], Guoying Zhao[a*], Peng Cui[ab],Jun Huang[b],SuojiangZhang[a*]**

[a]Institute of ProcessEngineering, Chinese Academy of Sciences, P. O. Box 353, Beijing, 100190,China

Fax : +86 10 82627080

[b]State Key Laboratory ofMaterials-Oriented Chemical Engineering, College of Chemistry and ChemicalEngineering, Nanjing University of Technology, Nanjing, 210009, China

Fax: +86 2583172263

**Abstract**

Smallamounts of ionic liquids or trifluoroethanol(TFEA), have been studied as promotersfor sulfuric acid catalyzed alkylation of isobutane/butene. Long-term studiesup to 40+ recycles indicated that the working life of $H_2SO_4$with suitable ionic liquid-based promoter was prolonged at least three timesthat of sulfuric acid. New catalytic active species formed in-situ by the abovepromoter and $H_2SO_4$ was detected by NMR and ESI-MS analytictechnology. In addition, the promoters are believed to lower the interfacial tensionof hydrocarbon feeds/sulfuric acids dispersion and thus improve the quality ofalkylate.

See more of this Session: High-Throughput and Compact Chemical Processing

# (589f) Comparison of Different Process Routes From Coal with Energy and Exergy Analysis

Yuehua Yao, Xiangping Zhang and **Suojiang Zhang**, Institute of Process Engineering, Chinese Academy of Sciences, Beijing, China

**Comparison of different process routesfrom coal with energy and exergy analysis**

Y. H. Yao[a,b], Y.J. Chen [a,b], X. P. Zhang[a,*], S. J. Zhang [a], B. Z. Ren[b] , J. H. Wang[b]

[a] *Institute of ProcessEngineering, Chinese Academy of Sciences, 100190, Beijing, China*

[b] *School of ChemicalEngineering and Energy, Zhengzhou University, 450001, Zhengzhou, China*

***Corresponding author:** *xpzhang@mail.ipe.ac.cn* **(X. P. Zhang)**

Chinahas plenty coal resource but there are many ways to utilize coal to productfuels and natural gas, et al. Facing the problem of energy shortage in theworld and the target of cutting CO<sub>2</sub> emission, thequestion of sifting the best one from existing process routes of coal utilizationshould be solved at first, it matters whether coal sector of China can satisfythe demand of low carbon economy and that the development form of Chinese coalresource in the future.

Inthis study, considering the existing process routes of coal, models of thedirect coal liquefaction process, the direct coal combustion process, thecoking process and coal gasification process for F-T synthesis oil and naturalgas production were developed. Based on simulation results, combining with theindustrial data, both the conservation of energy and the quality of energy forthe processes were investigated. Exergy loss, CO<sub>2</sub>emissions and yield of the four process routes were calculated and compared.For the four systems, the carbon and hydrogen flows were analyzed, and the mainreasons for carbon loss and the hydrogen atom utilization were discussed. Theresults can provide a roadmap for the future coal utilization with low globalwarming effect and high energy utilization.

**Acknowledgement**

This research was supported by the NationalBasic Research Program of China (973 Program, 2009CB219900) and State Science and Technology SupportProgram (2009BAC64B00)•.

**References**

Maria Sudiro,Alberto Bertucco. (2009). Production of synthetic gasoline and diesel fuel byalternative processes using natural gas and coal: process simulation and optimization.*Energy*, 34, 2206-2214.

Zhang, X.P.,Christian Solli, Edgar G. Hertwich, Tian, T. (2009). Exergy analysis of theprocess for dimethyl ether production through biomass steam gasification. *Ind.Eng.Chem.Res.* 48,10976-10985.

See more of this Session: High-Throughput and Compact Chemical Processing Technologies
See more of this Group/Topical: International Congress on Energy 2011

# (611a) Activated Carbon Process for Simultaneous SOx/NOx Removal From Concentrated CO2 At Elevated Pressures

**Monica Zanfir**, Minish Shah and Rahul Solunke, R&D H2 & Energy, Praxair, Inc., Tonawanda, NY

**Background**

Oxy-combustion is considered as one of the $CO_2$ capture options for the coal-fired power plants. The $CO_2$-rich flue gas from such a plant will have to be compressed and purified for sequestrtion. Praxair is developing a near zero emissions technology that will produce $CO_2$ virtually free of trace impurities. One key element of this technology is the removal of $SO_x$ and $NO_x$ from the flue gas as a result of the combustion of sulfur and nitrogen compounds contained in coal. The technology developed uses activated carbon to catalytically convert the $SO_x$ and $NO_x$ to their superior oxides $SO_3$ and $NO_2$ which are further adsorbed on the activated carbon material and removed from the gas phase. This process takes advantage of compression requirements of the $CO_2$ purification process and is conducted at elevated pressures. High pressure enhances the oxidation and adsorption process. Once the activated carbon became saturated with the contaminants it can be easily regenerated by simply washing it with water. During this stage the adsorbed oxides react with water to form the corresponding acids $H_2SO_4$, $H_2SO_3$, $HNO_3$ and $HNO_2$. In this manner the $SO_x$ and $NO_x$ contaminants are captured in the liquid phase. The waste water can be minimized by the recirculation of the wasted acid solution. The simplicity and good performance of this technology make it attractive and easy to implement.

**Experimental Testing and Scale-Up**

Comprehensive experimental investigation was conducted in a single bed batch unit containing activated carbon. Tests with synthetic flue gas with $SO_x$ only,

$NO_x$ only and mixtures of $SO_x$ and $NO_x$ assessed the process performance for a wide range of operating conditions. The effect of concentration level, ratio of $SO_x$/ $NO_x$, water concentration, temperature and pressure was investigated. For this purpose break-through tests for several adsorption-regeneration cycles were conducted. Process performance in terms of material capacity to adsorb a certain amount of contaminants, and the efficiency of each contaminant removal were determined.

The findings obtained from bench-unit experimental investigation created the bases for the first stage of scale-up. A dual bed continuous unit for the purification of 0.125 TPD $CO_2$ has been designed and it is being built. The continuous unit consists of two fixed activated carbon beds operating alternatively in adsorption and regeneration mode in order to ensure continuous purification of a synthetic flue gas stream. The main objectives of the continuous unit test program are to assess material longevity and long term process performance. At the same time it has been sought to optimize cycle steps duration.

**Results and Conclusions**

The single bed bench unit tests with synthetic flue gas at elevated pressures and ambient temperatures showed that activated carbon is able to remove $SO_x$ and $NO_x$ when fed individually or together. Investigation of various operating conditions showed that the process has better performance at ambient temperature (~20 $^oC$) compared to temperatures above ambient. The presence of moisture had a beneficial effect on the retention of $SO_x$. Higher operating pressure (220 psig vs. 50 psig) significantly improved the process performance, especially for $NO_x$ removal. The process achieved simultaneous removal of SOx and NOx from flue gas with the removal efficiency of > 99 % for $SO_x$ and > 96 % for $NO_x$.

The present paper will discuss in detail the influence of feed composition and operating conditions on $SO_x$ /$NO_x$ removal using activated carbon as adsorbent. Further, the results from the continuous unit and recommendations for the scale-up of the technology will be presented.

See more of this Session: Simultaneous Energy and Emission Reduction for

# (611b) Simultaneously Increasing Profit and Reducing Emissions Through Process Optimization for Integrated Petrochemical Plants

**Jian Zhang** and Qiang Xu, Dan F. Smith Department of Chemical Engineering, Lamar University, Beaumont, TX

Refineries and petrochemical plants are leading segment of the entire chemical process industry. It was reported that the value of the product shipments by refineries in U.S. accounted for 9.32% of the entire manufacturing sector of the U. S. economy in 2007. On the other hand, air emissions from refineries and petrochemical plants are significant. The carbon dioxide ($CO_2$) emissions from refineries account for about 4% of the global $CO_2$ emissions, close to 1 billion tons of $CO_2$ per year. Due to an increasing concern, specific policies for refinery emission control have been enacted, and these emission control polices will heavily affect U.S. petroleum refineries.

When an emission reduction strategy is implemented in a petrochemical plant, it most likely causes more conservative design and operations, and requires higher capital investment and manufacturing costs. However, emissions in a petrochemical plant generally mean the loss of raw materials and energy which could supposedly generate products. Thus, emission reduction may possibly bring opportunities for the plant profit increment. In this paper, a process optimization model and a methodology were developed, which enabled profitable emission reduction (PER).

The model includes major facilities of refinery and petrochemical plants. For each facility model, certain simplifications are made to reduce the model complexity, and meanwhile, keep the model accuracy. Facility models include material balance equations, property correlations, capacity constraints, and utility consumption correlations. Facility models are tuned and validated with actual operating data.

The model is a multi-objective mixed-integer nonlinear programming (MINLP) model. The integer variables and constraints enable the model to optimize both process flowsheet and operating parameters. The objective functions include profit, energy consumption, carbon dioxide ($CO_2$) emission, volatile organic compounds (VOC), and nitrogen oxides ($NO_X$). Pareto frontier method is employed to show relations of these objective functions. Optimization results give PER (profitable emissions reduction) region for various production schemes. When production schemes are in the PER region, the optimized flowsheet can reduce the $CO_2$, VOC and $NO_X$ emissions by 10%, as well as increase operating profit by 4%.

See more of this Session: Simultaneous Energy and Emission Reduction for Chemical Industries
See more of this Group/Topical: Process Development Division

# *Reducing Spray Dryer Bend Choking While Increasing Feed Rate*

Bedanta Krishna Bhuyan
Birla Institute of Technology & Science, Pilani
Rajasthan, India

## *Introduction*

Spray drying is a method of producing a dry powder from a liquid or slurry by rapidly drying with a hot gas. This is the preferred method of drying of many thermally-sensitive materials such as foods and pharmaceuticals. Air is the heated drying media; however, if the liquid is a flammable solvent such as ethanol or the product is oxygen-sensitive then nitrogen is used. All spray dryers use some type of atomizer or spray nozzle to disperse the liquid or slurry into a controlled drop size spray. The most common of these are rotary nozzles and single-fluid pressure swirl nozzles. Alternatively, for some applications two-fluid or ultrasonic nozzle are used. Depending on the process needs drop sizes from 10 to 500 micrometres can be achieved with the appropriate choices. The most common applications are in the 100 to 200 micrometre diameter range. The dry powder is often free-flowing.
The hot drying gas into the spray dryer can be passed as a co-current or counter-current flow to the atomizer direction. The co-current flow enables the particles to have a lower residence time within the system and the particle separator (typically a cyclone device) operates more efficiently. The counter-current flow method enables a greater residence time of the particles in the chamber and usually is paired with a fluidized bed system.

The spray dryer is the device used in spray drying to get the fine particles in Aditya Birla Chemicals Thailand Ltd (ABCTL). It takes a liquid stream and separates the solute or suspension as a solid and the solvent into a vapor. The solid is usually collected in a cyclone. The liquid input stream is sprayed through a rotary disk atomizer into a hot vapor stream and vaporized. Solids form as moisture quickly leaves the droplets. The atomizer is usually used to make the droplets as small as possible to get the desired particle size, maximizing heat transfer and the rate of water vaporization. Spray dryers can dry a product very quickly compared to other methods of drying. They also turn a liquid solution or slurry into a dried powder in a single step, which can be advantageous for profit maximization and process simplification. The spray dryers are designed efficiently to work at the specified working parameters. But the increased demand of products led to doubling the feed rate into the spray dryer. This gave rise to the problem of bend choking at the spray dryer outlet, in which the bend gets clogged/choked with solid particles. Bend choking is a serious problem in the spray drying process since it leads to high pressure build up inside the spray dryer chamber and affects the drying process inside it. Due to the choking of the bend, these deposited solids at the bend have to be manually removed and cleaned, because of which there is a huge loss of production time. Manual cleaning up also leads to loss of products.

# *Problem Analysis*

Spray drying produces solids in the form of fine particles which are free flowing. Due to the co-current drying process used in the spray dryer at ABCTL, the transportation of these fines and their subsequent separation in the cyclone is more efficient. Hence, these fine particles are not responsible to choke the bend. A closer look at the choking in the bend showed the presence of larger masses of solids. These masses of solid or lumps are not free flowing and hence fall to the lowermost place in the spray dryer chamber, the bend. Deposition of such lumps of solids usually entraps the fine particles on their way out through the bend and subsequently chokes up the whole bend giving rise to the problem of bend choking.

Earlier attempts were made at ABCTL to disperse the deposited solids using an air blast cylinder. A blast of air was released at certain intervals when the bend was reported to choke. But the sudden blast of air at the bend was shown to be responsible for reverse flow of these solid particles, which in turn entrapped more fine particles and enhanced the choking process.

The above attempt makes it evident that to reduce the bend choking, the lumps from the bend have to be reduced. Hence, reducing the presence of lumps at the bend is a viable solution to the problem of bend choking. These lumps are formed in the spray dryer due to insufficient drying of the atomizer droplets before they reach the spray dryer wall. So a thorough study of the process parameters inside the spray dryer was done to get a depth into the problem of bend choking at the spray dryer outlet. It includes temperature of hot air supply, atomizer RPM, particle residence time, additives in feed, hot air supplies, particle drying, wall temperature, guide plates and feed rate. These are discussed below in detail.

The following table shows the Mono Sodium Phosphate (MSP) condition of main plant from March to July 2010.

Table.1. Main Plant MSP conditions from March to July

| *date* | *Grade* | *RM* | *Liqor Na/P* | *Liqor pH* | *Liqor SG* | *Feed rate $M^3/h$* | *S/D inlet temp °C* | *S/D outlet temp °C* | *Atomizer frequency Hz* | *Cyclone diff pressure Mm WC* | *Remarks* |
|---|---|---|---|---|---|---|---|---|---|---|---|
| *6-7/3/2010* | *MSP feed* | *DSP+S/A+blend* | *1.005-1.008* | *4.58-4.66* | *1.58-1.62* | *4.8-7* | *345-360* | *139-142* | *BL 50* | *80-90* | *Smooth run* |
| *23/3/2010* | *MSP feed* | *DSP+blend* | *1.003-1.008* | *4.57-4.71* | *1.56-1.61* | *4.5-5.8* | *330-338* | *145-152* | *BL 50* | *70-78* | *Smooth run* |
| *27/3/2010* | *MSP HM* | *S/A* | *1.005-1.007* | *4.62-4.68* | *1.51-1.59* | *6-6.2* | *371-378* | *151-153* | *BL 50* | *70-75* | *Smooth run* |
| *2/4/2010* | *MSP feed* | *DSP+blend* | *1.004-1.008* | *4.58-4.72* | *1.6-1.65* | *4.5-5.2* | *330-350* | *147-152* | *BL 50* | *77-88* | *Atomizer vibration high* |
| *3/4/2010* | *MSP feed* | *DSP+blend* | *1.004-1.008* | *4.58-4.72* | *1.6-1.65* | *4.5-5.2* | *330-350* | *147-152* | *BL 45* | *76-88* | *Atomizer vibration high* |
| *4/4/2010* | *MSP feed* | *DSP+blend* | *1.004-1.008* | *4.58-4.72* | *1.6-1.65* | *4.5-5.2* | *330-350* | *147-152* | *BL 45* | *76-88* | *Atomizer vibration high* |
| *10/4/2010* | *MSP HM* | *DSP+blend* | *1.006-1.009* | *4.66-4.69* | *1.57-1.61* | *5.5-6.8* | *340-345* | *140-145* | *BL46* | *76-88* | *Smooth run* |
| *17/4/2010* | *MSP feed* | *DSP+S/A+blend* | *1.005-1.008* | *4.61-4.63* | *1.6-1.64* | *4.5-5* | *315-320* | *150-160* | *BL46* | *79-82* | *Atomizer vibration high* |
| *5/5/2010* | *MSP feed* | *DSP+S/A* | *1.004-1.007* | *4.57-4.69* | *1.56-1.62* | *4.5-5.5* | *315-330* | *149-153* | *BL46* | *76-87* | *Atomizer serviced* |
| *13/5/2010* | *MSP feed* | *DSP+S/A+blend* | *1.002-1.008* | *4.51-4.7* | *1.56-1.64* | *5-5.7* | *330-350* | *148-151* | *BL 46>50* | *76-88* | *Atomizer vibration high* |
| *14/5/2010* | *MSP feed* | *DSP+S/A+blend* | *1.002-1.008* | *4.51-4.7* | *1.56-1.64* | *5-5.7* | *330-350* | *148-151* | *BL 50* | *76-88* | *Atomizer vibration high/Bendchoke* |
| *15/5/2010* | *MSP feed* | *DSP+S/A+blend* | *1.002-1.008* | *4.51-4.7* | *1.56-1.64* | *5-5.7* | *330-350* | *148-151* | *BL 50* | *76-88* | *Atomizer vibration high/Bendchoke* |
| *16/5/2010* | *MSP feed* | *DSP+S/A+blend* | *1.002-1.008* | *4.51-4.7* | *1.56-1.64* | *5-5.7* | *330-350* | *148-151* | *BL 50* | *76-88* | *Atomizer vibration high/Bendchoke* |
| *21/5/2010* | *MSP* | *C/S* | *1.003-* | *4.55-* | *1.69-* | *5-6* | *320-335* | *140-147* | *BL 50>56* | *86-92* | *Atomizer vibration* |

| | | | | | | | | | | | |
|---|---|---|---|---|---|---|---|---|---|---|---|
| | SAPP | | 1.007 | 4.62 | 1.64 | | | | | | high/Bendchoke |
| 22/5/2010 | MSP feed | DSP | 1.003-1.008 | 4.53-4.7 | 1.6-1.65 | 5-5.5 | 320-335 | 148-154 | BL 50 | 70-92 | Atomizer vibration high/Bendchoke |
| 23/5/2010 | MSP feed | DSP | 1.003-1.008 | 4.53-4.7 | 1.6-1.65 | 5-5.5 | 320-335 | 148-154 | BL 50 | 70-92 | Atomizer vibration high |
| 24/5/2010 | MSP feed MSP HM | DSP | 1.003-1.008 | 4.53-4.7 | 1.6-1.65 | 5-5.5 | 320-335 | 148-154 | BL 50 | 70-92 | Atomizer vibration high |
| 31/5/2010 | MSP feed | DSP+S/A+blend | 1.002-1.007 | 4.5-4.68 | 1.56-1.6 | 4.5-6.2 | 315-330 | 146-150 | BL 50 | 76-81 | Atomizer vibration high |
| 1/6/2010 | MSP feed | DSP+S/A+blend | 1.002-1.007 | 4.5-4.68 | 1.56-1.6 | 4.5-6.2 | 315-330 | 146-150 | BL 50>48 | 76-81 | Atomizer vibration high |
| 2/6/2010 | MSP feed | DSP+S/A+blend | 1.002-1.007 | 4.5-4.68 | 1.56-1.6 | 4.5-6.2 | 315-330 | 146-150 | BL 48 | 76-81 | Atomizer vibration high |
| 10/6/2010 | MSP HM | DSP+S/A+blend | 1.003-1.005 | 4.55-4.7 | 1.59-1.63 | 5.8 | 340-352 | 145-152 | BL 50 | 67-77 | Atomizer vibration high |
| 11/6/2010 | MSP HM | DSP+S/A+blend | 1.003-1.005 | 4.55-4.7 | 1.59-1.63 | 6.0-6.5 | 340-352 | 145-152 | BL 50 | 67-77 | Atomizer vibration high |
| 18/6/2010 | MSP HM | S/A | 1.005-1.008 | 4.62-4.7 | 1.58-1.62 | 6.0-6.5 | 360-370 | 144-150 | BL 50>48>46>50 | 76-84 | Atomizer vibration high |
| 19/6/2010 | MSP HM | S/A | 1.005-1.008 | 4.62-4.7 | 1.58-1.62 | 6 | 360-370 | 144-150 | BL 50>AH40 | 76-84 | Atomizer vibration high/Bendchoke |
| 23/6/2010 | MSP HM | S/A | 1.005-1.008 | 4.6-4.7 | 1.57-1.62 | 6 | 360-370 | 145-152 | AH 40>46 | 84-100 | - |
| 3/7/2010 | MSP feed, MSP HM | DSP+blend | 1.003-1.007 | 4.54-4.72 | 1.57-1.62 | 5-6 | 350-362 | 147-151 | BL 50 | 69-77 | Atomizer vibration high |
| 4/7/2010 | MSP feed, MSP HM | DSP+blend | 1.003-1.007 | 4.54-4.72 | 1.57-1.62 | 5-6 | 350-362 | 147-151 | BL 50 | 69-77 | Smooth run |
| 12/7/2010 | MSP feed, MSP HM | DSP+ S/A | 1.003-1.005 | 4.56-4.72 | 1.58-1.63 | 5-5.6 | 330-340 | 144-151 | BL 50>52 | 80-85 | Atomizer vibration high |
| 13/7/2010 | MSP feed, MSP HM | DSP+ S/A | 1.003-1.005 | 4.56-4.72 | 1.58-1.63 | 5-5.6 | 330-340 | 144-151 | BL 52 | 80-85 | Smooth run |

### Hot Air Temperature

The hot air enters the spray dryer at a temperature of about 500 °C. This is a mixture of combustion air and dilution air. The trend of the inlet hot air temperature with the cyclone pressure difference was studied, based on the above data and the plot is represented below.

Legend: Temperature-Blue; Cyclone pressure difference-red; feed rate-green

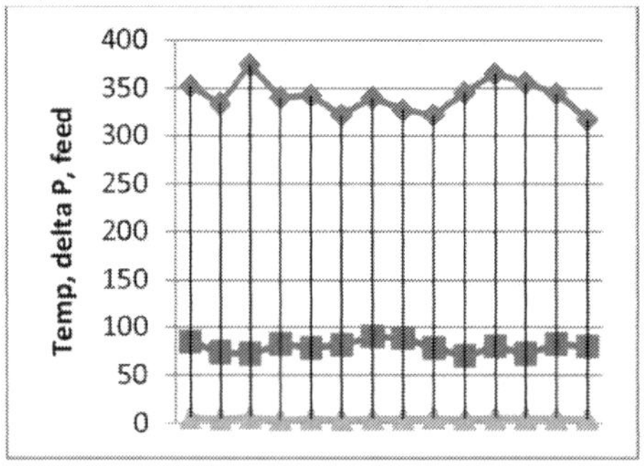


Fig 1. Plot of temperature & delta-P at different time intervals.

Based on the above plot, the best fit correlation was obtained as: press diff = -0.1285 T +123.7528. Hence it is to be noted that the evaporation rate will increase with the increase in inlet hot air temperature. This is because the temperature difference between feed and the hot drying air is the driving force for drying. But, in the existing case, temperature of hot air has to be limited to the current operating temperature for maintaining product quality since excessive high temperature can cause damage to the product particles.

***Atomizer RPM***

Atomizer RPM is an important parameter that controls the size of the droplets released into the drying environment. Increased atomizer RPM gives more fine droplets and hence increases the surface area for drying. Hence an increase in atomizer RPM is expected to increase the drying process of the inlet liquor feed. But in the existing case, the atomizer is already running at its maximum RPM according to its design specifications. So, in order to improve the drying process inside the spray dryer, a new atomizer with higher RPM can be installed and is expected to reduce the choking occurring at the bend of the spray dryer.

***Additives in feed***

The presence of blended phosphate in the feed was shown to cause high atomizer vibration alarm and bend choking. Taking note of the problem associated with blended as an additive, presently it has been removed completely from the feed and atomizer vibration as well as bend choking was seen to be reduced thereafter.

***Wall Temperature***

The inner wall of the spray dryer chamber is normally heated by the hot air that is striking against the wall. But if a wet particle strikes the wall with more than 10% moisture, then it sticks to the wall. Due to lack of heat supply from the wall, the particles aggregate with time to form lumps on the wall. Externally heating the wall may dry up the particles as they strike the wall, and help reduce lump formation. But, due to the presence of thick insulation on the spray dryer wall, it is difficult to provide external heating.

***Feed Rate***

For optimum working of the spray dryer, it must work at the specified design conditions of feed. The increase in feed rate was shown to have a greater impact towards bend choking. But due to demand for increased production rate, the feed rate was doubled. This has resulted in bend choking. The present objective is to reduce the bend choking at the current feed rate specified.

***Guide Plates***

The guide plates are present in the atomizer, just above the point of release of the feed droplets. The purpose of the guide plates is to guide the rotation of the drying air. The guide plates should normally form an angle with the vertical of $27^{\circ}$. When this cone angle is increased, the air is given a faster rotation and contrary when the angle is reduced, the air will rotate less. The angle should be no less than $25^{\circ}$ and no more than $30^{\circ}$. With greater angle than $30^{\circ}$ a tendency to form deposits on the ceiling may occur and with angles less than $25^{\circ}$, it may cause deposits to form on the cylinder and cone.

The guide plates of the atomizer are already aligned at $27^{\circ}$ with the vertical, which is the optimized angle to reduce deposits on the spray dryer walls.

***Particle residence time & drying***

The more time the feed droplet particles come in contact with the drying air, the more is the drying achieved in the spray dryer. Normally in a spray dryer with a rotary atomizer, the particle dries in two stages. In the first stage the droplets gets atomized and are thrown out of the atomizer as small droplets, hence giving rise to increased surface area for drying. After this in the second stage, the droplets come in contact with the hot air and the temperature difference drives the droplets to drying. But due to increased feed, not all droplets dry completely to give the fine particles. These particles often move out of the influence of the hot air flow and translate radial towards the spray dryer wall. During this period, there is no more medium to enhance the drying of these moisture laden particles. Hence particles with more than 10% moisture, when they strike the wall, stick to it and begin to form lumps due to subsequent deposition. These lumps when they grow bigger in size, falls down due to their weight and gets deposited at the bend of the spray dryer outlet and leads to choking of the bend, as explained earlier.

Hence reducing the moisture content of the particles before they strike the wall of the spray dryer will help to reduce the formation of lumps on the wall. This suggests that improving the drying process inside the spray dryer will help in reducing bend choking.

## Strategic Analysis

All the related process parameters in the spray drying process were studied in detail to sort out the problem of the choking at the bend. The above analysis exemplifies the following key strategies to the given problem:

1. Increase the residence time of the droplet particles, before it reaches the wall or prevent it from contacting with the wall.
2. Improve drying rate by allowing for the particles to dry even after it has gone out of the influence of existing hot air.

The above mentioned targets can be achieved by supplying a stream of hot air along the wall of the spray dryer. This stream of hot air will be fundamental in contacting with the particles near the spray dryer wall, drying them and at the same time increase their residence time by preventing them from reaching the wall. A similar stream of air is provided in an air-blast cylinder along its inner walls and it is called the co-flow air stream. But in the air-blast cylinder, the motive of this co-flow air to reduce the reverse flow near the walls, due to the sudden drop in pressure. The motive in the present case of the spray dryer is to supply a similar stream of co-flow air along the walls of the chamber with the motive to improve the drying and the residence time of the particles inside the spray dryer chamber.

## *Technical Feasibility*

Additional air supply into the spray dryer along its walls was supposed to improve the drying inside the spray dryer chamber. But the constraints in the addition of a new stream of air have to be studied for making the proper design and probable implementation. In this case, the spray dryer in the main plant of ABCTL, the exhaust is limited and under presently working at full capacity. So, this puts a constraint that additional volume of air cannot be used for the above mentioned proposal to solve the problem. But, our necessity in the above mentioned strategy is of a stream of co-flow air and not compulsorily additional air. The addition of co-flow air into the spray dryer without using additional air can be achieved in the following ways:

1. Diverting all existing drying air as co-flow air into the spray dryer.
2. Reducing the flow of existing drying air into the spray dryer and adding a stream of co-flow air from ambient air.
3. Dividing the stream of existing drying air and diverting one stream as co-flow air.

The feasibility of the above three proposed ways of adding co-flow air are now discussed in detail.

***Diverting all existing drying air as co-flow air into the spray dryer***

Diverting all the existing drying air as co-flow air will not increase the stages of drying. The purpose of the co-flow air stream is to provide an additional stage of particle drying after it has exited from the influence of the drying air near the atomizer region. Hence, this is not a feasible option to solve our purpose.

***Reducing drying air and adding co-flow air from ambient air.***

For an easy implementation of supply of co-flow air into the spray dryer chamber, the stream of co-flow air can be supplied from ambient air with a portion of existing drying air cut off. Although in this case, installation will be simpler but adding ambient air into the spray dryer chamber is shown to have negative impact on the drying process by bringing down the temperature inside the spray dryer chamber. These were proved on the basis of CFD simulations using the ANSYS Fluent CFD code. The results of this simulation are presented in the table below.

Table 2. Effect of addition of ambient air as co-flow stream on particle drying

| parameters | Existing drying air case | Dilution air reduced + ambient air |
|---|---|---|
| Evaporation rate. Kg/s | 0.00246 | 0.001835 |
| Lowest temperature, K | 375.58 | 344 |
| Wall film height (max), m | 1.356e-5 | 1.33e-5 |

Hence, the co-flow air stream taken from ambient air needs to be heated before adding into the spray dryer to maintain the temperature inside the chamber and facilitate efficient drying of particles. But this requires further installation of heaters, heat exchangers, etc for the purpose of heating the ambient air and again turns out to be not so feasible in terms of installation and cost effectiveness.

***Dividing the stream of existing drying air and diverting one stream as co-flow air***

The existing stream of drying air can be divided before entering the spray dryer and one stream can be added along the walls of the spray dryer as co-flow air. With the non-feasibility of the other two ways of adding the co-flow air proved as above, the most feasible way to introduce co-flow air is to divide the drying air using a damper and then fine tuning it to the required ratio based on the resistance in the flow paths.

The division ratio of existing air is now important for efficient drying to take place inside the chamber. On a preliminary note, the second stage of drying in presence of existing drying air is the stage where majority of evaporation takes place and a bulk of the moisture is lost and hence the volume flow of drying air should be higher than that of the co-flow supplied into the spray dryer chamber. But still a proper estimate of the ratio of division of drying air is required so that droplet drying inside spray dryer chamber can be optimized to give maximum efficiency and reduce the lump formation which will serve our initial purpose to reduce bend choking. This can be done by the use of various tools to verify the proposed strategy. The proposed solution is feasible enough to be implemented, only when it shows the desired results.

## Tools

***Experiments & Pilot Plant***

The conventional methods of testing the design conditions and checking the results before a final plant setup includes experimentation and a pilot plant scale model. The various proposed parameters are tested through experimentation and then the results are analyzed and tried in a pilot plant. After the success of the pilot plant, the actual plant is worked upon or changes to existing plants are applied. But this process of experimentation requires huge expenditure for even minimal tasks and at the cost of a lot of time.

***Process Simulation***

The use of Computational Fluid Dynamics (CFD) simulations has gained huge popularity in the chemical industries in the past few decades due to their accuracy in predicting the actual plant conditions through computer codes. CFD codes like ANSYS FLUENT software contains the broad physical modeling capabilities needed to model flow, turbulence, heat transfer, and reactions for industrial applications ranging from air flow over an aircraft wing to combustion in a furnace, from bubble columns to oil platforms, from blood flow to semiconductor manufacturing, and from clean room design to wastewater treatment plants. Special models that give the software the ability to model in-cylinder combustion, aeroacoustics, turbomachinery, and multiphase systems have served to broaden its reach.

Today, thousands of companies throughout the world benefit from the use of ANSYS FLUENT software as an integral part of their design and optimization phases of product development. Advanced solver technology provides fast, accurate CFD results, flexible moving and deforming meshes, and superior parallel scalability. User-defined functions allow the implementation of new user models and the extensive customization of existing ones. ANSYS FLUENT's interactive solver set-up, solution, and post-processing make it easy to pause a calculation, examine results with integrated post-processing, change any setting, and then continue the calculation within a single application.

These benefits with the extensive range of physical modeling capabilities and fast, accurate and inexpensive CFD results, ANSYS FLUENT is one of the most comprehensive software packages for CFD modeling available in the world today.

## Simulations using Fluent 6.3

The professional CFD code developed by ANSYS, Fluent 6.3.26 was used to verify that the proposed solution is helping our case in improving the drying process inside the spray dryer chamber. The simulations were carried out both using the 2-D and the 3-D model. The 2-D model was used to get an initial idea of the flow inside the spray dryer chamber and paved the way for further simulations using the 3-D models. The models of the spray dryer were created using the Gambit software which is integrated with the Fluent code. The atomizer was first modeled and simulations were carried out to get the results of its outflow which were then fed into the spray dryer simulations. The CFD simulation of the spray dryer included both discrete particle modeling (DPM) and spray injections to model the droplet injections from the atomizer into the spray dryer chamber and their subsequent drying. Swirl was introduced to model the swirling air of the process.

The initial results obtained from the 2-D simulation suggest that there was an improvement in the drying process of droplets inside the spray dryer chamber. These are depicted in the table and figures below.

Table3. Data showing improvement in drying process with introduction of Co-Flow air stream

| *parameters* | *Existing drying air case* | *Addition of co-flow air* |
|---|---|---|
| *Evaporation rate. Kg/s* | *0.003733* | *0.004051* |
| *Lowest temperature, K* | *380* | *380* |
| *Wall film height (max), m* | *3.55e-5* | *2.042e-5* |

Hence, it is evident that an increment in the drying rate and the maximum particle residence time was achieved due to the introduction of the co-flow air stream. But still a proper estimate of the ratio of division of drying air is required so that droplet drying inside spray dryer chamber can be optimized to give maximum efficiency and reduce the lump formation. For this, different ratios of drying air to co-flow air were to be tried in the existing model and simulated to find out the optimum condition. Due to time constraint relating to the project only three ratios of swirling air to co-flow air were simulated, which are 50:50, 60:40 and 70:30 respectively. The results of these simulations are shown below.

Table4. Study of different ratios of drying air & co-flow air to obtain the best drying condition (2-D)

| **parameters** | **Evaporation rate (kg/s)** | **Residence time (s)** | **Temperature range (K)** |
|---|---|---|---|
| **No Co-Flow air** | *0.0334* | *0.175* | *386.4-500* |
| **With Co-Flow air (swirl:co-flow = 50:50)** | *0.0364* | *0.182* | *387.76-500* |
| **With Co-Flow air (swirl:co-flow = 60:40)** | *0.0462* | *0.187* | *386.4-500* |
| **With Co-Flow air (swirl:co-flow = 70:30)** | *0.0550* | *0.189* | *387.2-500* |

To be noted is that the ratio of swirling air or drying air to the co-flow air that gives the best drying case is 70:30. In this case, there was a 65% increase in the drying rates of particles and 8%

increase in the particle residence time. The overall temperature inside the spray dryer was also observed to increase by a minor margin. Apart from these, it was also observed that the deposition of the solids was reduced (assuming the worst case in the simulations that any particle striking the wall, sticks to it), the reverse flow below the atomizer was suppressed and the particle diameters were uniform throughout the chamber.

Based on these results obtained from the 2-D simulations for different ratios of division of drying air, a 3-D model of the spray dryer was modeled and simulations were carried to divide the present 10 kg/s of drying air into swirl air and co-flow air. Two cases each of diverting 2.1 kg/s and 3.1 kg/s from the drying air were undertaken for the simulations and studied.

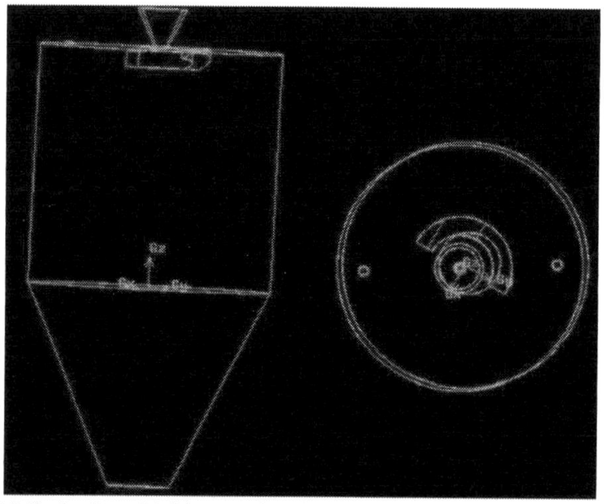

Fig2. The Gambit model of the spray dryer chamber showing the side view & top view

Table5. Study of different ratios of drying air & co-flow air to obtain the best drying condition (3-D)

| Parameters | No co-flow air(existing) | Co-flow @ 2.1 kg/s | Co-flow @ 3.1 kg/s |
|---|---|---|---|
| Evaporation rate (kg/s) | *2 e-3* | *2.5 e-3* | *1.5 e-3* |
| Temperature range (K) | *373-450* | *373-450* | *373-450* |
| Total number of droplets | *5.8 e9* | *5.12 e9* | *5.2 e9* |
| Total mass of feed (kg) | *3.015* | *3.024* | *3.136* |
| Wall film height (m) | *3 e-5* | *2 e-5* | *1 e-5* |
| Total wall film mass (kg) | *0.79* | *0.35* | *0.46* |
| Deposition on wall (%) | *26.2* | *11.5* | *15.4* |

Based on the above data, diverting 2.1 kg/s of drying air into the co-flow stream is the optimum case for improving the drying process and an increase of 0.1356 kg/s in productivity is expected with the same. An overall reduction of 50% in wall deposition has been observed in this case. This study of the spray dryer chamber using 3-D simulations is a more accurate representation of the actual industrial process and further study of the operating parameters and change in ratios can give even better results for the desired outcome.

# *Implementation Strategy*

The proposed co-flow air stream has to be added to the spray dryer chamber. The introduction can be done by the following two techniques:

1. Addition through certain spots in the existing design.
2. Civil changes in existing design to make a spot of introduction.

### *Addition through certain spots in the existing design*

A clear look at the spray dryer chamber was taken to find out any openings in the spray dryer, which are not in use and can be used to introduce the stream of co-flow air. Two ways of introducing the co-flow air stream was found. One is through the four holes that are present on the top wall of the spray dryer and the other is through the manhole on the cylinder wall of the spray dryer chamber which is at the same height as the atomizer inside.

There are two sets of closely spaced holes of about 0.5 m diameter on opposite sides of the atomizer on the top wall of the spray dryer. These sets of holes can be used as inlet points for the diverted co-flow air from the drying air. But on each side, if both the closely spaced holes are used for air inlet, it may cause interference and distorted turbulent flow which will affect the swirl and consequently the drying inside the dryer. Hence it is desired to use only one hole on each side of the atomizer.

There is also a manhole present on the wall of the spray dryer chamber, which is at the same height as the atomizer. It is at a distance of about two meters form the flow duct of hot drying air and also falls straight in the direction of required swirl of the co-flow air required. Hence, no additional bends or air guide plates needs to be installed to impart the swirl to the air.

### *Limitations & troubleshooting*

The holes present on the top wall of the spray dryer chamber are about 0.5 m high and hence just joining a duct from the diversion of existing drying air to these holes will be of no use. To give the co-flow a proper direction and the swirl inside the chamber, air vents or a 90 degree bend has to be provided inside the spray dryer, below the hole.

For even easier implementation of co-flow air, only one hole out of four can be used. But the use of two holes is expected to give better results because of the more wall surface the air will sweep. Based on the number of holes used and the air vents used the damper for the division of air flow from the drying air has to be fine tuned based on the resistances offered to their flow by their respective paths.

The manhole seems to be a better option for introduction of co-flow air into the spray dryer in terms of easy implementation. But since this manhole is presently being used for cleaning purposes along with the other manhole at a lesser height than this, a new manhole might be needed to be carved on the spray dryer for cleaning purposes.

***Civil changes in existing design to make a spot of introduction.***

Civil changes in an existing design are normally not desired by industries, especially when it is an equipment provided by some other organization. It is generally chosen as the last resort to tackle a problem. Co-flow air can be introduced at equally spaced points near the wall of the spray dryer with an imparted swirl.

If more capacity is expected and the problems are to be removed from the root, the current design and geometry of flow distribution must be reviewed.

# *Conclusions*

The spray drying process was carefully studied along with its relevance to the various plant processes. Every minute detail of the process was taken into consideration for the proposals for improvement. The suggestions are based on the viability of implementation in the present plant conditions. The suggestions for organizations that can help in further improvements have been mentioned in the presentation. The bend choking problem existing in the plant has a great impact on the overall productivity and also to the surrounding industries, hence they have to been dealt with urgency. Similar changes can be applied into any other industry involving Spray Drying to reduce the existing bend choking problem as well to reduce it after increasing feed rate. The above mentioned techniques will help in the same.

# (611d) Thermodynamics Studies of Metal Oxides Conversion for CO2 Sequestration In Steel Industry

**Hui Li**[1], De-Long Xu[2], Xiao Fan[3], Yong-Hua Duan[3] and Yong Min[1], (1)Xi'an University of Architecture &Technology, Institute of Powder Engineering, Xi'an, China, (2)Xi'an University of Architecture and Technology, Xi'an, China, (3) Material Institute, Xi'an University of Architecture and Technology,, Xi'an, China

Due to the concerns of "global warming", CO2 sequestration from steel industry are becoming more and more important for the environments FeCO3, one of component for steel, was selected to investigate in order to design suitable processing to sequestrate the CO2 at the steel plant. It was found that the raw material particle size, surface area as well as chemical components were important factors, which will effect on the CO2 absorption efficiency, cycle times, particle activities, and so on. Various analytical equipments such as SEM, XRD, FTIR, TGA, were used to analysis the processing and materials during the processing. Based on the above study,we designed an efficient chemical loop, which is not only sequestrate CO2 but also select suitable iron oar for the steel industry-B

See more of this Session: Simultaneous Energy and Emission Reduction for Chemical Industries
See more of this Group/Topical: Process Development Division

# (637a) Process Intensification of Multiphase Processes with a Rotor-Stator Spinning Disc Reactor

**John van der Schaaf**, Frans Visscher, Kevin M.P. van Eeten, Mart H.J.M. de Croon and Jaap C. Schouten, Chemical Engineering & Chemistry, Eindhoven University of Technology, Eindhoven, Netherlands

An overview of hydrodynamics and mass transfer for cocurrent gas-liquid flow in a rotor-stator spinning disc reactor is presented. The dependency of disc spacing and rotating speed on gas-liquid and liquid-disc mass transfer is discussed. Additionally results for multiple rotor-stator spinning discs mounted on a single axis are presented. Results will be shown for the application of the multiple spinning disc process for the production of a pharmaceutical by an ethynilation reaction. An outlook for further application to liquid-liquid, liquid-solid, and gas-liquid solid processes will be given.

See more of this Session: Advances In Process Intensification
See more of this Group/Topical: Process Development Division

# (637b) Achieving Process Intensification: A Phenomena-Based Synthesis/Design Methodology

**Philip Lutze**, Deenesh K. Babi, John M. Woodley and Rafiqul Gani, Department of Chemical and Biochemical Engineering, Denmark Technical University, Lyngby, Denmark

In recent years, Process Intensification (PI) has attracted considerable interest as a potential means of process improvements and to meet the increasing demands for sustainable production [1]. PI seeks to improve processes by increasing efficiency, reducing energy consumption, operational costs, volume and waste as well as simplifying the flowsheet. Thus, a variety of intensified operations and equipment have been developed by academia and industry but, to date, only a limited number of intensified technologies have achieved implementation, such as reactive distillation, dividing wall columns and reverse flow reactors [2]. One major reason for this is that the identification of the best PI option with the currently used techniques is neither simple nor systematic.

In previous work [3] we reported the development of a general computer-aided systematic synthesis/ design methodology. In order to manage the complexity of the potentially large number of intensified options, the optimization problem is solved by a decomposition approach. In this methodology, redundant intensified options were systematically removed by checking against predefined constraints. However, at present, this methodology is limited to already pre-defined PI unit operations which can be retrieved, together with all the necessary information for synthesis and design of each, from a knowledge base tool. In order to *invent* new unit operations and based on them, new process designs and flowsheets, it is necessary to investigate PI options at the phenomenological level [4, 5], which is the topic of this contribution. The basis of the phenomena-based process synthesis is phenomena building blocks consisting of mass, component, energy and momentum balances as well as connectivity rules between the phenomena building blocks. Two different connections between phenomena exist: A connection implying the simultaneous occurrence of phenomena or a connection implying the sequential occurrence of phenomena which depends on the operational window of the phenomena involved. Based on this approach, a general methodology has been developed for phenomena-based synthesis/design, similar to the one for pre-defined PI unit operations [3]. First, the process design problem is defined, in terms of a set of product qualities and quantities as well as additional process constraints and the desired (target) process improvement. Next, this information is analyzed to identify all potential phenomena building blocks and their corresponding operational windows. Based on this, the phenomena building blocks are joined together according to connectivity rules to match the defined process intensification targets and a superstructure of flowsheet options is generated. In the following step, the generated options are screened in terms of additional logical and structural constraints as well as performance specifications before unit operations are identified. For example, a sequence of at least three simultaneous ideal mixing and reaction phenomena can be identified as a plug flow reactor. In the penultimate step, a reduced optimization (NLP) problem is solved to identify the best intensified flowsheet. The advantage of the phenomena-based process synthesis and design is that it generates potentially *novel* process options because the initial search space is wider and allows space beyond those involving existing unit operations and is truly predictive. Also, the method provides simultaneous development of the necessary process models.

In this contribution the application of the phenomena-based process synthesis/ design methodology implemented within a computer-aided framework will be presented together with a selected example of industrial importance.

[1] Becht, B., Franke, R., Geißelmann, A., Hahn, H. An industrial view of process intensification Chem Eng Process 2009; 48, 1, 329–332.

[2] Harmsen, J. Process Intensification in the petrochemicals industry: Drivers and hurdles for commercial implementation. Chem Eng Process 2010; 49, 1, 70-73.

[3] Lutze, P., Gani, R., Woodley, J.M. Process Intensification: A perspective on process synthesis. Chem Eng Process 2010; 49, 6, 547-558.

[4] Freund, H., Sundmacher, K. Towards a methodology for the systematic analysis and design of efficient chemical processes: Part 1. From unit operations to elementary process functions. Chem Eng Process 2008; 47, 12, 2051-2060.

[5] Papalexandri, K.P., Pistikopoulos, E.N. Generalized modular representation framework for processs synthesis. AIChE J 1996; 42, 4, 1010-1032.

See more of this Session: Advances In Process Intensification
See more of this Group/Topical: Process Development Division

# (637c) Membrane-Assisted Processes for Energy-Efficient Separations: An Optimisation Based Design Approach Using Evolutionary Algorithms

**Katharina Koch** and Andrzej Grak, Department of Biochemical and Chemical Engineering, Laboratory of Fluid Separations, TU Dortmund University, Dortmund, Germany

The separation of azeotropic multicomponent mixtures is still a challenging issue in chemical process design. The replacement of conventional, energy-intensive separation methods such as extractive distillation with energy-efficient, membrane-assisted processes can lead to significant energy savings and cleaner processes and thus constitutes an example of how synergies can be used for process intensification. Nevertheless, this signifi-cant economic and ecological potential is hardly exploited in industry due to the lack of a general process design methodology and missing detailed process know-how.

To promote their industrial application, an optimisation based design approach for hybrid processes comprising distillation, pervaporation and vapour permeation is developed and verified for the separation of a ternary azeotropic mixture of acetone, isopropyl alcohol (IPA) and water.

The optimisation approach is based on a generic process model including rate-based models for distillation and membrane separation, which are essential to describe the increased complexity of hybrid processes and strong interactions between the unit operations involved. The parameters for prediction of trans-membrane flux for both membrane separations are determined in lab-scale experiments using different hydrophilic polymeric membranes. The results using different empirical and semi-empirical models for calculation of membrane permeances shows a very good agreement between calculated and measured permeate fluxes.

The optimal design of a membrane-assisted hybrid separation process involves the determination of several structural parameters, e.g. the number of apparatuses and their connection, and corresponding operational parameters, e.g. the flow rates of connecting streams. This complex mixed-integer non-linear design problem is optimised by incorporating the detailed models into a process superstructure, which represents several possible process configurations appropriate for the given separation task. The optimisation problem is solved by applying an evolutionary algorithm to minimize the annualized total production costs as objective function. This approach facilitates the simultaneous determination of the process configuration, dimension of apparatuses and operating conditions required for the optimal hybrid process.

This approach is demonstrated for the determination of the optimal process configuration for superstructures of membrane-assisted processes comprising up to four apparatuses. The results show that the developed methodology is applicable without detailed simulation studies at the early design phase. Finally, the results will be compared to a conventional distillation process. Major achievements concerning the energy efficiency can be reached by integration of membranes into the process.

*The support of the "Deutsche Forschungsgesellschaft" for the joint project „Optimisation based framework for mem-brane-assisted hybrid processes" (cooperation with Prof. Marquardt, RWTH Aachen) is gratefully acknowledged.*

See more of this Session: Advances In Process Intensification
See more of this Group/Topical: Process Development Division

# (637d) Isolation of High Grade Methyl Acetate Via Pervaporation

**Thomas Winkler**, Department of Chemical Engineering and Environmental Technology, Graz University of Technology, Graz, Austria, Susanne Lux, Institute of Chemical Engineering and Environmental Science, University of Technology Graz, Graz, Austria and Matthus Siebenhofer, Institute of Chemical Engineering and Environmental Technology, Graz University of Technology, 8010 Graz, Austria

**Isolation of high grade methyl acetate via pervaporation**

*T. Winkler, S. Lux, M. Siebenhofer*

*Graz University of Technology, Institute of Chemical Engineering and Environmental Technology, Graz/Austria*

Aim of process intensification is to overcome limitations of mass and heat transfer, product yield and selectivity of chemical reactions, through new or different combination of technologies and/or process steps. Thermodynamic limitations and challenges in purification of the desired product are unavoidable when dealing with esterification reactions. Speeding up reaction kinetics of equilibrium reactions and elimination of azeotropes by adaptive configuration of reaction and separation steps was the basic strategy of this research activity.

The synthesis of methyl acetate via esterification of acetic acid with methanol served as a welcome model process, as the basic type of reaction is applicable to a variety of esterification reactions. Methyl acetate forms low boiling azeotropes with methanol and with water. The bulk product methyl acetate is produced in large quantities based on the established Eastman Kodak process.

This project deals with the investigation of alternative paths in the synthesis and purification of methyl acetate. Selective permeation is a promising route of azeotrope breaking. This unit operation can handle higher water contents than reactive distillation technology. Side reactions can be suppressed since operation temperatures are moderate.

Esterification is carried out under acid limiting conditions. When using a hydrophilic membrane methanol and water can pass the membrane, whereas methyl acetate is kept in the retentate phase. In case of hydrophobic membranes methyl acetate is transferred into the permeate phase.

Separation selectivity and permeate flux of several membranes of hydrophilic and hydrophobic surface properties were investigated. The influence of the operation parameters on the membrane performance was quantified and modeled. Permeate flux was determined based on a solution-diffusion model. Process schemes for the acid limited reaction route were designed and validation experiments were conducted.

See more of this Session: Advances In Process Intensification
See more of this Group/Topical: Process Development Division

# (637e) Hydrotreating of Vegetable Oil In Microreactor

**Lalita Attanatho** and Goran N. Jovanovic, Department of Chemical, Biological and Environmental Engineering, Oregon State University, Corvallis, OR

Hydrotreatingreaction of vegetable oil is an alternative method for the production of renewableBiodiesel fuel. This reaction involves conversion of triglycerides into normalalkanes, leads to a deoxygenated and stable product, which is fully compatiblewith petroleum derived diesel fuel. The hydrotreating process uses hydrogen toremove oxygen from triglyceride molecules at elevated temperature in thepresence of a solid catalyst.

Inthis work, the catalytic hydrotreating of vegetable oil was studied in acontinuous flow microchannel reactor of inner diameter 508 [m] and of varied length; 0.5 [m] - 15 [m]. Themicrochannel reactor was fabricated from SS-316. The walls of the microreactorwere coated with a thin $Al_2O_3$ film, which was thenimpregnated with Ni-Mo catalyst containing phosphorus as promoter. Conversionof vegetable oil was experimentally determined in the temperature range of300-400 [°C], oil flow rate in the range of0.1-0.3 [ml/min], with hydrogen-to-triglyceride molar flow ratio in the rangeof 10-15, and at constant system pressure of 34 [bar]. The liquid products were analyzed in a gas chromatograph forits composition of triglyceride, diglyceride, monoglyceride, fatty acid andalkanes.

Detailedmathematical model was developed to represent this two-phase chemical reactionprocess. Mathematical model is entirely based on first principles, i.e. noadjustable or correlation parameters were used. *Figure 1* is schematicrepresentation of a section of the microchannel reactor used in numericalsimulation.

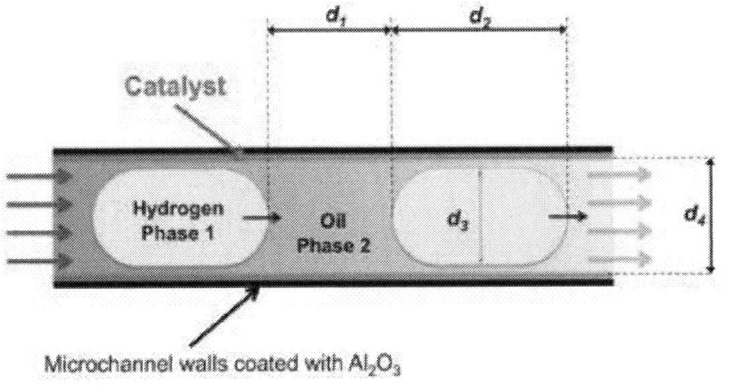


Bothexperimental and numerical results show that microchannel reactor is well suitedfor the hydrotreating process of vegetable oil. Hydrotreating of vegetable oilin a microchannel reactor provides complete elimination of hydrogen masstransport deficiency, which arises from low hydrogen miscibility in vegetableoil.

# (637f) A Novel Intensified Device for the Production of Highly Monodisperse Polymer Particles In the Low Nanometer Range Through Miniemulsion Polymerization

**Ins Souilem**[1], Ren Muller[2], Yves Holl[3], Michel Bouquey[2] and Christophe Serra[2], (1)LIPHT-ECPM Strasbourg, Strasbourg, France, (2)Lipht-Ecpm, LIPHT-ECPM Strasbourg, Strasbourg, France, (3)Ics, LIPHT-ECPM Strasbourg, Strasbourg, France

**A novel intensified device for the production of highly monodisperse polymer particles in the low nanometer range through miniemulsion polymerization**

Souilem Ins(1)*, Muller Ren(1), Holl Yves(2),
Bouquey Michel(1), Serra Christophe A(1)

* souilem@unistra.fr

*1 Group for the Intensification and Intrapolation of Polymer Processes (G2IP)*

*Laboratory of Polymer Engineering for High Technologies (LIPHT), Eac(CNRS) 4379*

2 Group of Collodes et Macromolecules / Charles Sadron Institut

*European Engineering School of Chemistry, Polymer and Material Science (ECPM)*

*University of Strasbourg (UdS)*

**Abstract** An original device called RMX was used to synthesize miniemulsions of poly(methyl methacrylate). This mixer/reactor is based on a strong elongational flow component contributing to a highly efficient dispersive mixing compared to existing laboratory mixers. It presents the following important technical features:

mixing volumes in the range of 5 to 45 cc.

easy feeding of components even for reactive systems.

tightness of liquids and gases is maintained during all steps, authorizing the processing of materials containing volatile components.

The principle of operation is based on the alternative displacement of two opposite pistons placed themselves in two different cylindrical chambers. Each piston pushes the material from one chamber to the other through a static mixing element, fitted with a given number of holes of diameter in the range 0.2 -1 mm (Figure 1).

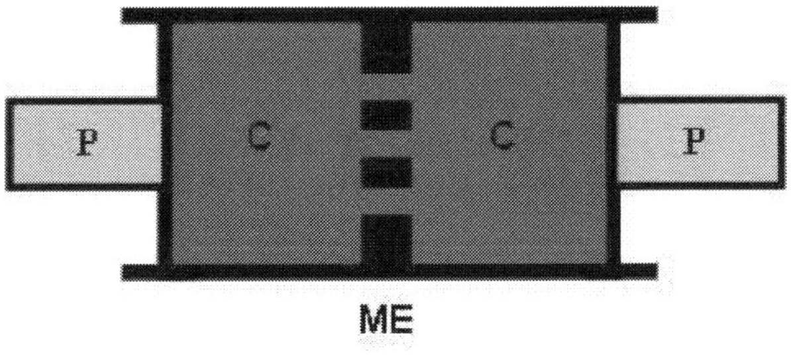


Figure 1: Schematicrepresentation of the RMX device: P pistons C mixing chambers MEmixing element

The first part of this work consists on testing theefficiency of this device as a mixer. That'swhy we worked on a direct emulsion (monomer in water) stabilized with a surfactant.Figure 2 shows the evolutionof the droplet size of monomer during the dispersion step at differentoperating pressures compared to the ultrasonification and ultraturrax. Theresults show that only the RMX allows to reach very small droplet size (lowerthan 80 nm) even with very low pressure.

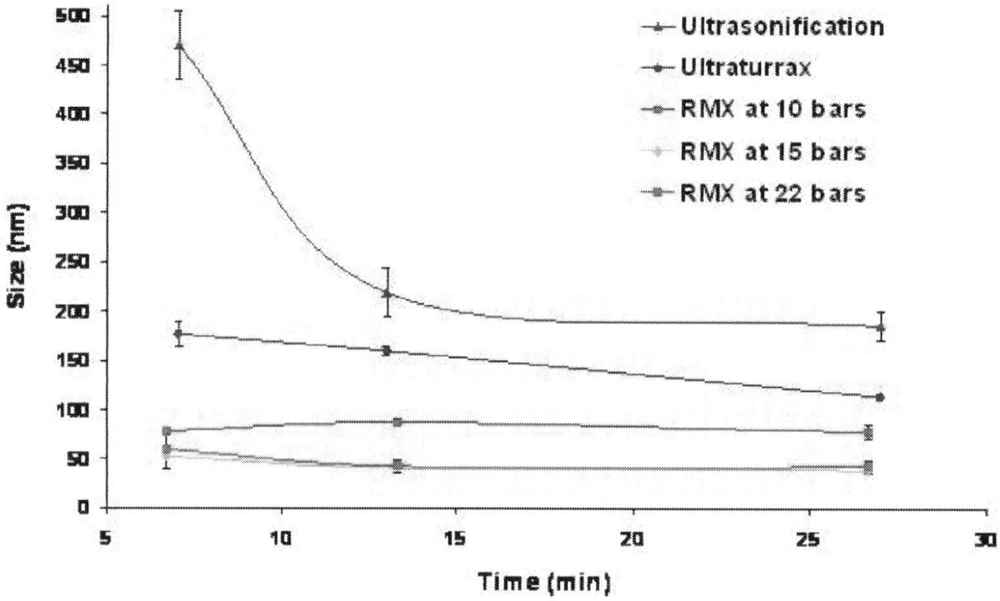


Figure 2: Evolutionof the droplet size of monomer in the emulsification step: comparison betweenultrasonification, ultraturrax and RMX

The second part consists on synthesizing miniemulsionsof poly(methyl methacrylate). After the dispersion of all reagents, thepolymerization of the monomer takes place by increasing the temperature.

Upon polymerization, it was found that polymerparticles with diameters below 100 nm could be obtained, thus differing fromconventional dispersion methods that have not yet reached this size domain. Theinfluence of both material parameters (monomer, surfactant and inhibitorconcentrations) and process parameters (pressure and flow rate, number and sizeof the holes) on the particle size distribution was also investigated.

For example, Figure 3 shows the evolution of particlesize and the particle size distribution (PSD) at different pressures.

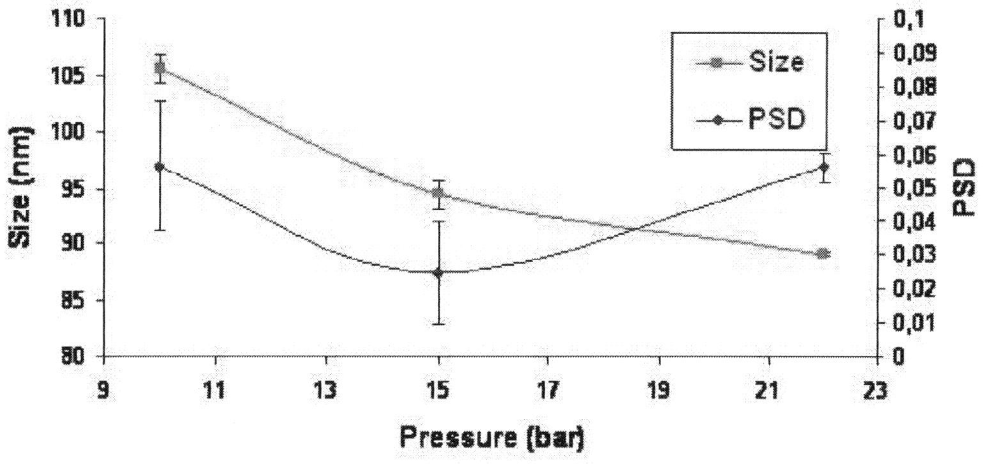


Figure 3:Evolution of the polymer particle size and the PSD as a function of thepressure

**Keywords:** miniemulsion, static mixer/reactor, particle size, dispersion,polymerization

See more of this Session: Advances In Process Intensification
See more of this Group/Topical: Process Development Division

# (641a) Direct Conversion of Lignocellulosic Biomass to Drop-In Hydrocarbon Fuel

**Xuejun Pan** and Li Shuai, Biological Systems Engineering, University of Wisconisn, Madison, Madison, WI

This research reports an innovative process (HDA) for producing fuel-grade hydrocarbon from lignocellulosic biomass. The process directly converts polysaccharides in lignocellulosic biomass to furan-based precursors for producing fuel-grade hydrocarbons in one step under mild conditions without prior pretreatment and saccharification of the feedstock biomass. Specifically, in the presence of acetone, halide salt, and small amount of mineral acid, the lignocellulosic biomass undergoes hydrolysis of polysaccharides (both cellulose and hemicellulose) to monosaccharides, dehydration of the monosaccharides to furfural (from pentoses) and hydroxymethylfurfural (HMF, from hexoses), and aldol-condensation of HMF and furfural with acetone, and ends up with the formation of furan-derived oxygenates (furfural- and HMF-acetone adducts) with chain length of C5–C15. The resultant furan-based precursors can be subsequently converted to fuel-grade hydrocarbons via hydrogenation. The process integrates the three reactions (Hydrolysis-Dehydration-Aldol condensation) in one step, and is abbreviated as HDA. At the conditions of 120-140°C and 1-2hour, cellulose and hemicellulose in lignocellulosic material (e.g., ground or chipped wood) were completely hydrolyzed into sugars, and more than 80% of the sugars were selectively converted to the furan-acetone adducts. Meanwhile, lignin in the biomass is extensively depolymerized and extracted. Preliminary evaluation indicated the HDA lignin has high purity, low molecular weight with narrow distribution, and good reactivity. The lignin is expected to have great potential in co-products development.

See more of this Session: Chemical and Catalytic Conversions and Processes for Renewable Feedstocks
See more of this Group/Topical: Sustainable Engineering Forum

# (641b) A Method for Rapid Screening of Catalysts for Hydrodeoxygenation of Biomass Intermediates

**David K. Johnson**, Biosciences Center, National Renewable Energy Laboratory, Golden, CO

The focus of biofuels development is shifting from alcohol fuels to advanced biofuels that would be drop-in replacements for petroleum-derived hydrocarbons. There is a wide range of intermediates that can be produced from biomass by chemical or biological routes, but virtually all require further deoxygenation before they can be used as fuel components. Potential routes for decreasing the oxygen content of biomass components include dehydration, hydrodeoxygenation and decarboxylation (possibly after controlled oxidation). Deoxygenation by HDO is preferred over dehydration and decarboxylation reactions, as it is likely to result in higher yields through conservation of carbon, retention of hydrogen, and the production of more energy dense products. Hydrodeoxygenation (HDO) is a hydrogenolysis process that removes oxygen by converting it to water, however, HDO is often performed at high temperatures and pressures that significantly increase the cost of the transformation. We need to develop HDO catalysts that more efficiently convert biomass-derived components into fuel products that are compatible with the existing fuel distribution infrastructure. The feedstocks for the chemical transformations can include sugars, lignins, bio-oils, and chemicals obtained by biological transformations of sugars and biomass. It is clear from previous research that in order to design commercially viable catalysts for HDO upgrading of biomass-derived intermediates, the catalyst system must (1) exhibit high activity under mild conditions (<200°C), (2) contain low-cost materials, (3) demonstrate high stability in an aqueous environment, and (4) have high selectivity for the desired products. To accomplish this, a microwave heated (to 250°C), pressurizable (to 250 psi), automated reaction system has been assembled to rapidly screen catalyst systems using model compounds. To allow automated operation, hydrogen is provided in-situ from a hydrogen donor material, such as formic acid, *i*-propanol or decalin. The rapid screening not only allows the rapid testing of catalysts, but also of reaction conditions, including temperature, reaction time, solvent, and hydrogen donor. The results from tests with model compounds such as acetovanillone and hydroxymethyl furfural, representing typical lignin and carbohydrate derived intermediates, with heterogeneous catalysts, will be described.

# A Novel Reactive Distillation Approach for Producing 1,3-Dihydroxyacetone from Glycerol

*Xi Hong, Carl T. Lira, and Dennis J. Miller*
*Department of Chemical Engineering and Materials Science, Michigan State University*
*East Lansing, Michigan 48824*

## Introduction

Glycerol is a viable renewable feedstock that has emerged as a consequence of triglyceride conversion to biodiesel. In addition to glycerol conversion to commodity chemicals such as propylene glycol and epichlorohydrin, formation of higher-value specialty chemical products such as 1,3-dihydroxyacetone (DHA) will provide income for the biorefinery.

DHA is the active ingredient in sunless tanning products. Currently, DHA is industrially produced by fermentation of glycerol, but this microbial approach is challenging because of low production rates and high recovery costs [1]. As a result, the high selling price of DHA limits access of sunless tanning products in the general population.

Recently, scientists have attempted conventional chemical oxidation processes to make DHA from glycerol using metal catalysts with air or molecular oxygen. The goal is to achieve direct oxidization of the secondary hydroxyl group in glycerol to a carbonyl bond. Unfortunately, this reaction is always accompanied with primary alcohol oxidation and a variety of products (i.e. glyceroaldehyde, glyceric acid, etc.) are produced. The highest yield obtained from this method is in the range of 40-60% [2-4].

We present here a novel chemical process to convert glycerol to DHA that greatly improves the efficiency and reduces cost. The essential feature of this process is the selective formation of a 6-membered acetal of glycerol via reactive distillation, followed by selective oxidation of the secondary alcohol to its corresponding ketone product using ruthenium metal and a chemical oxidant. The ketone product is then hydrolyzed to obtain DHA.

A continuous process of making DHA from glycerol has been designed and will be presented. Reactions involved in the process are rapid and can be carried out under mild conditions. Yields of individual process steps are all in excess of 90%, suggesting that this process has good potential to reduce DHA production costs.

## Experiments

### Materials

Anhydrous glycerol (99.5%), 1-propanol (99.9%) and 1-butanol (99.9%) were obtained from J.T. Baker. Ethanol (99.5%), 1,1-diethoxyethane (99%) and sodium hypochlorite (15% aqueous solution) were purchased from Pharmco and Aldrich, respectively. Sodium carbonate (> 99%), sodium bromate (> 99%), potassium bromate (> 99.8%), calcium hypochlorite (> 98%), sodium periodate (> 99%), sodium iodide (> 99%) were purchased from Sigma-Aldrich. Amberlyst-15 cationic exchange resin (16-50 mesh), purchased from Sigma-Aldrich, was investigated as the catalyst. Before being used in reactions, it was pretreated at 90°C in a vacuum oven for 2 hours. Ruthenium dioxide hydrate (99.9% trace metals basis, 28% water content) and ruthenium trichloride hydrate (reagent plus, 46% Ru content) were obtained from Aldrich. Hydrogen (99.999%) and helium (99.999%) were used for GC analysis; both were purchased from Linde Gas, Inc.

**Analysis**

<u>Gas Chromatography Analyses</u>

1-Butanol and 1-propanol were used as internal standard and solvent, respectively, for this analysis. To prepare samples for analysis, 0.16 ml 1-butanol and 0.2 ml of reaction sample were added to a volumetric flask and the mixture was diluted with 1-propanol to 10 ml of solution. A Perkin-Elmer Sigma 2000 gas chromatograph with FID and a DB-Wax capillary column (15m × 0.53mm × 1.0 μm) was used for analysis. Ultra high purity helium was used as the carrier gas; the column temperature was ramped from 40°C (3min) at a rate of 30°C/min to 220°C and held for 1 min. The total program time was 10 min. On column injection was applied to analyze the samples. Injector and detector chamber temperatures were set at 60°C and 300°C, respectively.

<u>GC-MS analysis</u>

The structure of some products was identified using GC-MS analysis (GC-Agilent 6890, MS-Agilent 5973). A 1μL sample was separated in a DB-wax capillary column (30m × 0.25mm × 0.25 μm). The initial column temperature was 40 °C (for 5min), the temperature was then increased at 20 °C/min to 250 °C and held for 10 min. Total program time was 24.5 min. Injector temperature was set at 240 °C. Electron impact (EI) mass spectra were monitored from 20 to 300 m/z. The MS source was set at 230 °C.

## Results and discussion

**Synthesis of glycerol acetal**

GA isomers were synthesized at 45 °C by mixing glycerol (3.1 kg, 33.9 mol), DEE (4.0 kg, 33.9 mol), and Amberlyst-15 cationic exchange resin (120 g) and stirring for 4 h. Following reaction and removal of volatile species (DEE and ethanol) by rotary evaporation, GC-MS analysis of the product mixture showed the presence of four different structures corresponding [5, 6] to the four GA isomers displayed in Table 1.

**Table 1.** Structure and IUPAC name of GA isomers

| | Structure | IUPAC Name | Designation |
|---|---|---|---|
| a) | 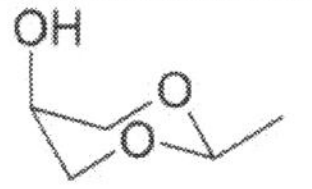 | cis-5-hydroxy-2-methyl-1,3-dioxane | Cis-6 |
| b) | 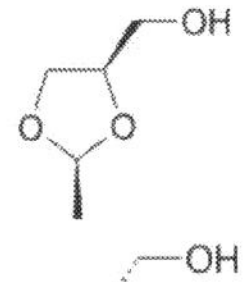 | cis-4-hydroxymethyl-2-methyl-1,3-dioxolane | Cis-5 |
| c) | 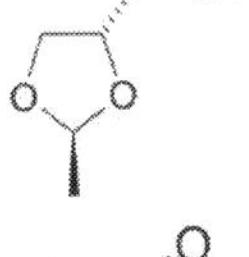 | trans-4-hydroxymethyl-2-methyl-1,3-dioxolane | Trans-5 |
| d) | 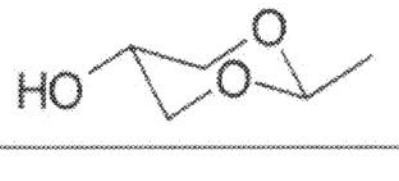 | trans-5-hydroxy-2-methyl-1,3-dioxane | Trans-6 |

**GA isomerization equilibrium**

In the presence of acid catalyst [7, 8], the four GA isomers interconvert and reach an equilibrated composition. To demonstrate this equilibration, three samples obtained from the distillation experiments below with widely differing isomer compositions (cis-6 from 0% to 84%, cis-5 from 2% to 54%, trans-5 from 2% to 41%, and trans-6 from 0% to 63%) were stirred for 5 hr at 25 °C in the presence of 5 wt% (relative to GA) Amberlyst-15 cationic exchange resin. Identical isomer compositions were observed in the products from the three samples: cis-6, cis-5, trans-5 and trans-6 were 37±1%, 17±1%, 13±1%, and 33±1%, respectively. The molar percentages of 6-membered isomers (cis-6 + trans-6) and 5-membered isomers (cis-6 + trans-5) are approximately 70% and 30%, respectively, at equilibrium.

Rapid isomerization requires vigorous agitation (to ensure mass transfer between solid catalyst and solution) and elevated temperature. For example, Sample 1 took 12 h to reach equilibrium at 0 °C, but only one hour at 80 °C. Increasing temperature also shifts the equilibrium composition. For Sample 1, increasing temperature from 0 °C to 80 °C decreased the sum of cis-6 + trans-6 isomers from 78% to 54%. However, the total quantity of the cis- form isomers (cis-6 + cis-5) and trans- form isomers (trans-6 + trans-5) isomers remained essentially constant over all temperatures at 55±2% and 45±2%, respectively.

**Production of the cis-6 GA isomer**

A glass vacuum distillation column was constructed to separate the four GA isomers. The column (height = 3.6 m, i.d. = 50 mm) was packed with 0.6 cm 316 Stainless Steel PRO-PAK distillation packing (Scientific Development Co.). The column reboiler, consisting of a two-liter round-bottom flask, was heated with a 750 W electric heating mantle. Vapor temperature was measured at the top of the column.

In the initial distillation, 896 g (800 ml) of a GA mixture containing 35% cis-6, 18% cis-5, 15% trans-5, and 32% trans-6 isomers was placed in the reboiler. To avoid isomer interconversion, the solution was maintained at a weakly basic state by adding 50 g of sodium carbonate to the mixture. This experiment was carried out at a pressure of 20.5 ± 1.0 mm Hg (except for Sample 1, where the pressure was temporarily reduced to 9.0 mmHg).

Twelve 60-mL distillate samples were collected over the course of the distillation and analyzed by GC. Ninety percent of the initial charge to the still pot was recovered as distillate in these samples (Table 2). The material remaining in the still pot consisted primarily of the trans-6 isomer and glycerol; inclusion of this material gave essentially complete (100±2%) closure of the material balance for the experiment.

**Table 2**. Summary of GA isomers recovered in the initial distillation

| | Initial GA charge | | Product recovery in distillate | | Percent recovered in distillate (%) |
|---|---|---|---|---|---|
| | Mass (g) | Composition (wt%) | Mass (g) | Percent of total | |
| Cis-6 | 311 | 35 | 288 | 36 | 93 |
| Cis-5 | 166 | 18 | 152 | 19 | 92 |
| Trans-5 | 132 | 15 | 120 | 15 | 91 |
| Trans-6 | 287 | 32 | 246 | 30 | 86 |
| Total | 896 | 100 | 806 | 100 | 90 |

Figure 1 gives the reboiler liquid temperature, vapor temperature at the top of the column, and composition (via GC) of the distillate samples taken during the distillation. The results clearly show the order of decreasing volatility of the GA isomers as cis-6, cis-5, trans-5, and finally trans-6. As the more volatile isomers are removed from the still pot, liquid and vapor temperature gradually increase.

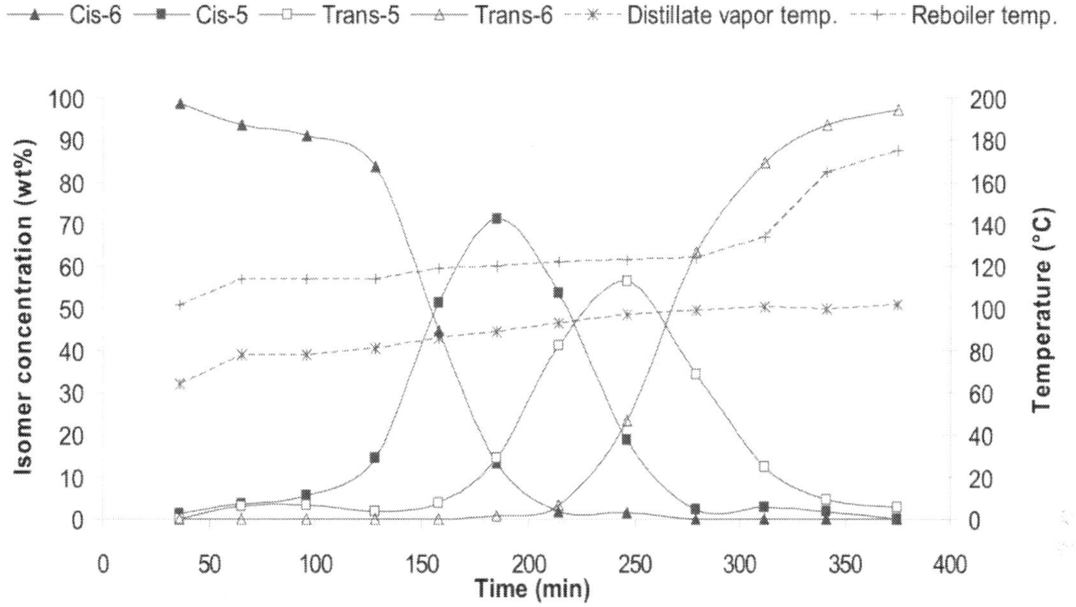


**Figure 1**. Initial distillation of GA isomer. Column pressure was 20.5±1.0 mmHg (except for the first sample, where the pressure was 9.0 mmHg).

Based on the result of the initial distillation and the fact that GA isomers can interconvert to each other and reach an equilibrated isomer composition in the presence of acid catalyst, a continuous reactive distillation experiment was then carried out to produce the cis-6 isomer as the sole product. In this run, 504 g (450 mL) of a mixture of 36 wt% cis-6, 16 wt% cis-5, 14 wt% trans-5, and 34 wt% trans-6 GA isomers was charged to the reboiler pot of the glass column along with 50 g of Amberlyst-15 cationic exchange resin. For this experiment, the column pressure was maintained at 2.9±0.2 mmHg. Once the distillate product started to condense at the top of the column, a continuous feed of the same GA isomer mixture was introduced to the column reboiler to maintain its liquid level constant. The feed rate was adjusted to match the distillate rate from the column, which was approximately 1.8 mL/min. During the distillation, 16 distillate samples were collected with an average volume of 59 mL.

As shown in Figure 2, this experiment produced samples with high concentration of cis-6 isomer (from 90% to 96%) continuously during operation of the column. The reboiler temperature and distillate vapor temperature were essentially constant during operation, indicating that the reboiler composition was constant and that the column was operating at steady state. The total quantity of distillate obtained was 1028 g, more than twice the amount of the initial charge to the reboiler, demonstrating that continuous production of the cis-6 isomer from the GA acetal mixture is feasible. Based on these results, it is fully expected that an appropriately designed commercial-scale column with sufficient separation capacity would provide essentially pure cis-6 isomer.

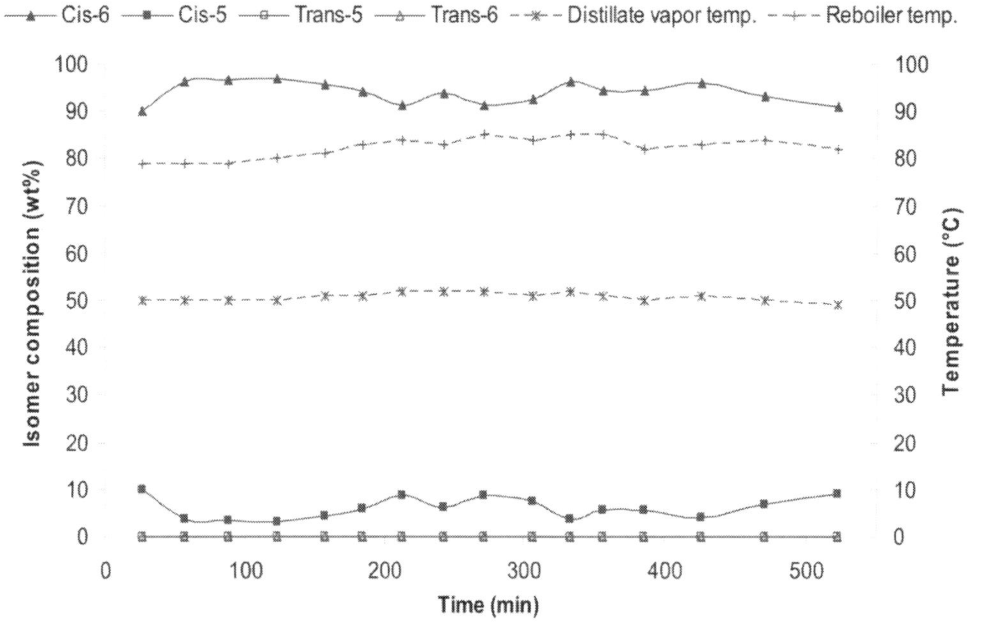


**Figure 2**. Continuous reactive distillation of GA isomers. Column pressure = 2.9±0.2 mm Hg.

**Oxidation of cis-6 to the precursor of DHA**

With the availability of cis-6 isomer, a new synthesis pathway can be developed to obtain valuable chemicals. For instance, by oxidizing the remaining free hydroxyl group in cis-6 and hydrolyzing the corresponding ketone product (2-methyl-1,3-dioxan-5-one, G-tone), 1,3-dihydroxyacetone (DHA) can be produced [7].

For this purpose, a variety of oxidation reagents including pyridinium chlorochromate (PCC), ruthenium tetroxide, hydrogen peroxide, molecular oxygen, and ozone were tested to convert cis-6 into its ketone product.

The most effective method is the ruthenium tetroxide-catalyzed oxidation. The reagents used in this reaction involves a ruthenium species (i.e. $RuCl_3$ and $RuO_2$) and another oxidizing salt (i.e. NaClO, $NaBrO_3$, $NaIO_4$, $Ca(ClO)_2$, $KBrO_3$, etc.) During reaction, the ruthenium species is first oxidized to $RuO_4$, and then $RuO_4$ oxidizes cis-6 to the corresponding ketone product. The co-product is a black precipitate of $RuO_2$, which can be re-oxidized to $RuO_4$ by the oxidizing salt in the aqueous solution. The regenerated $RuO_4$ then enters the organic phase again for another cycle of oxidation. In this way, the alcohol substrate can be continuously oxidized to the desired product with catalytic amounts of $RuO_4$ while the secondary oxidant is consumed. When the reaction is finished, the residue $RuO_4$ species in the organic phase can be easily recovered by adding isopropanol to make it $RuO_2$ precipitate [9].

Table 3 presents the results of cis-6 oxidation experiments catalyzed by ruthenium-based substances. It is clear that this type of reaction very efficient. In some reactions over 90% yield of G-tone was achieved within 30 min.

**Table 3**. Ruthenium tetroxide catalyzed oxidation of cis-6 at 25°C, 1.0 atm.

| Reaction | Cis-6 (g) | Water (g) | Catalyst ($RuO_2 \cdot xH_2O$) loading (g) | Secondary oxidant solution/mass (g)/concentration | Time (min) | G-tone yield (%) |
|---|---|---|---|---|---|---|
| 1 | 0.41 | 10 | 0 | NaClO/4/10% | 30 | 0 |
| 2 | 0.42 | 10 | 0 | $KBrO_3$/10/saturated | 1440 | 0 |
| 3 | 4.0 | 50 | 0.40 | NaClO/100/10% | 30 | 80 |
| 4 | 0.39 | 11 | 0.31 | NaClO/4/10% | 30 | 76 |
| | | | | NaClO/5/10% | 60 | 95 |
| 5 | 0.45 | 10 | 0.05 | $Ca(ClO)_2$/10/ saturated | 20 | 87 |
| 6 | 0.40 | 10 | 0.30[a] | $NaIO_4$/10/saturated | 75 | 88 |
| 7 | 0.42 | 10 | 0.26 | $KBrO_3$/10/ saturated | 45 | > 99 |
| 8 | 0.43 | 10 | 0.32[a] | $KBrO_3$/10/saturated | 75 | 95 |
| 9 | 4.1 | 10 | 0.40 | $KBrO_3$/100/saturated | 20 | 97 |
| 10 | 0.30 | 10 | 0.41 | $NaBrO_3$/10/10% | 20 | 96 |
| 11 | 10.0 | 10 | 0.51 | $NaBrO_3$/100/20% | 40 | 90 |
| 12 | 3.0 | 10 | 0.10 | $NaBrO_3$/12.5/20% | 20 | 95 |

a. $RuCl_3 \cdot xH_2O$ was used as catalyst.

**Hydrolysis of G-tone to DHA**

Hydrolysis of G-tone to DHA is very fast and can be carried out at mild conditions (Scheme 1). In one experiment, 5 wt% (weight to substrate) of Amberlyst-15 cationic exchange resin was added to the final mixture of Reaction 12 in Table 3. When the mixture was heated to 30°C, complete hydrolysis of G-tone was achieved in 10 min.

**Scheme 1**. Processes of catalytic conversion of glycerol into useful chemicals.

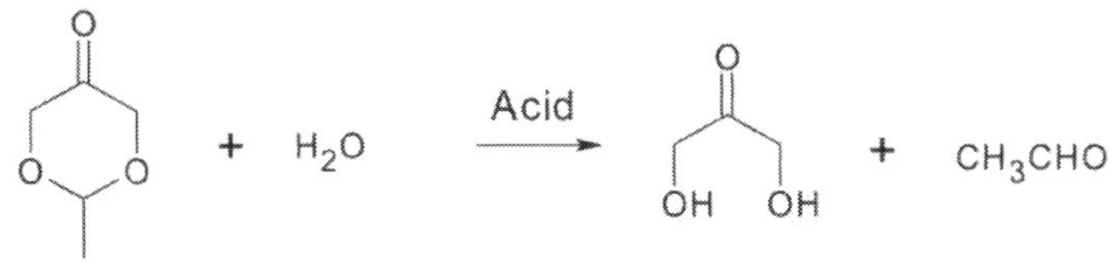


**A novel process of making DHA from glycerol**

Using cis-6 as an intermediate provides a relatively quick, environmentally safe and economical route to 1,3-dihydroxyacetone (DHA) from glycerol. Based on the experimental results obtained, a novel process of making DHA from glycerol is developed and described here.

Figure 3 presents the overall scheme of a novel DHA production process based on acetaldehyde or 1,1-diethoxyethane (DEE) as the glycerol acetal forming reagent. The process may also work with other aldehydes or ketones. Selection of the acetal forming reagent is dependent on price, availability and process feasibility.

In Figure 3, glycerol and acetaldehyde are introduced to Reactor 1 to form cyclic glycerol acetals. The reaction mixture, which contains four isomers of glycerol acetal, is then evaporated to get rid of water and unreacted acetaldehyde for recycling. After that, the mixture of glycerol acetal isomers is introduced to a vacuum distillation column, in which the cis-6 GA isomer is formed, distilled, and thus obtained from the top of the column. At the same time, other isomers, which have higher boiling points, react in the reboiler to re-establish isomerization equilibrium. The continuous recycling of the less-volatile glycerol acetal isomers through the equilibration-distillation procedure can transform all of the products into the desired 6-membered structure.

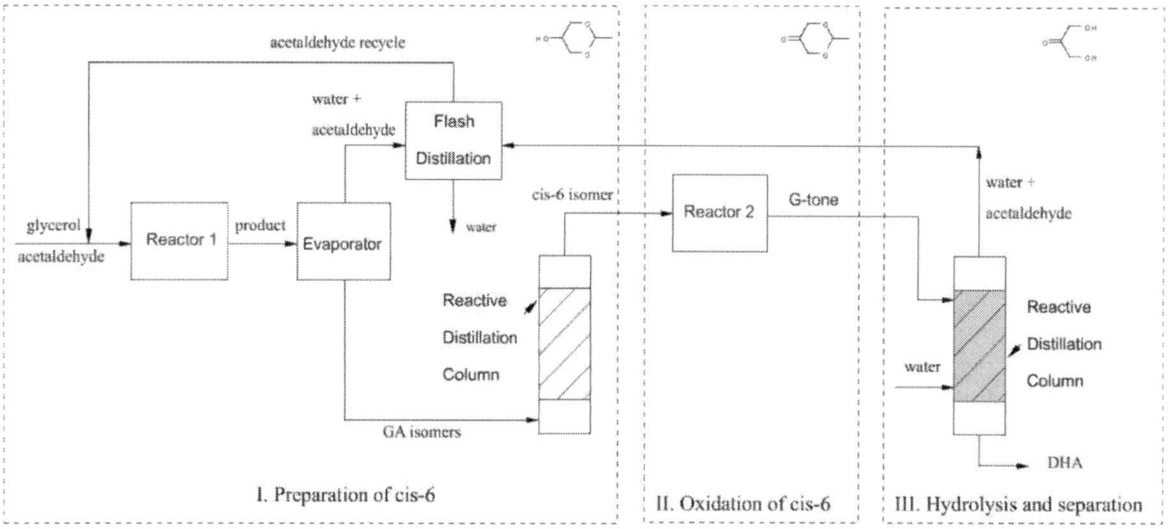


**Figure 3**. A novel process to produce DHA from glycerol.

In the second stage of this design, cis-6 GA isomer is introduced to the oxidation Reactor 2 to form 2-methyl-1,3-dioxan-5-one (G-tone). A preferred operation for this stage is to use ruthenium tetroxide-related oxidation.

In the last stage, G-tone and an excess amount of water are reacted, perhaps in a reactive-distillation column, to hydrolytically decompose G-tone to the desired product DHA and acetaldehyde, which can be recycled or reformed into DEE. The possible application of reactive distillation along with a large amount of water will overcome any chemical equilibrium limitations of this reaction, such that a high yield of DHA can be achieved from this process.

## Summary

A novel process to make 1,3-dihydroxyacetone (DHA) from glycerol is presented in this paper. The process involves three major steps: 1) formation of glycerol acetal (GA) and reactive distillation to get the 6-membered GA isomer 5-hydroxyl-2-methyl-1,3-dioxane (cis-6); 2) oxidation of cis-6 GA isomer to the corresponding ketone product 2-methyl-1,3-dioxan-5-one (G-tone); 3) conversion of G-tone to DHA via hydrolysis. Experimental results show that in each step, over 90% yield of the desired product can be achieved. Thus, the new method holds a good potential for commercial success.

## References


1. Mishra, R, Jain, S. and Kumar, A. (2008). Microbial production of dihydroxyacetone. Biotechnology Advances. 26, 293-303.
2. Garcia, R, Besson, M. and Gallezot, P. (1995). Chemoselective catalytic oxidation of glycerol with air on platinum metals. Applied Catalysis A: General. 127, 165-176.
3. Hu, W, Knight, D, Lowry, B. and Varma, A. (2010). Selective Oxidation of Glycerol to Dihydroxyacetone over Pt-Bi/C Catalyst: Optimization of Catalyst and Reaction Conditions. Industrial & Engineering Chemistry Research. 2010, 49, 10876–10882.
4. Kimura, H. (1993). Selective oxidation of glycerol on a platinum-bismuth catalyst by using a fixed bed reactor. Applied Catalysis A: General. 105, 147-158.
5. Da Silva Ferreira, A, Barbe, J. and Bertrand, A. (2002). Heterocyclic acetals f from glycerol and acetaldehyde in port wines: Evolution with aging. Journal of Agricultural and Food Chemistry. 50, 2560-2564.
6. Câmara, J, Marques, J, Alves, A. and Da Silva Ferreira, A. (2003). Heterocyclic acetals in Madeira wines. Analytical and Bioanalytical Chemistry. 375, 1221-1224.
7. Deutsch, J, Martin, A. and Lieske, H. (2007). Investigations on heterogeneously catalysed condensations of glycerol to cyclic acetals. Journal of Catalysis. 245, 428-435.
8. Aksnes, G, Albriktsen, P. and Juvvik, P. (1965). Studies of cyclic acetal and ketal isomers of glycerol. Acta Chemica Scandinavica. 19, 920-930.
9. Tojo, G. and Fernandez, M. (2006). Oxidation of alcohols to aldehydes and ketones. NY: Springer.

# (641d) Catalytic Oxidative Dehydration of Butanol Isomers: 1-Butanol, 2-Butanol and Isobutanol

**Ivan C. Lee**, US Army Research Laboratory, Adelphi, MD

The catalytic production of economically important four-carbon olefins was achieved through catalytic oxidative dehydration of 1-butanol, 2-butanol, and isobutanol using a millisecond contact time reactor. Both alumina foam and rhodium-alumina foam catalysts convert these four-carbon alcohols into four-carbon olefins, with contact time equal to 25 ms over a wide range of equivalence ratios (f) from 0.176 to 6.338 (or $C/O_{air}$ ratio from 0.0659 to 2.11) . The cracking of the carbon backbone was found to be minimal. A mixture of butene isomers was obtained due to the isomerization of the carbocation intermediates. The degree of this isomerization was determined by analyzing the ratios of butenes/ (total olefins), 1-butene/2-butene, cis-2-butene/trans-2-butene, and isobutene/ (linear butenes). In addition, a reaction mechanism of the catalytic oxidative dehydration of butanol in short contact times was proposed. In this mechanism, the C-O bond of the majority of butanol was broken to form olefinic species on the alumina surface. These surface olefinic species would undergo successive oxidation to form CO and/or $CO_2$, depending on the availability of oxygen.

See more of this Session: Chemical and Catalytic Conversions and Processes for Renewable Feedstocks
See more of this Group/Topical: Sustainable Engineering Forum

# (641e) Effects of Reaction Conditions On the Acid-Catalyzed Hydrolysis of Miscanthus Dissolved In An Ionic Liquid

**Sean J. Dee** and Alexis T. Bell, Chemical and Biomolecular Engineering, University of California - Berkeley, Berkeley, CA

Miscanthus, a perennial grass, is an attractive biofuel feedstock compared with corn and poplar due to its high energy yield per hectare and low agricultural inputs such as water and soil nutrients. Recent studies have shown the three primary components of miscanthus, cellulose, hemicellulose, and lignin, can be dissolved in ionic liquids (ILs) at levels approaching 20 wt%. An interesting question is then whether the cellulosic and hemicellulosic components of the dissolved miscanthus can be hydrolyzed to sugars in high yield. While acid-catalyzed hydrolysis of cellulose and hemicellulose has been studied extensively in ILs, the conversion of biomass is not as well understood. Recent literature has claimed that effective chemical conversions of biomass in ILs requires either thermochemical delignification which can destroy both hemicellulose and cellulose, or multistage processing that is inefficient and makes IL recycling energy intensive. This study investigated the acid-catalyzed hydrolysis of miscanthus dissolved in the ionic liquid 1-ethyl-3-methylimidazolium chloride ([Emim][Cl]). The effects of varying temperature, acid concentration, miscanthus concentration, and the gradual addition of water were undertaken. The similarities and differences between the rate of hydrolysis of microcrystalline cellulose (Avicel) and amorphous hemicellulose (Xylan from birchwood) compared to the cellulosic and hemicellulosic components of miscanthus were determined. The influence of miscanthus delignification on the hydrolysis rate was also explored. The present work has demonstrated that 90% biopolymer conversion from miscanthus could be achieved, with 93% selectivity to saccharine products in a single stage hydrolysis without pretreatment.

See more of this Session: Chemical and Catalytic Conversions and Processes for Renewable Feedstocks
See more of this Group/Topical: Sustainable Engineering Forum

# (641f) Secondary Thermochemical Xylo-Oligomer Hydrolysis of High-Solids Dilute-Acid Pretreatment Slurries

**Erik M. Kuhn**[1], Joseph Shekiro[1], Nick J. Nagle[2] and Richard T. Elander[1], (1) National Renewable Energy Laboratory, Golden, CO, (2)National Bioenergy Center, National Renewable Energy Lab, Golden, CO

Dilute-acid pretreatment is a critical step in the biochemical conversion of biomass to fuels and chemicals, releasing hemicellulose as oligomeric and monomeric xylose and rendering the cellulose fraction readily digestible by enzymes. It has been difficult to achieve very high yields of monomeric xylose during pretreatment, since acid hydrolysis also converts xylan to various oligomers and non-fermentable sugar degradation products. An alternative strategy for the production of high yields of monomeric xylose is to perform secondary thermochemical hydrolysis on the dilute-acid pretreated material. This secondary hydrolysis, typically performed at lower temperatures than the primary dilute-acid pretreatment step, liberates additional monomeric xylose while minimizing the yield of sugar degradation products. This work presents the results of laboratory and pilot scale primary and secondary thermochemical hydrolysis experiments over a wide range of conditions. Our results suggest that reduced severity conditions for primary dilute-acid pretreatment provide the highest overall yields of monomeric xylose after secondary thermochemical hydrolysis for the conditions tested. The results of our LC-MS investigations suggest that higher-molecular weight xylo-oligomers are converted to xylose during secondary thermochemical hydrolysis, but that a small fraction of the xylan in corn stover is converted to stable but non-fermentable compounds, such as xylo-oligomers, during both primary and secondary hydrolysis.

See more of this Session: Chemical and Catalytic Conversions and Processes for Renewable Feedstocks
See more of this Group/Topical: Sustainable Engineering Forum

# (675a) Microbial Oil Production by Oleaginous Fungi and Yeasts Using the Hydrolysate From the Dilute Sulfuric Acid Pretreatment of Wheat Straw

**Xiaochen Yu**, Jijiao Zeng, Yubin Zheng and Shulin Chen, Department of Biological Systems Engineering, Washington State University, Pullman, WA

**Microbial oil production by oleaginous fungi and yeasts using thehydrolysate from the dilute sulfuric acid pretreatment of wheat straw**

Xiaochen Yu, JijiaoZeng, Yubin Zheng and Shulin Chen

Department ofBiological Systems Engineering, WashingtonState University,Pullman, WA 99164-6120

### Abstract

Takinginto account the fundamental problems around the world: hunger, the lack ofenergy and the deterioration of the environment, biodiesel production fromlignocellulosic biomass has increasingly attracted much attention, since itdecreases the potential to spark conflicts between food and energy issues byusing non-starch, non-food biomass, and additionally it emits less green housegas to the environment than conventional fossil fuels. Generally lignocellulosicbiomass is firstly hydrolyzed either chemically or biologically to fermentablesugars derived from cellulose and hemicellulose, which will be then converted tosingle cell oil by oleaginous microorganisms, as the feedstock for biodieselproduction. Our work comprehensively explores the use of the hydrolysate fromthe dilute sulfuric acid pretreatment of wheat straw for microbial oilproduction. In this study, we investigated the performance of the oleaginousyeast and fungi on both detoxified and non-detoxified hydrolysate. The results showedthat all the strains could use the hydrolysate to produce lipids anddetoxification of the hydrolysate was not necessary for optimal cell growth andlipid production by the fungus and yeast strains. Among all the tested strains,the yeast *Cryptococcus curvatus* andthe fungus *Mortierella isabellina*achieved 34% and 50% lipid content respectively on the non-detoxifiedhydrolysate with up to 20% lipid yield. The overall mass balance shown in Fig.1indicated that lipid yield using the hydrolysatefrom dilute sulfuric acid pretreatment of wheat straw was 4.7 g lipids/100 g initial dry biomass. Our finding is quite significant as (1) itelucidates the feasibility of microbial oil production from lignocellulosicmaterials and suggests an alternative strategy for biodiesel production, (2) itprovides the possibilities for the further scale-up and demonstration, whichwill accelerate the development of biodiesel production towards thecommercialization, (3) it simplifies the hydrolysate fermentation process andcould probably reduce the cost since the detoxification step is not required.

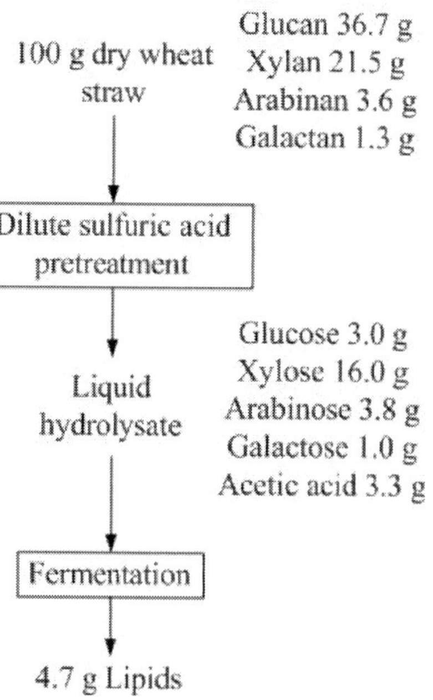


Fig. 1. The overall mass balance ofconverting lignocellulosic biomass into lipids

See more of this Session: Sustainable Forest Product and Process Innovation
See more of this Group/Topical: International Congress on Energy 2011

# BIOCORE - On a paradigm for the design and synthesis of real-life biorefineries

*Aikaterini D. Mountraki*[a], *Athanassios Nikolakopoulos*[a], *Konstantinos A. Pyrgakis*[a], *Bouchra Benjelloun Mlayah*[b], *Antonis C. Kokossis*[a]

[a] *School of Chemical Engineering, National Technical University of Athens, Greece*

[b] *Compagnie Industrielle de la Matičre Végétale, France*

**Abstract**

As oil is a fast depleting resource (BP, 2010), much is expected by the production of biofuels as valid options to replace fossil-oil transportation fuels. Biorefineries are not an entirely new concept. Many of the traditional biomass conversion technologies such as the sugar and paper industries are aspects of the bio-refining approach. By adapting some processes (such as fermentation of sugars to produce ethanol) these existing industries can develop into biorefineries. Following this skepticism, biomass is trying to act as a replacement for oil and chemicals (P. Nieuwenhuizen, A. D. Little, 2010). Lignocellulosic biomass includes agricultural and forestry residues, municipal wastes and greases. It does not require the cultivation of productive land and uses materials which are normally considered waste, such as wood and crop residues. This paper presents the efforts made through the work of BIOCORE for the transition from an existing paper industry (CIMV), to a sustainable biorefinery, by integrating and assessing diverse transformation processes. Tools used for this purpose, are systemic methods of composition and integration, optimization and process modelling. In design problems, for which there is prior knowledge of chemistry and products available, the synthesis process provides solutions at a conceptual level, defining the optimal portfolio of the products to be produced and enables a systematic partition of options by simultaneous optimization approach. Anticipating the development of new solutions and options, this paper outlines the major challenges for the development effort and the design of new biorefineries. The realization of biorefineries is not far from reality and can emerge from many conventional processes playing an important role in the energy game and the chemical market.

**Keywords:** Biofuels, Process Design, Flowsheeting, Synthesis, Biorefinery.

# (675c) Purification of Lignocellulosic Hydrolyzates by Nonionic Polymer Flocculation and Scale up Studies

**Rakesh Yasarla**, Paper & Bioprocess Engineering, State University of New York, Syracuse, NY and **Ramarao Bandaru**, Paper and Bioprcess Engineering, State University of New York, Syracuse, NY

Lignocellulosic hydrolyzates produced by hydrothermal treatment of wood chips contain hemicellulose sugars, acetic acid and significant quantities of colloidal material. These particles are mostly composed of lignin and its derivatives, which have a wide range of particle size distribution. Separation of these colloidal materials is necessary to improve fermentability of sugars into biofuels and other bioproducts. Flocculation of wood hydrolyzates prior to microfiltration improves their filterability. In this paper we investigated the separation of sugar maple wood extracts at bench and pilot scales ( of 1000 kg) by a non-ionic polymer Polyethylene Oxide (PEO) with two different molecular weights (100 and 1000 kDa).

The rate of flocculation was monitored by sedimentation of the suspensions. The optimal dosage of PEO depended on the temperature and extract concentrations and varied between 20 to 50 ppm. Although the pH of the extracts was varied from 2 through 10, the effect on suspension stability was minimal. The optimal temperature for flocculation at optimal polymer dosage was 21.5°C. The composition of hemicelluloses in the supernatants after flocculation were not altered from the raw extract showing that flocculation does not remove significant amount of fermentable sugars. Hence such pretreatment clarification is not likely to affect downstream bioproduct yields.

Flocculation and clarification with PEO were demonstrated on large pilot scale batches of 1000 kg of extracts. Further the flocculated extract was filtered and the filtrate showed a 99.5% reduction in turbidity, from 12000 NTUs in the raw extract to 50 NTUs in the filtrate.

See more of this Session: Sustainable Forest Product and Process Innovation
See more of this Group/Topical: International Congress on Energy 2011

# (675d) Revitalizing the Forest Bioproducts Industry: Steps Required to Overcome Innovation Fatigue

**Jeff Lindsay**, Asia Pulp and Paper, Shanghai, China, Cheryl Perkins, Innovation Edge, Neenah, WI and Mukund Karanjikar, NA

The challenges of the forest products industries, especially the pulp and paper industries, represent cases of "innovation fatigue" with several well-known "innovation fatigue factors" limiting progress and innovation. We discuss these factors and identify steps needed to revitalize innovation, including steps that can build culture and processes that will support innovation and new product development.

See more of this Session: Sustainable Forest Product and Process Innovation
See more of this Group/Topical: International Congress on Energy 2011

# REMOVAL AND RECOVERY OF DYE FROM PAPER MACHINE EFFFLUENTS

T. Utgata*, P. K. Mohanty**, S. Gupta, S. Chattopadhyay and A.K.Ray
IIT Roorkee, Saharanpur Campus Department of Paper Technology
Saharanpur, India
*Century pulp & Paper, Lalkua, Uttar Pradesh
**Ballarpur industries ltd., Orissa

## ABSTRACT

For production of various kinds of coloured paper, dyes are employed. The un-adsorbed dyes go with the effluents to the water bodies. The dyes used are usually not biodegradable and therefore they must be removed by any means to make the process environment friendly. Adsorption of dye on a suitable adsorbent is thus a plausible alternative of abating colour (dye) pollution from paper machine effluents. Experimental studies have been carried out using Methyl violet and Victoria blue dyes as adsorbents, on adsorbents like wood meal, bagasse pith, saw dust and crushed paper for both batch as well as continuous packed column. Parametric effects of particle size, agitation, initial dye concentration, time, and temperature on the rate of dye removal have been investigated in batch as well as continuous adsorption column.

Results indicate that a rapid initial uptake of dye takes place attributing to surface mass transfer and the rate of dye removal is found to be reduced greatly at the later stage as intra particle diffusion becomes the rate-controlling step. The surface mass transfer coefficients, $K_s$ correlated in terms of dimensionless $Sh/Sc^{0.33}$ term, against the particle variables and the intra particle rate parameter K as a function of agitation, initial dye concentration and particle size are simulated. (98.0-99.5 % of the dye adsorption has been possible within one hour. Regeneration of adsorbent and desorption of adsorbate have been tried. Experiments show promising results.

## KEYWORDS

Colour organic Dye, adsorption, wood meal, bagasse pith, sawdust, methyl violet, Victoria blue, rate kinetics, process parameters, batch reactor, models in trap article diffusion, surface mass transfer coefficients

## INTRODUCTION:

Effluents from paper machine section are colored due to the use of some colorants, such as colored pigments and dyes,organic direct, basic and sometimes acid dyes, the common dyes presently being employed by the Indian mills are: direct blue, C.I Basic green 4, C.I. acid yellow 36, aquamarine, acid orange, brilliant yellow, methylene blue, methyl violet, rhodamine, Victoria blue pergusol red, nigrosine etc.The commonly used coloured papers are coloured wove, coloured posters, envelope, matchbox etc.Their quantity, however, depends whether the end use is for cultural or decoration purposes.

The dyes are used in the manufacture of paper for a variety of reasons, including eye appeal, color-coding and brand identification. The selection of suitable colorants depends upon a number of factors such as end -use requirements, physical and chemical properties and handling characteristics. Coloring is done: (a) to produce a paper having given color or shade; and (b) to produce a white paper having a desired tint or "off-whiteness". To accomplish these ends the papermaker utilizes a number of different coloring materials that fall into one of five categories. These are non-biodegradable and cause pollution to the water bodies in terms of color, toxicity and loss of aesthetic value; they are stable to sunlight oxidizing and reducing agents and are resistant to aerobic or anaerobic digestion process. In addition to the above characteristics, color dye is one of the costliest items of paper mill inventory. Though various treatment techniques are available for removal of BOD and suspended solids from waste stream, no economically feasible process for dye color removal has been established till today. Of late stringent government legislations have enforced to spur the continuous research activities in order to remove and reuse dye from industrial effluent. Adsorption on solids appears one viable alternative to control the color displayed by dye.

The dye suspension is mixed in beaters/ mixing chests with pulp and sufficient retention is allowed for its stability and optical color pick-up by adsorption or by chemisorption mechanism. The unadsorbed dye goes with white water and gets diluted by fresh water showers. Actual color of the outgoing effluent or the extent of color deposition is rarely measured is the mills .A Major portion of white water is recirculated in well designed mill and a small portion is sent out as effluent.

The important parameters from industry viewpoint are: retention; the components and conditions of furnish (the furnish variables); the paper machine system and the interaction with other chemicals that affect color development obtained from the dye (e. g. Microbiol control agent). In most instances the papermaker desires to retain as much of an added dye as possible. This is possible by studying the equilibrium depending upon the shades-light or deep; heating the stock, adding salts to the dyed pulp and adding alum to the furnish after dye addition and adding a promoter i.e. a dye retention aid. The later is a low molecular weight, high charge density resins. Sources of interfering substances include fresh water and its constituents, pulping residuals, bleaching residuals and the broke.

It has been investigated that there are possibilities of measuring the color removal by adsorption on solid mass from both thick and effluents. Peat, wood meal/wood dust, saw dust, activated carbon, and bagasse pith and crushed pulp etc.; some of which are industrial solid wastes generating particulate pollution. The profitable utilization of the above solid particulate matters will also serve the development of an environmentally compatible process.

**Literature Survey:**

The data from paper mil survey clearly indicates that the most important paper products use dyes like methyl violet and Victoria blue. Their quantity also varies significantly.

There are innumerous information available about the removal of various pollutants including dye from waste water, either from many chemical process industries ,like sugar mill, paper mill, textile mill , dye industries , distilleries etc. or synthetically prepared solution for removal of acid dye ,indigo dye , Victoria blue,direct dye violet and orange dye, metanil blue,BOD, COD of waste water streams (1-27). The adsorbents used were either from natural sources like saw dust, fly ash, wollastonite, rice husk , waste tea leaves, lignite, peat, wood chip ,babul wood bark, wheat straw, wheat husk,coconut or activated carbon.Space limits to indicate all of them. Some are mentioned here.

Basic dye using natural absorbents (Poots at al., 1976,1978), spruce wood meal for Teflon blue-acid blue 25 (McConvey and McKay, 1977), activated carbon as adsorbents (McKay, 1982) have been tried. The correlation between mass transfer coefficients and intra-particle diffusion rate parameter as function of three variables, namely initial dye concentration, agitation and particle size has been developed (McKay, 1982). The mechanism for preferential adsorption of high and low molecular weight dyes on different preferred sites of granular activated carbon (Paschal, 1976) and drastic reduction of surface area and efficiency due to thermal regeneration were evaluated (Paschal, 1976). The fundamentals of mass transfer in fluid particle and intra-article in slurries has been studied (Furusawa and Smith, 1973) and a model approaching a real system has been proposed. The subsequent researchers adopted the same model for simulation purposes of many important systems.

Freundlichs isotherm, Langmuir isotherm and Lagergren kinetic models are mostly used to explain the phenomena for practical applications. **Deshmukh(2007)** studied on adsorption of basic dye in aqueous solution using activated carbon and coconut shell and corncob as adsorbents. Adsorption kinetics was studied using Lagergren and Helffrichsing Models. **Gautam et al.(2011)**worked for the removal of Victoria blue in simulated waste water using sand as an adsorbent. Equilibrium isotherms were analyzed by Freundlich, Langmuir, Temkin, Dubinin-Radushkevich equations and the goodness of fitting the curve inspected using a linear regression analysis and another two nonlinear error functions; some-of- square –error and Chi square test $\chi^2$. Vrious exponential parameters ,agitation contact times, initial dye concentration, dose of adsorbent, effect of cations and anions, pH, and temperature, of the solution affecting the adsorption behavior of dye on the adsorbent sand has been studied. The thermodynamic parameters, Gibbs free energy ($\Delta G_o$), change in enthalpies ($\Delta H_o$) and change in entropy($\Delta S_o$) have been calculated

In most of the study the parameter studied were adsorption materials, pH of the solution, adsorbent dose, particle size, initial concentration of dye, agitation , time, temperature and type of activation.

There is no data available regarding the removal of colour ( dyes) used by paper mill as a function of input parameters, specially on the adsorbents like hard wood meal, bagasse pith, and saw dust, some of which are industrial solid wastes generating particulate pollution. The purpose of profitable utilization of the above solid particulate matters will also serve the development of an environmentally compatible process. Till now no study has been made on the colour removal kinetic characteristics of two major dyes, namely Victoria blue and methyl violet which are largely used by Indian paper mill. No information is also available on the pretreatment of the adsorbents, which are necessary for accurate design estimates. In the present study the colour removal kinetics of the aforesaid adsorbate- adsorbent combination are attempted in a simple batch reactor as well as continuous reactor. However, results of only batch adsorption experiments are shown in this article.

## MATHEMATICAL ANALYSIS:

The wellknown ideal but approximate model of adsorption developed by **Furusawa and Smith ( 1973 )** have been used in the present investigation for estimating removal of colour dye of paper mill with some additional conditions as follows:

The adsorbent must be free from any solute including the adsorbate. Furthermore, the particle size of all adsorbents passing through a particular mesh is assumed to have same specific surface area and pores. Shape factor of all adsorbents are considered equal to 1. This is true in the sense that all fibrous mass is cylindrical in nature. The adsorbent must be sufficiently porous and possesses a very little surface resistance and Knudsen diffusion will not take place.

Based on the above assumptions the following equation relating transient concentration change can be written:

$$\partial C_t / \partial t = -K_s S_s (C_t - C_o); \text{ at } t = 0, C_t = C_o \quad .......(1)$$

Where, $C_o$ ,$C_t$ are concentrations of dye in bulk liquid initially and at any time t respectively and $S_s$, the outer surface of the adsorbent particles per unit volume of particle free slurry. The simplified form of the differential mass balance equation can be written in cylindrical coordinates with radius r of the adsorbent particles as:

$$D_v \left(1/r^2\, \partial^2 C_r / \partial r^2 + 2/r.\, \partial C_r/\partial r\right) - \rho_P\, \partial C_{nr} / \partial t = \varepsilon_P\, \partial C_r/\partial t \quad .....(2)$$

where $C_r$, the concentration in the liquid filled pores at radius r, $C_{nr}$, the concentration of adsorbed dye at radius at radius r of the particle, Dv, the total effective diffusivity, $\varepsilon_p$, the porosity of the adsorbent particle, $\rho_P$, the particle density. The boundary and initial conditions are:

$$D_v (\partial C_r/\partial_r) = K_s (C_t - C_o) \quad .....(3)$$

$$\text{and } \partial C_r/\partial r = 0 \text{ when } r = 0 \quad .....(4)$$

$$C_r = o \text{ at } t = o \quad .....(5)$$

$K_s$ being external mass transfer coefficient.

Since equilibrium is assumed during adsorption at interior site, $n_r$, concentration of adsorbed dye at radius r of particle filled pore at radius r are related by the isotherm

$$\partial n_r/\partial t = \partial/\partial C_r\, [K\, C_r /(1+m\, C_r)]\, (\partial C_r/\partial t) \quad .....(6)$$

Eq. 6 has been numerically solved to relate $C_t$ vs t curves to the rate coefficients $K_s$ and $D_V$.

When intraparticle diffusion resistance is negligible and when the isotherm is linear an analytical solution for $C_t$ vs t is possible in the case $C_r = C_s$ and $n_r$ is also uniform throughout. The equations representing the particle mass transfer, Eqs. 2 and 3 can then be replaced by

$$M\, \partial n_r/\partial t = K_s S_s (C_t - C_s) \quad .....(7)$$

and also Eq. 6 becomes

$$dn_r /dt = K dC_s/dt \quad .....(8)$$

Analysis of isotherm plotted in the form $C_o$ /n ($C_o$ M/X) against $C_0$ indicate the Langmuir constant K.

Small cylindrical particle size of saw dust, bagasse pith, and wood meal can be assumed to be spherical particles of diameter $D_P$ and the outer surface area of the particles is obtained from M as

$$S_s = 6M/ D_P \quad \rho_t (1-\varepsilon_P) \qquad .....(9)$$

Eqs. 1, 7 and 8 with initial conditions from eq. 2 and t = 0 can be solved analytically to give

$$C_t/C_o = 1/(1+mK) + mK /(1+mK)[(-1 +mK)/mK]K_s S_s t \qquad .....(10)$$

Where, the constant in Langmuir adsorption isotherm is m, K, being the adsorption equilibrium constant, m, mass of adsorbent particles per unit volume of particle free dye.

From the contact time results, the mass transfer coefficient between bulk liquid and outer surface of particles, $K_s$ can be obtained from initial slope of $C_t/C_o$ vs t curve. Alternatively a plot of In[$C_t$/ $C_o$-1/(1+mK] vs ln t will give a straight line at t = o with intercept ln [(-1+mk)/(1+mk)]$K_sS_s$ and slope Validity of the equation at t = o is based on the assumption that the influence of the intra particle diffusion does not affect the results and the isotherm becomes linear at t = o. Therefore from the initial value of (- 1+mK) / mK vs time, $K_s$ can be evaluated.

Diffusion rate processes are usually expressed in terms of $t^{0.5}$. A plot of fraction dye adsorbed against $t^{0.5}$, if shows distinctly different linear regions having different slopes, then the initial part represents the surface adsorption followed by the intra-particle diffusion. The slope of X/M against $t^{0.5}$ is defined as the rate parameter k characteristic of the rate of diffusion in the region where intr-aparticle diffusion is the rate-controlling step.

The film mass transfer coefficients $K_s$ can be presented in the dimensionless mass transfer Numbers $Sh/Sc^{0.33}$ where Sh is the Sherwood Number, $Y/D_v$.

## EXPERIMENTAL

### Materials

The adsorbents used in all experiments were Victoria blue (basic blue 26) and Methyl violet supplied by M/s Star paper mill Ltd. India in their original form. The adsorbents were hardwood meal ( Eucalyptus Tereticornis) collected from the chipper section of the said mill; bagasse pith obtained after separating the cellulose from bagasse by dry depithing method passing through 30 mesh screen ; saw dust collected from local carpentry shops ,crushed waste paper stock and activated charcoal.

### Experimental procedure

The adsorbents were pretreated with hot water and cold-water solubility using Tappi specification (TIS - 59) to remove the water-soluble coloring matter, extractives and resinous materials to ensure that they should not interfere with the colour used during invesigation.

The physico-chemical characteristics of the adsorbents like density, solubility and viscosity were measured. The determined final concentration of dye in the sample was determined spectrophotometrically using an ELICO spectrophotometer ((Model-ClL-125). First standard curves for different dye concentrations at a wavelength 550nm were draw. Size distributions of different adsorbents were determined using Tyler mesh screen sizes. The densities of all the adsorbents were evaluated.

The experiments were conducted in a series of batch reactors to investigate the effect of agitation at atmospheric conditions (303K) with 36 mesh fraction as particle size for both the absorbents, each having 200 ml volume. For bagasse and 100 ml for other adsorbents, the effect of initial dye concentrations with time were studied for all particle sizes, for wood meal, saw dust and bagasse pith for both Victoria blue and methyl violet as enumerated in Table 1.

The characteristics of adsorb ate- adsorbents and the conditions maintained are shown in Table 1.

**Table 1 Operating conditions of Batch reactor**

| Parameters | Details |
|---|---|
| Adsorbents | Wood meal; 5 gms each |
| | Bagasse pith |
| | Sawdust |
| Particle Size, mesh | 36, 44, 52,60, 100 |
| Adsorbents | Methyl violet, 100ml / 200ml |
| | Victoria blue |
| Initial dye conc., gpl, | 0.5, 1, 2 |
| Time interval, hrs. | 0.25-18.75 |
| Temperature, K | 303,318,348 |
| Agitation speed, r.p.m. | 300,450,550 |

The concentrations were measured at different time intervals from 3, 5.5, 6, 12 and 18 hours spectrophotometrically. The effect of temperature was studied for the same amount of absorbents for wood meal and bagasse pith only from three levels or temperature, 303,318 and 348K at constant volume of adsorbents and particle size of 36 mesh. Concentrations of dye were measured at three time intervals, namely 0.25, 0.75 and 1.5 hrs.

## RESULT AND DISCUSSIONS

The experimental results obtained from the laboratory were analysed and some typical representative figures 1- 13 are shown to indicate the effect of some important parameters: **The above figures are interpreted as under:**

### Effect of time:

**Fig. 1** is drawn d**imensionless concentration $C_t/ C_o$ for bagasse pith of particle size 36 mesh** as a function of time and degree of agitation for both methyl violet and Victoria blue with bagasse pith as adsorbent. The figure also compares the behavior blue and Victoria blue with speed of the agitator as a parameter.

Figure 2 indicates effect of time on concentration profiles for both victoria blue and methylene blue for bagasse pith of 36 mesh at 550 rpm at 2 gpl concentration. It is evident from the plot that up to 15 min's time, nearly 83% for the input dye is removed in case of Victoria blue whereas 50-60% for the case of methyl violet. At the initial stage there is a sharp decrease of $C_t/ C_o$ value followed by a much slower dye removal rate in subsequent stage. For 88% dye removal approximately 45 min's are required for Victoria blue in comparison to 70% for methyl violet.

**Effect of agitation:**

Fig 1 also reflects the influence of agitation on the dye removal kinetics. The results indicate that for the methyl violet dye, the rate of colour removal from the bulk solution is controlled by the degree of agitation. The effect is not so much sensitive in case of victoria blue. However in both the cases it is found that higher the degree of agitation more is the rate of dye removal. The effect of increasing stirring rate is to decrease the resistance to mass transfer in the film surrounding the adsorbent particles. It is evident that there are two distinct regions to each curve. When the solid adsorbent is added to the solution at the start of each adsorption reaction, there is an initial rapid rate of dye removal. The initial region of the plot is attributed to adsorption of dye particles on to the surface of adsorbent particles. At about 15 minutes after start of each run, the rate of dye removal starts to fall off to give a smooth curve. This part of the curve is probably due to intraparticle diffusion. As the entire process is in unsteady state large time period is required to reach 99.9% dye removal.

**Effect of initial dye concentration**

The results of a series of contact time- concentration experiments for varying initial dye concentration are shown in **Figure 3** as dimensionless concentration, $C_t/C_o$ vs time with initial dye concentration as a parameter.

Figure reveals that the values of $C_t/ C_0$ with time for both Victoria blue adsorbate with initial concentrations of 0.5(gpl) kg/m$^3$ and 2 kg/m$^3$ respectively. In other words, higher the concentration better is the initial rate of dye removal. But this rate as expected is much lower than those with agitation. Influence of agitations is therefore is very much clear. Further increase of initial concentration beyond 2 kg/m$^3$ is impractical. As explained earlier, that plot of dimensionless concentration of effluents as a function of reaction time at different rpm, the two regions exist for each curve. The initial region of rapid adsorption on particles lasts for a longer period of time than with lower initial dye concentrations before intraparticle diffusion fully controlling the process. This observation is further explained by the fact that for the same particle size, in each run, the greater the initial dye concentration the greater is the driving force between the bulk solution and particle surface.

**Effect of varying particle sizes**

Figure 5. Value of concentration of 5 range particle size using Victoria Blue (VB)
Figure 6. Value of concentration of 5 range particle size using Methyl Violet (MV)
Figure 7. Effect of varying particle size for wood meal and saw dust
Figure 8. Effect of different temperatures on Ct/Co profiles for Victoria Blue (VB) and Methyl Violet (MV)
Figure 9. Logarithm of rate parameter versus particle size, Dp, for Victoria blue, for bagasse pith
Figure 10. Rate parameter versus particle size, Dp, for Victoria blue, for bagasse pith
Figure 11. Rate parameter versus particle size, Dp, for Victoria blue, and for bagasse pith
Figure 12.Degree of adsorption as a function of time with RPM as a parameter
Figure 13. Effect of Particle size on rate parameter, Ks, Sherwood no. and $Sh/Sc^{0.33}$ for bagasse pith.

**Figs. 4-7** indicate the value for concentrations for various particle sizes used in the experiment for Victoria blue and Methyl violet at the same initial concentration, 1 gpl and 2 gpl.for various types of adsorbent.

It is quite clear that with decrease in particle size the value of concentration decreases with time exponentially. This can be explained by the fact that the surface area exposed for adsorptions increased as the particle size is decreased; thereby more mass transfer is inevitable. However the net gain is not much as anticipated for bagasse pith as adsorbent though initially little advantage is obtained for smaller particle sizes. But finally 36 mesh sizes scores over the other. This fact is due to the agglomeration of smaller size particle by prolonged exposure which inhibits mass transfer by increasing the resistance.

**Comparison of nature of adsorbents**

**Figures** also compare the relative efficiency of all the adsorbents i.e. bagasse pith, saw dust and wood meal. It is evident that for bagasse the dye removal rate is much faster than those of saw dust and wood meal for both the adsorbents. Wood meal possesses the slowest rate. 90-95% dye removal is accomplished in case of bagasse at about 3 hours' time in comparison to 78-80% for sawdust and maximum 50% for wood meal at the same particle size, and temperature and for methyl violet adsorbent. For victoria blue 95% is achieved within 1.25 hours' time for bagasse whereas at 3 hours 97% for sawdust and maximum 85% in case of wood meal are estimated. A similar behavior is observed for all particle sizes. For cases of sawdust and wood meal, however 36-mesh size gives better performance than other smaller size particles. It may be the fact that the extractives (alcohol- benzene, ethyl ether etc.) or resins present in these adsorbents have the tendency to agglomerate the particles in presence of dye, transforming them to be impervious to mass transfer.

**Effect of temperature**

The effect of temperature on the dye removal kinetics is shown in **Fig. 8 for** three temperatures, namely, 303K, 318K and 348K for both Victoria blue and methyl violet respectively at three time intervals using bagasse pith of particle size of 36 mesh only.

As it is expected higher the temperature more is the rate kinetics. The increase in rate is due to lowering of viscosity and hence reducing the film resistance.The resistance in the pore diffusion is unaltered which is considered to impose maximum resistance. Therefore not very significant improvement can be achieved. The results at still higher temperature 363K did not yield satisfactory performance because of significant vaporization of solvent at atmospheric pressure and the adsorption equilibrium became disturbed. Hence it becomes unsuitable for prediction or for design use.
The degree of adsorption by the two dyes were more clearly shown in Figs. 9-10.

**Evaluation of various parameters**

The density, surface area per unit volume and diffusivity of wood meal are 92 $kg/m^3$, 5.5$cm^{-1}$, 2 $X10^{-5}$ $cm^2/s$ respectively. Whereas the density and viscosity of methyl violet and Victoria blue are 1040,1004.9 $kg/m^3$ and 0.99 and 1.16 mPas respectively. Using the Furusawa and Smith adsorption model and based on McKay's suggested procedure as already explained in the section of mathematical analysis, the following parameters were estimated graphically: Langmuir Isotherm analysis (from $C_o$ M/X and $C_o$ relationship at various particle diameter; the Langmuir Constant is of the order of 217 $cm^3g^{-1}$.); the value of mass transfer coefficients between the bulk solution and particle surface from the initial slope of the In ($C_t/C_o$-1 /(1+mK) against t plot; the effect of diffusion from X/$C_o$ i.e. the fraction of dye adsorbed and $t^{0.5}$ at varying particle sizes; The rate parameters determined from the slopes of the linear parts of graph of X/M against $t^{0.5}$ and its dependence on particle diameter was determined.A plot of log k vs log of ( $d_{pm}$, the mean particle diameter was also made. A straight line is obtained. $K_s$ was determined through the plot of ln [ vs t is a$C_t/C_o$- 1/(1+mK)]. Effect of intra-particle diffusion with varying degree of agitation and the resultant effect on mass transfer coefficient was estimated. It is found that increasing the mean particle diameter results in increase in $K_s$. The various parameters are shown in the following table.

| $D_P$,mm | $K_S$ | -log $K_S$ | Sh | Log Sh | Sc | $Sh/Sc^{0.33}$ | Log of $Sh/Sc^{0.33}$ | Log $D_P$ |
|---|---|---|---|---|---|---|---|---|
| 0.417 | 0.38 | 0.42 | 792 | 2.8937 | 57717 | 21.2566 | 1.3 | -1.37 |
| 0.244 | 0.377 | 0.4236 | 459 | 2.662 | 57717 | 12.319 | 1.09 | -1.61 |
| 0.208 | 0.333 | 0.47753 | 343 | 2.535 | 57717 | 9.20 | 0.96 | -1.68 |

This can be explained by the fact that a large surface area is presented to the dye molecules in solution. The opposite effect occurs for large adsorbent particles. The film mass transfer coefficients, $K_s$ were transformed in the dimensionless mass transfer form $Sh/Sc^{0.33}$.

**Fig.11-12 indicate the effect of particle size on rate parameter, mass transfer Sherwood No. and $Sh/Sc^{0.33}$ for bagasse pith.** All give distinct linear relationship. The dependence of rate parameter $K_S$ on particle diameter shows that the intra-particle diffusion is predominant rate controlling step during a large part of adsorption process.

**Simulation study:**

The model predicted data were simulated with practical data obtained from a model solution and the conditions of the mill.As there is no effluent available from any Indian mill , the above solution were prepared by pulping rice straw at the conditions of chemical cooking.The conditions maintained during the preparation of solution were as under:°SR:35,Rosin,2%, Alum,4%,Talcom,8-10%, Methyl Violet: 2 gpl which is required for manufacturing coloured poster paper. Trials were made for both methyl violet and Victoria blue with crushed paper as adsorbent. Adsorption of Victoria blue is found to be much higher than that for methyl violet. Approximately 99.9% is adsorbed in case of Victoria blue within 15 minutes time whereas 98.5% is adsorbed in case of methyl violet within the same time for a 5 ch slurry height and 95% is adsorbed at 3 cm slurry height. At most one hour is required for achieving 99% adsorption of the dye on the adsorbent.

## RECOVERY OF DYE

Some experimental trials have been carried out to desorb the adsorbed dye from the exhausted adsorbent using several solvents. The solvent chosen were Methanol, Acetone, 1,4 Dioxane and Acetic Acid. Acetone gives best results whereas the acetic acid has poor performance. Almost all the dye was removed by the first three solvent treatment process and very minor amount remained with the adsorbent, which may be chemically reacted (due to chemi-adsorption mechanism) and they could not be removed inspite of use of larger amount of solvent. It may be explained by the fact that the adsorbent microstructure might have formed strong bonds with the dye molecules. The desorption capacity of acetic acid has been found quite low as compared to other solvents.

## CONCLUSION

Adsorption rate decreases with an increase of time. Higher initial concentration, temperature, degree of agitation, and particle size improve the dye removal rate. However, the effect of initial concentration and agitation has pronounced effect than temperature and particle size. Without agitation the rate becomes much slower and hence time required will be much more.

Adsorption rate of Victoria blue is found much higher than that of Methyl violet. Approximately 88-90% of Victoria Blue added to the adsorbate is adsorbed within 15 minutes' time with proper agitation, (for the same degree of dye removal without agitation near about 75 minutes are required) using bagasse pith of 36 mesh size whereas 55% is adsorbed in the case of Methyl violet within the same time. Total treatment and regeneration time probably will not cross 2.5 hours' time.

Bagasse pith has been found to be the best adsorbent compared to sawdust and wood meal. Saw dust will be the next preferred one..

Higher particle size than 36 mesh is found to decrease the dye removal rate in case of sawdust and wood meal.

The influence of rate parameter shows that the intraparicle diffusion is predominant rate controlling step during a large part of adsorption process. The equilibrium data follows Langmuir adsorption isotherm.

**REFERENCES:**


1. Ottamayer, W. (1930). The use of Peat and Wood for the treatment of textile effluent-Gesundhrits- Ingenieur, 53, 185.**(1930).**
2. Poots, V.J.P., McKay, G. and Healy, J.J. (1976). The removal of acid dye from effluents using natural adsorbents-II Wood,Water Research,10,1067(**1976).**
3. **McKay G. and Poots V.J., Kinetics and diffusion in color removal from effluent using wood as adsorbent,J.Chem.Tech. Biotechnol, Vol.30,Pp 279-281,1980**
4. McConvey I.F. and McKay, G. (**1977)** Dimensionless mass transfer correlations for colour removal from effluent using Wood meal as adsorbent. Indian Chemical Engineer,April-June,**1981**, Vol.. XXIII, No.2, PP.33-37
5. McKay, G. (**1982)** Intraparticle diffusion and surface mass transfer processes during colour removal from effluent dye carbon adsorption- Colourage 29,11
6. Paschal, B.D. (**1976**). Proceeding of 31$^{st}$. Industrial Waste Water Conference, Purdue University, 175.
7. Poots, V.J.P., McKay, G. and Healy, J.J.,Removal of Basic Dye from effluents using Wood as an Adsorbent, J.W.P.C.F.,50,926(**1978**).
8. Furusawa T.,and Smith J.M.(**1973),** Fluid particle and Intraparticle mass transfer in slurries, Industrial Engineering Chem. Fundamentals,12,197
9. Amin Nevine Kamal, Removal of reactive dye from aqueous solutions by adsorption in to activated carbon prepared from sugar cane bagasse pith,Desalination 223(**2008**) 152-161
10. G. McKay, J.F. Porter and G.R. Prasad, The removal of dye colours from aqueous solutions by adsorption on low-cost materials, Water, Air Soil Pollut., 114 **(1998)** 423–438.
11. W.T. Tsai, C.Y. Chang, M.C. Lin, S.F. Chien, H.F. Sun and M.F. Hsieh, Adsorption of acid dye onto activated carbons prepared from agricultural waste bagasse by $ZnCl_2$ activation, Chemosphere,45 **(2001)** 51–58.
12. S.P. Raghuvanshi, R. Singh and C.P. Kaushik, Kinetics study of methylene blue dye bioadsorption on bagasse, Appl. Ecol. Environ. Res., 2 **(2004)** 35–43.
13. G. McKay, G. Chl Wal Hul and G. Buning Chen, Film-pore diffusion modeling and contact time optimization for the adsorption of dyestuffs on bagasse pith, Chem. Eng. J., 84 **(2001)** 77–94.
14. G. McKay, M. El-Geundi and M.M. Nassar, Adsorption model for the removal of acid dyes from effluent by bagasse pith using a simplified isotherm, Adsorp. Sci. Technol., 15 **(1997)** 737–752.
15. G. McKay, M. El-Geundi and M.M. Nassar, Pore diffusion during the adsorption of dyes onto bagasse pith, Process Safety Environ. Protect., 74 **(1996)**277–288.
16. G. McKay, M. El Geundi and M.M. Nassar, Equi- librium studies during the

removal of dyestuffs from aqueous solutions using bagasse pith, Water Res., 21 **(1987)** 1513–1520.
17. N. Kannan and M.M. Sundaram, Kinetics and mechanism of removal of methylene blue by adsorp- tion on various carbons — a comparative study, Dyes Pigments, 51 **(2001)** 25–40.
18. V.K. Garg, R. Gupta, A.B. Yadav and R. Kumar, Dye removal from aqueous solution by adsorption on treated saw dust, Bioresour. Technol., 89 (**2003**)121–124.
19. P. Nigam, G. Armour, I.M. Banat, O. Singh and R. Marchant, Physical removal of textile dyes from effluents and solid-state fermentation of dye- adsorbed agricultural residues, Bioresour. Technol.,72 **(2000)** 219–226.
20. S. Senthilkumaar, P. Kalaamani, K. Porkodi, P.R. Varadarajan and C.V. Subburaam, Adsorp- tion of dissolved reactive red dye from aqueous phase onto activated carbon prepared from agri- cultural waste, Bioresour. Technol., 97 **(2006)**1618–1625.
21. V.J.P. Pools, G. McKay and J.J. Hearly, Removal of basic dye from effluent using wood as an adsorbent, Water Pollut. Control, 50 **(1978)** 926.
22. Vineeta Gautam , Karan Pratap singh, sanjay Kumar and Vijay Laxmi Yadav,Removal of Victoria Blue by using Ganga River Sand as a low cost Adsorbent, Proc. Of 13th Conference of International Academy of physical sciences , held at University of petroleum and energy studies on 14-16th June,**2011** pp.222
23. McKay G,M.J.Bino and A.R.Altamami, The absorption of of various pollutants from aqueous solution on to activated carbon, Water Research,19(**1985**),491-495
24. Pragya Tiwari and N.P.Shukla, Water Pollution Control-Color removal from Textile Effluent, ibid,V 12 –V-14-26.
**25.** Mall I.D. and K.Mangesh Kishore, Treatment of Metanil Yellow bearing wastewater using coal flyash, Proc. Of the Eleventh National Convention of Chemical engineers, held at Roorkee between Sept.28-29,1995, PP.V-37-V-41,**1995**
**26.** Khare S.K., Pandey K.K., srivastava R.M. and Singh V.N., Removal of Victoria blue from aqueous solution by flyash, Chem. Tech. Biotechol 38,99-104,**1987**
27. Deshmukh, S.K.,Kinetic Performance and parameters evaluation for decolorization of industrial waste water using low cost adsorbent,Proceedings of 23rd National Convention of Chemical Engineers on recent trends in Chemical Engineering, Oct.,5-7 ,**2007,** pp.340-350

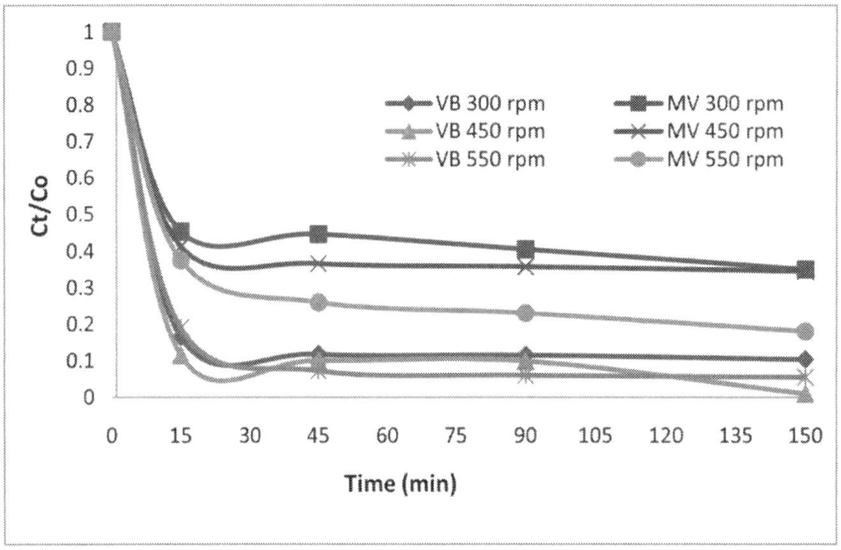


Figure 1. Effect of time and degree of agitation on dimensionless concentration for dyes, Victoria Blue (VB) and Methyl Violet (MV) with bagasse pith as adsorbent.

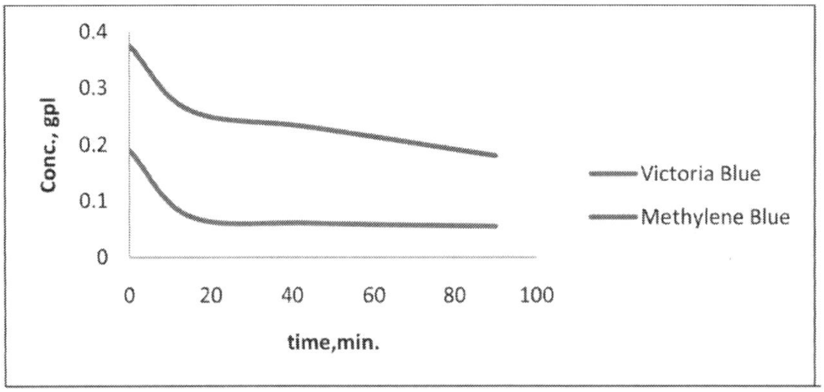


Figure 2 Effect of time on concentration profiles for both victoria blue and methylene blue for bagasse pith at 550 rpm at 2 gpl concentration

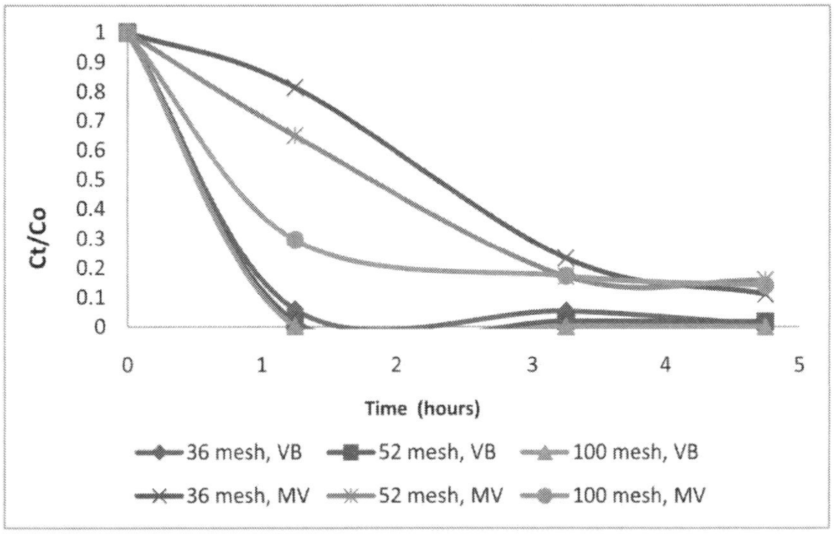

Figure 3. Effect of initial dye concentration at 2gpl on $C_t/C_o$ profiles at various particle sizes

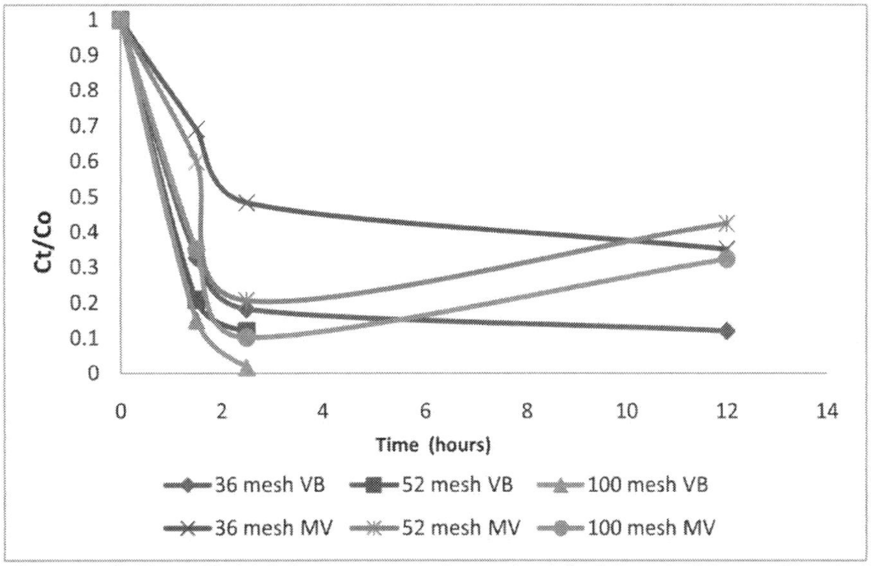

Figure 4. Effect of initial dye concentration at 1gpl for Victoria Blue and Methyl Violet.

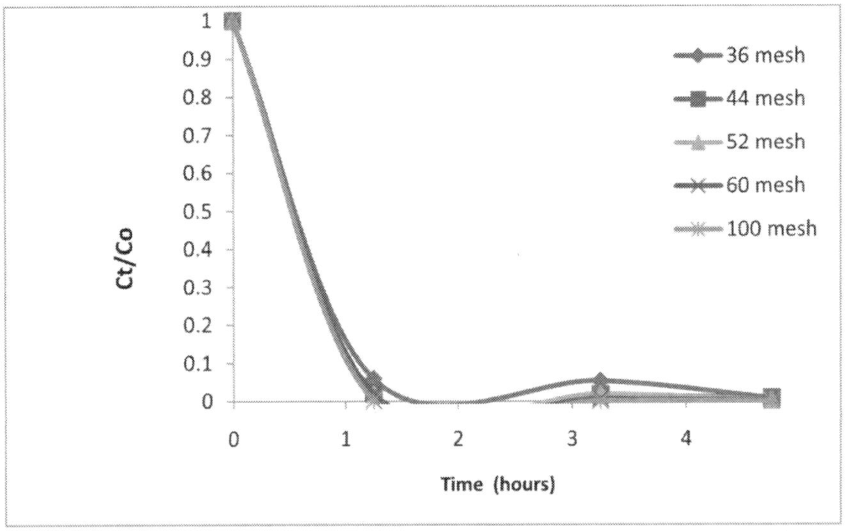

Figure 5. Value of concentration of five range particle size using Victoria Blue (VB)

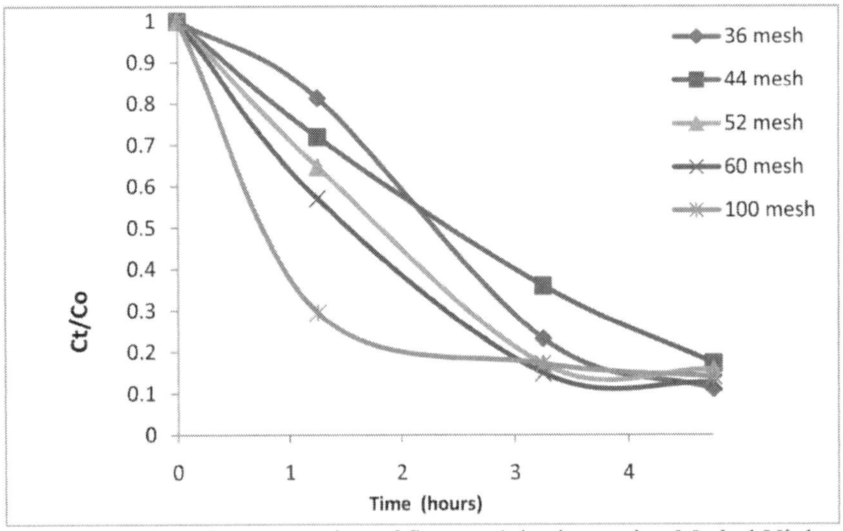

Figure 6. Value of concentration of five particle sizes using Methyl Violet (MV)

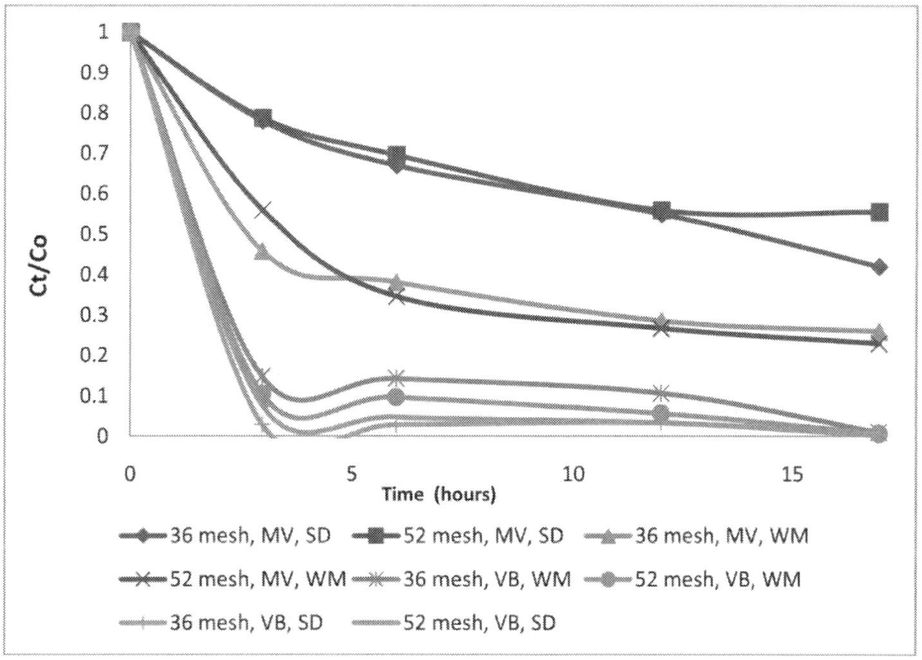


Figure 7. Effect of varying particle size for wood meal and saw dust

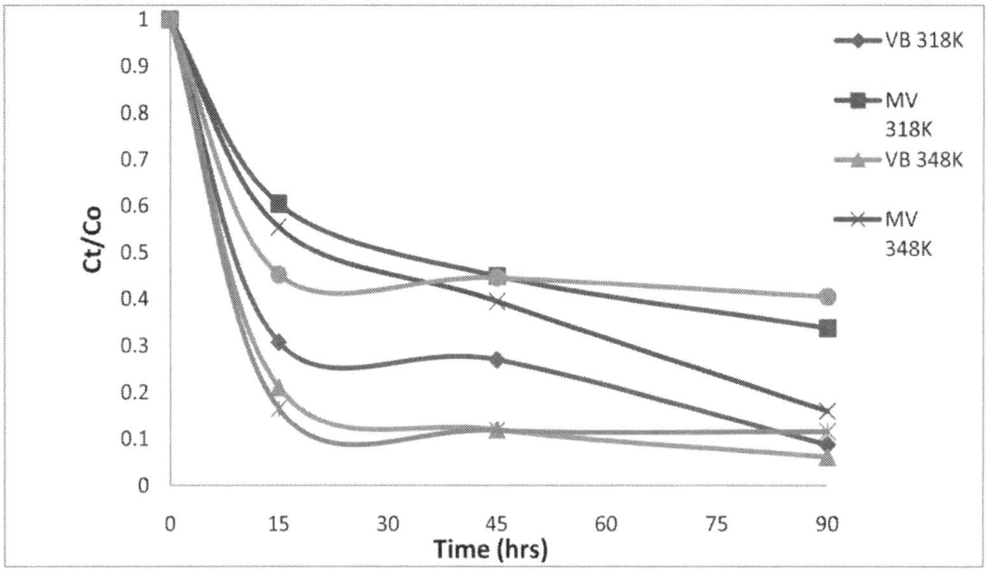

Figure 8. Effect of different temperatures on Ct/Co profiles for Victoria Blue (VB) and Methyl Violet (MV)

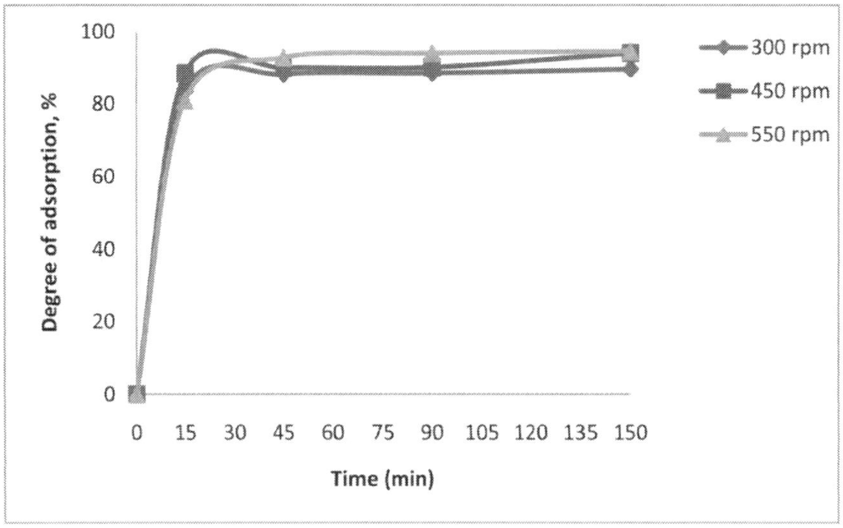

Figure 9.Degree of adsorption as a function of time with RPM as a parameter

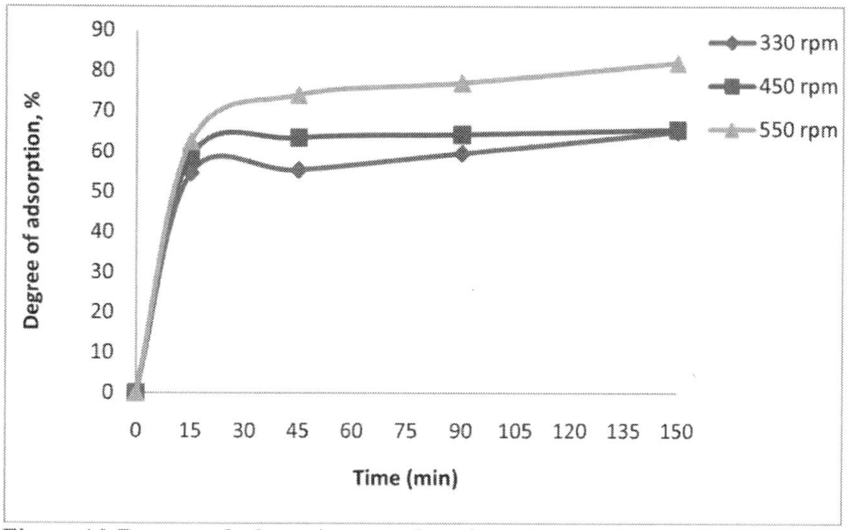

Figure 10.Degree of adsorption as a function of time with RPM as a parameter

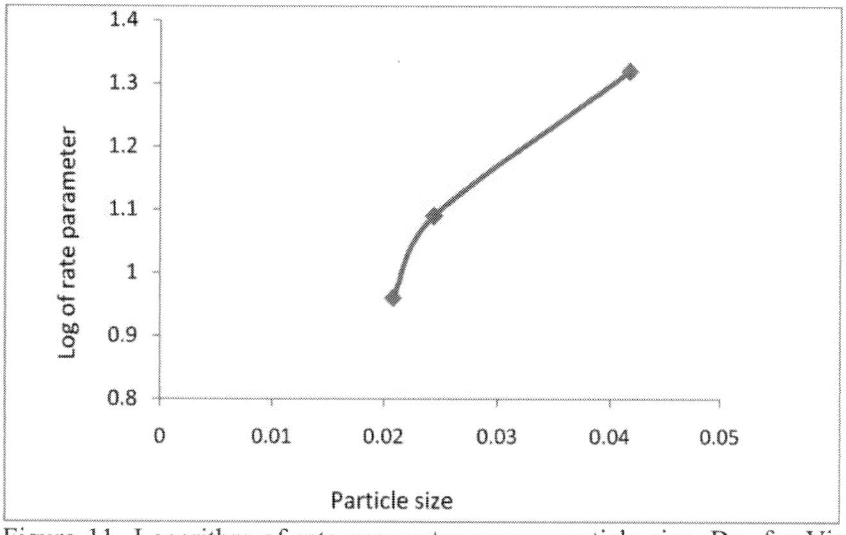

Figure 11. Logarithm of rate parameter versus particle size, Dp, for Victoria blue, for bagasse pith

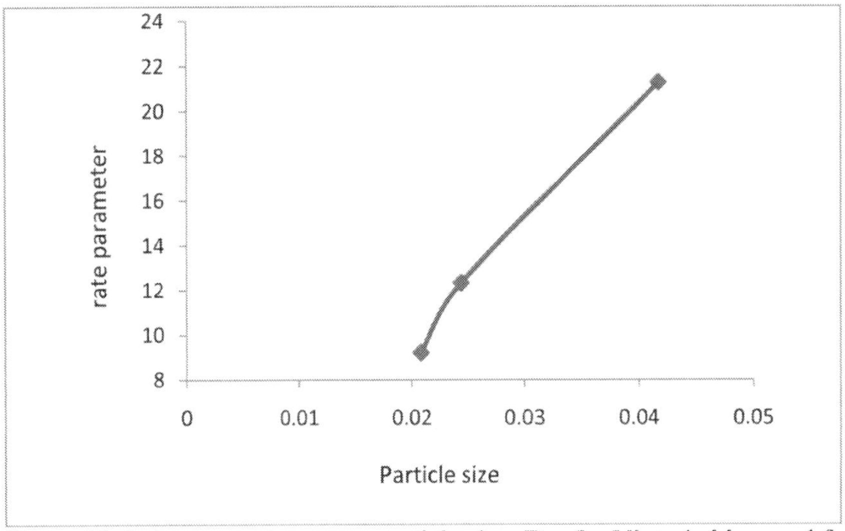


Figure 12. Rate parameter versus particle size, Dp, for Victoria blue, and for bagasse pith

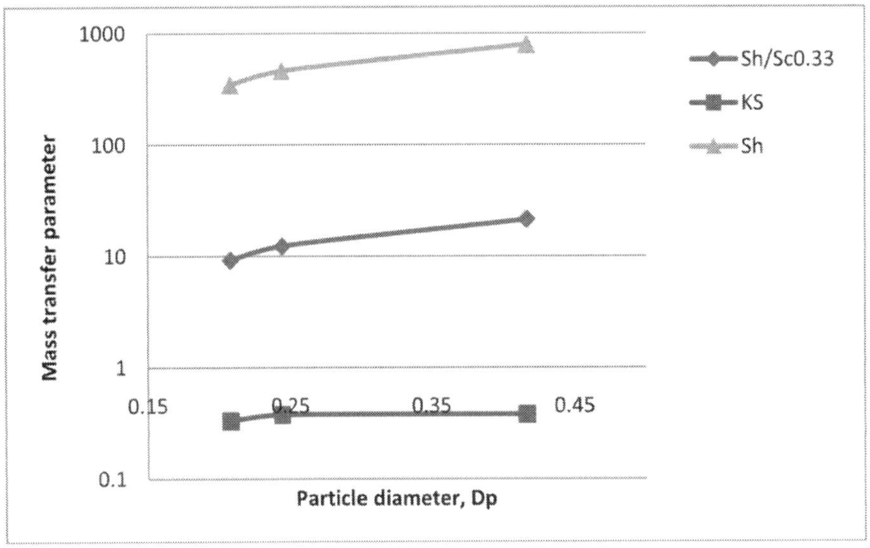


Figure 13. Effect of Particle size on rate parameter, Ks, Sherwood no. and $Sh/Sc^{0.33}$ for bagasse pith.

# INFLUENCE OF CALENDERING PARAMETERS ON COATING LAYERS OF AGRI-RESIDUE BASE PAPER FORMED BY DIFFERENT COATING PROCESSES.

*Sanjay Tyagi 1 #* , ***A K Ray*** [2] *& R M Mathur* [3]

[1] *Scientist C, Paper Testing and Energy Management Division, Central Pulp & Paper Research Institute, Saharanpur, India*
[2] *Professor, Department of Paper Technology, IIT Roorkee, Saharanpur , India*
[3] *Director, Central Pulp and Paper Research Institute, Saharanpur, India*

## ABSTRACT

The quality of coated paper is dependent on coating color composition, consisting of binders and pigments, coating process followed by quality of base paper and its source and other external process and design conditions such as type of coating equipments etc. One of the most important parameter next to coating color composition is coating process (such as blade coating, roll coating, air knife coating etc.). This is of particular importance for non wood based pulps which are relatively shorter in size and impart unevenness on the surface, thus causing the problem of fiber rising and fluff generation. It thus becomes very necessary to examine the response of coating process to non wood agro residue based pulp for growing Indian paper industry. Further, effect of calendering variables on the quality of coated paper made from agro residue needs to be examined. Unfortunately, extremely limited literature is available in this regard. In the present work, blade coating and film transfer coating processes were compared on paper made from the blend of agriculture residue pulps and that of softwood through exhaustive experimental trials made on pilot coater. Results indicated that blade coating process reduces the fiber and flock scale unevenness more efficiently than film transfer coating process resulting in improved paper gloss and print gloss mainly after the first pass only. Sheet opacity decreases more gradually in film transfer coating and occurs throughout the calendering nip. Calendering temperature plays more important role for blade coated paper and readily improved the sheet and print gloss. Reduction in PPS roughness of GCC coated paper is more during subsequent calendering trials as compare to china clay and blend of two. The effectiveness of calendering as described by "calenderability" showed increase in gloss and print gloss with increasing number of nips. The value of calenderability also founds to be in good correlation with calendering temperature and with formulations of different coating pigments.

## INTRODUCTION

The main aim of coated paper calendering is to make the paper structure more suitable for printing. Coated papers, which are intended mainly for high quality products like magazines, fliers, catalogs and books are often printed in multicolor printing processes where low roughness and uniform pore structure are essential for print result. The quality of colour print is generally related to the uniformity of the coating layer which is affected significantly by the base paper characteristics as well as by the interaction between base and coating, by coating formulation, the coating process and drying and calendering conditions.

The surface of the coated paper is formed during the coating process through consolidation of the coating color and subsequent drying of the coating layer and the base paper. The coated paper surface properties as such, mainly depend on coating color formulation applied on the surface of paper. The coating color properties affects water retention in coating color and its transfer to base paper during coating consolidation at blade pressure and drying of paper [1]. Beside this, it also depends on the type of coating process used (i.e. blade coating, film transfer coating & surface sizing process), coat weight, number of coating layers and intensity of drying. In long dwell time blade coating the time between application and metering normally long enough for structural changes in the base paper to occur before the blade [2]. Because of the base paper roughness, the coating layers thickness after the blade is uneven. The paper becomes rougher during the consolidation process and during drying of the coating layer when water is removed from the coating color. Depending on blade pressure, evenness of coating layer and hence coated surface smoothness decreases after the blade due to spring back effect of under – the – blade compressed base paper areas. It is not very clear how different coating processes influence the formation of coated layer structure on paper and their consequent coated paper properties. In film transfer coating, there are less coating layer thickness variation than in blade coating. This is due to the nature of application process where an even layer of wet coating is applied to the surface of the base paper [3]. Also, the behavior of paper, coated with different coating process during calendering is not very clear. There is also no mention of studies on the mill made paper based on agricultural residues whose surface characteristics are entirely different than those papers made from hardwood and softwood pulps. There is a need to examine these aspects for the base paper indigenously manufactured from non-wood agricultural residue raw materials.

In the present study, an attempt has thus been made to explain how coating process affects the different properties of coated paper made from non-wood agricultural residue raw materials . The influence of calendering variables like number of nips and calendering temperature on coated paper properties were also studied. Printing properties were tested in terms of print gloss. The quantification of the calendering effect of gloss and print gloss of coated using delta (Δ) gloss is meaningful only if the initial uncalendered gloss values of coating are equal. Therefore, effectiveness of calendering process on gloss and print gloss of paper coated with different coating process was studied using a new term called "calenderability" (κ).

## EXPERIMENTAL METHODS

All blade coating and film transfer coating experiments in this investigation were performed on pilot scale coater supplied by DT Paper Science, Finland. Pilot coating trials were carried out on the commercial base paper procured from the mill in reel form and made from agricultural residue pulp blended with soft wood pulp in the ratio of 75: 25. The coating pigments and other chemicals were also procured from the mill. The coating formulations used are given in Table I. The base paper was coated on single side with the coat weight of 17 $g/m^2$ at the speed of 30m/min. The distance between the applicator roll and blade was kept 250mm. The metering blade measuring 84 by 310 mm was used in this trial. The blade angle was kept at 31°. The blade load was adjusted by pneumatic cylinders at the pressure of 1.3 bar. For drying of paper two IRT- Monocassettes of infrared units were used to give a temperature of 60 – 70°C. The temperature of air drying unit was kept at 50°C. For film transfer coating, the coating solution was metered by applicator roll before applying to paper at the pressure of 2.1 bar. The web was then dried by IR dryer and air dryer.

**I: Coating formulation used for blade coating and film transfer coating trials on pilot coater**

| **Coating Ingredients** | **Parts in formulation** | | |
|---|---|---|---|
| China clay | 70 parts | 100 parts | - |
| GCC | 30 parts | - | 100 parts |
| Latex | 15 parts | 15 parts | 15 parts |
| Thickener | 0.5 parts | 0.5 parts | 0.5 parts |
| Dispersant | 0.5 parts | 0.5 parts | 0.5 parts |

All calendering experiments were carried out on laboratory single nip calender (supplied by DT Paper Science, Finland) running at the speed of 22 m/min at the linear load of 95kN/m. For calendering soft nip calendering mode was used. Since, the laboratory calender has only single nip, the paper was passed several times to increase the number of nips. (i.e. for 3 nips we have passed same sheet three times, for 5 nips we have passed paper 5 times and so on). All printing trials were made on laboratory IGT printability tester. (AIC 2 -5). Different paper and printing properties were evaluated using different ISO standard test methods.

## RESULTS AND DISCUSSION

### (a) Response of calendering nips on coated paper surface properties

The responses of calendering nips are expressed in terms of variation of surface, opacity and printing properties such as sheet roughness (PPS), printing opacity, sheet gloss and print gloss as a function of number of passes in the laboratory calender. The responses of multiple passes in a single nip calender do not exactly emulate the responses from multiple nips in actual calender, the results shown here can only used for comparison purposes between two kinds of coating processes. The results are depicted in Figures 1 to 4. From the detailed study the following note-worthy remarks can be drawn.

- In five pass or nip calendering process approximately 50% of total roughness decrease was occurred in the first nip. The reduction of roughness in calendering was larger for film transfer coated paper compared to those from blade coated paper (Fig.1). In both cases, the main roughness decrease occurred after first pass only. (i.e. about 50% in case of blade coating and 46% in case of film transfer coating). In subsequent 3 and 5 passes the roughness reduction was lower.
- The improvement in paper gloss was more for blade coated paper. Similarly, to roughness, the improvement in gloss was more after first nip (Fig. 2). The improvement in gloss in subsequent nips or passes was less for both cases. Both roughness and gloss results indicated that the deformation of blade coated paper structure occurred more easily in calendering. For blade coated paper, the roughness decrease (and consequently the gloss increase) occurred more during the first three passes and further change in gloss were small.
- The decrease in sheet opacity occurred in both cases, but decrease in opacity was more gradual in case of film transfer coating process and it occurred throughout the calendering sequence, while for blade coated paper maximum decrease in opacity occurred in first pass and there were little change in subsequent stages. (Fig. 3)

- The increase in print gloss of coated paper followed the same trend as the paper gloss. The maximum improvement in print gloss occurred after first pass in both cases. The print gloss improvement in subsequent nips was less in both cases. (Fig. 4)

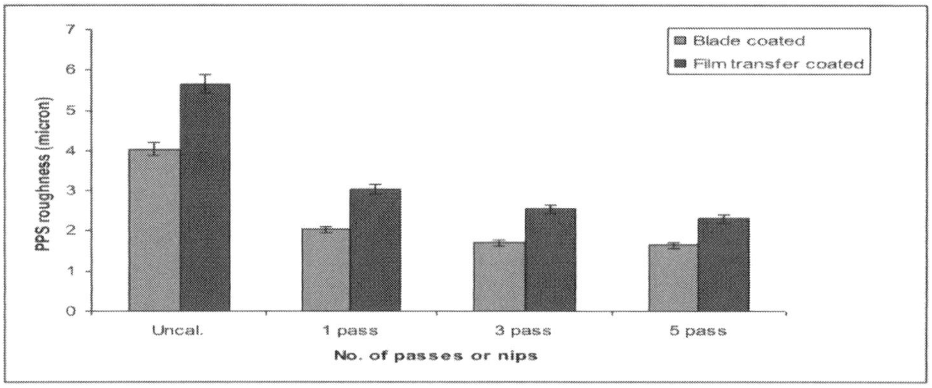


**1. Response of calendering nips on PPS roughness of blade coated/ film transfer coated paper**

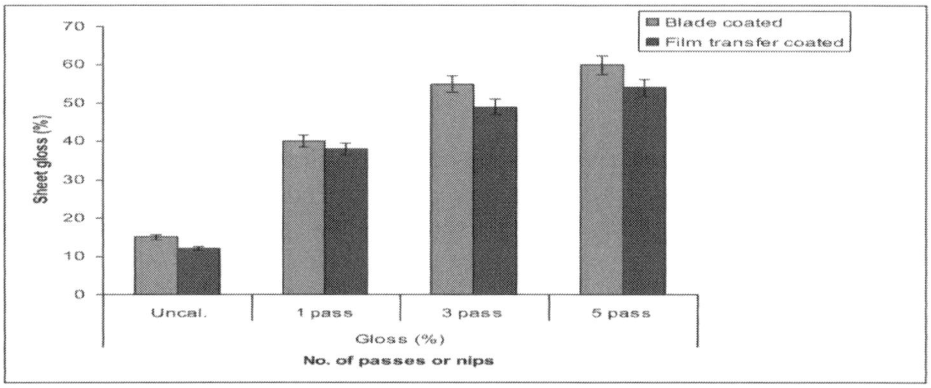


**2. Response of calendering nips on sheet gloss of blade coated/ film transfer coated paper**

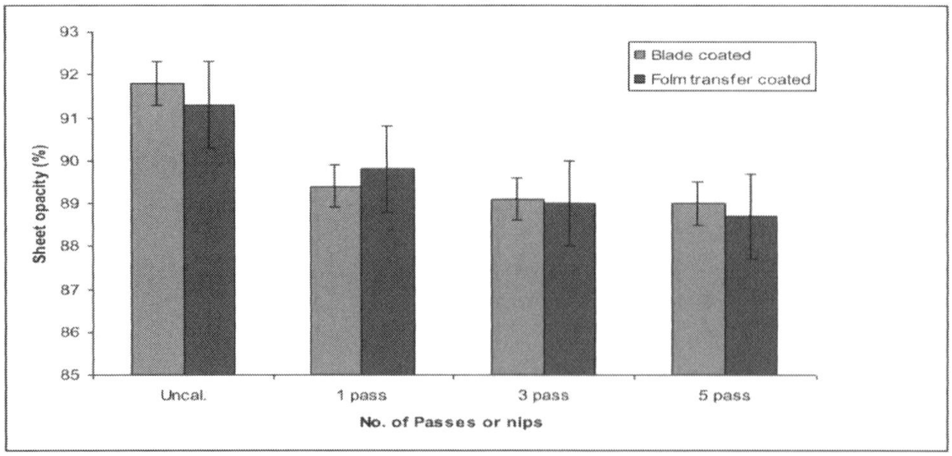


**3. Response of calendering nips on sheet opacity of blade coated/ film transfer coated paper**

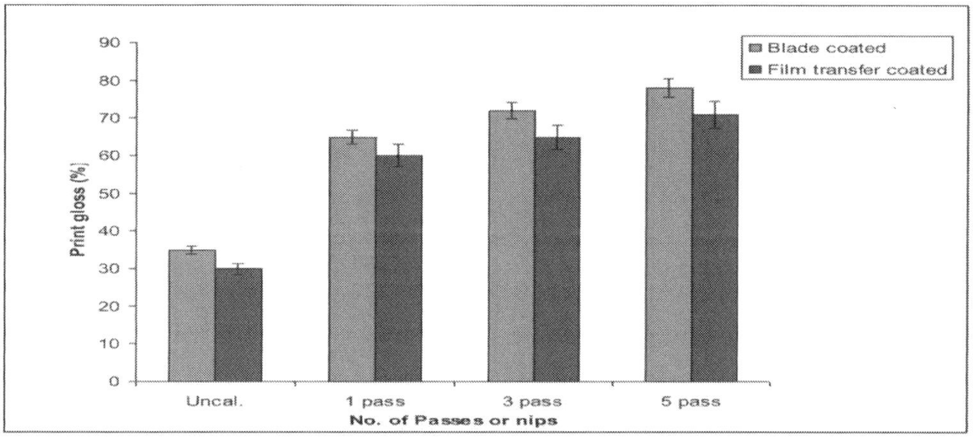


**4. Response of calendering nips on print gloss of blade coated/ film transfer coated paper**

**(b) Effect of calendering temperature on the coated paper surface properties**

The effect of calendering temperature on the gloss of paper coated through blade coating process and through applicator roll process is shown in Figures 5 through 8.

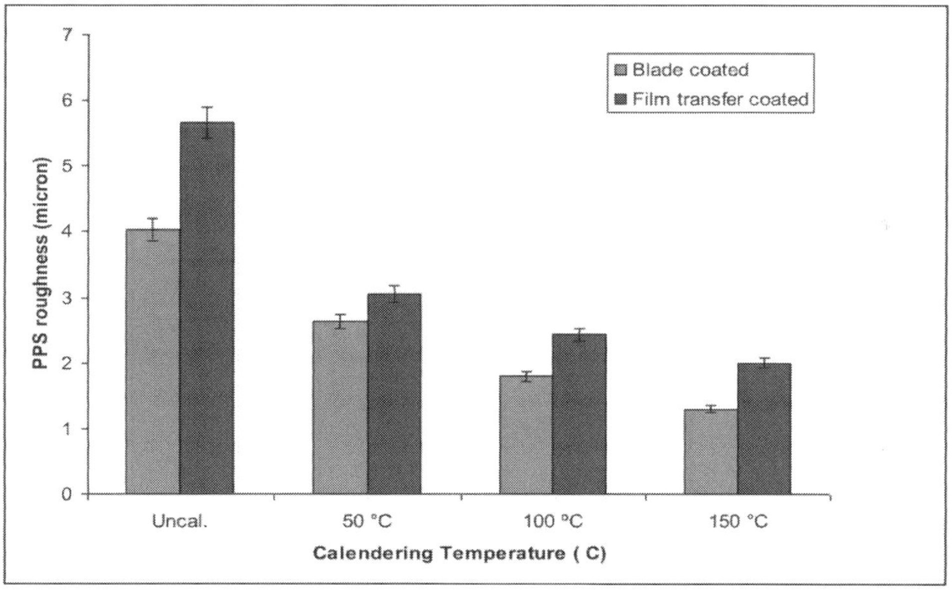


**5. Response of calendering temperature on PPS roughness of blade coated/film transfer coated paper**

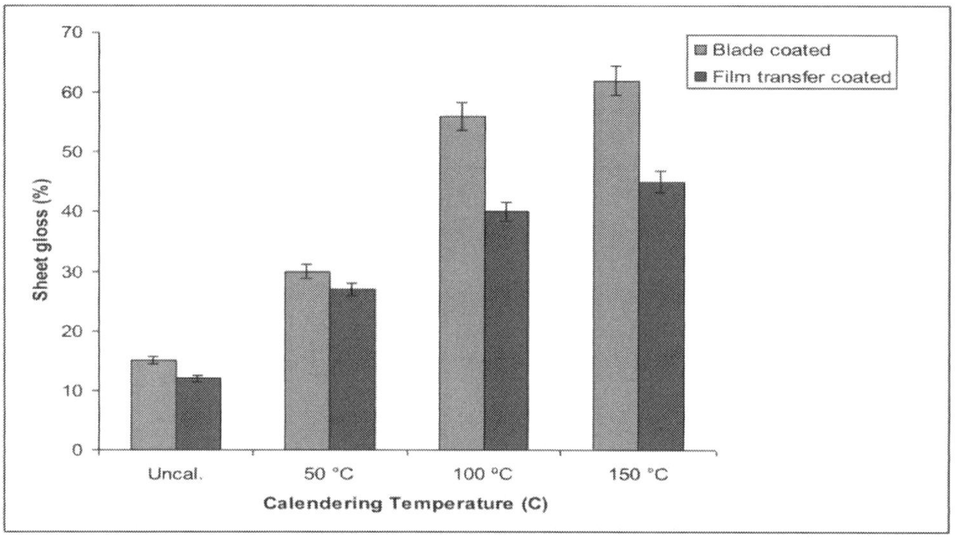


**6. Response of calendering temperature on sheet gloss of blade coated/film transfer coated paper**

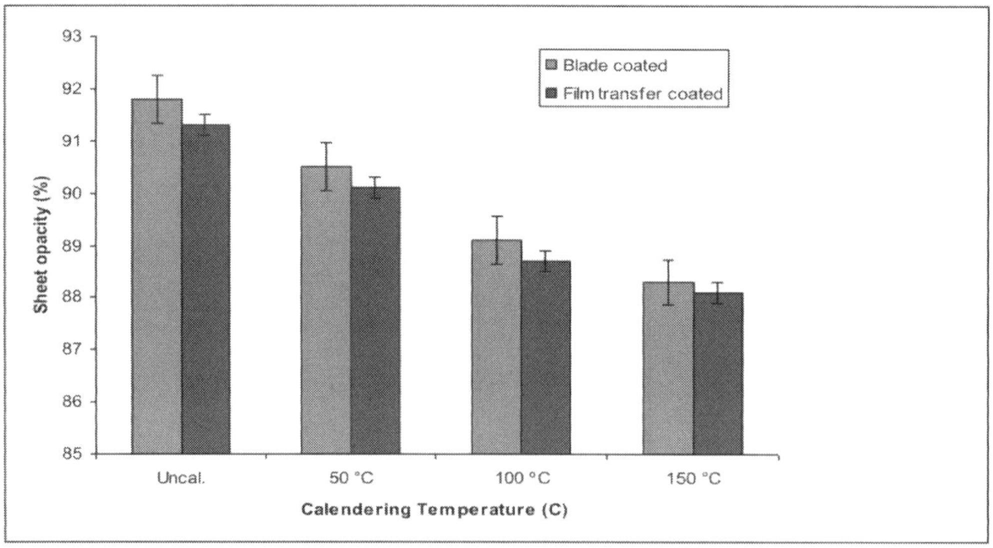


**7. Response of calendering temperature on sheet opacity of blade coated/film transfer coated paper**

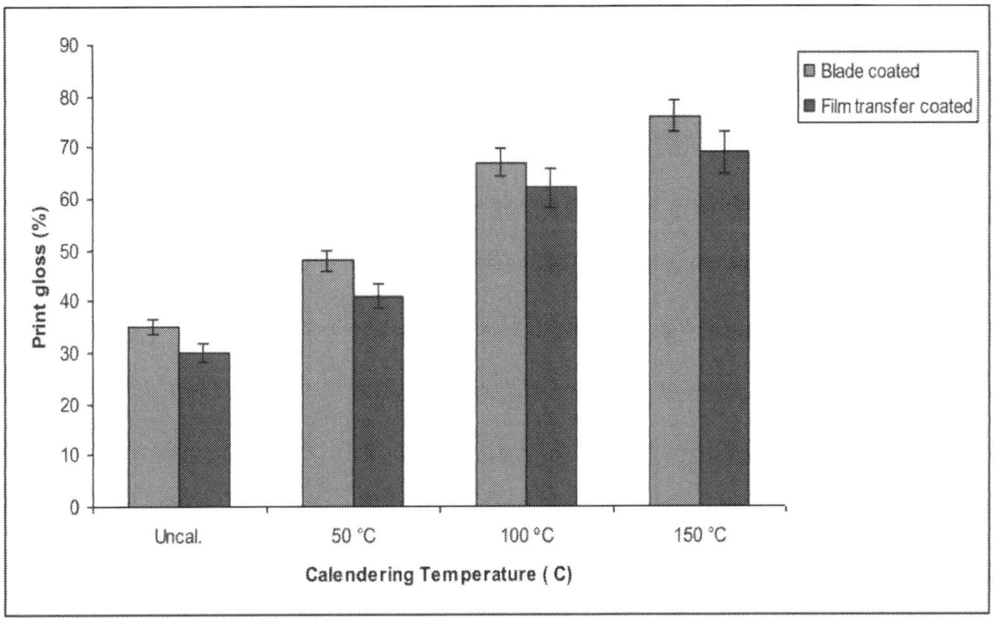


**8. Response of calendering temperature on print gloss of blade coated/film transfer coated paper**

The analysis of above figures revealed that an increase in calendering temperature increased the gloss and print gloss of coated sheet but reduced the sheet roughness. The main change in paper properties during calendering had occurred at lowest calendering temperature. The subsequent use of higher temperature had only a minor influence on properties of paper. The blade coated paper shows more improvement in gloss and print gloss (Fig. 6 & 8). The improvement in gloss is more upto 100°C in both cases. On further increasing the temperature there was not much improvement in gloss value. There was only a minor influence of temperature on roughness in both cases with film transfer coated paper showing more change as compared to blade coated paper. The effect of calendering temperature on opacity is given in Figure 7. It revealed that the decrease in opacity was nearly same for both blade coated and film transfer coated paper and occurs gradually throughout the calendering process

**(c) Effect of base paper roughness on paper coated with different pigments**

Roughness of base paper results in higher roughness of coated paper. It increases the coating non uniformity on paper surface and decreases coating coverage. This sometimes lead to print non-uniformity. The effect of calendering conditions of number of passes or nips is represented in figures 9 to 12. The main influence of the roughening of the base paper on the calendering behaviour of the paper was seen in the different roughness reduction of the clay coated and calcium carbonate coated paper. Clearly for uncalendered paper, the roughness of clay coated papers was lower as compared to ground carbonate coated papers. The roughness of paper coated with the blend of fine clay/GCC in the ratio of 70:30 lies in between that of fine GCC and clay. The experiment squeal of five calendering passes and temperature of 50 °C was although taken arbitrary, it is more practically at mill condition. It simulates more practically the temperature condition maintaining at industry of 50°C and number of passes limited to five with six roll calender stack. However, to reduce the reduction of bulk many industries attempted for five passes whereas one or three passes are the common cases of soft nip calenders.

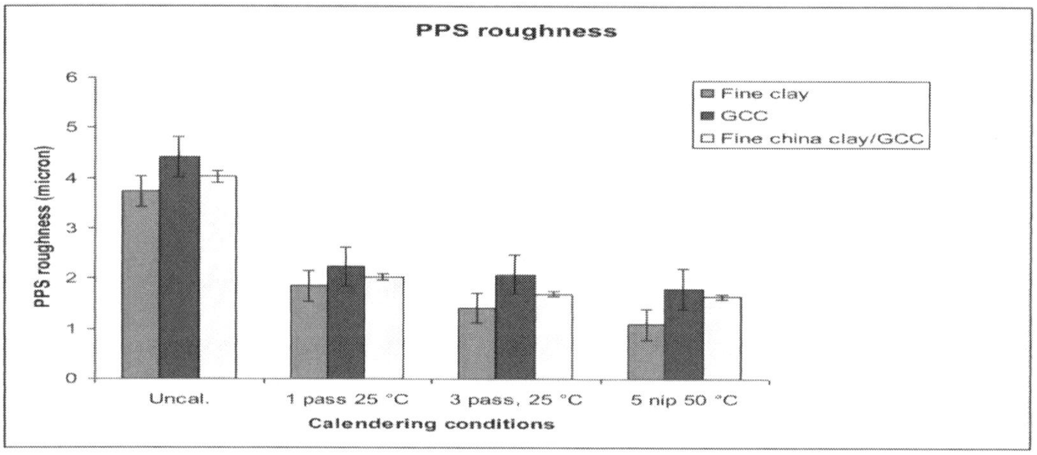


**9. PPS roughness of calendered /uncalendered blade coated paper with different coating pigments**

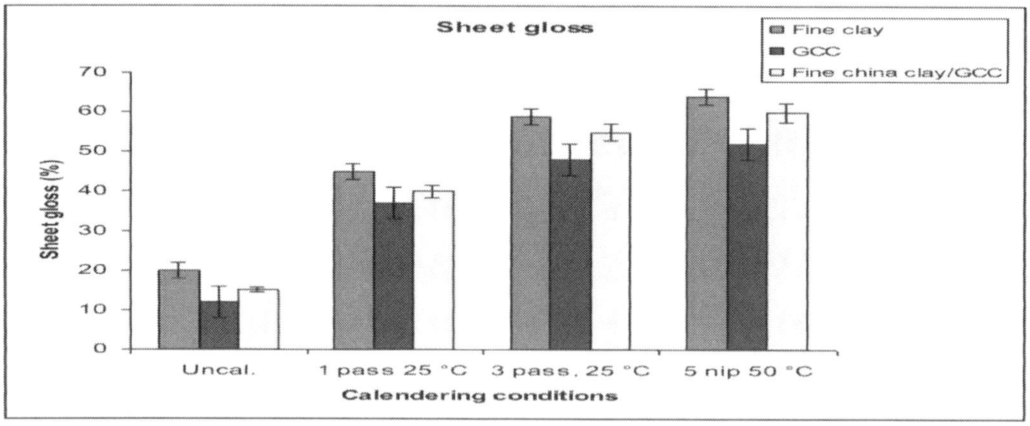


**10. Sheet gloss of calendered /uncalendered blade coated paper with different coating pigments**

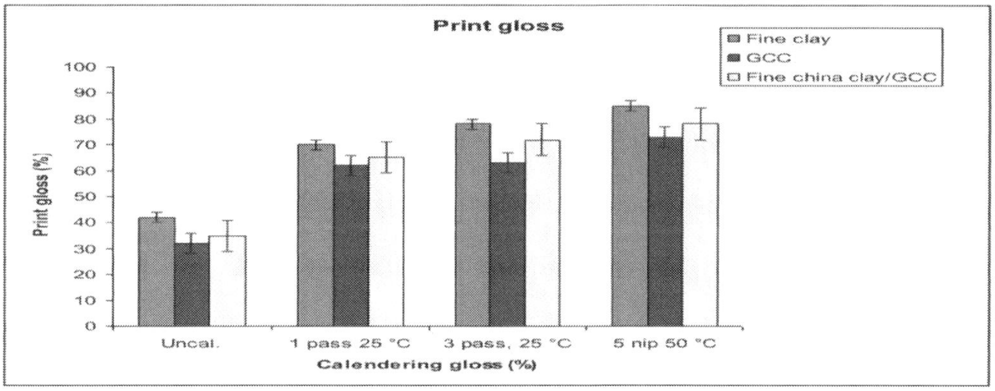


**11. Print gloss of calendered /uncalendered blade coated paper with different coating pigments**

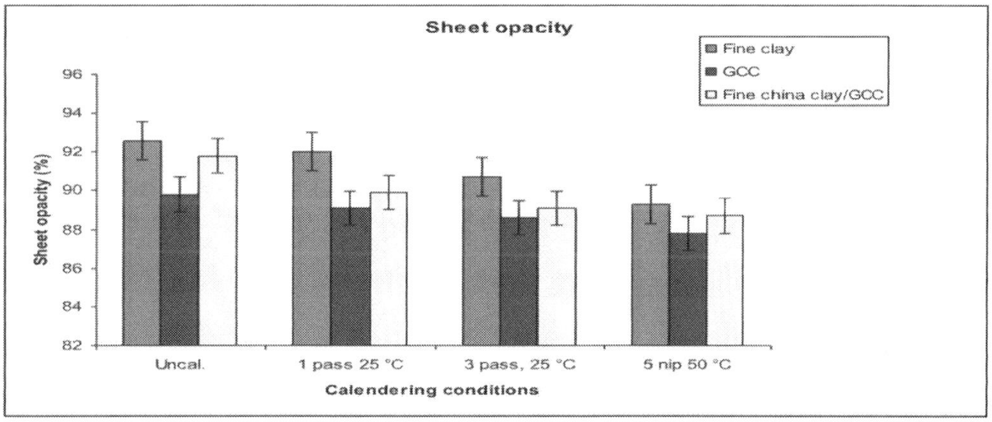


**12. Sheet opacity of calendered /uncalendered blade coated paper with different coating pigments**

An examination of above plots inferred that at higher temperature, the difference in roughness reduction between clay and carbonate coated papers was small (Fig 9). The papers coated with clay/carbonate blend have values in between fine clay and GCC. The fiber network was bulkier because of roughening and so more prone to deformation during calendering. The difference in paper gloss and print gloss was most likely due to difference in particle size and shape and only partly due to the difference in calendering behaviour of the bulk structure of the paper. (Fig 10 & 11)

The opacity of both fine clay and GCC coated paper decreased as the consequent increase in gloss. The reduction in subsequent nip was more in case of fine clay/GCC blend coated sheets as compared to fine clay and GCC coated sheets alone (Fig. 12).

**Calenderability and gloss of paper**

It is obvious from above that gloss is an important optical property of coated paper that affects end use performance. The effect of calendering on gloss is quantified using delta (Δ) gloss i.e. difference between calendered gloss $G_i$ and uncalendered gloss $G_o$. However, when comparing the gloss improvement for different coatings Δ gloss is meaningful, if the initial uncalendered gloss values of coating are identical. The potential increase in gloss is larger for coatings with low uncalendered gloss as compared to coatings with high uncalendered gloss. With different coating processes, applying coating formulations of different pigments and binders quantification using Δ gloss makes the problem still more complicated, so the effectiveness of calendering is described by a quantity given by $(G_i - G_0)/(100 - G_0)$, where $(G_i - G_0)$ is Δ gloss and $(100 - G_0)$ is the potential for gloss improvement. The quantity $(G_i - G_0) / (100 - G_0)$, ranges from zero where calendering has no effect on gloss to 1, when $G_i$ has increased to 100.

In this study calenderability κ is defined as

$\kappa = (100 - G_0) / (100 - G_i)$ where $G_0$ = uncalendered gloss & $G_i$ = calendered gloss after i nips

The value of κ ranges from 1 where calendering has no effect on gloss to infinity. The quantity $(G_i - G_0)/(100 - G_0)$ is related to κ by $1 - (1/\kappa)$

i.e. $(G_i - G_0)/(100 - G_0) = 1 - (1/\kappa)$

Thus the effect of blade coating and film application coating process on gloss and print gloss, gloss increment due to calendering temperature with respect to calenderability was studied. The results are given in figures 13 to 18. The effect of different coating pigments on gloss and print gloss of coated papers was also studied with respect to calenderability. The results are given in Table II. The values of calenderability shows increasing trend with an increase in the gloss values due to increased number of nips.

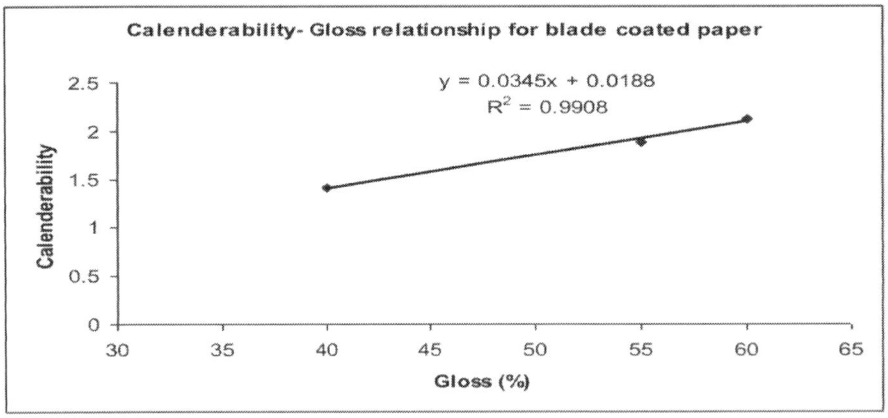


**13. Calenderability- gloss relationship for blade coated paper**

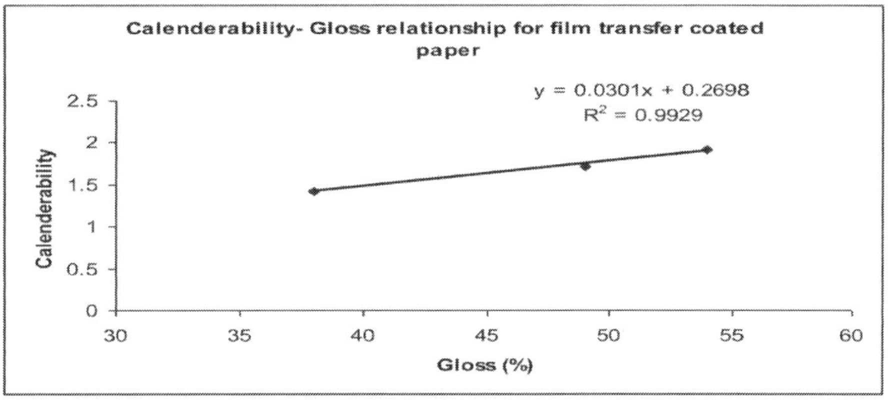


## 14. Calenderability- gloss relationship for film transfer coated paper

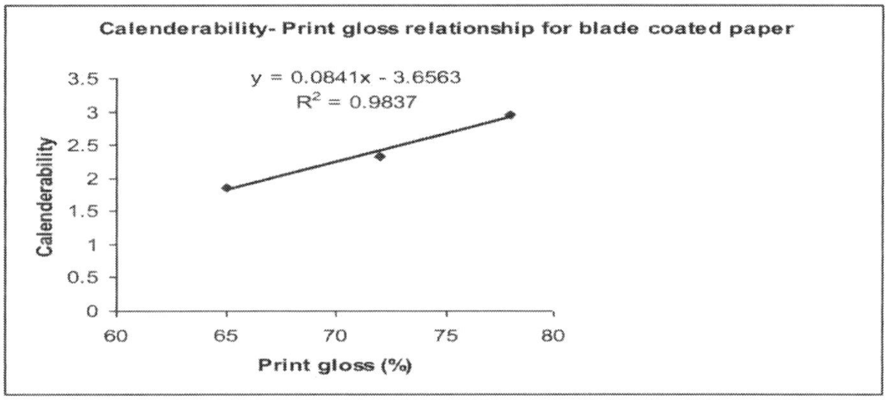


## 15. Calenderability- print gloss relationship for blade coated papers

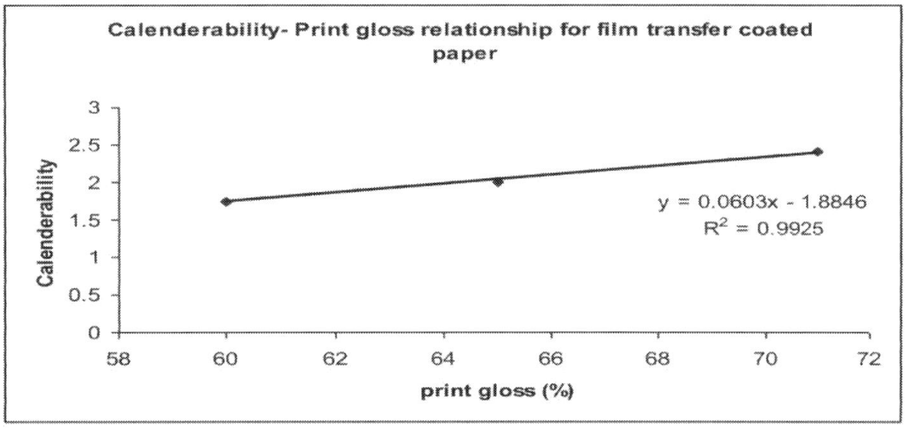


## 16. Calenderability- print gloss relationship for film transfer coated papers

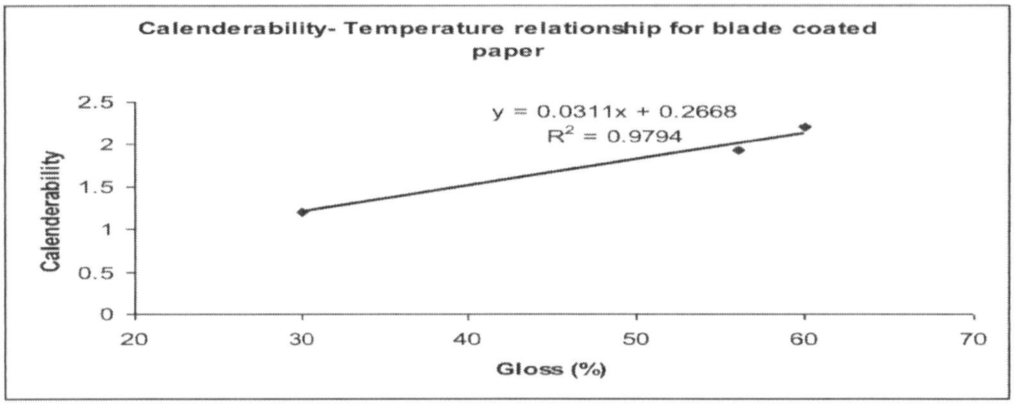


**17. Calenderability- Temperature relationship for gloss development for blade coated papers**

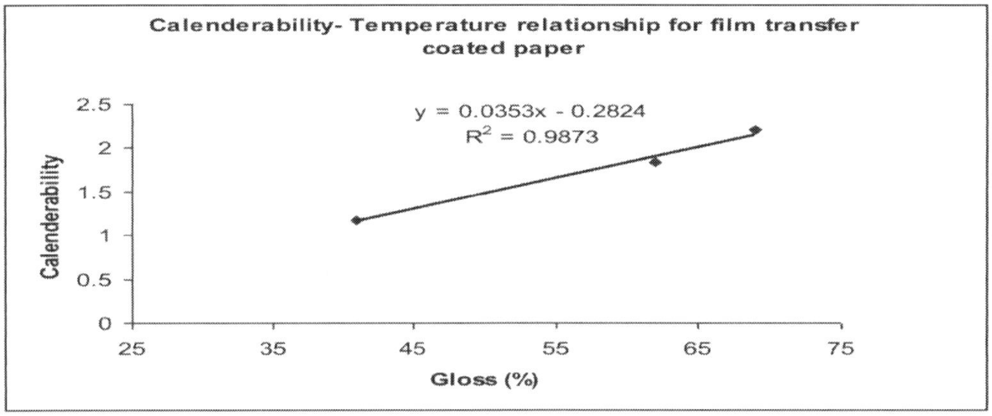


**18. Calenderability- Temperature relationship for gloss development film transfer coated papers**

## II: Calenderability of paper coating – effect of coating pigments

| | Sheet Gloss | | | | | | | | | |
|---|---|---|---|---|---|---|---|---|---|---|
| | Uncal. | 1 pass | 3 pass | 5 nip 50 °C | Δ 1 pass | Δ 3 pass | Δ 5 pass | κ after 1 pass | K after 3 pass | K after 5 pass |
| Fine clay | 20 | 45 | 59 | 64 | 25 | 39 | 44 | 1.45 | 1.6 | 2.22 |
| GCC | 12 | 37 | 48 | 52 | 25 | 36 | 40 | 1.39 | 1.39 | 1.83 |
| Fine china clay/GCC | 15 | 40 | 55 | 60 | 25 | 40 | 45 | 1.41 | 1.89 | 2.12 |
| | Print gloss (%) | | | | | | | | | |
| Fine clay | 42 | 70 | 78 | 85 | 28 | 36 | 43 | 1.93 | 2.63 | 3.86 |
| GCC | 32 | 62 | 63 | 73 | 30 | 31 | 41 | 1.79 | 1.84 | 2.5 |
| Fine china clay/GCC | 35 | 65 | 72 | 78 | 30 | 37 | 43 | 1.85 | 2.32 | 2.95 |

Δ = difference between final value and initial value

The figures 13 to 18 and data given in Table II reveal that the value of calenderability increases as the value of gloss increases with increment in number of passes or nips. The same is true for print gloss also. This is well in accordance of results shown earlier in this work where both gloss and print gloss increases with number of calendar nips and calendar temperature. Also the vale of calenderability increases with an increase in calendering temperature increases from 50°C to 150°C as expected. The behaviour of calenderability on different coating pigments reveals that the value of calenderability is more in case of paper coated with china clay in comparison to GCC. For the blend of GCC/Clay the value lies in between two. It also in well accordance earlier findings in this chapter showing higher gloss for fine clay coated paper as compared to GCC. This is in excellent agreement with the results put forward by Kan et. al. [4]

**CONCLUSIONS**

The blade coating process reduces the fiber and flock scale unevenness more efficiently than film transfer coating process. Blade coating process produces better smooth sheets than film transfer coating process. The improvement in paper gloss and print gloss was found to be more for blade coated paper and that too after the first pass only. This indicates that deformation of blade coated paper occurs more easily in calendering. The decrease in sheet opacity in more gradual in film transfer coating and occurs throughout the calendering nip. The blade coated paper more readily showed improvement in sheet and print gloss as compared to film transfer coated paper with increasing calendar temperature. PPS roughness of blade coated paper shows only a minor change with increasing calendering temperature when different pigments are used in coating formulation. The roughness of GCC coated paper reduces more during subsequent calendering trials. At high temperature difference in roughness reduction between clay and carbonate coated paper becomes small. The reduction in roughness in subsequent calendering nip is more in case of fine clay as compared to GCC and fine clay/GCC blend coated sheets. The concept of calenderability founds to be in excellent correlation with gloss and print gloss values of coated paper and its value increases like gloss value with increasing number of nips. The value of calenderability also founds to be in good correlation with calendaring temperature. The results of calenderability for gloss value of paper coated with formulations of different coating pigments also correlates well and are in well accordance with results found earlier.

**References**:


1. Singh, S., De, A., Gupta, R., Krishnan, S. V. R and Umashankar, R., "Effect of Blade on Blade Coating on The print Results of High Quality Packaging Boards", *IPPTA convention Issue,* 61-64, (2003).
2. Linnonmaa J., and Trefz M., "*Pigment coating techniques*", Chapter 24 in *Pigment Coating and Surface Sizing of Paper.* Lehtinen E (editor), Fapet Oy, Helsinki , Finland, 415– 486, (2000)
3. Trefz, M. – "Film coated offset grades- theoretical aspects and practical experience", *Tappi Coating conference proceedings,* TAPPI PRESS, Atlanta, p. 435 (1995).
4. Kan, Charles S., Kim Linda H, Lee, Do I. & Gilder, Ronald L. Van, "Viscoelastic properties of paper coatings: Relationship between coating structure –Viscoelasticity and end use performance", *Tappi Journal*, 80 (5): 191 – 201,(May 1997).

# (709a) Methyl Stearate Transesterification with n-Butanol: Process Intensification Using Distillation with External Side Reactors

**Venkata KS Pappu**, Chemical Engineering and Material Science, Michigan State University, East Lansing, MI, Aspi Kolah, Michigan State University, East Lansing, MI, Carl T. Lira, Chemical Engineering & Material Science, Michigan State University, East Lansing, MI and Dennis J. Miller, Chemical Engineering and Materials Science, Michigan State University, East Lansing, MI

Fatty acid butyl esters (FABE) are commercially important chemicals used in a wide range of applications such as lubricants, emulsifiers, detergents, and as biofuel constituents. An alternate approach to direct esterification of fatty acids for producing higher alcohol esters is to first make the fatty acid methyl esters (FAME) via traditional base catalysis and then transesterify FAME with other alcohols using heterogeneous catalysts to produce the desired esters. In previous work, we have characterized kinetics of methyl stearate transesterification with n-butanol using Amberlyst 15; this reaction is slow at 90°C even with a very large excess of butanol. Above 90°C or at high catalyst loadings, ether formation takes place, liberating water which leads to hydrolysis of the esters.

Conventional reactive distillation will likely be uneconomical for this reaction system because of the slow reaction rate and temperature limitations. We present here a study that evaluates the potential of carrying out reactive distillation in a column with external side reactors for continuous transesterification of FAME with n-butanol. Process simulation using AspenPlus describes column performance as a function of operating conditions, the number of side reactors utilized, and location of side draws and re-entry points in the column. The column configuration that maximizes conversion of the methyl ester to its butyl counterpart is described.

See more of this Session: Process Intensification by Process Integration
See more of this Group/Topical: Process Development Division

# (709b) A Comparison of the Exergetic and Economic Optimization of the Catalytic Distillation, Based On a Rigorous Model. Case Study: ETBE Synthesis

Jorge M. Gomez, Department of Chemical Engineering, Universidad de los Andes, Bogota, Colombia, **Laura Marcela Fonseca**, Department of Chemical Engineering, Universidad de los Andes, Bogot, Colombia, Jean-Michel Reneaume, Laboratoire de Thermique, Energtique et Procds (LaTEP - EA 1932), cole Nationale Suprieure en Gnie des Technologies Industrielles, Pau, France and Sabine Sochard, Universit de Pau et des Pays de l'Adour (UPPA), Pau, France

**Abstract**

With finite natural resources andhigh levels of consumption, due to the incredible growing speed of globalpopulation, it is really important to think about sustainable development *[1]*.One of the highest consumption resources is energy, not in vain it is the mainconstraint to reach sustainable development. There are many possible causes ofhigh energy consumption, perhapsthe most important one is the lack of industrial processes focused on energysaving *[2]*. As a result, exergeticoptimization arises as an interesting alternative solution to mitigate thisproblem. Exergy is well-known as the potential tocause change in terms of available work and useful energy *[3]*.

In this work a comparison betweenan exergetic and economic optimization, of acatalytic distillation (CD) process to obtain Ethyl *Tert*-butyl Ether (ETBE) ispresented. In order to achieve this, a Mixed Integer Non-Linear Programming (MINLP) problem was formulated, aiming to optimize not only continuousvariables but integer ones too ***[4]***. The strategy of solution was acombination between the algorithms: Reduced Sequential Quadratic Programming(r-SQP) and Simulated Annealing (SA). The exergeticoptimization was carried out through the minimization of the exergetic objective function, defined as the second-lawbalance around the catalytic distillation column. The main purpose of this exergetic optimization was to reduce the exergy losses by optimizing design and operating variablesof the process *[5]*. In addition an economic optimization was presented andreached by the minimization of the economic objective function, aiming to reducethe costs of the process and increase its profitability.

In order to assure the correct performance of the column, both, exergetic and economic objective functions, were subjectedto a group of constraints, composed by: model constraints, operationalconstraints, and hydraulic constraints. The first group of constraints was definedby mass balances, energy balances, mole fraction summations, and phaseequilibrium (MESH equations); the second was established by the purity of theproduct (ETBE) and the third ones avoids entrainment flooding, down -flow flooding and weeping -dumpling problems.Finally, a comparison between both, the results obtained with an economic optimizationand the results achieved in an exergetic optimizationare presented.

**Bibliography**

*[1]* J. Aronsand H. van der Kooi. Efficiency and Sustainabilityin the Energy and Chemical Industries. New York: Marcel Dekker. (2004)

*[2]* Gong Mei. Using Exergy and Optimization Models to Improve Industrial EnergySystems towards Sustainability. Linkoping University.Linkoping, Sweden, pages: 1-10.(2004)

*[3]* Cerci Yunus.The minimum work requirement for distillation processes. Exergy,an International Journal, **2**, pages:15-23. (2002)

*[4]* Gmez Jorge, Reneaume Jean-Michel, Roques Michel,Meyer Michel and Meyer Xuan. A "MINLP" formulation for optimaldesign of a catalytic distillation column based on a generic non equilibriummodel. 20, pages: 925-930 (2005)

*[5]* Kaiser Victor and Gourlia Jean-Paul. The Ideal-Column concept: applying Exergy to Distillation. Chem. Eng, **19,** pages: 41-53 (1985)

See more of this Session: Process Intensification by Process Integration
See more of this Group/Topical: Process Development Division

# (709c) Reactive Dehydration for Bioethanol Purification

James R. Kittrell[1], **Carl R. Dupre**[1], Michael F. Malone[2] and Wei Qi[2], (1)KSE Inc, Amherst, MA, (2)Chemical Engineering, University of Massachusetts, Amherst, MA

The development of a new Reactive Dehydration Technology will be discussed, with the goal of substantial reduction of the energy consumption for purifying alcohol-water mixtures, e.g., in the manufacture of fuel grade ethanol from corn. Although the immediate application of the new technology is fuel grade ethanol manufacture, the cost, energy consumption, and carbon footprint of evaporation of water are major impediments to commercialization of many projects. This goal of reduced energy consumption is to be accomplished by adding a novel dehydrating reagent to the wet ethanol, which reacts with water using a catalyst. The dehydrating agent is subsequently regenerated and reused in a closed chemical cycle. In addition, the use of process simulations for the conceptual design will be discussed.

This research is partially funded by the the U.S. Department of Energy, Office of Science and the Office of Energy Efficiency and Renewable Energy.

See more of this Session: Process Intensification by Process Integration
See more of this Group/Topical: Process Development Division

# (709d) Heat Integrated Reactive Distillation for the Indirect Hydration of Cyclohexene to Cyclohexanol

**Aspi K. Kolah**[1], Venkata K.S. Pappu[1], Carl T. Lira[1], Dennis J. Miller[1], Emmanuel A. Dada[2], C.B. Panchal[3], Richard D. Doctor[3] and John C. Prindle[4], (1) Chemical Engineering and Materials Science, Michigan State University, East Lansing, MI, (2)ChemProcess Technologies (CPT), LLC, League City, TX, (3) Energy System Division, Argonne National Laboratory, Argonne, IL, (4) Department of Chemical and Biomolecular Engineering, Tulane University, New Orleans, LA

Cyclohexanol is an important bulk chemical feedstock for the polymer industry, primarily as a precursor to nylon, but also as a plasticizer component and solvent. The conventional process for cyclohexanol synthesis is either cyclohexane partial oxidation, hydrogenation of phenol, or the direct hydration of cyclohexene. The direct hydration route of cyclohexene to form cyclohexanol, which is the simplest in principle, suffers from limited solubilities of the cyclohexene-water pair, thus requiring large amounts of catalyst and high residence times with intense mixing, and process chilling.

An alternative to the direct hydration of cyclohexene is the two-stage indirect hydration process employing heat-integrated reactive distillation. The justification for a two-stage operation employing (1) esterification and (2) hydrolyzation builds upon the earlier studies by Sundmacher, et al., [1] Here acetic acid is used as the reactive entrainer in a reactive distillation operation. In the first stage of this process cyclohexene from the inlet mixture of cyclohexene and cyclohexane reacts with acetic acid to form cyclohexyl acetate, which is then hydrolyzed with water to form cyclohexanol in the second stage. In this study, the heat-integration will improve yields from the operation.

We present the results of a reaction of cyclohexene in a mixture of cyclohexene and cyclohexane with acetic acid to form cyclohexyl acetate. This reaction is carried out using the MSU pilot scale reactive distillation column packed with Katapak-SP11 structured packing. The effect of flow rates, reflux ratio, and column geometry has been experimentally investigated. Column performance is predicted using RADFRAC in Aspen Plus process simulation software with appropriate kinetic parameters derived from experimental batch kinetic studies.

Kinetic parameters have been estimated using a LHHW model. Reactive residue curve maps have been presented along with plant scale simulations.

[1] Steyer, Frank and Kai Sundmacher, Cyclohexanol Production via Esterification of Cyclohexene with Formic Acid and Subsequent Hydration of the Esters Reaction Kinetics, Ind. Eng. Chem. Res. 2007, 46, 1099-1104

See more of this Session: Process Intensification by Process Integration
See more of this Group/Topical: Process Development Division

# NEW REACTIVE DISTILLATION PROCESS FOR THE PRODUCTION OF THE FUEL ETHER TAME USING DIFFERENT PRESSURE THERMALLY COUPLED TECHNOLOGY

*Hong Li*[a, b, *], *Xin Gao*[a, b], *Xingang Li*[a, b]
[a] *School of Chemical Engineering and Technology, Tianjin University, Tianjin, China*
[b] *National Engineering Research Center for Distillation Technology, Tianjin, China*

## Introduction

Distillation columns and chemical reactors represent the backbone of what distinguishes chemical engineering from other engineering disciplines. Reactive distillation (RD) combines chemical reaction and distillation into a single process unit. RD has economic advantages in some systems over conventional reactors, separation and recycles flowsheets, particularly for reversible reactions in which chemical equilibrium constraints limit conversion in a conventional reactor [1]. The addition of an in-situ separation function in a reactor offers the promise of improved conversion and selectivity, and for this reason, there is considerable academic and industrial interest in the area of reactive distillation. The integration of reaction and separation in one single process unit shows several advantages compared to the separately performed processes. Process intensification and energy integration represent ways of economical efficiency, as well as environmental friendly operating conditions [2-4].

The conventional RD column consists of a reactive section in the middle with non-reactive rectifying and stripping sections at the top and bottom. The task of the stripping section is to recover reactant from product stream in bottom. In the rectifying section, the reactant is reclaimed from the product stream in the top of column. The TAME reactive distillation system with two reactants and one product is irreversible reaction. Isoamylenes is very volatile, and TAME is very heavy, so the product is removed from the bottom of the column. The lighter isoamylenes leave in the distillate stream along with a significant amount of methanol [5].

In this paper, the process of catalytic distillation for *tert*-amyl methyl ether (TAME) synthesis from isoamylenes (2M1B and 2M2B) and methanol has been studied. A brief description of equilibrium stage model developed is given for simulation of the catalytic distillation process of the synthesis of *tert*-amyl methyl

[*] Corresponing author
E-mail : lihong.tju@163.com

ether (TAME) from isoamylenes (2M1B and 2M2B) and methanol. The model predicted the reaction conversion and the energy consumption of process. Validation based on pilot plant experiments proved that the model accuracy is satisfactory. A new process technology was designed by means of different pressure thermally coupled technology for energy-saving.

## Novel Thermally Coupled Technology

The conventional RD column for producing TAME consists of rectifying section, reactive section and stripping section. The column is operated at an overpressure of 0.4MPa ~ 0.5Mpa [6]. Increasing column pressure causes increased reaction rates and a shift of chemical equilibrium duo to higher boiling point temperatures in the reactive distillation section. On the other hand, increasing column pressure reduces the relative volatility. In order to providing a good recovery of 2M2B and improving the TAME purity in bottom product, the stripping section have to increasing the energy consumption of reboiler in catalytic distillation column bottom. So increasing the pressure of the stripping section is unnecessary.

Because both reaction and separation occur in a single vessel operating, the temperatures and pressures of reaction and separation are not independent. Therefore, reactive distillation is limited to systems in which the temperatures and pressures conductive to reaction are compatible with temperatures and pressures conductive for distillation separation. Pressure in conventional distillation design is usually set by a minimum temperature in the reflux drum (so that cooling water can be used) or a maximum temperature in the reboiler (to prevent fouling or thermal decomposition). Establishing the optimum pressure in a reactive distillation column is more complex because of the interplay between reaction and phase separation. Most VLE relationships show an increase in volatility with decreasing pressure. On the other hand, reaction rates decrease with decreasing pressure. If the chemical equilibrium constant increases with decreasing pressure or temperature, low operating pressure or temperature, which facilitates the phase separation, may require lots of catalyst or liquid tray holdup to compensate for the low reaction rates.

Based on the analysis above, different pressure thermally coupled technology could be applied to the catalytic distillation process for producing TAME. In this process technology, the RD column is divided to two columns with different pressure. This unit is a highly integrated configuration as it contains a reactive part placed in higher pressure column, products separation in the lower pressure column. The process scheme used in our design study is shown in Fig.1. The flow scheme consists of a pre-reactor followed by a higher pressure RD column and a lower pressure stripping column. The top stream from the higher pressure RD column is used as the teat energy for the reboiling of the lower pressure stripping column, therefore the thermally coupled process is realized.

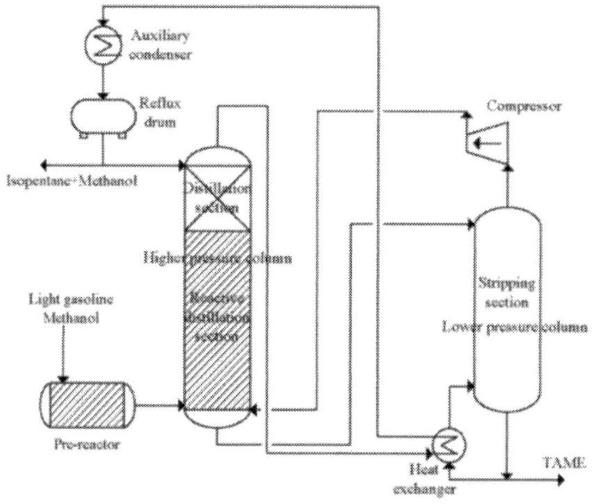


Fig.1 The scheme of catalytic distillation process applied to different pressure thermally coupled technology for producing TAME

## Results and Discussions

A comparison of conventional catalytic distillation process for producing TAME and novel catalytic distillation process applied to different pressure thermally coupled technology for producing TAME shows that the energy consumption of new process could be reduced about 46.7%. The purity of TAME obtained as bottom product of lower pressure stripping column is high, and the excess of methanol can be recirculated to the pre-reactor allowing a good overall yield of the process. Finally, the equilibrium stage model is used to optimal effect parameters and design factors of the new process technology on conversion and energy consumption. The catalytic distillation process integrated by different pressure thermally coupled technology seems to be a very attractive and challenging process.

**Keywords:** catalytic distillation, TAME, different pressure thermally coupled, process intensification

**ACKNOWLEDGEMENT**

The authors are grateful for the financial support from the National Basic Research

Program of China (973 Program, 2009CB219905). Supported by Program for Changjiang Scholars and Innovative Research Team in University (No. IRT0936).

**Reference**

1. Gao X., Li X., Li H., Hydrolysis of methyl acetate via catalytic distillation: Simulation and design of new technological process. Chemical Engineering and Processing 49, 1267-1276 (2010).
2. Malone M.F., Doherty M.F., Reactive distillation. Industry and Engineering Chemistry Research 39, 3953-3957 (2000).
3. Taylor R., Krishna R., Modelling reactive distillation. Chemical Engineering Science 55, 5183-5229 (2000).
4. Tuchlenski A., et al., Reactive distillation – industrial applications, process design & scale-up. Chemical Engineering Science 56, 387-394 (2001).
5. Baur R., Krishna R., Hardware selection and design aspects for reactive distillation columns. A case study on synthesis of TAME. Chemical Engineering and Processing 41, 445-462 (2002).
6. Luyben W.L., Comparison of pressure-swing and extractive-distillation methods for methanol-recovery systems in the TAME reactive-distillation process. Industrial and Engineering Chemistry Research 44, 5715-5725 (2005).

# (736a) Process Integration and Analysis of Chemical Looping Based Power and Fuel Generation Systems

**Liang Zeng**[1], Fanxing Li[2], Deepak Sridhar[3], Hyung Rae Kim[3], Andrew Tong[4], Siwei Luo[1] and L.-S. Fan[5], (1)William G. Lowrie Department of Chemical and Biomolecular Engineering, Ohio State University, Columbus, OH, (2)Chemical and Biomolecular Engineering, North Carolina State University, Raleigh, NC, (3) Chemical and Biomolecular Engineering, The Ohio State University, Columbus, OH, (4)Chemical and Biomolecular Engineering, Ohio State University, Columbus, OH, (5)William. G. Lowrie Dept. of Chemical and Biomolecular Engineering, The Ohio State University, Columbus, OH

Chemical looping gasification (CLG) process can flexibly convert carbonaceous fuel such as coal and biomass into hydrogen, liquid fuels and electricity while effectively capturing the $CO_2$ byproduct. The unique features of the CLG process are in light of its indirect conversion scheme via redox cycles of metal oxide based oxygen carriers. In the first step, fuel is converted into $CO_2$ and $H_2O$ by high oxidation state oxygen carrier in one reactor. The reduced metal oxide can be reoxidized in a subsequent reactor by either steam for hydrogen production, or air for heat and power generation. This study first illustrates the CLG process configurations. Then we will concentrate on two process intensification cases based on the CLG process. The integration between the CLG and the SOFC technology provides a closed loop between the chemical looping oxidizer and the SOFC anode through the circulation of the gaseous mixture of steam and $H_2$, which can significantly enhance the electricity generation. Another novel process that integrates CLG process with $CO_2$ hydrogenation is proposed for liquid fuel synthesis. Both exergy analysis and process simulation studies confirm the advantages of the proposed processes. The results indicate that, through process intensification, the chemical looping gasification strategies have the potential to notably increase process efficiencies in carbonaceous fuel conversions.

See more of this Session: Energy Efficiency by Process Intensification
See more of this Group/Topical: Process Development Division

# (736b) Theory of Process Design Based On Self-Heat Recuperation

**Yasuki Kansha**[1], Akira Kishimoto[1], Muhammad Aziz[2] and Atsushi Tsutsumi[1], (1)Institute of Industrial Science, The University of Tokyo, Tokyo, Japan, (2) Advanced Energy Systems for Sustainability, Tokyo Institute of Technology, Tokyo, Japan

In chemical industries, physical conditions of process streams must be adjusted to satisfy the condition of the following processes such as separation and reaction processes. This required physical condition change can be explained by Helmholtz and Gibbs free energies with the second law of thermodynamics. However, many energy saving technologies recently developed are only considered on the basis of the first law of thermodynamics, energy conservation. Simultaneously, some researchers have been paid attention to the analysis of process exergy and irreversibility of the process under the consideration of the second law of thermodynamics. However, many of these investigations only show the calculation results of exergy analysis and the possibility of energy saving of the processes.

Recently, Kuchonthara and Tsutsumi have proposed the energy recuperative integrated gasification power generation system and developed the design way of the system in their papers. Based on exergy recuperation, Kansha et al. have developed the self-heat recuperation technology, applied it to several chemical processes and showed the energy saving possibility as compared with the conventional counterparts.

In this research, the minimum energy required for the thermal processes was derived theoretically according to the irreversibility point of view. This minimum energy required was compared with the energy required for the self-heat recuperative thermal processes and conventional processes by using process simulator. From these results, we show the direction of energy saving process designs from irreversibility point view with fundamental developments.

See more of this Session: Energy Efficiency by Process Intensification
See more of this Group/Topical: Process Development Division

# (736c) Novel Technology for Producing Bio-Butanol

Xiuqin Dong, Minli Tao, Shenghua Qian, Yingzhe Yu and **Minhua Zhang**, R & D Center for Petrochemical Technology, Tianjin University, Tianjin, China

Butanol is an important organic solvent and industrial intermediate. It is widely used in plastic, organic synthesis, paint and pharmacy industry. Moreover, Butanol's Brit-ish thermal units (BTU) is almost equal to that of gasoline and higher than ethanol. Therefore, butanol has been recently identified as a good candidate of next-generation clean and renewable biofules to address the concerns for the availability of fossil feed stocks and climate change.

Recent advances in the field of biotechnology have resulted in a renewed interest in the fermentation production of butanol (coproducts are acetone and ethanol) by using raw materials like starch, sugar and lignocelluloses. However, the traditional acetone butanol ethanol (ABE) fermentation process has many disadvantages such as high energy and material consumption and serious water pollution. For instance, the steam consumption is more than thirteen tons/per ton of solvent.

In the past few years, our team has developed a series of new technologies to facilitate an effective and energy-efficient product removal in the traditional ABE fermentation process. The multi-effect thermal coupling technique was employed in the distillation unit which consumes the largest amount of energy in the entire process. With the heat integration and optimization, more than 35% energy was saved and the steam consumption was reduced to 6.5 tons/per ton of solvent while the product quality was also improved with the high-efficient distillation process. Based on the whole process modeling and optimization, both the heavy and light enrichment areas were identified for bleeding the impurities respectively. Accordingly, the total product yield was increased by 2% to reach 99%. In the meantime, the water flow was also optimized to minimize the water usage as well as the waste water.

The production cost was greatly reduced in this improved ABE fermentation process. Currently, a 70,000 MTA bio-butanol plant has been built up using this technology.

# (736d) Optimal Synthesis of Refrigeration System for Natural Gas Liquefaction

**Meiqian Wang**, Jian Zhang, Qiang Xu and Kuyen Li, Dan F. Smith Department of Chemical Engineering, Lamar University, Beaumont, TX

Liquefaction of natural gas (LNG) is an energy-intensive process. Energy efficiency maximization is the major objective in LNG process design and operation. A lot of efforts have been done to minimize the energy consumption of the LNG process, most of which use heuristic methods or nondeterministic search methods and thus could not guarantee the global optimal results. In this paper, a novel methodology for LNG process optimization on energy consumption minimization is presented. This methodology is based on mathematical programming and deterministic global optimization. The difficulties of this systematic study include: i) develop process synthesis model for optimal design and operation of LNG process; ii) solving the mixed integer nonlinear programming (MINLP) optimization problem. iii) develop rigorous process simulation model for optimal results validation. Also presented are the technologies for obtaining reliable thermodynamic property formulas based on rigorous simulation and the methods for MINLP model simplification.

LINDOGlobal solver in GAMS is employed to solve the optimization problem. The LINDO global optimization procedure employs branch-and cut methods to break an NLP model down into a list of subproblems. Each subproblem is either solved to find the optimal solution or split into two or more subproblems. Given appropriate tolerances, after a finite, though possibly large number of steps a solution provably global optimal to tolerances is returned. The results obtained from GAMS are validated through rigorous simulation. Based on the methodology, the optimal design and operation conditions of LNG process can be simultaneously achieved. A case study about a C3MR LNG process is employed to demonstrate the efficacy of the developed methodology. Comprehensive thermodynamic analysis is also conducted in this paper to give deep understandings of the LNG refrigeration system and optimization procedure.

See more of this Session: Energy Efficiency by Process Intensification
See more of this Group/Topical: Process Development Division

# (736e) Simulation and Optimization of Natural Gas Liquefaction Process for Energy Saving

**Wonsub Lim**[1], Kyungjae Tak[1], Inkyu Lee[1], Il Moon[1], Kwangho Choi[2] and Daeho Ko[2], (1)Chemical and Biomolecular Engineering, Yonsei University, Seoul, South Korea, (2)GS Engineering & Construction, Seoul, South Korea

Oil is the most dominant source of energy in the global energy market over the past decades but the global energy market structure has been changing sharply because of environmental restrictions. Natural gas is a clean source of energy and is preferred because it offers a number of significant environmental benefits against other fossil fuels such as oil and coal. Pipelines remain the dominant method of transportation of natural gas, because the large volume necessary for use makes most other forms of transportation too costly. Another method of transporting natural gas is to be compressed, refrigerated and supplied as a liquid known as liquefied natural gas(LNG). When there is a considerable distance involved in transporting natural gas, LNG is the preferred method of natural gas supply because of technical, economic, safety and political reasons. Therefore, LNG is expected to play a major role as a clean energy. The global use of LNG is increasing sharply. Worldwide LNG demands expand from 159 million ton in 2006 to 500 million ton in 2030, achieving 3.6-fold growth.

This study focuses on the design of liquefaction process in LNG plant. The simulation and optimization of the liquefaction process are carried out to obtain the improved process design for energy saving. The optimal design of processes requires less power and cost at given specific plant capacity. Designing cost-effective process of high efficiency becomes a key contribution for optimal LNG plant. The conditions and configurations of liquefaction process are suggested in comparison with several commercial liquefaction processes. An efficient natural gas liquefaction process is developed in this research.

This research was supported by a grant from the GAS Plant R&D Center funded by the Ministry of Land, Transportation and Maritime Affairs (MLTM) of the Korean government.

See more of this Session: Energy Efficiency by Process Intensification

# REDUCING COSTS IN BIOETHANOL PURIFICATION USING A HYBRID PROCESS LIQUID-LIQUID EXTRACTION - EXTRACTIVE DISTILLATION

*Adriana Avilez-Martínez[a], Jaime Saucedo-Luna[a], Juan Gabriel Segovia-Hernandez[b], Salvador Hernandez[b], Fernando Israel Gomez-Castro[b] and Agustin Jaime Castro-Montoya[a]*
*[a] Universidad Michoacana de San Nicolás de Hidalgo, Facultad de Ingeniería Química, Michoacán, México*
*[b] Universidad de Guanajuato, Campus Guanajuato, División de Ciencias Naturales y Exactas, Guanajuato, Guanajuato, México*

## Introduction

Ensuringthe energy supply for the future is nowadays an important concern for humankind. Traditionally, energy requirements have been satisfied using fuels obtained from petroleum. However, due to the observed reduction in petroleum production and the high environmental impact that the use of fossil fuels causes, a variety of efforts have been undertaken to find alternative sources of energy. Biofuels represents an important example of such clean, renewable sources. In the transportation industry the most promising candidate to replace gasoline as fuel for gasoline engines is ethanol. This alcohol has a higher octane number, broader flammability limits and higher heats of vaporization than gasoline[1]. Furthermore, ethanol can be used mixed with gasoline, and thus no additional engine modification investment is required. To these performance advantages it can be added the lower greenhouse gas emissions of ethanol, as compared to gasoline. It has been established that if ethanol replaces gasoline as automotive fuel, greenhouse gas emissions will be reduced by more than 85% when considering the entire fuel life cycle[2]. In recent years, the countries with the highest production of ethanol worldwide have been Brazil, the United States and Canada[3]. Moreover, it has been predicted that, by 2022, a third of energy requirements in the United States will be satisfied by fuels obtained from biomass, and at least half of that proportion will be fulfilled by ethanol.[4]

Bioethanol can be obtained from sources with a high content of sugars, or materials that can be converted into sugars, such as corn and sugar cane. However, since these crops are used for human food, alternative sources are required. A potential solution to this problem is to use lignocellulosic compoundsas raw material. Using such non-edible biomass to produce ethanol also reduces the cost of production. The production of bioethanol from lignocellulosic materials basically consists of four steps: delignification, depolymerization, fermentation and purification[5]. The delignification step involves the separation of cellulose, hemicellulose and lignin, so the molecules of cellulose and hemicelluloses can be further treated to obtain sugars. In the depolymerization step, the lignocellulosic biomass is hydrolyzed with enzymes or acids as catalysts, obtaining sugars such as glucose, xylose and arabinose[6]. Sugars are then fermented, usually with yeast or bacteria, to obtain mainly ethanol and water. In the purification step, ethanol must be separated from water. The final product must contain less than 0.5 wt% of water in order to be useful as automotive fuel in combination with gasoline, or as oxygenate for gasoline. Nevertheless, purification by conventional distillation is not possible because the ethanol-water mixture forms an azeotrope when a purity of 96 wt% of ethanol is reached. Thus, non-conventional methods are necessary to achieve the required ethanol purity. One alternative to obtain pure ethanol is to use conventional distillation to pre-concentrate the ethanol to purity levels close to the azeotrope, and then reach the desired purity by using extractive distillation. The main purpose of extractive distillation is to modify the relative volatilities of the components of a given mixture by using an entrainer. One entrainer used to separate ethanol-water mixtures is ethylene glycol, and recently the use of glycerol has been proposed to separate such mixtures using extractive distillation[7]. It has been reported that the industrial oil company Sasol Ltd. uses n-dodecane as a solvent to extract light alcohols, such as ethanol and 2-propanol, from water[8]. Liquid-liquid extraction could thus be an alternative to separate ethanol from water. However, liquid-liquid extraction alone is unable to produce the purity levels required for bioethanol. Therefore, in this study we propose the use of hybrid systems of liquid-liquid extraction and extractive distillation to reduce energy consumption and total annual costs for the purification step in the bioethanol production process. The liquid-liquid extraction is done using n-dodecane as entrainer, reducing the concentration of water; in a second separation stage, extractive distillation is used to reach the desired ethanol purity using glycerol as entrainer. The performance of the proposed separation strategy is then compared to that of the purification process involving conventional distillation and extractive distillation.

## Design methodology

To analyze the processes under consideration, simulations have been performed in the commercial simulator Aspen Plus. The feed stream for all the alternatives analyzed is 100 lbmol/h with a composition of 90 mol% of water and 10 mol% of ethanol. This stream enters the purification process at 30°C and 1 atm. The vapor-liquid equilibrium has been calculated using the NRTL equation for determination of the activity coefficients, using the binary parameters provided by the simulator.

Three different processes have been analyzed. First, a conventional separation sequence (CS-1) is considered. This separation method involves the pre-purification of ethanol using conventional distillation until a maximum feasible purity is achieved. This purity will be lower than the composition of the azeotrope. The bottoms stream consists on water with a purity of 99.99mol%. The concentrated ethanol is then treated in an extractive distillation column, using glycerol as entrainer, aiming for a purity of 99.99 % in mole fraction of ethanol. Once the ethanol has been purified, an additional conventional distillation column is used to purify the glycerol so it can be reused. For the extractive distillation column, an initial design is supposed. This initial design must allow reaching the desired purity for ethanol, with a recovery higher than 98 mol%. Once the design specification for ethanol has been fixed, the degrees of freedom of the column are modified one by one, aiming to reduce the energy requirements for the separation. The distillation column where entrainer is purified is designed aiming for a purity of 99 mol% for glycerol.

For the two alternative schemes proposed, the possibility of pre-purifying the ethanol by using liquid-liquid extraction with n-dodecane as entrainer is considered. To prove the ability of n-dodecane to separate water from ethanol by liquid-liquid extraction, Figure 3 shows the ternary diagram for the ethanol/water/n-dodecane mixture at 1 atm. From the liquid-liquid envelope, it can be seen that, if n-dodecane is used as entrainer, almost pure water can be obtained in one phase, and in the other phase a mixture of ethanol, n-dodecane and some water is obtained.

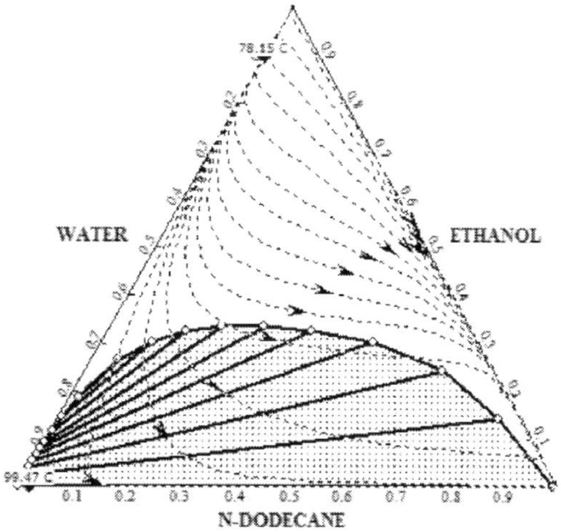

Figure 1. Ternary map for the ethanol/water/n-dodecane mixture

In the context of the ideasdiscussed above, Figure 2 presents the two proposed purification sequences.

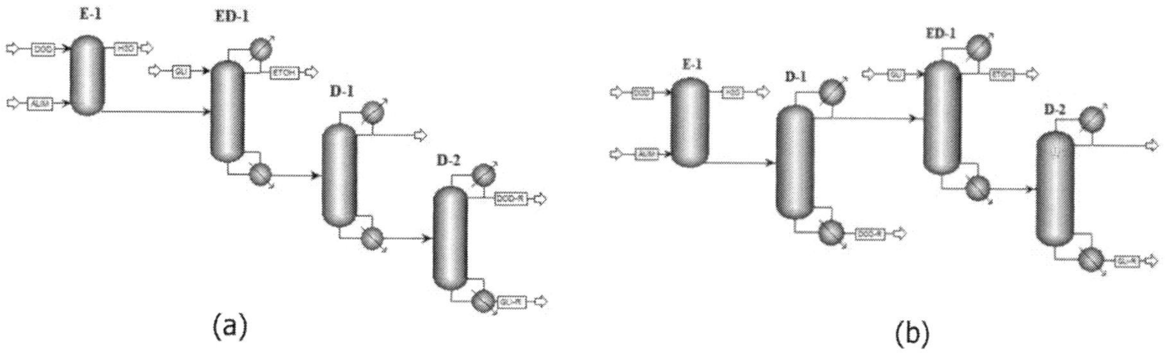

Figure 2. Proposed separation sequences (a) AS-1, (b) AS-2.

In the AS-1 approach, the original mixture enters a liquid-liquid extraction column, where a portion of the water is separated from ethanol using n-dodecane as extraction agent. Then, in an extractive distillation column, the rest of the water is separated from ethanol using glycerol as entrainer, aiming for a purity of 99.99 mol% of ethanol in the distillate stream. The bottoms stream leaving the extractive distillation column enters a distillation column where water is separated from the glycerol and n-dodecane. In a second distillation column, the entrainers are recovered so they can be reused. To design the liquid-liquid extraction and the extractive distillation columns, an initial guess is first considered, aiming for a global recovery for ethanol higher than 98 mol% and a purity of 99.99 mol% for ethanol at the top of the extractive distillation column. Once the design specifications have been reached, degrees of freedom are manipulated to reduce energy consumption in the extractive distillation column. The bottoms stream of the extractive distillation column contains n-dodecane, glycerol and some water. The extraction agents are purified in a conventional distillation train, separating first the water and, in a second column, the n-dodecane and the glycerol. Both columns are designed aiming for a purity of 99 mol% for n-dodecane and glycerol.

The second purification alternative (AS-2) is shown in Figure 6. In this separation scheme, ethanol is pre-purified in a liquid-liquid extractive column using n-dodecane as extraction agent. Then, the stream containing ethanol, n-dodecane and some water enters a distillation column, where n-dodecane is recovered. Ethanol is then obtained with a purity of 99.99 mol% in an extractive distillation column, using glycerol as entrainer. In a final separation stage, glycerol is recovered to be reused as entrainer. To design the liquid-liquid extraction column and the extractive distillation column, an initial guess is required, aiming for a recovery higher than 98 mol% for ethanol and a purity of 99.99 mol% at the top of the extractive distillation column. The conventional distillation column following the liquid-liquid extraction column is designed aiming for a purity of 99 mol% for n-dodecane in the bottoms stream. Once the design specifications are reached, an analysis of the design variables is performed to reduce energy consumption in the conventional distillation column and the extractive distillation column. The next piece of equipment is a conventional distillation column, where glycerol is recovered with a purity of 99 mol%.

## Results and discussion

The first analyzed system is the sequence AS-1. The impact of the number of stages in the liquid-liquid extraction column on the energy requirements of the extractive distillation column has been found to be almost insignificant. The impact of the molar feed ratio for n-dodecane and ethanol/water feed to the liquid-liquid extraction column on the heat duty will be first discussed. It has been found that a low flow rate for n-dodecane is required to perform the pre-purification of ethanol from the ethanol/water mixture with a low heat duty in the extractive distillation column. The number of stages in the extractive distillation column has an insignificant impact on the energy consumption. When analyzing the effect of the location of the feed streams on energy requirementsit has been found that, to reach a low heat duty for the extractive distillation column, the glycerol stream must be located in a stage close to the top of the column and the ethanol stream must be located in a stage close to the bottom of the column. It has also been observed that aglycerol/ethanol molar ratio between 1 and 1.3 is required to obtain lower energy consumption for the extractive distillation column. The main design characteristics of the columns involved in the AS-1 purification scheme are shown in Table 1.

We will now consider the results for the AS-2 sequence. The design for the liquid-liquid extraction column has been considered to be the same as that obtained for the AS-1 sequence, since said design has proven to be appropriate to perform the separation with the lowest energy consumption for the extractive distillation column. Similar trends to the results obtained for the sequence AS-1 have been observed on the optimization procedure of the sequence AS-2, thus the results will only be discussed. The influence of the number of stages on the extractive distillation column's energy requirements is quite small. When the effect of feed stream locations on the extractive distillation column is considered, it has been observed that the glycerol stream must be located on the top stages of the column, and the ethanol feed stream must be located near to the bottoms of the column. When the effect of glycerol/ethanol molar feed ratio is analyzed, it has been found that the glycerol molar flow required to perform the separation with a low heat duty is relatively low, with a glycerol/ethanol ratio of around 0.8. Design characteristics of the columns in the AS-2 purification scheme are shown in Table 2.

Table 1. Design characteristics of columns in the AS-1 sequence.

| | E-1 | ED-1 | D-1 | D-2 |
|---|---|---|---|---|
| Number of stages | 20 | 20 | 10 | 12 |
| Feed stage | 20 | 18 | 6 | 7 |
| Entrainer feed stage | 1 | 3 | - | - |
| Feed molar flow (kmol/h) | 45.36 | 15.44 | 29.14 | 28.03 |
| Entrainer molar flow (kmol/h) | 9.98 | 18.14 | - | - |
| Top stage flow (kmol/h) | 39.90 | 4.44 | 1.11 | 9.80 |
| Bottoms stage flow (kmol/h) | 15.44 | 29.14 | 28.03 | 18.23 |
| Design Pressure (kPa) | 101.35 | 101.35 | 101.35 | 101.35 |
| Reflux ratio | - | 4.4556 | 0.9551 | 1.4898 |
| Heat duty (GJ/h) | - | 1.39 | 0.51 | 1.12 |

Table 2. Design characteristics of columns in the AS-2 sequence.

| | E-1 | D-1 | ED-1 | D-2 |
|---|---|---|---|---|
| Number of stages | 20 | 16 | 18 | 15 |
| Feed stage | 20 | 13 | 15 | 7 |
| Entrainer feed stage | 1 | - | 3 | - |
| Feed molar flow (kmol/h) | 45.36 | 15.44 | 5.39 | 4.60 |
| Entrainer molar flow (kmol/h) | 9.98 | - | 3.63 | - |
| Top stage flow (kmol/h) | 39.90 | 5.39 | 4.42 | 0.9349 |
| Bottoms stage flow (kmol/h) | 15.44 | 10.05 | 4.60 | 3.6654 |
| Design Pressure (kPa) | 101.35 | 101.35 | 101.35 | 101.35 |
| Reflux ratio | - | 0.023 | 1.4498 | 1.9999 |
| Heat duty (GJ/h) | - | 0.99 | 0.32 | 0.22 |

The total energy requirements for the two analyzed processes and the conventional process are presented in Table 3, together with the energy savings taking the conventional separation scheme as a basis. It can be seen that, when the two entrainers enter the extractive distillation column (AS-1), the total energy requirements in the process are higher than those of the conventional separation sequence. On the other hand, if the n-dodecane is separated from the ethanol/water mixture before entering the extractive distillation column, energy savings of around 30% over the conventional sequence are obtained.

Table 3. Total energy requirements for the analyzed sequences.

| | $Q_T$ (GJ/h) | $\%_{savings}$ |
|---|---|---|
| Conventional | 2.19 | 0% |
| AS-1 | 3.02 | -37.15% |
| AS-2 | 1.54 | 30.03% |

## Conclusions

Alternatives have been proposed for the purification of a typical mixture of ethanol-water obtained from fermentation of biomass. These new schemes involve the use of n-dodecane as entrainer in a liquid-liquid extraction system to pre-concentrate the ethanol from the ethanol-water mixture, followed by extractive distillation with glycerol as entrainer. According to the results, among the alternatives under analysis, the most promising is the system involving pre-concentration of ethanol in a liquid-liquid extraction column using n-dodecane as entrainer, followed by separation of the n-dodecane from the mixture containing ethanol and the remaining water, and finally purification of the ethanol, using extractive distillation with glycerol as entrainer. This process requires less energy than the traditional separation sequence. The other proposed separation method resulted in higher energy consumption, because a higher molar flow is processed in the extractive distillation column.

## References


1. Balat, Mustafa, Havva Balat and Cahide Öz (2008), "Progress in Bioethanol Processing", *Progr. Energ. Combust. Sci.*, 34, pp. 551-573.

2. Bergeron, Paul (1996)"Environmental Impacts of Bioethanol",in*Handbook on Bioethanol: Production and Utilization,*Charles E. Wyman, ed. Washington, D.C., Taylor & Francis.

3. Chiaramonti, David (2007) "Bioethanol: Role and Production Technologies", in *Improvement of Crop Plants for Industrial End Uses*, Paolo Ranalli, ed. Dordrecht, The Netherlands, Springer.

4. Eggeman, Tim and Carrie Atiyeh (2010), "The Role for Biofuels", *Chem. Eng. Prog.*, 106, pp. 36-38.

5. Champagne, Pascale (2008), "Bioethanol from Agricultural Waste Residues", *Environ. Prog.*, 27, pp. 51-57.

6. Elnashaie, Said and Parag Garhyan (2004), "Bioethanol Production – Solving the Efficiency Bottleneck", *The Chemical Engineer*, October, pp. 30-32.

7. Garcia-Herreros, Pablo, Jorge M. Gomez, Iván D. Gil and Gerardo Rodriguez (2011), "Optimization of the Design and Operation of an Extractive Distillation System for the Production of Fuel Grade Ethanol Using Glycerol as Entrainer", *Ind. Eng. Chem. Res.*, 50, pp. 3977-3985.

8.Bhownath, Rinay (2008), "The Use of n-Dodecane as a Solvent in the Extraction of Light Alcohols from Water", Master of Sciences in Engineering Thesis, University of Kwa-Zulu Natal, Durban, South Africa.

# OPTIMIZATION OF BLAST FURNACE STEELMAKING PROCESS FROM A PROCESS INTEGRATION PERSPECTIVE

Hamid Ghanbari[1,2*], Mikko Helle[1], Frank Pettersson[1], Henrik Saxén[1]
[1]*Thermal and Flow Engineering Laboratory, Department of Chemical Engineering, Åbo Akademi University, Åbo, Finland*
[2]*Center for Advanced Process Decision-making, Department of Chemical Engineering, Carnegie Mellon University, Pittsburgh, PA, USA*
*Corresponding author: ghanbari@cmu.edu

## Abstract

In this paper, mathematical modeling and optimization have been used to study the process integration feasibility of a steelmaking plant with methanol production to suppress carbon dioxide emissions. A blast furnace simulator with options to describe new technologies, such as cold oxygen injection combined with top gas recycling and natural gas injection is used as the core of the analysis. An optimization problem for the whole industrial complex is stated, where the specific steelmaking cost is minimized under process and resource constraints. The results show the potential of combining the technologies for different scenarios of costs of emission and carbon capturing and storage, and also illustrate how the steel production rate influences the performance of the system.

*Keywords*: Primary Steelmaking, $CO_2$ Emissions, Methanol Production, Process Integration

## Introduction

Primary steelmaking systems are one of the major carbon dioxide emitters due to large production volumes and operation based on fossil fuels (mainly coal). Recently, many efforts have been launched to increase the energy efficiency and to suppress the emissions from the unit processes of the steelmaking plants, e.g., in the ULCOS research program which aims at carbon dioxide emission reductions of 50% compared to today's best available technology.[1]

One proposed solution is to apply new technologies, such as full pellet charging, top gas recycling, cold oxygen injection, and/or alternative fuel injection in the blast furnace, and some of these concepts have been studied in experimental scale and implemented in industrial practice. Another approach is to focus on using renewable energy, such as biomass, which could reduce fossil-fuel based emissions dramatically. However, process integration also shows a great potential to increase the efficiency and reduce the emissions of the steelmaking systems: Large amount of off-gases are available from different unit processes, such as coke oven gas, blast furnace top gas and basic oxygen furnace gas, and these could be utilized in a polygeneration system rather than as fuels in the combined heat and power (CHP) unit producing (district) heat and electricity. Production of methanol is an option, where the methanol could be considered as a fuel or as a raw material for other chemical processes. Methanol production could also yield more flexibility to the plant due to problems caused by seasonal and day-time changes of the demand for district heat and electricity.

The carbon dioxide emissions from primary steelmaking mainly stem from the operation of the ironmaking blast furnace. Numerous of efforts to increase the energy efficiency and decrease the emissions from the BF have been reported. Replacing metallurgical coke as a reducing agent with hydrogen bearing materials, such as natural gas (NG), has been studied and implemented.[2-4] The results show a good replacement ratio, increased furnace productivity and, depending on the NG price, a potential for decreased production costs. Nogami et al.[5] analyzed material and energy balances of an ironmaking system with natural gas injection, metallic charging and top gas recycling with carbon dioxide capturing, where the residual gases were consumed in other unit processes of the steel plant. The study showed that natural gas injection decreased the $CO_2$ emissions from the ironmaking system, but the decrease in the energy input was small. On the other hand, the top gas recycling operation increases the $CO_2$ emissions due to the requirements of the gas scrubbing operation, so the stripped $CO_2$ should be stored. Another option to decrease the $CO_2$ emissions is efficient blast furnace operation with natural gas injection and methanol synthesis based on the off gases:[6,7] Natural gas injection leads to a top gas with decreased concentrations of carbon oxides and an increased $H_2$ concentration, which are advantageous for the methanol synthesis.

In the present study, a combination of natural gas injection, hearth injection of recycled top gas together with oxygen enriched hot blast is compared with top gas recycling and cold oxygen injection in an integrated steel plant with a polygeneration system that produces methanol as a chemical by-product. Nonlinear programming is used to find the optimal states of the integrated plant with respect to the specific cost of liquid steel. The sensitivity of the optimal states to changes in the specific costs and in the steel production rate is also analyzed.

## System and Its Models

The integrated primary steelmaking system producing liquid steel (ls) from the basic oxygen furnace (BOF) is presented in Fig. 1. The most complex mathematical model of the system is that of the blast furnace (BF) with options to implement different technologies.

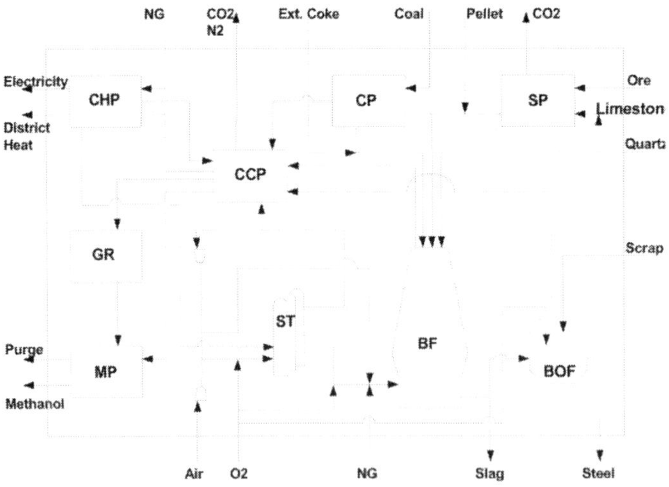


**Figure 1** Integrated steel plant. CP: Coke Plant, SP: Sinter Plant, ST: Hot Stoves, CCP: $CO_2$ Capturing Plant, BF: Blast Furnace, BOF: Basic Oxygen Furnace, CHP: Combined Heat and Power Plant, GR: Gas Reforming unit, and MP: Methanol Plant.

The nonlinear BF model has six input variables: The hot metal production rate, the volume flow rates of recycled top gas and oxygen injected, the specific natural gas injection rate, the temperature of the injected tuyere gases (here termed "blast temperature"), and the specific pellet rate. Table 1 lists the inputs and some outputs of the BF model, as well as their corresponding lower and upper limits.[8] In the present study, three alternative ways of operating the blast furnace are considered: (1) Conventional Blast Furnace operation (CBF): The blast is heated and no top gas is recycled. (2) Top Gas Recycling Blast Furnace (TBF): The top gas is recycled and is injected cold together with hot blast in hearth tuyeres. (3) Oxygen in Blast Furnace (OBF): Recycled top gas is heated and injected together with cold oxygen in hearth tuyeres. It should be noted that (2) and (3) assume carbon capture and storage (CCS) of 95% of the $CO_2$ of the recycled top gas[9], and also that moisture is removed before injection. The excess nitrogen is purged from the system by the non-recycled top gas. For all three cases, both a single steelmaking plant and a plant integrated with methanol production (CBF-Int, TBF-Int and OBF-Int) are studied. Mass and energy balances for the other process units, including the coke plant, sinter plant, and basic oxygen furnace, are based on empirical coefficients and efficiency factors derived from historical data of a reference steel plant. The maximum sinter and coke production rates of the plant were set at 160 t/h and 55 t/h, respectively. The carbon dioxide emission was calculated by a carbon balance equation, including all (fossil) carbon bearing inputs and outputs from the system. For more details the reader is referred to Helle et al.[10] and Ghanbari et al.[11]

**Table 1** Blast furnace input (first six rows) and some of the output variables and their constraints. BF production (reported in tons of hot metal, $t_{hm}$) was selected to give a steel production in [150,180] $t_{ls}$/h.

| Variable | Range |
|---|---|
| BF production rate | 131-157 $t_{hm}$/h |
| Recycled top gas | 0-220 $km^3n$/h |
| Blast oxygen content | 21-99 vol-% |
| Specific natural gas rate | 0-120 kg/$t_{hm}$ |
| Blast temperature | 250-1200 °C |
| Specific pellet rate | 0-600 kg/$t_{hm}$ |
| Specific coke rate | $\geq 0$ kg/$t_{hm}$ |
| Flame temperature | 1800-2300 °C |
| Top gas temperature | 115-250 °C |
| Bosh gas volume | 150-220 $km^3n$/h |
| Solid residence time | 6.0-9.5 h |
| Slag basicity | 1.0-1.2 |
| Slag rate | $\geq 0$ kg/$t_{hm}$ |
| Top gas volume | $\geq 0$ $km^3n$/h |
| Top gas CO content | $\geq 0$ vol-% |
| Top gas $CO_2$ content | $\geq 0$ vol-% |
| Top gas $H_2$ content | $\geq 0$ vol-% |
| Top gas $N_2$ content | $\geq 0$ vol-% |
| Top gas heating value | $\geq 0$ MJ/$m^3n$ |

## Optimization Problem

The optimal operational state of the system was investigated by minimizing the specific costs of liquid steel production. The economic objective function, $F$, expressed in specific terms (i.e., per ton of liquid steel) is given by

$$\frac{F}{€/t_{ls}} = \left\{ \left( \frac{\dot{m}_{ore}}{t/h} \cdot \frac{C_{ore}}{€/t} + \frac{\dot{m}_{pel}}{t/h} \cdot \frac{C_{pel}}{€/t} + \frac{\dot{m}_{coal}}{t/h} \cdot \frac{C_{coal}}{€/t} + \frac{\dot{m}_{coke,ext}}{t/h} \cdot \frac{C_{coke,ext}}{€/t} + \frac{\dot{m}_{NG}}{t/h} \cdot \frac{C_{NG}}{€/t} + \frac{\dot{m}_{lime}}{t/h} \cdot \frac{C_{lime}}{€/t} + \frac{\dot{m}_{scrap}}{t/h} \cdot \frac{C_{scrap}}{€/t} + \frac{\dot{V}_{O2}}{t/h} \cdot \frac{C_{O2}}{€/t} + \frac{\dot{m}_{CO2}}{t/h} \cdot \frac{C_{CO2}}{€/t} + \frac{\dot{m}_{CO2,CCS}}{t/h} \cdot \frac{C_{CCS}}{€/t} - \frac{\dot{m}_{MeOH}}{t/h} \cdot \frac{C_{MeOH}}{€/t} - \frac{P}{MW} \cdot \frac{C_{el}}{€/MWh} - \frac{\dot{Q}_{dh}}{MW} \cdot \frac{C_{dh}}{€/MWh} \right) / \left( \frac{\dot{m}_{ls}}{t_{ls}/h} \right) \right\}$$

where $\dot{m}_i$, $\dot{V}_i$, and $C_i$ are mass flow rates, volume flow rates and unit cost terms, respectively. The total inflow of oxygen considers the pure oxygen flows to the BF and BOF, while the (net) outflows of methanol (subscripts MeOH), power (el) and district heat (dh) from the system appear in the last three credit terms. The nonlinear programming problem, $\min(F)$, was solved numerically subject to constraints of the variables (Table 1) and to the reported limits of the raw material flows, using the cost factors of Table 2. The system was studied for a set of different liquid steel production rates in the range [150, 180] $t_{ls}$/h, capturing cost in the range [10, 50] €/$t_{CO2}$ and carbon dioxide emission cost in the range [0, 50] €/$t_{CO2}$ for different technologies of blast furnace operation in a single steel plant or in a steel plant integrated with a polygeneration system.

**Table 2** Cost factors in the objective function

| $C_{ore}$ | $C_{pel}$ | $C_{coal}$ | $C_{coke,ext}$ | $C_{NG}$ | $C_{lime}$ | $C_{quartz}$ | $C_{O2}$ | $C_{scrap}$ | $C_{MeOH}$ | $C_{el}$ | $C_{dh}$ |
|---|---|---|---|---|---|---|---|---|---|---|---|
| €/t | €/t | €/t | €/t | €/t | €/t | €/t | €/km$^3$n | €/t | €/t | €/MWh | €/MWh |
| 80 | 120 | 145 | 300 | 200 | 30 | 30 | 50 | 100 | 250 | 50 | 10 |

## Results and Discussions

The optimal operational conditions for the system were first investigated at a constant steel production rate of 170 $t_{ls}$/h for different blast furnace operating routes (CBF, TBF, OBF) considering a CCS cost of $C_{CCS}$ = 30 €/$t_{CO2}$ for a single steel plant and for a steel plant integrated with methanol production. Figure 2a shows how the specific emissions of the system, expressed per tons of liquid steel, vary with the emissions costs, while the corresponding specific costs of liquid steel are depicted in Fig. 2b. Lines without circular markers correspond to a steel plant, while lines with markers apply to the steel plant integrated with methanol production.

In all scenarios, process integration is seen to reduce the specific emissions and decrease the specific costs, and for low emission costs CBF-Int and TBF-Int show the lowest steel costs. Cold oxygen injection decreases the emission of the system considerably, and for an emission cost exceeding 25 €/$t_{CO2}$ its steelmaking costs becomes the lowest. At increased emission costs, the optimal operational condition of OBF-Int is shifted towards higher recycling degree of the top gas, yielding decreased emissions. For an emission price exceeding 35 €/$t_{CO2}$, the emissions of OBF-Int are lower than those of OBF.

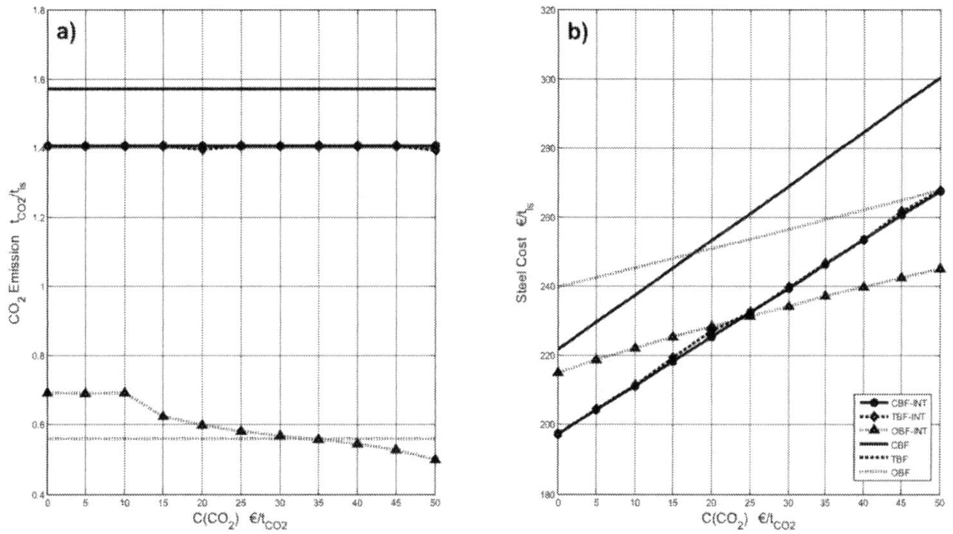


**Figure 2 a)** Specific emissions and **b)** specific costs of steel production for a steel plant with (lines with circles) or without (lines without circles) methanol production for a steel production rate of 170 $t_{ls}$/h and a CCS cost of 30 €/$t_{CO2}$. Cases optimized: Conventional (CBF, solid lines), top gas recycling with hot (TBF, dashed lines) or cold oxygen (OBF, dotted lines) blast.

Figure 3a shows the production of methanol from residual gases of CP, BF and BOF in an integrated steelmaking plant, while Fig. 3b illustrates the top gas recycling ratio for the different steelmaking routes. The results demonstrate that in the considered scenarios with a combination of natural gas injection and oxygen (enriched) blast, maximum methanol production is obtained without top gas recycling. The top gas recycling degree for integrated OBF increases as increasing emission costs, which will decrease the methanol production. However, the ratio of methanol production to substitue natural gas used in utility is estimated to 3-4, which can be compared with a ratio of 2.5-3.3 for heavy oil used in utility in the same system, and with a ratio of 1.2 in a methanol plant with heavy oil gasification.[11] This shows the potential of integrating methanol production with a steelmaking plant.

Figure 4 shows the specific emissions of carbon dioxide and specific cost of liquid steel for different steel production rates and fixed emission (35 €/$t_{CO2}$) and CCS (30 €/ $t_{CO2}$) costs. The former decrease and the latter increase slightly with the steel production in the interval. The best performance is presented by the integrated OBF steelmaking which yields the lowest costs and low emissions: The estimated steel costs for an integrated plant are around 20 €/$t_{ls}$ lower than those of a single steel plant. Thus, upgrading the steelmaking gases to methanol must be considered a promising concept.

The external coke required by the system and the methanol production from it for different steel production rates are depicted in Fig. 5a. The results show no need to buy coke for the OBF steelmaking concept, but for production rates > 160 $t_{ls}$/h, the other options need a considerable extra supply of coke. Figure 5b illustrates a weak dependence between the methanol and steel production rates. The methanol production rate is about twice for CBF-Int and TBF-Int compared to the OBF-Int route in the optimal states. This shows the complexity of the system and stresses the importance of decision making, where multiple aspects (production rates, costs, environmental aspects, etc.) have to be taken into account.

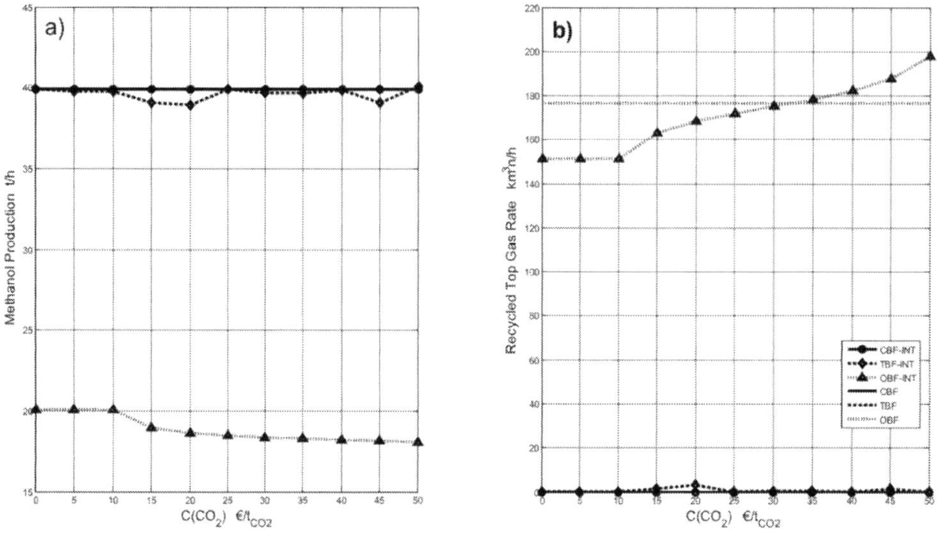


**Figure 3 a**) Methanol production rate and **b**) Recycled top gas rate in a steel plant integrated with methanol production for a steel production rate of 170 $t_{ls}$/h and a CCS cost of 30 €/$t_{CO2}$. Cases optimized: Conventional (CBF, solid lines), top gas recycling with hot (TBF, dashed lines) or cold oxygen (OBF, dotted lines) blast, for single steel plant or a plant with integrated methanol production.

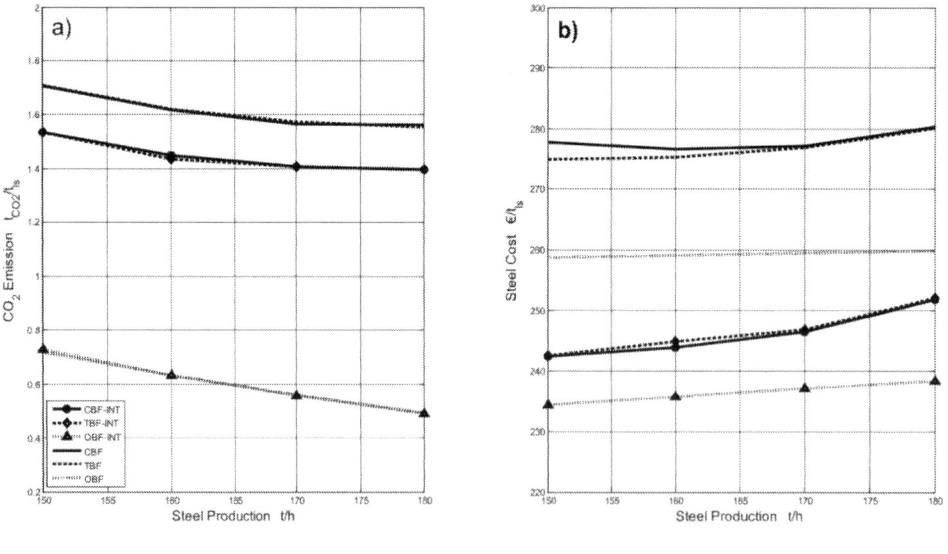


**Figure 4 a)** Specific emissions and **b)** specific costs of steel production for a steel plant with (lines with circles) or without (lines without circles) methanol production for a $CO_2$ emission and a CCS cost of 35 and 30 €/$t_{CO2}$, respectively. Cases optimized: Conventional (CBF, solid lines), top gas recycling with hot (TBF, dashed lines) or cold oxygen (OBF, dotted lines) blast.

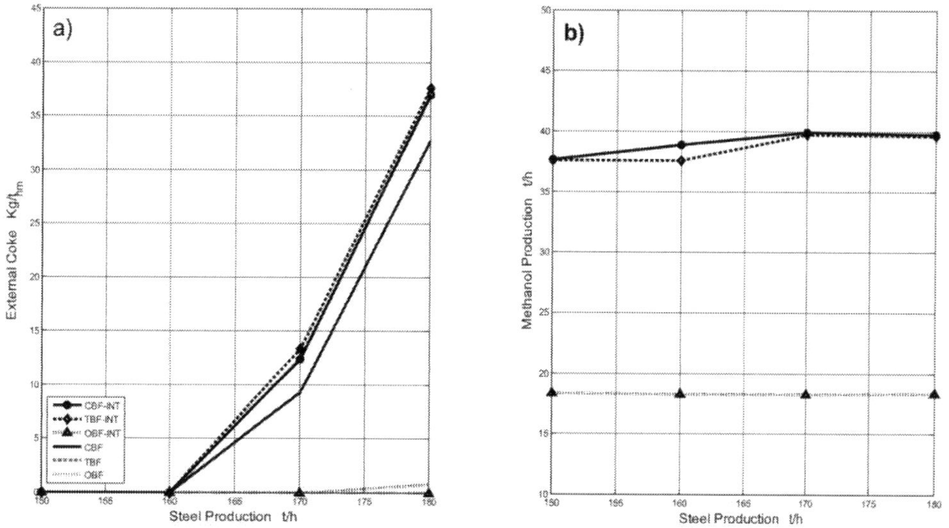


**Figure 5 a)** Feed flow of external coke and **b)** methanol production rate for a steel plant with (lines with circles) or without (lines without circles) methanol production for a $CO_2$ emission and a CCS cost of 35 and 30 €/$t_{CO2}$. Cases optimized: Conventional (CBF, solid lines), top gas recycling with hot (TBF, dashed lines) or cold oxygen (OBF, dotted lines) blast.

## Conclusions

In the work reported in this paper, a mathematical model has been applied to investigate the effect of integration of a polygeneration system with steelmaking on steel cost and carbon dioxide emissions. The system was optimized with respect to steelmaking costs in order to yield states that have a potential of being commercially exploited. A complex model of the blast furnace with options to implement different technologies, such as top gas recycling, natural gas injection, cold oxygen injection and hot oxygen enriched blast, was applied as the core of the system. The optimal operational conditions for different liquid steel production rates and scenarios for CCS and emission costs were studied. The oxygen blast furnace with top gas recycling may be a suitable option for suppressing $CO_2$ emission from the system in that the costs of steelmaking were kept reasonable, while the use of oxygen-enriched hot blast was found less suitable to be combined with top gas recycling. However, the optimal methanol production rate was lower than in the other integrated plants. In summary, natural gas injection combined with methanol production holds promise as an alternative to decrease the emissions from the steelmaking system, still keeping the costs competitive.

## References

1. Danloy, G.; Berthelemot, A.; Grant, M.; Borlee, J.; Sert, D.; van der Stel, J.; Jak, H.; Dimastromatteo, V.; Hallin, M.; Eklund, N.; Edberg, N.; Sundqvist, L.; Skoeld, B. -.; Lin, R.; Feiterna, A.; Korthas, B.; Mueller, F.; Feilmayr, C.; Habermann, A. Rev. Metall. -Cah. Inf. Techn. 2009, 1.

2. Feshchenko, S. A.; Pleshkov, V. I.; Lizunov, B. N.; Lapshin, A. A.; Soveiko, K. N.; Loginov, V. N.; Vasil'ev, L. E. Metallurgist 2007, 11-12, 605-611.

3. Ryzhenkov, A. N.; Yaroshevskii, S. L.; Zamuruev, V. P.; Popov, V. E.; Afanas'eva, Z. K. Metallurgist 2006, 5-6, 235-240.

4. Halim, K. S. A. Mater Lett 2007, 14-15, 3281-3286.

5. Nogami, H.; Yagi, J.; Kitamura, S.; Austin, P. R. ISIJ Int 2006, 12, 1759-1766.

6. Muramatsu, A.; Sato, H.; Akiyama, T.; Yagi, J. ISIJ Int 1993, 11, 1144-1149.

7. Yagi, J.; Akiyama, T. Process integration of ironmaking blast furnace with methanol synthesis; Efficiency, Costs, Optimization, Simulation and Environmental Impact of Energy Systems (ECOS 2000), pp.1401-1410, The Netherlands.

8. Helle, H.; Helle, M.; Saxén, H. Chem Eng Sci 2011, submitted.

9. Tobiesen, F. A.; Svendsen, H. F.; Mejdell, T. Ind Eng Chem Res 2007, 23, 7811-7819.

10. Helle, H.; Helle, M.; Saxén, H.; Pettersson, F. ISIJ Int 2010, 7, 931-938.

11. Ghanbari, H.; Helle, M.; Pettersson, F.; Saxén, H. Ind Eng Chem Res 2011, submitted.

# (763c) A Novel Conceptual Design for Liquefied Natural Gas Receiving Terminals

**Meiqian Wang**, Jian Zhang and Qiang Xu, Dan F. Smith Department of Chemical Engineering, Lamar University, Beaumont, TX

Natural gas mainly consists of methane, inert gas, and some heavy hydrocarbons, such as ethane, propane, and butane etc. which are referred as natural gas liquids (NGL). The distribution pipeline companies and natural gas buyers usually specify their desirable ranges of natural gas components and heating values. These specifications may vary widely with respect to different markets. In U.S., for instance, the relatively low heating value is required for the natural gas. Generally, there are two ways to adjust the heating value: adding inert gas or extract NGL. Since NGL can be feedstock for many chemical processes, it has more economic values than that as part of the natural gas stream. Thus, it would be more desirable to lower heating value via NGL recovery. However, a major concern for building NGL fractionator system is the huge amount of cooling and heating duties requested by the condenser and reboiler.

As known, shale gas as one of the most promising unconventional gases is a raising star in the global energy market. Currently, treated shale gas is usually burnt to generate electricity after simply treatment. However, if NGL recovery could be conducted for shale gas, the overall profit will be significantly increased. The problem about NGL recovery for shale gas is that it has to be liquefied first before sent to the fractionation system, which will require a lot of cold energy.

On the other hand, the liquefied natural gas that shipped to receiving terminals needs re-gasification before distribution. A major concern here is the heating energy consumption because natural gas liquefaction needs a great deal of heating energy. In this work, a new conceptual design of natural gas receiving terminals is proposed, which integrates re-gasification with shale gas liquefaction and NGL recovery. The cold energy from re-gasification could be partly recovered by shale gas liquefaction and NGL fractionators' condensers. Meanwhile, recovered NGL can be easily transported to nearby chemical plants because natural gas receiving terminals normally are built near chemical industry concentrated coast regions. The optimal design and operation of the proposed receiving terminal are identified via coupling rigorous process simulation and simplified optimization models.

See more of this Session: Advances in Process Intensification and Integration
See more of this Group/Topical: Process Development Division

**Compact Heat Exchange Reactors for Process Intensification**

**Abstract**

Many chemical processes are operating in reactors that are a compromise between productivity and control of the reaction. Consequently the reaction rate can be limited by the mixing or heat transfer capability of the reactor, rather than the actual reaction kinetics. This compromise results in equipment that is larger than it needs to be and decreased operating efficiency.

Chart's Compact Heat Exchange Reactor (CHER) combines a heat exchanger and reactor into a single device which are used to improve chemical processes using the philosophy of Process Intensification (PI). The role of mixing and heat transfer in controlling chemical reactions will be presented, as well as techniques used for identifying when compact heat exchange reactors can be applied to chemical processes, along with examples of potential and actual applications and includes case studies of their use in the fine chemicals, pharmaceutical, and fuel processing fields. In many applications, the processing rate achieved was many times higher than that anticipated.

Chart Energy & Chemicals manufactures compact heat exchangers and chemical reactors that are used for PI using either aluminum or stainless steel or special metals. Chart has been manufacturing compact brazed aluminum plate-fin heat exchangers for over 50 years, with these being used as chemical reactors as early as 1979, known as FinTec® reactor. Over the last 15 years a shim-based, diffusion bonded heat exchanger and reactor, known as ShimTec®, has been developed and widely applied.

# MAGNETIC HEAT CIRCULATOR BASED ON SELF-HEAT RECUPERATION

*Yui Kotani[1], Muhammad Aziz[2], Yasuki Kansha[1], Chihiro Fushimi[1], Atsushi Tsutsumi[1]*
*[1]The University of Tokyo, Tokyo, JAPAN*
*[2]Tokyo Institute of Technology, Tokyo, JAPAN*

## Introduction

Generally, a process that involves heating begins and ends at ambient temperature. Thus if the enthalpy of inlet and outlet fluid is equal, from the conservation law of energy, all provided heat is discharged. The net energy consumption is equal to the energy discharged. To reduce the energy consumption, heat recovery technology represented by pinch technology [1] has been used. Because temperature difference is needed for heat transfer, not all heat can be recovered and additional heating is needed (Fig. 1 a).

Recently, Kansha et al. proposed a heat circulating method; self-heat recuperation [2]. In this method, no heat is provided to the process.

## Heat Recuperation Technology

The schematic image of heat recuperation is presented in Fig. 1b [2]. The fluid that is heated in HEX (1 → 2), is adiabatically compressed (3 → 4) in order to create the temperature difference needed for heat transfer. The hot fluid gives heat back to cold fluid in HEX (4 → 5). Part of the compression work is recovered at the turbine (5 → 6), the remaining heat is discharged (6 → 7) so that the fluid temperature is back to the inlet temperature. From the conservation law of energy, the heat discharged is equal to the necessary work that is provided to the system, i.e. the compression work minus the work recovered by the turbine. With this heat recuperation technology, it is possible to circulate all heat. It is estimated to reduce the energy consumption to 1/5 - 1/20 compared to conventional energy conservation technology with a furnace as shown in Fig. 1a [2].

In order to apply the heat recuperation technology, adiabatic compression at the highest temperature of the process fluid is essential. The existing compressor can only be applied to temperature up to around 250 $^{o}$C. Also, heat elevation by adiabatic compression is only valid for gas practically. In this research, heat circulation using magnetocaloric effect (MCE) instead of compression is proposed so that 1) heat of solid and liquid, and/or 2) the heat above 250 $^{o}$C, can be recuperated.

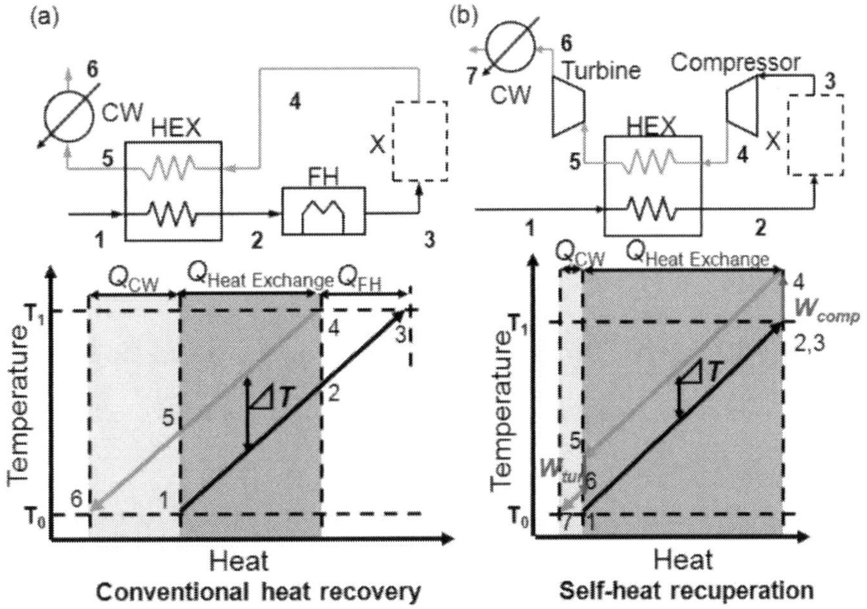


Fig. 1 Comparison between (a) conventional heat recovery and (b) self-heat recuperation technology [2]

## Magnetic Heat-Circulator

### *Magnetocaloric Effect*

By exposing magnetic material to magnetic field, the magnetic spin faces the same way. The lattice of the magnetic material absorbs the heat that is generated by the entropy difference and elevates the temperature of the magnetic material. Brown [4] successfully created 14 °C temperature difference using Gadolinium (Gd) as working material. Many magnetic heat pumps have been proposed [4]-[6] since the Brown's study.

### *Magnetic Heat-Circulator*

A schematic image of a magnetic heat-circulator (MHC) is shown in Fig. 2. Magnetic material is used as working material. The temperature of the working material rises when it is magnetized (a → b), and decreases when it is demagnetized (c → d). The fluid is heated in HEX1 (1 → 2), and via the following process, it is cooled back to the inlet temperature in HEX2 (4 →5). By giving all heat to the working material temporarily, it is possible to circulate all heat provided to the fluid.

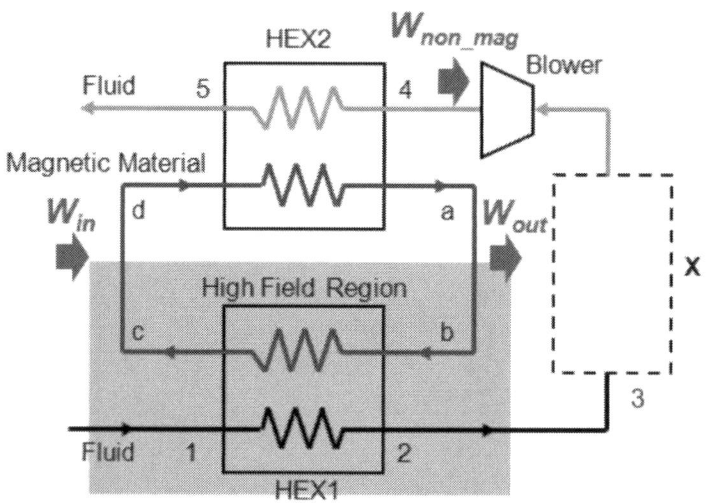


**Fig. 2** Schematic image of MHC

The theoretical work that is necessary to drive the MHC ($W_{net}$) is shown by eq.(1);

$$W_{\mathrm{net}} = W_{\mathrm{in}} - W_{\mathrm{out}} + W_{\mathrm{non\_mag}} \quad (1)$$

where $W_{in}$ is the work that is needed to pull out the magnetic material from the magnetic field, $W_{out}$ is the work that the magnetic field pulls in the magnetic material, and $W_{non\_mag}$ is the work needed to circulate the fluid. $W_{in}$ and $W_{out}$ is equivalent to the difference of magnetic potential and can be calculated by integration of strength of magnetization ($M$) with the strength of magnetic field (H).

Temperature-entropy (T-S) diagram of the MHC is shown in Fig. 3.The area surrounded by the isomagnetic line of the working material is equivalent to the magnetic work in net.

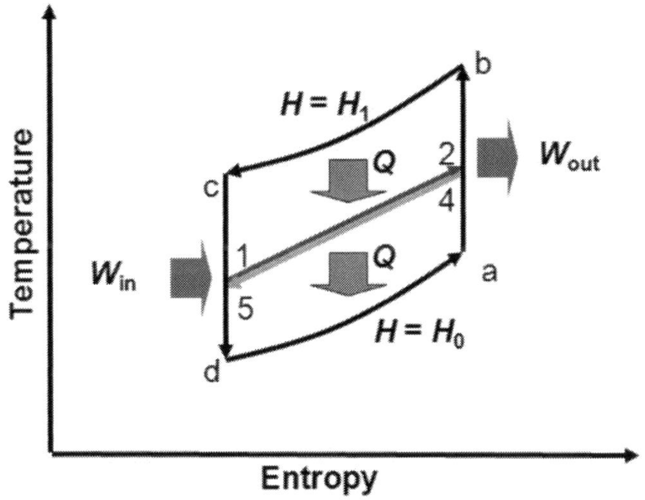

Fig. 3 T-S diagram of MHC

Table 1 shows the result of calculation of coefficient of performance (COP) using Gd (Curie temperature of 20 °C) as working material under 1 and 2 T magnetic fields when its temperature was raised from 275 K to 290 K, provided that no irreversible heat is produced during magnetization and demagnetization. $W_{in} - W_{out}$ was calculated by the area surrounded by the isomagnetics line of Gd. The heat circulated ($Q$) is the multiple of raised temperature and Gd's specific heat.

**Table 1** COP of MHC using Gd as working material

| $\mu_0 H$ [T] | $T_{2,4}$ [K] | $T_{1,5}$ [K] | $Q$ [J/kg] | $\Delta T_{ad}$ [K] | $W_{in}$-$W_{out}$ [J/kg] | COP |
|---|---|---|---|---|---|---|
| 1 | 290 | 275 | 3450 | 3.0 | 36.7 | 94.0 |
| 2 | | | | 4.8 | 60.7 | 56.9 |

## Conclusion

By using magnetocaloric effect, magnetic heat-circulator without compressor is proposed. Under 2 T magnetic field, COP of 56.9 was gained. The MHC can be applied to process in temperature above 250 °C and to process that uses solid or liquid. By applying self-heat recuperation, reduction of energy consumption can be obtained.

### *References*


[1] B. Linhoff, V. Sahdev , (2000), "Pinch Technology," Ullman's Encyclopedia of Industrial Chemistry

[2] Y. Kansha, N. Tsuru, K. Sato, C. Fushimi, A. Tsutsumi, (2009), "Self-Heat Recuperation Technology for Energy Saving in Chemical Process," *Industrial & Engineering Chemical Research*, **48**, pp. 7682-7686

[3] G.V. Brown, (1976), "Magnetic heat pumping near room temperature," *Journal of Applied Physics*, **47**, pp. 3673-3680.

[4] N. Hirano, S. Nagaya, M. Takahashi, T. Kuriyama, K. Ito, S. Nomura, (2002), "Development of Magnetic Refrigerator for Room Temperature Application," *Advances in Cryogenic Engineering*, **47**, pp. 1027-1034

[5] C. Zimm, A. Boeder, J. Chell, A. Sternberg, A. Fujita, S. Fujieda, K. Fukamichi, (2006), "Design and performance of a permanent-magnet rotary refrigerator," *International Journal of Refrigeration*, **29**, pp. 1302-1306

[6] T. Okamura, K. Yamada, N. Hirano, S. Nagaya, (2006), "Performance of a room-temperature rotary magnetic refrigerator," *International Joural of Refrigeration*, **29**, pp. 1327-1331

# (763f) Continuous Flow Reactors: Process Intensification of An Industrial Case

**Alessandra Vizza**, Roland Guidat and Yi Jiang, Reactor Technologies, Corning, Avon, France

Continuous Flow Reactors: process intensification of an industrial case

Reactor Technologies, Corning SAS, Avon, France

**Alessandra Vizza**, Roland Guidat, Berengere Chevalier, and Yi Jiang, Reactor Technologies, Corning SAS, Avon, France

This presentation will highlight the performance benefits for an industrial chemical process through the use of Corning' Advanced Flow Reactor.

Benefits of continuous flow processes over batch chemistry including the rapid synthesis of organic compounds, particularly under hazardous reactions conditions; rapid numbering up from milligram to kilogram amounts. A process performed in continuous flow allows a more efficient and safe process keeping a constant quality of the final product. Implementing continuous technology requires a different approach compared to traditional batch processes. The work described here shows how to fill this gap in the transition from traditional to continuous technologies and from feasibility test to an optimization. Finally the development of a customised set up will be described.
This presentation will underline the process intensification and ease of process scale-up for a continuous Beckmann Transposition reaction related to an industrial lactame production.

The purpose is to show how to effectively apply a process intensification to overcome the limitation in the actual process, increasing the reaction yield and enhancing the thermal control in a reaction involving dangerous chemicals. It will be shown how precise control of reaction conditions has a significant impact on the yield.

We will emphasize how the use of a milliscale continuous flow reactor in the area of fine chemical synthesis allows getting more efficiency and increased safety.

See more of this Session: Advances in Process Intensification and Integration
See more of this Group/Topical: Process Development Division

# (767a) Exploring Protein-Excipient Interactions for Optimal Design of Protein Stabilizers

Brock C. Roughton[1], **Anthony I. Pokphanh**[2], E. M. Topp[3] and K. V. Camarda[2], (1)Bioengineering Graduate Program, University of Kansas, Lawrence, KS, (2) Department of Chemical and Petroleum Engineering, University of Kansas, Lawrence, KS, (3)Department of Industrial and Physical Pharmacy, Purdue University, West Lafayette, IN

The number of proteins being considered as therapeutic candidates is rapidly increasing. Many obstacles are faced in the formulation of drug products containing proteins and a system approach to designing formulations is needed. A common and significant problem in protein therapeutic formulation is aggregation. Aggregation occurs through protein-protein interactions that lead to the formation of dimers, trimers, or larger aggregates. Protein aggregation can be caused by and also result in unfolding and loss of native conformation. Protein aggregates are often no longer effective as drug molecules and can elicit an immune response *in vivo*. Lyophilization, or freeze-drying, is a common method employed to minimize aggregation by limiting protein mobility. However, even in lyophilized formulations, aggregation can still occur in significant measures. Excipients, or additives, can be included in a formulation to stabilize the protein. Several excipients have been shown to minimize aggregation in lyophilized formulations, but the mechanisms are not well understood. Two basic hypotheses exist: protein aggregation is limited when protein mobility is limited by formation of a glass or amorphous solid and protein aggregation is limited by direct stabilizing interactions between an excipient and a protein. The current work explores direct excipient interactions with the surface of folded proteins and incorporates the information in the computational molecular design (CMD) of optimal excipients for aggregation minimization.

Protein-excipient interactions were studied using computer docking simulations to provide information for design of excipients using the CMD methodology. Several proteins were considered (calmodulin, lysozyme, myoglobin, and b-lactoglobin) along with carbohydrate (mannitol, sucrose, trehalose, and raffinose) excipients. The excipients chosen are commonly used as stabilizers in protein formulations. Blind docking simulations were performed to determine the most likely regions of direct interaction between excipients and proteins. In the blind docking experiments, the excipient was allowed to be flexible while the amino acid resides were held rigid. The amino acids that interacted with an excipient's docked conformation were recorded for fifty total docked conformations for each protein-excipient pair. The regions were compared to computationally predicted aggregation prone regions and experimental hydrogen/deuterium exchange (HDX) data. Regions where excipients exhibited protective effects in the HDX experiments were chosen as sites for direct docking simulations with the excipients. In the directed docking simulations, both the ligand and amino acid residues were allowed to be flexible.

Aggregation prone regions, or "hotspots", are areas on a protein which are likely to participate in aggregation reactions. The hotspots were calculated based on the amino acid sequence of the protein using two previously published algorithms available online: Aggrescan (http://bioinf.uab.es/aggrescan/) and PASTA (http://protein.cribi.unipd.it/pasta/). The calculated hotspot regions were compared to the regions of frequent protein-excipient interaction determined by the blind docking simulations. Hot spot regions calculated by the algorithms did not compare well with the docking results for most protein-excipient pairs. The amino acid sequence alone may not be sufficient to accurately predict regions prone to aggregation.

In HDX experiments, direct interaction between an excipient and protein limits the exchange of hydrogen on the protein backbone for deuterium. The use of peptic digests in HDX provides region specific information on exchange. Regions of reduced exchange in HDX data for protein-excipient pairs matched well with regions of frequent protein-excipient interaction in the blind docking simulations, indicating that the docking simulations gave reasonable estimates of regions of protein-excipient interaction that can lead to stabilization or protection of the protein. Directed docking simulations were performed for the regions of limited exchange for each protein-excipient pair. The docked free energies of the excipients were compared to the percent deuterium incorporated by the regions. The docked free energy provided an indication of how favorable the protein-excipient interaction was. An optimal stabilizer will bind preferentially to vulnerable regions on the protein's surface, minimizing contact between the protein and other protein molecules in the formulation. The docked free energy correlated well with the percent deuterium incorporated, indicating that the docking simulations could provide quantitative information about protein-excipient interaction. Regions with low deuterium uptake bound excipients with lower docked free energies. A quantitative structure-property relationship (QSPR) was developed to relate the docked free energy to the molecular topology of the excipients. The QSPR was then utilized in a computational molecular design framework to generate candidates that would bind protein regions with a minimum free energy. Prediction intervals at 95% were used to compare the candidates and estimate the uncertainty in binding energy prediction. By designing an excipient candidate that will optimally bind to a region on a protein, the region will be protected from interaction with other proteins and aggregation will be minimized. For the proteins considered, several such candidates have been designed using the computational molecular design methodology.

See more of this Session: Methodologies for Product Design
See more of this Group/Topical: Computing and Systems Technology Division

# (767b) Comprehensive Computer-Aided Molecular Design Framework for Pure Component Design

**Apurva Samudra** and Nick Sahinidis, Chemical Engineering, Carnegie Mellon University, Pittsburgh, PA

Computer-aided molecular design (CAMD) has become a powerful tool to design novel molecules to meet product target property criteria. Over the years, the field has matured from neighborhood-based searches to fathoming the chemical space using optimization and enumeration. The techniques used in this field have been rooted in the basic group contribution (GC) property methods and thus more generic design problems have either been decomposed or projected onto GC properties to make them compatible with CAMD methods. These techniques have also been plagued by the exponential search-space of molecular structures which inhibit a generic enumeration based design approach.

We have developed an optimization-based framework for basic molecular design aimed to automate, augment and accelerate the steps in CAMD. This framework employs integer optimization efficiently and effectively to cut down on the computational cost of CAMD. Using property decomposition and model transformation, this framework achieves the flexibility required for extensive molecular design along with accuracy for basic group contribution properties. The framework is based on GC+ [1] property models for thermodynamic properties like (Tc,Pc,Tm,Tb,Hv etc.). The framework is split into three stages. In the first stage, mixed-integer linear programming (MILP) is used to generate molecular candidates that fit a relaxed range of defined property targets with approximate GC + models. This approach tackles the exponentially increasing search-space very effectively by exploiting the linearity in property models. The second stage of the framework screens the solutions further by using structure-dependent corrections to the approximate property values. Detailed chemical structures are developed using a graph theory approach. Different descriptors in different GC+ models are used by estimating their contribution via connectivity indices. The last stage is used to validate the model using molecule-specific property techniques and to integrate the CAMD problem with many involved property prediction techniques that extend its applicability to diverse problems.

We present various extensions to the framework to include other property models for various applications. These extensions range from simple groups contributions for solvent design, e.g. Hansen solubility parameter, UNIFAC, to complex property models and simulations for other diverse applications. We demonstrate the simplicity of use of the framework. Simple methodology to extend the framework for desired application for the use is presented. We conclude by presenting many case studies illustrating efficiency, wide application domain and versatility of proposed framework.

[1] R. Gani, P. M. Harper, and M. Hostrup. Automatic Creation of Missing Groups through Connectivity Index for Pure-Component Property Prediction. Industrial & Engineering Chemistry Research, 44:7262–7269, 2005.

See more of this Session: Methodologies for Product Design
See more of this Group/Topical: Computing and Systems Technology Division

# (767c) Systematic Molecular Design of Biofuel Additives Using Hybrid Characterization and Group Contribution Based Techniques

**Subin Hada**, Charles C. Solvason and Mario Richard Eden, Department of Chemical Engineering, Auburn University, Auburn, AL

Any type of feedstock that contains free fatty acids and/or triglycerides such as vegetable oil, waste oil, animal fat, and waste grease can be converted into biodiesel. Because of its biological origin, biodiesel has many advantages compared to its petroleum derived counterparts. However, it does exhibit several limitations like oxidative stability and low temperature operability that have impaired its wide use and commercialization. Simultaneous solution of these problems has proven difficult as improvements in one area tend to impair another. As a result, design of an optimal fuel for all environments can be a quite difficult task. In addition, feedstock origin will impact the final quality of the biodiesel product. Therefore, to produce fuel grade biodiesel, the characteristics of the feedstock are very important during the initial research and production stage.

Several approaches are possible for improving the fuel properties of biodiesel. Modification of the fatty acid composition through physical processes or uses of additives are the most prevalent. Fuel additives, such as antioxidants, cetane enhancers, or cold-flow improvers have become indispensable tools not only to alleviate the drawbacks described above, but also to assure that any fuel blend will meet international and regional standards regardless of origin. However, an additive solution to one problem often aggravates another problem. Furthermore, the questions of additive compatibility, required additive levels, the effect on other properties, and whether these additives function as designed for biodiesel fuels with differing fatty acid profiles still remain challenges and that need further investigation.

Therefore, it is desired to molecularly design biodiesel additives that simultaneously account for the unintended effect on other fuel properties in the neat and the blend fuel in order to achieve the performance properties of the petroleum based fuel. In this way, biofuels can be formulated that are adaptable to a range or blend of feedstocks and the desirable fuel characteristics like oxidative stability and wide operating temperature range.

In this work, we aim to identify all possible compounds which, when added to off spec biodiesel, results in a fuel that satisfies performance standards such as ASTM D6751 in the US and EN14214 in Europe. To meet this end, multivariate characterization data obtained from IR spectroscopy of common additives were combined with decomposition and property clustering techniques in a reverse problem formulation. In this approach, the fuel additive property targets are identified in the first reverse problem followed by molecular design to match the targets. The characterization data consists of multitude of properties of interest (such as cetane number, melting point, and kinematic viscosity) to ensure adequate performance. To facilitate an efficient design we consolidated these various properties into a latent property domain using principle component analysis (PCA) techniques. Finally, characterization based molecular design using group contribution parameters are then used to build novel additives that match the fuel specifications in the latent property space.

See more of this Session: Methodologies for Product Design
See more of this Group/Topical: Computing and Systems Technology Division

# (767d) Multiscale Approach to Colloidal Systems (emulsions) Design

**Andrea Nieto**, Diego Pradilla, Natalia Murcia and Oscar Alvarez, Universidad de Los Andes, Bogota, Colombia

Multiscale Approach to Colloidal Systems (emulsions) Design

A. Nieto, D.Pradilla, N. Murcia, O. Alvarez

Group of Process and Product Design (GDPP)

Department ofChemical Engineering

Universidad de losAndes, Bogota-Colombia

Preparation of emulsions (generally mixing ofimmiscible liquids, water and oil) involves several variables (emulsificationconditions, system formulation and composition), being the final properties aconsequence of the established level for each of these variables. To analyzeseparately the effect of a given process condition, formulation or compositionon emulsion's properties, facilitates the establishment of a relation betweenthe study variable and the changes in measured properties. However, thestatement above limits the comprehension of the existent interactions betweenformulation, composition and process.

In this paper, molecular interactions of thesystem water/mineral oil/non ionic surfactant used in the preparation of inverseand direct emulsions were studied, evaluating the effect on the microscopic(particle size) and macroscopic (apparent viscosity, elastic and loss moduli)properties of the system. To achieve this, different agitation geometries andemulsification processes were used in order to establish the influence of theamount of energy added during emulsion preparation. The molecular scalebehavior is evaluated using near-infrared spectroscopy NIR (FOSS 5000Smartprobe Analyzer), size distribution and particle size are settled using dynamiclight scattering technique (Zetasizer nano ZS), and finally the rheologicalbehavior is measured from steady state and oscillatory tests using a controlledstress rheometer (TA Instruments AR-G2).

The study of the phenomena that occur duringthe preparation of these emulsions at three different scales: macroscopic,microscopic and molecular, simultaneously with the effect of the amount ofenergy added, results in the understanding of the relations between process(emulsification conditions), product (formulation and composition) andproperties (rheology), allowing us to design emulsions with specificcharacteristics. For instance, it has been established for concentrated inverseemulsions (volumetric fraction of dispersed phase > 0.9) that elasticmodulus is independent of impeller type used during emulsification.

See more of this Session: Methodologies for Product Design
See more of this Group/Topical: Computing and Systems Technology Division

# (767e) Teaching Chemical Product Design

**Ed Cussler**, Department of Chemical Engineering and Materials Science, University of Minnesota, Minneapolis, MN and **Geoff D. Moggridge**, Chemical Engineering and Biotechnology, University of Cambridge, Cambridge, CB2 3RA, United Kingdom

Teaching Chemical Product Design

E L Cussler and G D Moggridge

Universities of Minnesota and Cambridge

Since its inception around a century ago, the chemical industry has focussed on the manufacture of commodities. A commodity chemical, produced at over 1000 tons per year, is sold into a world market where the products are differentiated by price. Benzene, polypropylene and titanium dioxide are examples. This industry had its Golden Age from 1940 to 1980, with growth equivalent to that of the modern software industry.

More recently, as market growth slowed, chemical companies have moved towards higher value added products. These higher value added products, often produced at less than 10 tons per year, are often made of ingredients which cost a tenth or less of their selling price. They gain their value from a molecular or microstructure which gives them better performance. However, many chemical professionals are trained largely, if not exclusively, to serve the commodity chemical industry. These professionals now focus largely on the traditional question, "How should we make this commodity product?" These professionals must also consider what is increasingly the more relevant question, "What high value added products are we going to make?"

This paper describes teaching chemical product design, and so attempts to fill part of the educational gap between the commodity chemical industry and the new, more fragmented, high value added chemical product industry. Filling this gap has two parts. First, we present a four-step template for chemical product design of customer needs, product ideas, selection of the best idea, and product manufacturing. Second, we explore which step in this template is most important for which kind of product. As expected, the manufacturing step is the most important for commodities. Less obviously, the selection step is the most important for many non-commodity products.

We teach product design using two different strategies. Sometimes, we teach a separate course of lectures and smaller recitations, where the specific problems are discussed. The students develop ideas as teams, but write six to ten individual reports. Alternatively, we teach twelve classes which supplement to chemical process design. In this stategy, each student writes three individual essays. Both of these strategies work better for us than imitating the process design course.

We are interested that different experiences lead to trouble with different parts of product design. Inexperienced students are wonderful at generating new product ideas, but they have trouble making estimates which let them quickly select among possible alternative products. Experienced chemists and engineers have no trouble making quick estimates and sensible selections, but they are less effective at suggesting new ideas. Both groups have benefited from and enjoyed their efforts to get better at chemical product design. We challenge you, either as a professor or as a student: while this material is hard to teach and hard to learn, it can be the most satisfying part of your education. Have fun with your designs.

See more of this Session: Methodologies for Product Design
See more of this Group/Topical: Computing and Systems Technology Division

# AUTHOR INDEX

# AUTHOR INDEX

# AUTHOR INDEX

# AUTHOR INDEX

# AUTHOR INDEX

# AUTHOR INDEX